Mean-Field-Type
Games for Engineers

Mean-Field-Type Games for Engineers

Julian Barreiro-Gomez
Hamidou Tembine

CRC Press
Taylor & Francis Group
Boca Raton London New York

CRC Press is an imprint of the
Taylor & Francis Group, an **informa** business

First edition published 2022
by CRC Press
6000 Broken Sound Parkway NW, Suite 300, Boca Raton, FL 33487-2742

and by CRC Press
2 Park Square, Milton Park, Abingdon, Oxon, OX14 4RN

CRC Press is an imprint of Taylor & Francis Group, LLC

ISBN: 9780367566128 (hbk)
ISBN: 9780367566135 (pbk)
ISBN: 9781003098607 (ebk)

ISBN: 9781032128047 (eBook+ Enhancements)

DOI: 10.1201/9781003098607

Typeset in CMR10 font
by KnowledgeWorks Global Ltd.

Access the Support Material: www.routledge.com/9780367566128

To God.
To Mayerly, Margarita, Luis, and Juan
for their permanent support.
To Jacobo for being my biggest motivation.

Julián Barreiro Gómez

To Yandai, Pama, Marie-Claire,
Jean-Pierre and Florence
for their unconditional support

Tembine Hamidou Doumbodo

Contents

List of Figures

List of Tables

Foreword

Mean field was first studied in physics for the behavior of systems with large numbers of negligible individual particles. Recently mean-field game theory was introduced in the economics and engineering literature to study the strategic decision-making by small interacting agents of huge populations. Typically a mean-field game is described by a Fokker-Planck equation, and solved by a Hamilton-Jacobi-Bellman equation, which requires the number of agents approaches infinity. This assumption limits the practical usage of mean-field game theory in engineering fields.

Thanks to my friend, professor Hamidou Tembine, who also mentored my former Ph.D. student majored in mean-field game theory, his team and collaborators including Dr. Julian Barreiro-Gomez have pioneer works to introduce mean-field-type game theory to engineering scenarios. Mean-field-type games differ from mean field games since it takes into account higher-order statistics, it can be employed when dynamic programming cannot be applied, the number of interacting agents is not necessarily large, and it can handle non-symmetry non-negligible effect of individual decision on the mean field term. Those significant advantages of mean-field-type game theory open a whole gate for solving complex engineering problems that cannot be handled by classic methods.

With such a demand from engineering audiences, this book is very timely and provides a thorough study of mean-field-type game theory. The strenuous protagonist of this book is to bridge between the theoretical findings and engineering solutions. The book introduces the basics first, and then mathematical frameworks are elaborately explained. The engineering application examples are shown in detail, and the popular learning approaches are also investigated. Those advantageous characteristics will make this book a comprehensive handbook of many engineering fields for many years, and I will buy one when it gets published.

Zhu Han, IEEE/AAAS Fellow
John and Rebecca Moores professor
University of Houston

Preface

If you have picked this book, you are probably already aware about how powerful and suitable the mean-field-type control and game theory is in order to solve risk-aware problems in the engineering framework, and that a large variety of control and dynamic game problems can be set as particular cases of the mean-field-type games.

Our main goal in this textbook is to provide a quite comprehensive and simple treatment of the mean-field-type control and game theory, which can also be interpreted as risk-aware optimal interactive decision-making techniques. To this end, we exclusively focus on the so-called direct method either in continuous or discrete time. Our experience indicates that other existing methods reported in the literature to solve the class of stochastic problems we address in this book, such as partial-differential-equation-based methods, chaos expansion, dynamic programming, or the stochastic maximum principle, are not appropriate to start teaching beginner students in the field neither early-career researchers. We recommend to focus on understanding this book prior to moving on the study of other research manuscripts using other theoretical directions. In this regard, the contents of this book comprises an appropriate background to start working and doing research in this game-theoretical field.

To make the exposition and explanation even easier, we first study the deterministic optimal control and differential linear-quadratic games. Then, we progressively add complexity step-by-step and little-by-little to the problem settings until we finally study and analyze mean-field-type control and game problems incorporating several stochastic processes, e.g., Brownian motions, Poisson jumps, and random coefficients.

This smooth trip, starting with a scalar-valued state optimal control problem in continuous and discrete time, passes through the scalar-valued deterministic differential games and mean field games, the stochastic state-and-control-input independent diffusion differential games and mean field games, until we finally address the mean-field-type games with state-and-control-input dependent diffusion terms and incorporating Poisson jumps and random coefficients by means of switching regimes. On the other hand, we go beyond the Nash equilibrium, which provides a solution for non-cooperative games, by analyzing other game-theoretical concepts such as the Berge, Stackelberg, adversarial and co-opetitive equilibria. For the mean-field-type game analysis, we provide several numerical examples, which are obtained from a MatLab-based user-friendly toolbox that is available for the free use of the readers of this book.

We devote a whole part of the book to discuss about some learning approaches that guarantee converge to mean-field-type solutions. In particular, we present the constrained and static mean-field-type games where optimization algorithms may be applied such as distributed evolutionary dynamics, the receding horizon mean-field-type control also know as risk-aware model predictive technique, and the data-driven mean-field-type games motivating the use of artificial intelligence tools such as machine learning with either neural networks or simple linear regression. Finally, we present several engineering applications in both continuous and discrete time. Among these applications we find the following: water distribution systems, micro-grid energy storage, stirred tank reactor, mechanism design for evolutionary dynamics, multi-level building evacuation problem, and the COVID-19 propagation control.

Julian Barreiro-Gomez
Hamidou Tembine

Acknowledgments

We gratefully acknowledge support from the US Air Force, and the New York University in the US campus (NYU) and the UAE campus (NYUAD), for the research conducted at the Learning & Game Theory Laboratory (L&G Lab) and at the Center on Stability, Instability and Turbulence (SITE). This material is based upon work supported by Tamkeen under the NYU Abu Dhabi Research Institute grant CG002.

We also acknowledge our friends, faculty members, and researchers with whom we have had several scientific discussions about mean-field-type control and game theory, and also regarding its potential for engineering applications. We specially thank Prof. Tyrone E. Duncan and Prof. Bozenna Pasik-Duncan from the mathematics department at Kansas University in the US, and Prof. Boualem Djehiche from the mathematics department at Royal KTH in Sweden. We finally acknowledge all our co-authors with whom we have published several articles in the mean-field-type field.

Author Biographies

Julian Barreiro-Gomez received his B.S. degree (cum laude) in Electronics Engineering from Universidad Santo Tomás (USTA), Bogota, Colombia, in 2011. He received the M.Sc. degree in Electrical Engineering and the Ph.D. degree in Engineering from Universidad de Los Andes (UAndes), Bogota, Colombia, in 2013 and 2017, respectively. He received the Ph.D. degree (cum laude) in Automatic, Robotics and Computer Vision from the Technical University of Catalonia (UPC), Barcelona, Spain, in 2017; the best Ph.D. thesis in control engineering 2017 award from the Spanish National Committee of Automatic Control (CEA) and Springer; and the EECI Ph.D. Award from the European Embedded Control Institute in recognition to the best Ph.D. thesis in Europe in the field of Control for Complex and Heterogeneous Systems 2017. He received the ISA Transactions Best Paper Award 2018 in Recognition to the best paper published in the previous year. Since August 2017, he has been a Post-Doctoral Associate in the Learning & Game Theory Laboratory (L&G-Lab) at the New York University in Abu Dhabi (NYUAD), United Arab Emirates, and since 2019, he has also been with the Research Center on Stability, Instability and Turbulence (SITE) at the New York University in Abu Dhabi (NYUAD). His main research interests are: risk-aware control and games, mean-field-type games, constrained evolutionary game dynamics, distributed optimization, stochastic optimal control, and distributed predictive control.

Hamidou Tembine received the M.S. degree in applied mathematics from Ecole Polytechnique, Palaiseau, France, in 2006 and the Ph.D. degree in computer science from the University of Avignon, Avignon, France, in 2009. He is a prolific Researcher and holds more than 150 scientific publications including magazines, letters, journals, and conferences. He is an author of the book on Distributed Strategic Learning for Engineers (CRC Press, Taylor & Francis 2012), and Coauthor of the book Game Theory and Learning in Wireless Networks (Elsevier Academic Press). He has been co-organizer of several scientific meetings on game theory in networking, wireless communications, smart energy systems, and smart transportation systems. His current research interests include evolutionary games, mean-field stochastic games and applications. Dr. Tembine received the IEEE ComSoc Outstanding Young Researcher Award for his promising research activities for the benefit of the society in 2014. He received the best paper awards in the applications of game theory.

Symbols

Symbol Description

x	Scalar-valued system state	ℓ	Running cost (control case)
x_i	Scalar-valued system state of the ith decision-maker	h	Terminal cost (control case)
		L_i	Cost functional of the ith decision-maker
u	Scalar-valued control input		
u_i	Scalar-valued control input of the ith decision-maker	ℓ_i	Running cost of the ith decision-maker
X	Matrix/vector-valued system state	h_i	Terminal cost of the ith decision-maker
U	Matrix/vector-valued control input	$\mathbb{E}[\cdot]$	Expected value
		$\text{var}[\cdot]$	Variance
U_i	Matrix/vector-valued control input of the ith decision-maker	$\text{cov}[\cdot]$	Co-variance
		H	Hamiltonian (control case)
b	Drift	f	Guess functional (control case), or fitness functions in a population game
σ	Diffusion		
B	Standard Brownian motion	H_i	Hamiltonian of the ith decision-maker
N	Jump process		
$\tilde{N}$	Compensated jump process	F_i	Guess functional of the ith decision-maker (matrix-valued problems)
Θ	Set of jump sizes, or set of operating point in a gain-schedule strategy		
		f_i	Guess functional of the ith decision-maker (scalar-valued problems), or the fitness of the ith strategy in a population game
ν	Radon measure over Θ		
s	Regime switching		
$\mathcal{S}$	Set of regime switching		
$\tilde{q}_{oo'}$	Jump intensity from regime switching s to s'	BR_i	Best response of the ith decision-maker
W	Discrete-time noise		
m_x	Mean-field term of x	$\mathcal{N}$	Set of decision-makers
m_u	Mean-field term of u	$\mathcal{N}_0$	Set of risk-neutral decision-makers
m	Strategic distribution		
ϕ	Probability measure of the system state x	$\mathcal{N}_+$	Set of risk-averse decision-makers
L	Cost functional (control case)	$\mathcal{N}_-$	Set of risk-seeking decision-makers

$\mathcal{X}$ Feasible set of system state

$\mathcal{P}(\mathcal{X})$ Space of the probability measure of x

$\mathbb{U}$ Feasible set of control inputs

$\mathcal{U}$ Feasible control strategy

$\mathbb{U}_i$ Feasible set of control inputs for the ith decision-maker

$\mathcal{U}_i$ Feasible control strategy of the ith decision-maker

$\langle x, y \rangle$ Inner product for vectors x, y

$\langle A, B \rangle$ Trace $\text{tr}(A, B)$ for matrices A, B

Part I

Preliminaries

1

Introduction

We truly live in a more and more interconnected and interactive world. In recent years, we have seen emerging technologies such as internet of everything, collective intelligence including Artificial Intelligence (AI), blockchains, next-generation wireless networks, among many others. The quantities-of-interest in these systems involve both *volatilities and risks*.

A typical example of risk concerns in the current online market is the evolution of prices for the digital and cryptocurrencies (e.g., bitcoin, litecoin, ethereum, dash, and other altcoins (alternatives to bitcoin, etc.). The variance plays a base model for many risk measures. From random-variable perspective (probability theory) the volatility can be captured by means of the variance, which is a *mean-field term* comprising the second moment and the square of the mean. Another example concerns the variations of wireless channels in multiple-input-multiple-output systems. Non-Gaussianity of wireless channels has been observed experimentally and empirically, and its variability affects the quality of the communication.

The term *mean-field* has been referred to as a physics concept that attempts to describe the effect of an infinite number of particles on the motion of a single particle. Researchers began to apply the concept to social sciences in the early 1960s to study how an infinite number of factors affect individual decisions. However, the key ingredient in a game-theoretic context is the influence of the distribution of states and/or control actions onto the payoffs of the decision-makers. Notice that *there is no need to have a large population of decision-makers*. A mean-field-type game is a game in which the payoffs and/or the state dynamics coefficient functions involve not only the state and actions profiles but also the distributions of state-action process (or its marginal distributions).

Games with distribution-dependent quantity-of-interest such as state and/or payoffs are particularly attractive because they capture not only the mean, but also the variance and higher order terms. Such incorporation of these mean and variance terms is directly associated with the paradigm introduced by H. Markowitz, 1990 Nobel Laureate in Economics. The Markowitz paradigm, also termed as the mean-variance paradigm, is often characterized as dealing with portfolio risk and (expected) returns [2–4].

In this book, we address variance reduction problems when several decision-making entities take place. When the decisions made by the agents/players/decision-makers influence each other, the decision-making is

DOI: 10.1201/9781003098607-1

said to be interactive (interdependent). Such problems are known as *game-theoretical* problems.

"Interactive decision theory would perhaps be a more descriptive name for the discipline usually called *Game Theory*"

Robert Aumman
[5, page 47]

In this book we study the mean-field-type game theory, which can be also named as *risk-aware interactive decision-making theory*.

Next, we present some basic definitions corresponding to the structure of a particular class of games, which is addressed throughout this book.

1.1 Linear-Quadratic Games

We start by defining a particular class of either deterministic or stochastic differential games determined by a specific structure that the system dynamics and cost functional have.

Definition 1 (Linear-Quadratic Deterministic Games) *Game problems, in which the state dynamics is given by a linear deterministic system and a cost functional that is quadratic in the state and in the control inputs, are often called the Linear-Quadratic (LQ) games.*

Definition 2 (Linear-Quadratic-Gaussian Games) *Game problems, in which the state dynamics is given by a linear stochastic system with a Brownian motion and a cost functional that is quadratic in the state and in the control inputs, are often called the Linear-Quadratic Gaussian (LQG) games. Such games also belong to the family of stochastic linear-quadratic games.*

1.1.1 Structure of the Optimal Strategies and Optimal Costs

For generic LQG game problems under perfect state observation, the optimal strategy of the decision-maker is a linear state-feedback strategy, which is identical to an optimal control for the corresponding deterministic linear-quadratic game problem where the Brownian motion is replaced by the zero process. Moreover, the equilibrium cost only differs from the deterministic game problem's equilibrium cost by the integral of a function of time.

However, when the diffusion (volatility) coefficient is state and/or control-dependent, the structure of the resulting differential system as well as the equilibrium cost vector are modified. These results were widely known in both dynamic optimization, control and game theory literature.

In this book, several structures are studied from simple ones up to cases where the stochastic processes are not only dependent on both the system states and control inputs, but also on the distribution of the states and/or the control inputs.

1.1.2 Solvability of the Linear-Quadratic Gaussian Games

For both LQG control and LQG zero-sum games, it can be shown that a simple square completion method provides an explicit solution to the problem. It was successfully developed and applied by Duncan et al. [6–11] in the mean-field-free case (games in the absence of the distribution of the variables of interest). Moreover, Duncan et al. have extended the direct method to more general noises including fractional Brownian noise and some non-quadratic cost functionals such as on spheres and torus.

Here, we follow the same method in order to solve a large variety of mean-field-type control and game problems in both continuous and discrete time, making the solution of this complex problem accessible for early-career researchers and engineering students.

1.1.3 Beyond Brownian Motion

Inspired by applications in engineering (e.g., internet connection, battery state, etc) and in finance (e.g., price, stock option, multi-currency exchange, etc) where not only Gaussian processes but also jump processes (e.g., Poisson, Lévy, etc) play important features, the question of extending the framework to linear-quadratic games under state dynamics driven by jump-diffusion processes were naturally posed. Adding a Poisson jump and regime switching (random coefficients) may allow to capture in particular larger jumps which may not be captured by just increasing diffusion coefficients. Several examples such as multi-currency exchange or cloud-server rate allocation on blockchains are naturally in a matrix form.

Throughout this book, we discuss about several game-theoretical solution concepts and different structures as it has been pointed out in Section 1.1.1 including the analysis when processes beyond Brownian motion are taken into consideration, e.g., see Chapter 6 where mean-field-type game problems with jumps and regime switching (random coefficients) are studied.

1.2   Linear-Quadratic Gaussian Mean-Field-Type Game

According to the Definitions 1 and 2 in Section 1.1 corresponding to deterministic LQ and stochastic LQG games, respectively, we introduce next in Definition 3 a class of stochastic differential games that additionally involve the distribution of the variables of interest such as the system states and/or the control inputs.

Definition 3 (Linear-Quadratic Mean-Field-Type Games) *Game problems, in which*

- *the state dynamics is given by a linear stochastic system driven by a Brownian motion with a mean-field term (such as the expectation of the system state and/or the expectation of the control actions/control inputs) , and*

- *the cost functional is quadratic in the state, control, expectation of the state and/or the expectation of the control actions/control inputs, are often called LQG games of mean-field type, or Mean-Field-Type Linear-Quadratic Gaussian MFT-LQG games.*

The incorporation of mean-field terms into the stochastic differential game problems allows taking into consideration risk terms such as the variance, and higher order terms as introduced next. As a motivation to continue studying this book, notice that the incorporation of risk terms into the engineering problems solution and systems design can potentially enable the enhancement of the systems performance.

1.2.1   Variance-Awareness and Higher Order Mean-Field Terms

Most studies illustrated mean-field game methods in the linear-quadratic game with *infinite number of decision-makers* [12–16]. These works assume indistinguishability (also associated with homogeneity) within classes and the cost functionals were assumed to be identical or invariant per permutation of decision-makers indexes. Note that the indistinguishability assumption is not fulfilled for many interesting problems such as variance reduction and/or risk quantification problems, in which decision-makers have different sensitivity toward the risk, i.e., not all the decision-makers involved in the strategic interaction perceive, measure or interpret the risk in the same manner. One typical and practical example is to consider a multi-level building in which every resident has its own comfort zone temperature and aims to use the Heating, Ventilating, and Air Conditioning (HVAC) system to be in its comfort temperature zone, i.e., maintain the temperature within its own comfort zone. This problem clearly does not satisfy the indistinguishability assumption used in the previous works on mean-field games (the homogeneity of decision-makers is not satisfied). Therefore, it is reasonable to look at the problem beyond the *indistinguishability assumption*.

Here we drop these assumptions and deal with the problem directly with *arbitrarily finite number of decision-makers*, which can clearly be heterogeneous. In the LQ-mean-field game problems the state process can be modeled by a set of linear stochastic differential equations of McKean-Vlasov and the preferences are formalized by quadratic or exponential of integral of quadratic-cost functions with mean-field terms. These game problems are of practical interests and a detailed exposition of this theory can be found in [17–22]. The popularity of these game problems is due to practical considerations in (but not limited to) consensus problems, signal processing, pattern recognition, filtering, prediction, economics, and management science [23–26].

To some extent, most of the risk-neutral versions of these optimal controls are analytically and numerically solvable [6, 7, 9, 11, 19]. On the other hand, the linear quadratic robust setting naturally appears if the decision makers' objective is to minimize the effect of a small perturbation and related variance of the optimally controlled nonlinear process. By solving a linear quadratic game problem of mean-field type, and using the implied optimal control actions, decision-makers can significantly reduce the variance (and also the cost) incurred by this perturbation. The variance reduction and min-max problems have very interesting applications in risk quantification problems under adversarial attacks and in security issues with interdependent infrastructures and networks [26–30]. In order to provide few examples, in [27, 31], the control for the evacuation of a multi-level building is designed by means of mean-field games and mean-field-type control, and in [32], electricity price dynamics in the smart grid is analyzed using a mean-field-type game approach under common noise which is of diffusion type.

This book investigates how the *Direct Method* (sometimes also known as the verification method or, for linear-quadratic case, as square-completion method) can be used to solve different types of mean-field-type game problems which are non-standard problems [17], e.g., Non-cooperative solutions, Cooperative solutions, Co-opetitive solutions, Adversarial solutions, Stackelberg solutions, Berge solutions; and for scalar, vector and/or matrix-valued states and control inputs, and both in continuous and discrete time. Besides, we study the potential applications that this class of games have in engineering problems. To this end, the following section discusses about the role that risk terms play in the engineering field.

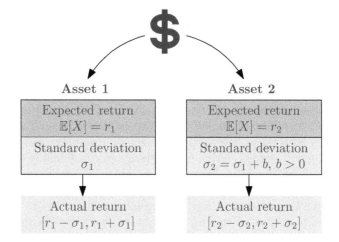

FIGURE 1.1
Diversification problem with two assets.

1.2.2   The Role of the Risk in Engineering Applications

Before introducing the mean-field-type control and game theory, we present the role of risk terms from the economical perspective. Let us consider two different assets where we can invest a total economic resource that is denoted by "$\$$": Asset 1 and Asset 2, as presented in Figure 1.1. The ith asset is characterized by two main features, i.e., the expected return that is denoted by $\mathbb{E}[X_i]$, and its volatility, which is expressed by means of the standard deviation that is denoted by σ_i. Let us assume that the assets have expected returns denoted by $r1, r_2 \in \mathbb{R}$, and volatilities $\sigma_1 < \sigma_2$. Thus, the return that can be earned from assets 1 and 2 belongs to the ranges

$$[r - \sigma_1, r + \sigma_1], \text{ and } [r - \sigma_2, r + \sigma_2],$$

respectively. For the sake of simplicity, letting $r_1 = r_2$, the asset 2 can potentially return more money than asset 1. Nevertheless, asset 2 is riskier.

This problem is known as a diversification optimization introduced by Harry Markowitz (Nobel Memorial Prize in Economic Sciences 1990) in [2,3]. The objective consists in maximizing the total expected return while minimizing the associate risk. For n assets the problem is formally stated as follows:

$$\underset{X_1, \ldots, X_n}{\text{maximize}} \sum_{i=1}^{n} \left(\mathbb{E}[X_i] - \sqrt{\text{var}(X_i)} \right),$$

being X_i the price of the ith asset. Figure 1.2 shows a two-asset portfolio-problem example. In this example, we can observe in the diagram that the asset 1 has an expected return $\mathbb{E}[X_1] = r_1$ and volatility $\sqrt{\text{var}(X_1)} = \sigma_1$, and

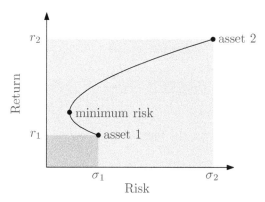

FIGURE 1.2
Risk vs Return plot in a portfolio problem with two assets.

for the asset 2 $\mathbb{E}[X_2] = r_2$ and $\sqrt{\mathrm{var}(X_2)} = \sigma_2$, respectively. The interesting result relies on the fact that there exists an optimal resource allocation (investment split) in the portfolio problem such that the risk (volatility) is minimized.

> "We next consider the rule that the investor does (or should) consider expected return a *desirable* thing and variance of return an *undesirable* thing."
>
> Harry Markowitz
> [33, page 15]

Once the concept of the volatility has been introduced from the economics perspective, we extend the use of this concept to the engineering field, and more precisely, in the automatic control and game theory context.

Mean-field-type game theory studies how to manage with risk terms, i.e., it concerns about quantifying and minimizing risk terms. In order to illustrate the risk management *in the context of engineering applications*, let us consider a temperature control example. The control objective consists in maintaining stable the temperature inside a room, which is represented by a system state that is denoted as $x(t)$, around a desired reference denoted by $r = 20$ (in Celsius degrees) by means of an actuator signal that is denoted by $u(t)$. Moreover, let us consider a disturbance affecting the dynamical behavior of the temperature, e.g., there are windows and doors opening and/or closing, error in the measurements of the current temperature, or any agent perturbing the evolution of the temperature. This control scheme is presented in Figure 1.3.

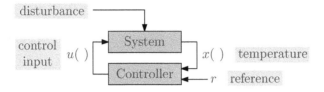

FIGURE 1.3
Feedback control scheme for the temperature control system.

The first option in order to design a temperature controller consists of look-
ing at the expected value of the current temperature, i.e., controlling $\mathbb{E}[x(t)]$
to meet the desired reference r. As an example, consider the performance
of two different scenarios/controllers presented in Figure 1.4(a). Clearly, Scenario 1 performs much better than Scenario 2 since it reaches a steady state
faster, i.e., shorter settling time; the temperature meets the desired target,
i.e., null steady-state error; and it does not oscillate too much, i.e., it has a
small over-shoot behavior.

Nevertheless, notice that the expectation of the temperature does not allow evaluating the risk (or volatility behavior), i.e., it does not allow observing
the real behavior of the temperature. Now, let us check the evolution of the
actual temperature $x(t)$ for both scenarios in Figure 1.4(b). Now, it is clear
that the performance of the controller in Scenario 2 is much better than the
one exhibited in Scenario 1 since it has less variations along the time, i.e.,
Figure 1.4(b) shows the risk terms along the time for the two different scenarios. A second option in order to design the temperature controller consists of
evaluating the variance of the variable-of-interest along the time (risk-aware
approach), i.e., $\mathbb{E}(x(t) - \mathbb{E}[x(t)])^2$. Figure 1.4(c) compares the variance of the
variable-of-interest for the two scenarios. It can be seen that Scenario 2 performs better due to the fact that the variance of $x(t)$ is smaller.

The risk-awareness analysis that was illustrated by means of the temperature controller can be extended to a large number of other engineering
applications in which uncertainties take place. Next, we discuss about such
uncertainties and point out several engineering problems of current research
interest to motivate the study of the mean-field-type control and game theory
that is presented in this book.

1.2.3 Uncertainties in Engineering Applications

Networked engineering systems are becoming of larger-scale nature (involving
a large number of system states, control inputs, and/or variables) and highly
interconnected (several sub-system coupled to each other). Game theory, or
the interactive decision-making theory [5, 34, 35], has become a quite powerful tool to analyze, design, and control these types of systems. Applications

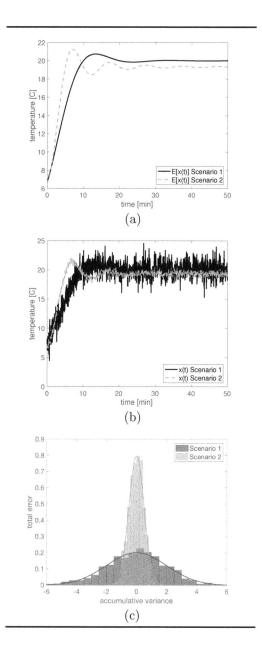

FIGURE 1.4
Evolution of the temperature for two different control scenarios. (a) evolution of the expected temperature. (b) evolution of the temperature. (c) variance comparison of the two scenarios.

FIGURE 1.5
Some engineering applications involving uncertainties.

include wireless networks [25, 36], water systems [37–39], power systems [32,39–41], traffic systems [42], blockchain [43], temperature control [39,44], among many others.

All the previously mentioned engineering applications, and not limited to them, might be affected by multiple uncertainties that can be taken into consideration. Figure 1.5 shows a summarized diagram of some engineering applications incorporating uncertainties where the risk-awareness control and game-theoretical techniques can be implemented. To illustrate how relevant the role of uncertainties is in the engineering applications, we describe next how these uncertainties appear and affect the systems performance.

- In social networks, the uncertainties are associated with the spreading of the information along the network such as fake news, uncertainties in the number of followers, likes, and/or the uncertainty in the reputation of the participants. Also, there can be malicious and spiteful agents perturbing and manipulating the overall behavior of the network. Indeed, opinion dynamics is quite related to the social networks where several uncertainties can affect the evolution of the overall average opinion.

- In drainage and drinking water networks, the objective consists of op-timally manage some of the inflows and outflows such that the drinking water demand is satisfied while economical costs are minimized, or alterna-tively, to manage the wastewater flows in a way that the pollution and/or the overflows are reduced or even driven to zero. In the drinking water networks, there exist uncertainties associated with the resource demand and to the weather conditions that might considerably affect the operation of the system. In wastewater networks, there is uncertainty associated with the precipitations that directly affect the risk to have overflows throughout the system.

- The uncertainties in the power systems can be interpreted in a similar way as for the water networks. There exists a demand of power with an associated uncertainty and the weather conditions also directly affect the performance of the renewable resources such as the wind turbines and the solar panels.

- In the transportation systems, two important goals are: (i) the road safety and (ii) the congestion management throughout the network while mini-mizing the travel time. There are some uncertainties associated with the weather conditions and road incident states, which might affect the traffic flow.

- In the context of blockchains and cryptocurrencies, two important goals are: (i) the incentives design for users, developers, producers, and verifiers; and (ii) the development of a consensus algorithm for the verification and validation of the transactions in a distributed fashion. As in traditional currencies, cryptocurrencies have many uncertainties as the exchange rates along the time. Besides, they add extra uncertainties related to network security and the proof-of-verification.

- In temperature control problem such as the air conditioning system many people use at home, car, or office; the system is continuously dealing with uncertainties when doors and windows are opened for a while, or simply by the fact the temperature measurements might have some errors related to the sensor devices.

Extensive modeling of these systems under *uncertainties* has been considered. Most often, the risk-neutral and the expected utility approach have been used. However, the expected utility which is a linear first-order performance metric may not capture the risk. One possible way to incorporate risk-awareness is through the variance of the performance as studied throughout this book.

1.2.4   Network of Networks/System of Systems

Engineering systems are becoming every time more complex, of large-scale nature (some of them are represented by means of networks), and highly interconnected. The aggregated overall model composed of such interconnected systems is known as *network of networks* or *system of systems*. In order to clarify this concept, let us consider the existing interdependence among the water, traffic, energy, district heating, and communication systems as shown in Figure 1.6. It can be seen that, each system has many components represented by nodes.

First, let us consider the drinking water transportation system. The control objectives consist of providing water to supply the demand while minimizing the operational economic costs. Such management is made by means of valves and pumps, whose operation requires energy, creating an interconnection with the power system. There is an operational cost related to the economic cost of energy to operate the active actuators in the system, and the other operational cost is related to the economic cost of water depending on the source from which it is gotten. Likewise, water is required in hydroelectric systems to produce energy. This fact establishes a relationship between the water and power networks. The interdependence is presented with links connecting the components of each network in Figure 1.6.

Now, let us consider the heat network, also known as district heating, and it is in charge of providing comfort temperature. Similarly, this system

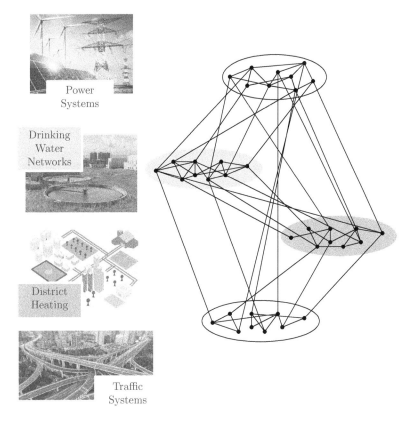

FIGURE 1.6
Network of networks. Interdependence among the water, traffic, energy, district heating and communication systems.

requires both water and power to operate the actuators that transport the thermo-fluids (water) throughout the network.

Regarding the traffic system, there is a large number of agents moving along the transportation network. Notice that there is a direct relationship between the agents (people) motion and the geographical distribution of the demands for the water, the electrical power, and the comfort temperature. Thus, we can observe the interdependence of the traffic system with the water, power, and heat networks illustrated in Figure 1.6. Hence, power systems are required in order to operate the traffic lights, which are the main actuators that regulate the traffic.

Finally, all the aforementioned systems work utilizing a communication network to exchange the information related to the actuators and measurements. Such communication network is represented in Figure 1.6 as the connectivity among all the different components belonging to the same or other systems.

1.2.5   Optimality Systems

For game problems considering mean-field terms, various solution methods such as the Stochastic Maximum Principle (SMP) (see [17]) and the Dynamic Programming Principle (DPP) with Hamilton-Jacobi-Bellman-Isaacs and Fokker-Planck-Kolmogorov equations have been proposed [17,45,46]. Alternatively, to address this class of stochastic control and game problems, this book focuses on the direct method highlighting its advantages among which we can find the following:

- It is easy to apply

- It does not require to solve Partial Differential Equations (PDEs) and

- It allows computing either explicit or semi-explicit solutions (i.e., solutions in terms of ordinary differential equations).

Figure 1.7 shows a time-dependent scheme on some of the recent literature review on different methods to solve mean-field-type control and game problems with *finite number of decision-makers*, i.e., in the context of *atomic games* where the decisions or strategic selections cannot be neglected.

In [47] and [48], optimal control problems under systems governed by mean-field-type Stochastic Differential Equations (SDE) are studied by applying the SMP. Similarly, in this work, both the system dynamics and cost functional are allowed to be of mean-field type. Further analysis on the SMP applied to mean-field-type problems has been made in [49], where authors consider second-order adjoint processes. Later on, solutions for mean-field-type control problems have been discussed by using methods either based on Hamilton-Jacobi-Bellman and Fokker-Planck coupled equations or based on SMP in [17]. Other authors have contributed by adding variations, extensions and also by using other different methods. For instance, [50] studies discrete-time mean-field-type control problems, and presents necessary and sufficient conditions for their solvability. In [51], the SMP is applied to solve jump-diffusion mean-field problems involving mean-variance terms, i.e., of mean-field type.

Other considerations have been added into these classes of problems. For instance, information related issues have been studied. In [52], it is proposed to solve mean-field-type control problems by applying dynamic programming, and two examples are presented: (i) a portfolio optimization, and (ii) a systemic risk model. In [53], the SMP is used to solve mean-field-type control problems with partial observation and the results are applied to financial engineering. The work in [54] addresses a partially observed optimal control problem whose cost functional is of mean-field type and the authors solve it by means of a maximum principle using Girsanov's theorem and convex variation.

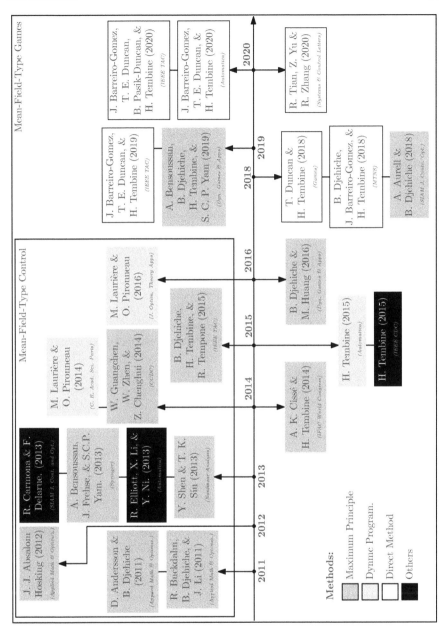

FIGURE 1.7
Brief recent literature review on different methods to solve mean-field-type control and game problems with finite number of decision-makers.

The well-posedness of mean-field-type forward-backward stochastic differential equations is studied in [55] motivated by the work previously reported in [56], where the same problem was discussed but in the context of mean field (not of mean-field-type). In [24] the SMP has been extended to the risk-sensitive mean-field-type control problem, which involves not only first and second moment terms, but also higher terms. In [57], a stochastic robust (H_2/H_∞) control of mean-field-type with state and control-input-dependent noise is studied. In [58], the well-posedness and solvability of an indefinite linear-quadratic mean-field-type control problem in discrete time is discussed, developing over previous discussions made in [50] by making further generalizations on the settings. The kind of partial-observation mean-field-type control problems, as those addressed by authors in [53] and [54], have been extended to the risk-sensitive case in [59] and [60], and by applying the backward separation method in [61] and [62]. Besides, the continuation of the research performed in [52] on dynamic programming has been reported in [63]. The dynamic programming approach has been later studied in an adaptive manner for discrete-time problems. Hence, a new consideration over the mean-field-type control problems by incorporating common noise has been studied in [64].

Different from the aforementioned works, which study and analyze mean-field-type control problems (i.e., with a unique decision-maker), the work in [65] presents these ideas for game theoretical problems (i.e., with multiple decision-makers), e.g., cooperative mean-field-type games. Thus, several developments on mean-field-type theory involving both control and game approaches have been reported in the literature. Regarding the mean-field-type game theoretical results, uncertainty quantifications have been studied in [66] in the context of mean-field-type teams and by using the Kosambi-Karhunen-Loeve expansion (chaos expansion approach), which allows representing the stochastic process as a linear orthogonal-functions combination. In [67], mean-field-type games are characterized with time-inconsistent cost functionals and the SMP is applied. Hence, new results on risk-sensitive mean-field-type games have been reported in [26] by extending the risk-sensitive control approach previously analyzed in [24], and sufficient optimality equations are established via infinite-dimension dynamic programming principle. Other classes of games have been studied in the mean-field-type area. The two-player game scenario has been widely studied for both the non-zero-sum and zero-sum (robust) cases in [68] and [69], respectively. In contrast to the different methods used before to solve mean-field-type control and game problems, e.g., dynamic programing or SMP, authors in [70] use the so-called *direct method* in order to compute explicit and semi-explicit solutions for mean-field-type game problems for the linear-quadratic case. The work in [70] served as a motivation for other results using the same method, e.g., the most recent results related to the direct method on mean-field-type games are [71] and [72], discussing non-linear continuous-time problems, and discrete-time problems involving different information issues, respectively. Regarding the most recent research in the field of mean-field-type theory, mean-field-type games with jump and regime switching are studied by using dynamic programing and SMP principles in [73] and following the direct method in [74]. Other game-theoretical solution concepts such as partial cooperation, competition, partial altruism, etc are studied by means of the co-opetitive mean-field-type games in [75] and by following the direct method for linear-quadratic problems.

Then, results on mean-field-type games started being applied over concrete engineering applications. For instance, mean-field-type games have been applied to distributed power networks with prosumers in [64], to the design of filters for big data assimilation in [76], to the network security analysis as a public good problem in [77], and to the demand-supply management in smart grid in [78], where also constraints have been taken into account.

The direct method has allowed designing risk-aware control and game techniques for a large number of engineering applications. For instance, in [79] a mean-field-type-based pedestrian crowd model is presented. In [32], applications on electricity price in smart grids and blocked-based power networks are studied by following the direct method. In [80], a class of control input constraints are considered for linear-quadratic mean-field-type games and an application of water distribution system is presented. Hence, pedestrian motion has been studied in the context of games in [81], and other applications in traffic, water, energy, block-chains, power systems, among others, have been developed by using mean-field-type theory, e.g., see the works in [82], [83], and [37].

The following section is in charge of presenting one of the methods to solve mean-field-type control and game problems consisting in the master system comprising the backward Hamilton-Jacobi-Bellman and forward Fokker-Plank-Kolmogorov equations. This method focuses on satisfying optimality conditions presented by means of a Partial-Integro Differential Equations (PIDE) system. The next section seeks to motivate the reader by pointing out that this book intends to considerably reduce the complexity to solve the underlying problems avoiding to solve PDEs.

1.3 Game Theoretical Solution Concepts

In the context of decision-making theory, there are several problem statements, information configurations, and/or strategic scenarios that can be studied. Perhaps, the most popular solution concept in game theory is the Nash equilibrium, which provides a solution for a non-cooperative game problem. Nevertheless, there is an enriched variety of solution concepts that we present next, i.e., adversarial, Stackelberg, Berge, among others, and that are of interest depending on the engineering problem we are working on.

Let us consider a general scalar-valued system involving n decision-makers from the set $\mathcal{N} = \{1, \ldots, n\}$, whose dynamics are given by a SDE as follows:

$$dx(t) = b(x, u_1, \ldots, u_n)dt + \sigma(x, u_1, \ldots, u_n)dB(t), \qquad (1.1)$$

where $x \in \mathbb{X}$ denotes the scalar system state, $u_i \in \mathbb{U}_i$ denotes the scalar control input for the decision-maker $i \in \mathcal{N}$, and B denotes a standard Brownian motion. Moreover, $b : \mathbb{X} \times \mathbb{R} \times \prod_{j \in \mathcal{N}} \mathbb{U}_j \to \mathbb{R}$ and $\sigma : \mathbb{X} \times \mathbb{R} \times \prod_{j \in \mathcal{N}} \mathbb{U}_j \to \mathbb{R}$ are state and control-dependent drift and diffusion terms, respectively. Each decision-maker has a cost functional of mean-field type as follows:

$$L_i(x, u_i) = h_i(x(T)) + \int_0^T \ell_i(x(t), u_i(t))dt,$$

where $h_i : \mathbb{X} \to \mathbb{R}$ denotes the terminal cost and $\ell_i : \mathbb{X} \times \mathbb{U}_i \to \mathbb{R}$ denotes the running cost. Next, we present the different game-theoretical problems (i.e., the different game theoretical solution concepts) that we are going to address along this book.

1.3.1   Non-cooperative Game Problem

The first problem is given by a non-cooperative situation involving two decision-makers, i.e., considering the set $\mathcal{N} = \{1, 2\}$. Each decision-maker is interested in minimizing the magnitude of the system state x while reducing the used energy determined by the control input.

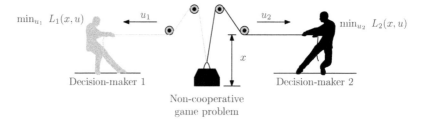

Non-cooperative
game problem

FIGURE 1.8
Two decision-makers illustrating a non-cooperative game problem.

Figure 1.8 shows an illustrative example for a non-cooperative situation. Each decision-maker solves the following problem:

$$\underset{u_i \in \mathcal{U}_i}{\text{minimize}} \ \mathbb{E}[L_i(x, u_i)],$$

subject to (1.1) and a given initial condition for the system state $x(0) \triangleq x_0$. The aforementioned problem represents a dilemma or a conflict since the decision-makers want to minimize the magnitude of the state but without making any effort (or applying the least effort as possible).

1.3.2 Fully-Cooperative Game Problem

Let us now suppose that the two decision-makers cooperate with each other to minimize the magnitude of the system state x while reducing the aggregative weighted effort to do so, i.e., $q_1 u_1 + q_2 u_2$ with $q_1, q_2 > 0$. In this regard, decision-makers optimize jointly as a team. The fully-cooperative game problem is given by:

$$\underset{\{u_j \in \mathcal{U}_j\}_{j \in \mathcal{N}}}{\text{minimize}} \sum_{j \in \mathcal{N}} \mathbb{E}[L_j(x, u_j)],$$

subject to (1.1) and a given initial condition $x(0) \triangleq x_0$.

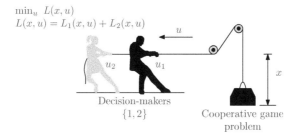

FIGURE 1.9
Two decision-makers illustrating a cooperative game problem.

Figure 1.9 shows an illustrative example of cooperation between two decision-makers. They act jointly pursuing the same objective. It is worth to mention that the cooperative game solution corresponds to the control problem solution when the control input in the control problem is of the dimension of the joined control inputs for all decision-makers in the game problem.

1.3.3 Adversarial Game Problem

There are other situations where the interests of the decision-makers are opposite i.e., what a decision-maker is pursuing goes against what its opponent wants. In this case, we have an adversarial situation. There is a decision-maker acting against the other. Let $u \in \mathbb{U}$ be the control input of the first decision-maker and let $v \in \mathbb{V}$ denote the control input of the second decision-maker. On one hand, the first decision-maker pursues to minimize the magnitude of the system state together with its applied energy. In contrast, the second decision-maker pursues to maximize the magnitude of the system state while minimizing the applied energy.

The adversarial game problem is given by:

$$\underset{v \in \mathcal{V}}{\text{maximize}} \ \underset{u \in \mathcal{U}}{\text{minimize}} \ \mathbb{E}[L(x, u, v)],$$

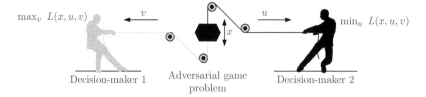

$$\max_v \ L(x, u, v)$$ v u $$\min_u \ L(x, u, v)$$

Decision-maker 1   Adversarial game   Decision-maker 2
problem

FIGURE 1.10
Two decision-makers illustrating an adversarial game problem.

subject to

$$\mathrm{d}x(t) = b(x, u, v)\mathrm{d}t + \sigma(x, u, v)\mathrm{d}B(t),$$

and a given initial condition $x(0) \triangleq x_0$. Figure 1.10 illustrates by means of an example the adversarial situation. It can be seen that one decision-maker pursues a contrary objective with respect to the other.

1.3.4 Berge Game Problem

Let us now analyze an altruistic behavior known as a mutual support situation. The idea is as the saying: *"I help you, and you help me."* First, it is necessary to highlight that this situation differs from the cooperative scenario where the decision-makers had a common interest. Let us consider only two decision-makers in the set $\mathcal{N} = \{1, 2\}$. The Berge game problem is given by

$$\underset{u_j \in \mathcal{U}_j}{\text{minimize}} \ \mathbb{E}[L_i(x, u)], \ i \in \mathcal{N} \setminus \{j\},$$

subject to (1.1) and a given initial condition $x(0) \triangleq x_0$. Notice that, it is also possible to consider a strategic interaction with multiple decision-makers with a set $\mathcal{N} = \{1, \dots, n\}$. Under such scenario, we can still study Berge solutions using a particular criterion. One alternative consists of minimizing the cost of the decision-maker whose cost functional is the highest. In other words, supporting the decision-maker who needs it the most, i.e.,

$$\underset{u_j \in \mathcal{U}_j}{\text{minimize}} \ \mathbb{E}[L_i(x, u)], \ i \in \arg \underset{a \in \mathcal{N} \setminus \{j\}}{\max} \{L_a(x, u)\},$$

subject to (1.1) and a given initial condition $x(0) \triangleq x_0$.

Figure 1.11 shows a mutual support situation. The first decision-maker acts minimizing the second decision-maker's cost functional and vice versa. This problem is quite different to the non-cooperative game problem due to the fact that the control input of a decision-maker is not incorporated in the cost functional of the other, what significantly modifies the optimal solution.

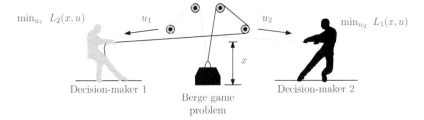

FIGURE 1.11
Two decision-makers illustrating a Berge game problem.

1.3.5 Stackelberg Game Problem

We analyze now a strategic interaction that is played in sequence with respect to the time. The game is composed of two stages comprising two different time instants known as action and reaction. Alternatively, for the two decision-maker game problem, it is interpreted that there is a leader denoted by j and a follower denoted by i. The first decision-maker, known as leader, plays; and then, the second decision-maker, known as follower, reacts against the leader's selection. This configuration has gotten special importance to model situations in which a hierarchical scheme emerges. For instance, the government imposing rules and the population reacting, or a huge and powerful company deciding the prices for the products and small competition companies reacting strategically to compete in the market. Hence, it is interesting to determining if the leader has any advantage over the follower and under which conditions it occurs.

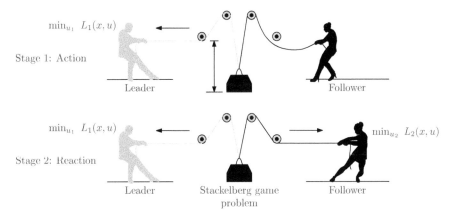

FIGURE 1.12
Two decision-makers illustrating a Stackelberg game problem.

The Stackelberg game solution is characterized by

$$u_j^* \in \arg \min_{u_j \in \mathcal{U}_j} \{L_j(x, u) : u_i \in \mathrm{BR}_i(u_j)\},$$

$$u_i^* \in \mathrm{BR}_i(u_j),$$

where $\mathrm{BR}_i(u_j) = \{u_i : u_i \in \arg \min L_i(x, u_i, u_j), \text{given } u_j\}$ denotes the best-response or *best reaction* against the decision made by the leader. Figure 1.12 presents the sequential strategic situation corresponding to a Stackelberg game problem. It can be seen that the leader selects its strategy and then the follower reacts to it.

1.3.6 Co-opetitive Game Problem

The mixture of both the cooperation and competition is known as co-opetition. When having a larger number of decision-makers beyond two, several combinations may emerge. In real life, some people might help others while behaving indifferent, or even in detriment, of the others. To provide a concrete example, consider some elections with three candidates: *a*, *b*, and *c*. Any candidate, let us say *b*, can act during the campaign against a candidate, let us say *a*, in order to help the other candidate, let us say candidate *c*. In this case, we cannot really claim that the decision-maker *b* is a non-cooperative or cooperative one since its behavior is a combination of both.

In order to demonstrate this concept, let us consider a strategic interaction involving three decision-makers. Figure 1.13 shows an illustrative example of a co-opetitive situation.

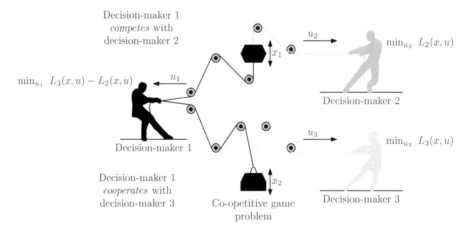

FIGURE 1.13
Three decision-makers illustrating a co-opetitive game problem.

The interaction of the decision-makers 1 and 2 is given by an adversarial situation, whereas the interaction between the decision-makers 1 and 3 is characterized by a non-cooperative situation. Such heterogeneity of the behavior with respect to different decision-makers in the same strategic interaction is known as a co-opetitive game problem.

1.3.7 Partial-Altruism and Self-Abnegation Game Problem

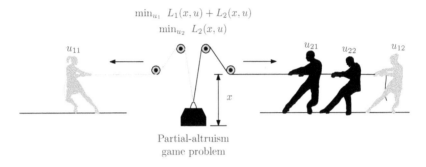

FIGURE 1.14

Four decision-makers illustrating a Partial-Altruism game problem.

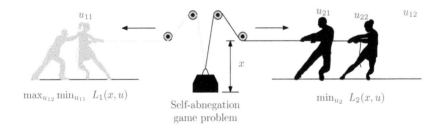

FIGURE 1.15

Four decision-makers illustrating a Self-Abnegation game problem.

Once the co-opetitive game problem has been presented in the previous section, we observe that there is a large variety of possibilities and situations to analyze with game theory. To motivate such flexibility we consider a two-team interactive decision-making. Figures 1.14 and 1.15 present a partial altruistic scenario and self-abnegation case, respectively. On one hand, we observe an altruistic behavior when one team helps the other. On the other hand, we observe a self-abnegation case when a member of the team is sabotaging (pushing those who are pulling).

1.4 Partial Integro-Differential System for a Mean-Field-Type Control

Let us consider a control problem as the one presented in Section 1.3.2 corresponding to the fully-cooperative game problem but with a unique decision-maker as shown in Figure 1.16. This control problem can be solved by

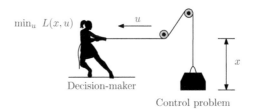

$\min_u \ L(x, u)$

u

x

Decision-maker

Control problem

FIGURE 1.16
Control problem. Equivalently, a single decision-maker decision-making problem.

computing the solution of the Hamilton-Jacobi-Bellman and Fokker-Plank-Kolmogorov equations, which compose a Partial-Integro-Differential Equation (PIDE) system for mean-field-type control. These equations describe optimality conditions for the solution of the mean-field-type problem.

Let us consider a general scalar-valued system whose dynamics are given by a SDE as follows:

$$dx(t) = b(t, x, u, \phi)dt + \sigma(t, x, u, \phi)dB(t), \tag{1.2}$$

where $x \in \mathcal{X}$ denotes the scalar system state, $u \in \mathbb{U}$ denotes the scalar control input, $\phi \in \mathbb{R}$ is the probability measure of x, and B denotes a standard Brownian motion. Moreover, the mappings $b : [0, T] \times \mathcal{X} \times \mathbb{U} \times \mathcal{P}(\mathbb{R}) \to \mathbb{R}$ and $\sigma : [0, T] \times \mathcal{X} \times \mathbb{U} \times \mathcal{P}(\mathcal{X}) \to \mathbb{R}$ are state and control-dependent drift, and diffusion terms, respectively. The control objective consists of minimizing the following cost functional of mean-field type as follows:

$$L(t, x, u, \phi) = h(x(T), \phi(T)) + \int_0^T \ell(t, x(t), u(t), \phi(t))dt,$$

where $h : \mathcal{X} \times \mathcal{P}(\mathcal{X}) \to \mathbb{R}$ denotes the terminal cost and $\ell : [0, T) \times \mathcal{X} \times \mathbb{U} \times \mathcal{P}(\mathcal{X}) \to \mathbb{R}$ denotes the running cost. The risk-aware control problem is:

$$\underset{u \in \mathcal{U}}{\text{minimize}} \quad \int_{\mathcal{X}} L(t, x, u, \phi)\phi(t, dx), \tag{1.3a}$$

subject to

$$dx(t) = b(t, x, u, \phi)dt + \sigma(t, x, u, \phi)dB(t), \tag{1.3b}$$

$$x(0) \triangleq x_0. \tag{1.3c}$$

The problem in (1.3a)–(1.3c) is of mean-field type, or equivalently, it is a risk-aware control problem involving mean-field terms, which are computed by using the probability measure of the system state as follows:

$$\mathbb{E}[x(t)] = \int_{\mathbb{R}} y \, \phi(t, dy),$$

$$\mathbb{E}[u(t)] = \int_{\mathbb{R}} u(t, y, \phi)\phi(t, dy),$$

since the optimal control input is expressed in terms of the system state and its probability measure. The solution of the risk-aware control problem is found by solving the following backward-forward partial-integro-differential system presented next:

$$\phi_t = -(\phi \, b)_x + \frac{1}{2}(\phi\sigma^2)_{xx}, \tag{1.4a}$$

$$\phi(0) = \phi_0, \tag{1.4b}$$

$$0 = V_t(t, \phi) + \int_{\mathcal{X}} H(t, x, \phi, V_{x\phi}, V_{xx\phi})\phi(t, dx). \tag{1.4c}$$

$$V(T, \phi) = \int_{\mathcal{X}} h(y, \phi)\phi(T, dy). \tag{1.4d}$$

The initial boundary condition for ϕ makes the equation go forward, and terminal boundary condition over V makes the equation go backward. Please notice that for simplicity in the notation, in the system (1.4a)-(1.4d) subindexes denote partial derivatives. For instance, $[\cdot]_t$ denotes the partial derivative with respect to time, and $[\cdot]_{xx}$ denotes the second partial derivative with respect to x. Hence,

$$V(t, \phi) = \underset{u \in \mathcal{U}}{\text{minimize}} \int_{\mathcal{X}} L(t, x, u, \phi)\phi(t, dx)$$

denotes the optimal cost from time t up to T in the running cost, and

$$H(t, x, \phi, V_{x\phi}, V_{xx\phi}) = \underset{u \in \mathcal{U}}{\text{minimize}} \left\{ \ell(t, x, u, \phi) + b(t, x, u, \phi)V_{x\phi} \right.$$

$$\left. + \frac{\sigma(t, x, u, \phi)^2}{2} V_{xx\phi} \right\} \tag{1.5}$$

denotes the *integrand Hamiltonian*. In (1.4a)–(1.4d), The Fokker-Planck-Kolmogorov equation is given by (1.4a), and the Hamilton-Jacobi-Bellman equation is the one in (1.4c). The solution of the system in (1.4a)–(1.4d) is complex and normally requires from the implementation of numerical methods. In fact, when problems involving stochastic processes beyond Brownian motion such as Poisson jumps, the Hamiltonian (1.5) becomes more difficult involving another integration over the jump set in the integrand Hamiltonian (see e.g., [73, 74, 84]) making more challenging to compute the solution. Besides, the system presented in (1.4) corresponds to the unique-decision-maker problem and it must be extended to multiple decision-makers for solving game problems.

Let us consider a mean-field-type games with $i \in \mathcal{N} = \{1, \ldots, n\}$ decision-makers where $n \geq 0$, $n \in \mathbb{N}$, corresponding to a non-cooperative game problem as shown in Section 1.3.1. The Hamilton-Jacobi-Bellman system corresponding to this game problem incorporates as many Hamiltonian as decision-makers as follows:

$$0 = V_{i,t}(t, \phi) + \int_{\mathcal{X}} H_i(t, x, \phi, V_{x\phi}, V_{xx\phi}) \phi(t, dx),$$

$$V_i(T, \phi) = \int_{\mathcal{X}} h_i(y, \phi) \phi(T, dy),$$

where the integrand Hamiltonian is

$$H_i(t, x, \phi, V_{x\phi}, V_{xx\phi}) = \underset{u_i \in \mathcal{U}_i}{\text{minimize}} \Big\{ \ell_i(t, x, u, \phi)$$

$$+ b(t, x, u, \phi) V_{i,x\phi} + \frac{1}{2} \sigma(t, x, u, \phi)^2 V_{i,xx\phi} \Big\}.$$

In the Hamiltonian, the functions have the following mappings:

$$b, \sigma : [0, T] \times \mathcal{X} \times \prod_{j \in \mathcal{N}} \mathbb{U}_j \times \mathcal{P}(\mathcal{X}) \to \mathbb{R},$$

$$\ell_i : [0, T) \times \mathcal{X} \times \prod_{j \in \mathcal{N}} \mathbb{U}_j \times \mathcal{P}(\mathcal{X}) \to \mathbb{R},$$

$$h_i : \mathcal{X} \times \mathcal{P}(\mathcal{X}) \to \mathbb{R}.$$

An important issue to study and analyze consists in the existence and uniqueness for the system in (1.4a)-(1.4d). Such problems have been studied, for instance, in [73], [85] and [86].

The following important remark clarifies/emphasizes that the approach this book follows pursues to make the mean-field-type game theory accessible to engineers and early career researchers in the field.

Important Remark

The motivation in this book consists in solving the mean-field-type control and game problems **by avoiding** to compute the solution of the *Backward-Forward Partial-Integro-Differential System for Mean-Field-Type either Control or Games*. Instead, this book proposes to follow the direct method either in continuous or discrete time in order to find semi-explicit solutions for the mean-field-type problems.

This book intends to be oriented to engineers, beginners in the mean-field-type control and games field, and early career researchers.

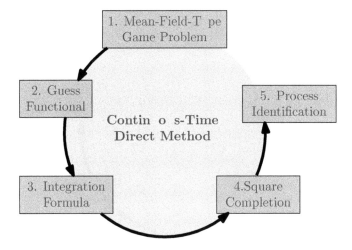

FIGURE 1.17
General scheme corresponding to the Continuous-Time Direct Method

1.5 A Simple Method for Solving Mean-Field-Type Games and Control

The mean-field-type game problems presented in this book are solved in a semi-explicit way by means of the so-called direct method. This method can be implemented in either continuous or discrete time. Next, we show the general steps corresponding to the direct method.

1.5.1 Continuous-Time Direct Method

Figure 1.17 presents the general scheme of the continuous-time direct method. In the first step, we have the mean-field-type game problem statement. Then, inspired by the structure of both the cost functional and the system dynamics of the mean-field-type game problem exhibited in the first step, a guess functional is proposed in the second step. Afterward the integration formula is applied, which is given by the Itô's formula. It follows to perform square completion (or optimization over the control-input-dependent terms) to determine the optimal control inputs in the fourth step. It could be identified that the direct method has a tight relationship with the HJB equation. Thus, this fourth step corresponds to the optimization of the Hamiltonian with respect to the control inputs. Finally, the process identification allows to deduce the optimality for both the control inputs and the cost functional.

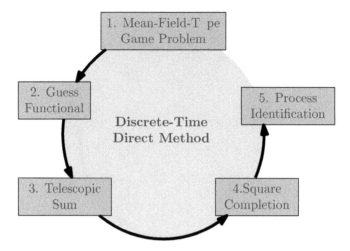

FIGURE 1.18
General scheme corresponding to the discrete-time direct method.

1.5.2   Discrete-Time Direct Method

In this book we also address discrete-time mean-field-type game problems, corresponding to the first step of the discrete-time direct method presented in Figure 1.18. It is worth to highlight that the discrete approach might be more suitable for implementation considerations in real engineering applications. Then, following the same reasoning as in the continuous-time direct method, a guess functional is also proposed in the second step. As a third step, and different from the continuous-time direct method, we apply the telescopic sum. Finally, steps four and five allow identifying the optimal control input by performing square completion (or optimization over the control-input-dependent terms), and the respective recursive equations associated with the Riccati equations. Throughout the book, the reader will observe that the complexity to get the solution for discrete-time problems is not high, but in contrast, the computation becomes long.

1.6   A Simple Derivation of the Itô's Formula

One of the main ingredients of the direct method is the integration formula, which is given by the Itô's formula. Here, we present a simple derivation of the Itô's formula by means of a power series or Taylor expansion. Let us consider the same SDE as in Section 1.4 in (1.2), i.e., $\mathrm{d}x = b\mathrm{d}t + \sigma\mathrm{d}B$, where $b : \mathbb{X} \times \mathbb{R} \times \mathbb{U} \to \mathbb{R}$ denotes the state and control-dependent drift term, and

$\sigma : \mathbb{X} \times \mathbb{R} \times \mathbb{U} \to \mathbb{R}$ is the state and control-dependent diffusion term. Let $f(t, x)$ be a twice-differentiable function and its Taylor expansion is as follows:

$$df = \frac{\partial f}{\partial t}dt + \frac{\partial f}{\partial x}dx + \frac{1}{2}\frac{\partial^2 f}{\partial x^2}dx^2 + \text{higher order terms}.$$

Now, replacing dx yields

$$df = \frac{\partial f}{\partial t}dt + \frac{\partial f}{\partial x}(bdt + \sigma dB) + \frac{1}{2}\frac{\partial^2 f}{\partial x^2}(bdt + \sigma dB)^2 + \text{higher order terms},$$

and

$$df = \frac{\partial f}{\partial t}dt + \frac{\partial f}{\partial x}bdt + \frac{\partial f}{\partial x}\sigma dB + \frac{1}{2}\frac{\partial^2 f}{\partial x^2}b^2 dt^2 + \frac{1}{2}\frac{\partial^2 f}{\partial x^2}\sigma^2 dB^2$$
$$+ \frac{\partial^2 f}{\partial x^2}b\sigma dtdB + \text{higher order terms}.$$

Considering the fact that $\mathbb{E}[B^2(t)] = t$ by the property of a Brownian motion, it follows that

$$df = \frac{\partial f}{\partial t}dt + \frac{\partial f}{\partial x}bdt + \frac{\partial f}{\partial x}\sigma dB + \frac{1}{2}\frac{\partial^2 f}{\partial x^2}\sigma^2 dt.$$

Finally, since df should be linear in $[dt, dB]$, by identification one obtains

$$df = \left(\frac{\partial f}{\partial t} + \frac{\partial f}{\partial x}b + \frac{1}{2}\frac{\partial^2 f}{\partial x^2}\sigma^2\right)dt + \frac{\partial f}{\partial x}\sigma dB,$$

which is the Itô's formula for standard Brownian process.

1.7 Outline

This book is divided into six main parts:

1. First part is devoted to the introduction of preliminary works on the field of mean-field and mean-field-type control and game theory. The Direct Method is introduced for both continuous-time and discrete-time versions. Then, we highlight the purpose and objectives of this book oriented to engineers and early career researchers. Moreover, this first part also motivates the study of mean-field-type control and games, and shows the outline of this book.

2. In the second part, we introduce both mean-field-free and mean-field games in continuous time. For the sake of simplicity in the explanation, this part of the book assumes the system state and control inputs to be scalar values. These approaches are solved by

using the Direct Method. These two type of games allow to highlight the main differences with respect to the mean-field-type games,
which is the core of this book. In addition, it is important to mention to the reader that the two main branches in this book, i.e.,
the continuous-time and discrete-time mean-field-type games, can
be studied independently. Therefore, the reader can feel free to omit
the continuous-time approach and go directly over the discrete analysis within Part II and Part V.

3. Afterward, the third part continues focusing on the one-dimensional
    case. We introduce the simplest mean-field-type game problem,
    which is solved by means of the Direct Method. In addition, we
    study different solution concepts, i.e., the non-cooperative, fully-
    cooperative, and the co-opetitive scenarios. The reader can also re-
    fer to the introductory section presented in Part I related to the
    different game-theoretical solution concepts.

    Finally, other two game-theoretical solution concepts are studied,
    i.e., the Stackelberg (leader-follower) mean-field-type game consist-
    ing in a sequential strategic interaction, and the Berge mean-field-
    type game consisting in a mutual support consideration.

4. The study of matrix-valued mean-field-type games is presented in
    the fourth part, where the proposed problems are semi-explicitly
    solved by using the Direct Method. In addition, this part also dis-
    cusses about an alternative to consider multiple coupled input con-
    straints by means of auxiliary variables and without affecting the
    method to obtain semi-explicit solutions.

5. Part five presents the discrete-time version of both the scalar-valued
    and matrix-valued cases. This part also studies the cooperative and
    non-cooperative solution concepts that are presented in the con-
    tinuous counterpart. Note that this part does not discuss about
    the co-opetitive, Stackelberg, and Berge problems in discrete time
    given that, although they are interesting problems, the authors of
    this book consider they do not add extra difficulty.

6. Finally, the last part focuses on showing other problem settings such
    as the stationary case, risk-aware model predictive control approach,
    data-driven mean-field-type games, and also presents a data-driven
    mean-field-type game perspective by using machine learning and
    massive data about the dynamical behavior of an unknown system
    by means of a simple linear regression. This part also presents sev-
    eral engineering applications.

Figure 1.19 shows the outline of this book divided into the main six parts,
and presents the suggested interdependence among the chapters to read. Thus,
we suggest some prerequisites for each chapter.

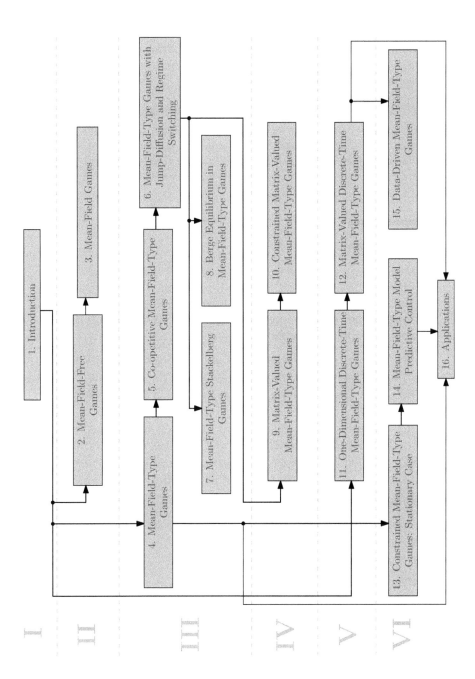

FIGURE 1.19
Outline of the book. Arrows show the interdependence among the chapters.

Important Notation

To improve the readability and ease the understanding of the proposed method presented throughout this book, we have omitted some arguments in the functions.

For instance, a drift function in a stochastic differential equation depending on time t, regime switching s, a system state x, the expectation of the system state $\mathbb{E}[x]$, a control input u, and the expectation of the control $\mathbb{E}[u]$, which is given by $b(t, s, x, \mathbb{E}[x], u, \mathbb{E}[u])$, we simply denote it by b.

1.8   Exercises

1. Mention an application from your research or study field of interests (different from the ones discussed in Section 1.2.3), in which the quantification and minimization of risk is (or could be) important to enhance the desired performance. Once you select your application, answer the following questions:

   (a) What is the control objective associated with your selected application?
   (b) What kind of uncertainties are involved in your selected application?
   (c) What kind of control strategies have been implemented to your selected application?
   (d) Which other system could your engineering problem be coupled with?, please see Section 1.2.4 as a guidance.

2. Define the following terms:

   (a) Mean
   (b) Standard deviation
   (c) Variance
   (d) Covariance
   (e) Skewness
   (f) Kurtosis

3. Define the following type of decision-makers:

   (a) Risk-aware decision-maker
   (b) Risk-averse decision-maker
   (c) Risk-neutral decision-maker
   (d) Risk-seeking decision-maker

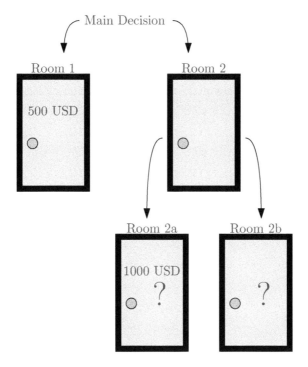

FIGURE 1.20
Risk doors exercise. Would you risk for a bigger reward?

    (e) Risk-sensitive decision-maker

4. Let us suppose you have the following offer. You can select between two rooms: 1 and 2. Once you have selected a door you cannot move back or change your selection.

    • Room 1: You will find 500 USD

    • Room 2: You will have the option to enter any of two rooms: 2a and 2b. In one of them, you will find 1000 USD and in the other there is nothing.

The situation is summarized in Figure 1.20. Answer the following questions:

    (a) What would you do?, i.e., to take the 500 USD for sure, or pursue to get 1000 USD?

    (b) Do you consider yourself a risky person?

    (c) What is the expected reward for rooms 1 and 2?

5. Let us consider the same scenario as in the previous exercise but with a reward xUSD in Room 1. Response then the following questions:

(a) For which values of x, a risk-averse decision-maker would select Room 1?

(b) For which values of x, a risk-neutral decision-maker would select randomly between Rooms 1 and 2?

(c) For which values of x, a risk-seeking decision-maker would select Room 2?

Part II

Mean-Field-Free and Mean-Field Games

2

Mean-Field-Free Games

In the context of mean-field-free games, the expected payoff functional is linear with the respect to the probability measure of the (individual) state. Game problems, in which the state dynamics is given by a linear system and a cost functional that is quadratic in the state and in the control inputs, is often called the linear-quadratic (LQ) differential games. Then, these are referred to as *linear-quadratic mean-field-free games*, see [20]. When, in addition, the linear system dynamics incorporates a diffusion term with a Brownian motion, the game is called a linear-quadratic (LQ) stochastic differential game, or equivalently, we refer to it as a *linear-quadratic stochastic mean-field-free games*.

In this chapter, we present the linear-quadratic game problems in which there is no mean-field term, i.e., the distribution of the variables-of-interest is part neither the cost functional nor the system dynamics. We first introduce a deterministic game problem involving arbitrary number of players in continuous time, i.e., a differential linear-quadratic game. This will serve as a basic example in order to understand the transition from optimal control theory to differential game theory. Moreover, we point out that the direct method used in this first part of the book is going to be followed throughout the entire book. In addition, the reader will be also introduced the discrete-time linear-quadratic game counterpart, which is solved by following the same method. We suggest the reader to follow the semi-explicit computation of the solutions while checking the steps diagram presented in Chapter 1, Section 1.5.

In order to continue with a natural extension of these differential (continuous time) and difference (discrete time) game problems, we continue by introducing the stochastic problem involving a Brownian motion or Wiener process. Indeed, these stochastic scenarios are equally solved by following the same direct method and pointing out the main difference in the application of the integration formula. Once both the deterministic and stochastic problems are solved in a semi-explicit manner, then we discuss about the relationship between their solutions. In this regard, the reader can observe the implication of adding a state and control-input-independent diffusion term.

DOI: 10.1201/9781003098607-2

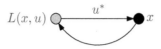

FIGURE 2.1
A basic scalar-valued control scheme.

2.1   A Basic Continuous-Time Optimal Control Problem

Let us consider the following dynamical system with a unique control input
as the one illustrated in Figure 2.1, which is mean-field-free, i.e.,

$$\dot{x} = b_1 x + b_2 u, \tag{2.1}$$

where $x \in \mathbb{R}$ corresponds to the system state and $u \in \mathbb{R}$ denotes the unique
control input. Hence, $\dot{x}$ denotes the time derivative of the state x. Finally,
$b_1, b_2 \in \mathbb{R}$ are the system state parameters, which might be considered time-
dependent, i.e., these parameters can be considered as $b_1, b_2 : [0, T] \to \mathbb{R}$
without this implying a modification in the results exposed in this chapter,
different from keeping considering $b_1(t)$ and $b_2(t)$ time-dependent terms. The
mean-field-free cost functional to be minimized is as follows:

$$L(x, u) = \frac{1}{2} q(T) x(T)^2 + \frac{1}{2} \int_0^T \left(qx^2 + ru^2 \right) dt. \tag{2.2}$$

The interpretation of the cost functional is explained as composed by
two main elements (also named control objectives) involving both the system states and the control input. The cost presented in (2.2) pursues to make
the system state vanish to minimize the quadratic penalization over the states,
but at the same time it is desired to do so by using the least energy as possible
by minimizing the quadratic penalization over the control input. The weight
parameters q and r determine the prioritization that is assigned over these
two control objectives, i.e., the weight parameters allow tuning in the cost
function which control objective is the most important one. Details about the
conditions of these parameters are presented in Problem 1.

Problem 1 (Mean-Field-Free Optimal Control Problem) *The   optimal control problem is given by*

$$\begin{cases} \underset{u \in \mathcal{U}}{\text{minimize}} \; L(x, u), \\[2mm] \text{subject to} \\[2mm] \dot{x} = b_1 x + b_2 u, \\ x(0) := x_0, \end{cases}$$

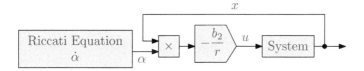

FIGURE 2.2
Feedback scheme for the linear-quadratic mean-field-free optimal control.

where the cost-functional parameters $q \in \mathbb{R}_{\geq 0}$, $r \in \mathbb{R}_{>0}$, the system state parameters $b_1, b_2 \in \mathbb{R}$, and $\mathcal{U}$ denotes the set of measurable and feasible real-valued control inputs.

Now, Proposition 1 presents the semi-explicit solution for the mean-field-free optimal control problem with scalar-valued parameters, i.e., being the system state and control input scalar values.

Proposition 1 *The linear-quadratic optimal control for Problem 1, and the corresponding optimal cost are given by:*

$$u^* = -\frac{b_2}{r}\alpha x,$$

$$L(x, u^*) = \frac{1}{2}\alpha(0)x_0^2,$$

where α solves the following differential equation:

$$\dot{\alpha} = -q - 2b_1\alpha + \frac{b_2^2}{r}\alpha^2, \tag{2.3}$$

with the terminal boundary condition established by

$$\alpha(T) = q(T).$$

Proof 1 (Proposition 1) *This proof is straight-forwardly obtained from the proof of Proposition 3, which is presented later on, and by considering a unique decision-maker, which corresponds to the solution for a non-cooperative differential game.*

Notice that, the solution of the Riccati equation in (2.3) determines the optimal control input law u^* and it is of state-feedback form, i.e., the optimal control input depends also on the system state. The solution of (2.3) is discussed in Remark 1, and it might have a blow-up as presented later on in Remark 2.

Remark 1 *The expression in (2.3) is known as a Riccati equation, which has been extensively studied in the literature. Here we analyze such Riccati* *equation and discuss about some important facts related to it. First, let us consider the following auxiliary variables:*

$$\tilde{a}(t) = \frac{b_2^2(t)}{r(t)},$$

$$\tilde{b}(t) = -2b_1(t),$$

$$\tilde{c}(t) = -q(t),$$

then, the equation in (2.3) can be re-written as follows:

$$\frac{d\alpha}{dt} = \tilde{a}\alpha^2 + \tilde{b}\alpha + \tilde{c}. \tag{2.4}$$

Let $\tilde{\alpha}$ be a particular solution of (2.4) and let y be any function of time t. Thus, let $\alpha = \tilde{\alpha} + y$ and

$$\frac{d\tilde{\alpha}}{dt} + \frac{dy}{dt} = \frac{d\alpha}{dt},$$

$$= \tilde{a}(\tilde{\alpha} + y)^2 + \tilde{b}(\tilde{\alpha} + y) + \tilde{c},$$

$$= \tilde{a}\tilde{\alpha}^2 + 2\tilde{a}\tilde{\alpha}y + \tilde{a}y^2 + \tilde{b}\tilde{\alpha} + \tilde{b}y + \tilde{c},$$

since $\tilde{\alpha}$ is a particular solution of (2.4), then from (2.4) we have that

$$\frac{d\tilde{\alpha}}{dt} = \tilde{a}\tilde{\alpha}^2 + \tilde{b}\tilde{\alpha} + \tilde{c},$$

and replacing back the term $d\tilde{\alpha}/dt$ yields

$$\tilde{a}\tilde{\alpha}^2 + \tilde{b}\tilde{\alpha} + \tilde{c} + \frac{dy}{dt} = \tilde{a}\tilde{\alpha}^2 + 2\tilde{a}\tilde{\alpha}y + \tilde{a}y^2 + \tilde{b}\tilde{\alpha} + \tilde{b}y + \tilde{c},$$

$$\frac{dy}{dt} = 2\tilde{a}\tilde{\alpha}y + \tilde{a}y^2 + \tilde{b}y. \tag{2.5}$$

The expression in (2.5) is known as a Bernoulli equation, i.e.,

$$\dot{y} - (2\tilde{a}\tilde{\alpha} + \tilde{b})y = \tilde{a}y^2,$$

which can be addressed by the substitution $z = 1/y$ as follows:

$$\frac{dz}{dt} = -\frac{1}{y^2}\frac{dy}{dt},$$

and (2.5) becomes

$$\tilde{a}y^2 = \frac{dy}{dt} - (2\tilde{a}\tilde{\alpha} + \tilde{b})y,$$

$$\tilde{a} = \frac{1}{y^2}\frac{dy}{dt} - \frac{1}{y}(2\tilde{a}\tilde{\alpha} + \tilde{b}),$$

$$-\tilde{a} = \frac{dz}{dt} + (2\tilde{a}\tilde{\alpha} + \tilde{b})z, \tag{2.6}$$

which is a linear differential equation. The equation in (2.6) *is easily solved by considering the terms* $\tilde{p} = (2\tilde{a}\tilde{\alpha} + \tilde{b})$, *and* $\tilde{q} = -\tilde{a}$, *i.e.,*

$$\frac{dz}{dt} + \tilde{p}(t)z = \tilde{q}(t), \tag{2.7}$$

where

$$\alpha = \frac{\tilde{\alpha}z + 1}{z}, \tag{2.8}$$

obtaining the solution of the Riccati equation shown in (2.4).

From Remark 1, one can determine a singularity of the Riccati equation in (2.4). This singularity is explained in Remark 2.

Remark 2 *If the linear differential equation in* (2.7), $\dot{z} = -\tilde{p}(t)z + \tilde{q}(t)$, *reaches zero for a time t within the interval* $[0, T]$, *then the Riccati equation in* (2.4) *has a blow-up. This claim can be easily observed from the expression of* α *in terms of z presented in* (2.8).

Given that the solution of the Riccati equation in (2.4) is expressed in terms of the variable z, then it is necessary to study the solution for the differential equation in (2.7). Thus, the solution of the z equation mentioned in Remark 2 is discussed next in Remark 3.

Remark 3 *The linear differential equation in* (2.7) *can be solved by multiplying to the exponential factor*

$$\exp\left(\int \tilde{p}(t)dt\right)$$

on both sides, i.e.,

$$\frac{dz}{dt}\exp\left(\int \tilde{p}(t)dt\right) + \exp\left(\int \tilde{p}(t)dt\right)\tilde{p}(t)z = \tilde{q}(t)\exp\left(\int \tilde{p}(t)dt\right),$$

$$\frac{dz}{dt}\left[\exp\left(\int \tilde{p}(t)dt\right)\right] + \frac{d}{dt}\left[\exp\left(\int \tilde{p}(t)dt\right)\right]z = \tilde{q}(t)\exp\left(\int \tilde{p}(t)dt\right),$$

$$\frac{d}{dt}\left(z\exp\left(\int \tilde{p}(t)dt\right)\right) = \tilde{q}(t)\exp\left(\int \tilde{p}(t)dt\right),$$

resulting in

$$z = \frac{\int\left[\tilde{q}(t')\exp\left(\int \tilde{p}(t)dt\right)\right]dt'}{\exp\left(\int \tilde{p}(t)dt\right)},$$

and finally obtaining the solution of (2.7).

Figure 2.2 shows a scheme corresponding to the state feedback control law for the mean-field-free optimal controller according to Proposition 1. It can be seen that it is necessary to solve the Riccati equation α in (2.4) to compute the optimal control input.

Sometimes, the time window for which we are designing an optimal controller is not defined. In other cases, we require to implement our controllers for all time t. In such cases, it is appropriate to design optimal solutions in the long term, i.e., considering an infinite time horizon. In addition, another reason to pursue an infinite-time horizon has to do with the simplicity to compute the optimal control input that is independent from differential equations. Next, we illustrate this by solving the same problem for the long term. Let us consider the infinite-horizon cost function as follows:

$$L^\infty(x, u) = \frac{1}{2} \int_0^\infty \left(qx^2 + ru^2 \right) \mathrm{dt}. \tag{2.9}$$

Notice that in (2.9) there cannot be a terminal cost due to the fact that the time horizon is infinite. The new infinite-horizon problem is stated next.

Problem 2 (Infinite-Horizon Mean-Field-Free Optimal Control)
The infinite-horizon optimal control problem is given by

$$\begin{cases} \underset{u \in \mathcal{U}}{\text{minimize}} \, L^\infty(x, u), \\[2mm] \text{subject to} \\[2mm] \dot{x} = b_1 x + b_2 u, \\ x(0) := x_0, \end{cases}$$

where the weight parameters are $q \in \mathbb{R}_{\geq 0}$, $r \in \mathbb{R}_{>0}$, and the system parameters are $b_1, b_2 \in \mathbb{R}$. Again, all these parameters can be function of time t as it has been discussed before.

Proposition 2 presents the explicit solution for the infinite-horizon optimal control problem where it can be observed that it is no longer needed to solve a differential equation to compute the optimal control input. Instead, it is necessary to solve an algebraic equation.

Proposition 2 *The infinite-horizon linear-quadratic optimal control Problem 2 is given by:*

$$u^* = -kx,$$

where

$$k = \frac{b_2}{r} \alpha^\infty,$$

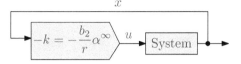

FIGURE 2.3
Feedback scheme for the infinite-horizon linear-quadratic mean-field-free optimal control.

and the quadratic equation

$$0 = \frac{b_2^2}{r}[\alpha^\infty]^2 - 2b_1\alpha^\infty - q \tag{2.10}$$

is solved by α^∞, being the parameter α^∞ the unique positive root of the latter polynomial in (2.10) denoted by $p(\alpha^\infty)$.

Figure 2.3 shows the diagram corresponding to the state-dependent feedback for the infinite-horizon linear-quadratic mean-field-free optimal control according to Proposition 2. In this regard, it is not necessary to solve the differential equation α presented in Proposition 1 but the value for the parameter α^∞ in (2.10), which is analyzed below.

Proof 2 (Proposition 2) *This proof is straight-forwardly obtained from the proof of Proposition 3 and it can be found in the following section where multiple decision-makers are incorporated to the problems. First, the solution is simplified by considering a unique decision-maker, and then by computing the steady state solution of the Riccati equations. The polynomial in (2.10) has two roots that we denote by r_1^o and r_2^o, i.e.,*

$$p(\alpha^\infty) = \frac{b_2^2}{r}(\alpha^\infty - r_1^o)(\alpha^\infty - r_2^o),$$

$$= \frac{b_2^2}{r}[\alpha^\infty]^2 - \frac{b_2^2}{r}(r_1^o + r_2^o)\alpha^\infty + \frac{b_2^2}{r}r_1^o r_2^o.$$

From the latter expression and comparing with the polynomial in (2.10), one observes that the product of the roots $r_1^o r_2^o$ is negative, which means that there exist a unique negative and a unique positive root in this case (recall $r > 0$ and notice that the term b_2 appears to the second power). Then,

$$\alpha^\infty = r\left(b_1 + \sqrt{b_1^2 + b_2^2\frac{q}{r}}\right)b_2^{-2} > 0,$$

and

$$k = \left(b_1 + \sqrt{b_1^2 + b_2^2\frac{q}{r}}\right)b_2^{-1}.$$

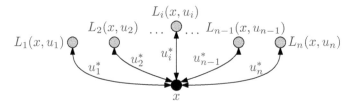

FIGURE 2.4

General scheme of the non-cooperative differential game. Dark gray nodes represent the n players, whereas the black node represents the common system state.

By replacing the optimal control input $u^* = -kx$ in the system dynamics in (2.1), one obtains that

$$\dot{x} = b_1 x - \left(b_1 + \sqrt{b_1^2 + b_2^2 \frac{q}{r}} \right) x,$$

$$= - \left(\sqrt{b_1^2 + b_2^2 \frac{q}{r}} \right) x.$$

Since $q \geq 0$, and $r > 0$ by assumption. Then, if $b_1 \neq 0$, it is concluded that the system is exponentially asymptotically stable, i.e., $x(t) \to 0$ as $t \to \infty$.

After having studied the optimal control problem with a unique control input and explained the direct method in these simple setups, we extend such problem statement by incorporating multiple control inputs, i.e., multiple decision-makers. In the coming sections, we start to study game-theoretical problems for the mean-field-free case.

2.2 Continuous-Time Differential Game

Different from Problem 1, we now consider a set of $n \geq 2$, $n \in \mathbb{N}$, decision-makers denoted by $\mathcal{N} = \{1, \ldots, n\}$ as shown in Figure 2.4. The dark gray nodes symbolize the players. All the players affect the same system state, which is symbolized by the black node in Figure 2.4. The arrows in the diagram symbolize how the information flow is. In other words, we can see that the unique information the players can observe is the system state, and we see that the evolution of the system state depends on all the strategies players apply.

These decision-makers (also known as agents) strategically interact to each other by means of the following deterministic dynamical system:

$$\dot{x} = b_1 x + \sum_{j \in \mathcal{N}} b_{2j} u_j, \tag{2.11}$$

where $x \in \mathbb{R}$ is still a scalar-valued variable and it corresponds to the system state, and $u_i \in \mathbb{R}$ denotes the control input of the decision-maker $i \in \mathcal{N}$. Finally, $b_1, b_{2j} \in \mathbb{R}$, for all $j \in \mathcal{N}$ are the system state parameters. Recall that these parameters could be time-dependent. Each decision-maker has an associated cost functional with a quadratic structure as follows:

$$L_i(x, u_1, \ldots, u_n) = \frac{1}{2} q_i(T) x^2(T) + \frac{1}{2} \int_0^T \left(q_i x^2 + r_i u_i^2 \right) dt. \tag{2.12}$$

According to the cost functional in (2.12), all the decision-makers are interested in minimizing the system state x with heterogeneous prioritization, which is determined by q_i, for all $i \in \mathcal{N}$. Besides, at the same time, all the decision-makers pursues to minimize the energy they apply into the system (2.11) according to the cost that with the quadratic term in u_i, for all $i \in \mathcal{N}$.

The interesting fact is that the system state is common for all the decision-makers and there exists a dynamical coupling through (2.11) creating a strategic interaction, *"everyone wants to minimize the magnitude of the state, but nobody wants to spend energy on it,"* illustrating a non-cooperative behavior. Such problem is formally stated next and it is known as a non-cooperative game problem. The reader may recall the illustration for the non-cooperative game problem presented in Chapter 1.

Problem 3 (Non-Cooperative Differential Game Approach) *The non-cooperative linear-quadratic differential game problem is given by*

$$\begin{cases} \underset{u_i \in \mathcal{U}_i}{\text{minimize}} \, L_i(x, u_1, \ldots, u_n), \\[2mm] \text{subject to} \\[2mm] \dot{x} = b_1 x + \sum_{j \in \mathcal{N}} b_{2j} u_j, \\[2mm] x(0) := x_0, \end{cases}$$

where the cost functional weight parameters are $q_i \in \mathbb{R}_{\geq 0}$, $r_i \in \mathbb{R}_{>0}$, and the system parameters are $b_0, b_1, b_{2j} \in \mathbb{R}$, for all $j \in \mathcal{N}$. Note that these parameters can also be considered dependent of time t.

Each decision-maker $i \in \mathcal{N}$ does its best given the circumstances, i.e., given the selection made by all the other decision-makers (players or agents) from the set $\mathcal{N} \setminus \{i\}$. This best action that the decision-makers can do is known as a best-response strategy, which is defined next.

Definition 4 (Differential Best-Response Strategy) *Any feasible strategy $u_i^* \in \mathcal{U}_i$ such that $u_i^* \in \arg\text{minimize}_{u_i \in \mathcal{U}_i}\, L_i(x, u_1, \ldots, u_n)$ is a Differential Best-Response strategy of the decision-maker $i \in \mathcal{N}$ against the strategies*

$$u_{-i} = (u_1, \ldots, u_{i-1}, u_{i+1}, \ldots, u_n)$$

selected from the other decision-makers $\mathcal{N} \setminus \{i\}$. The set of Best-Response strategies is given by $\mathrm{BR}_i : \prod_{j \in \mathcal{N} \setminus \{i\}} \mathcal{U}_j \to 2^{\mathcal{U}_i}$, where $2^{\mathcal{U}_i}$ is the power set of all the possible subsets of $\mathcal{U}_i$.

The set of Differential Nash equilibria, which is the solution concept for the non-cooperative game problem (see the statement in Problem 3), is defined by using the differential best-response as shown next.

Definition 5 (Differential Nash equilibrium) *Any feasible strategy profile, given by $[u_1^* \ \cdots \ u_n^*]^\top \in \prod_{j \in \mathcal{N}} \mathcal{U}_j$, such that the optimal $u_i^* \in \mathrm{BR}_i(u_{-i}^*)$, for all $i \in \mathcal{N}$, is a Nash equilibrium of the linear-quadratic differential game.*

Let us interpret the Definition 5. The condition describing the Nash equilibrium represents a situation in which all the decision-makers do their best. Therefore, they do not have incentives to modify their selected strategy in an unilateral manner. This definition of the Nash equilibrium as there is no incentives to change strategy is quite standard and it corresponds to the same one studied even in matrix static games, or closely, to the dynamic games. With these concepts in hand, we proceed to present the solution for the differential game in Proposition 3.

Proposition 3 *The feedback-strategies linear-quadratic differential Nash equilibrium for the non-cooperative problem 3, and the optimal costs are given by:*

$$u_i^* = -\frac{b_{2i}}{r_i}\alpha_i x,$$

$$L_i(x, u_1^*, \ldots, u_n^*) = \frac{1}{2}\alpha_i(0)x_0^2,$$

where α_i solves the following differential equations:

$$\dot{\alpha}_i = -q_i - 2b_1\alpha_i + \frac{b_{2i}^2}{r_i}\alpha_i^2 + 2\alpha_i \sum_{j \in \mathcal{N} \setminus \{i\}} \frac{b_{2j}^2}{r_j}\alpha_j,$$

with the terminal boundary condition for all the decision-makers given by

$$\alpha_i(T) = q_i(T).$$

Figure 2.5 shows the scheme corresponding to the optimal control inputs for the decision-maker $i \in \mathcal{N}$ in the mean-field-free differential game according

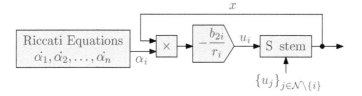

FIGURE 2.5
Feedback scheme for the ith decision-maker in the linear-quadratic mean-field-free differential game.

to Proposition 3. It can be seen that, similar as it occurs in the optimal control problem, the Riccati equation α_i should be computed, which is coupled with the other ordinary differential equations α_j, for all $j \in \mathcal{N} \setminus \{i\}$. Therefore, all the Riccati equations for all the decision-makers should be solved simultaneously in order to find the Nash equilibrium point.

The proof of Proposition 3 is presented next by following the direct method according to Figure 1.17 shown in Chapter 1.

Proof 3 (Proposition 3) *Inspired by the structure of the cost functional, we propose a quadratic guess functional, also known as ansatz, of the following form:*

$$f_i(t,x) = \frac{1}{2}\alpha_i x^2, \tag{2.13}$$

this guess functional corresponds to the optimal cost for the decision-maker $i \in \mathcal{N}$ from time t up to T, and it is tightly related to the function V_i used in the PIDE in Chapter 1, Section 1.4. The term α_i in the ansatz is a deterministic function of time. Now, we compute the integration formula obtaining the following:

$$f_i(T,x(T)) - f_i(0,x(0)) = \int_0^T \left(\frac{1}{2}\dot\alpha_i x^2 + \alpha_i x \dot x\right) dt,$$

$$= \int_0^T \left(\frac{1}{2}\dot\alpha_i x^2 + \alpha_i x \left(b_1 x + \sum_{j \in \mathcal{N}} b_{2j} u_j\right)\right) dt,$$

$$= \int_0^T \left(\frac{1}{2}\dot\alpha_i x^2 + \alpha_i x \left(b_1 x + b_{2i} u_i + \sum_{j \in \mathcal{N}\setminus\{i\}} b_{2j} u_j\right)\right) dt,$$

$$= \int_0^T \left(\frac{1}{2}\dot\alpha_i x^2 + b_1 \alpha_i x^2 + b_{2i}\alpha_i x u_i + \alpha_i x \sum_{j \in \mathcal{N}\setminus\{i\}} b_{2j} u_j\right) dt.$$

Therefore, the procedure is followed by computing the gap

$$L_i(x, u_1, \ldots, u_n) - f_i(0, x(0))$$

*by using the latter equation and the cost functional of the problem in (2.12).
The term $f_i(0, x(0))$ represents then the optimal cost from time $t = 0$. It
follows that*

$$L_i(x, u_1, \ldots, u_n) - f_i(0, x(0)) = \frac{1}{2}(q_i(T) - \alpha_i(T))x^2(T) + \frac{1}{2}\int_0^T \left(q_i x^2 + r_i u_i^2\right) dt$$

$$+ \int_0^T \left(\frac{1}{2}\dot{\alpha}_i x^2 + b_1 \alpha_i x^2 + b_{2i}\alpha_i x u_i + \alpha_i x \sum_{j \in \mathcal{N} \setminus \{i\}} b_{2j} u_j \right) dt.$$

Now, by grouping terms as:

- *Those terms at terminal time,*

- *Those terms common factor of x^2,*

- *Those terms depending on the control input u_i, and*

- *Those terms independent from x and u_i,*

yields the following:

$$L_i(x, u_1, \ldots, u_n) - f_i(0, x(0)) = \frac{1}{2}(q_i(T) - \alpha_i(T))x(T)^2$$

$$+ \frac{1}{2}\int_0^T x^2 \left(q_i + \dot{\alpha}_i + 2b_1 \alpha_i\right) dt$$

$$+ \frac{1}{2}\int_0^T r_i \left(u_i^2 + 2\frac{b_{2i}}{r_i}\alpha_i x u_i\right) dt$$

$$+ \frac{1}{2}\int_0^T \left(2\alpha_i x \sum_{j \in \mathcal{N} \setminus \{i\}} b_{2j} u_j \right) dt. \qquad (2.14)$$

*We optimize the latter expression over the control input by performing a square
completion procedure for the u_i-dependent terms. This procedure corresponds,
or is equivalent to, optimizing the Hamiltonian in the HJB equation (see Section 1.4), obtaining*

$$\left(u_i + \frac{b_{2i}}{r_i}\alpha_i x\right)^2 = u_i^2 + 2\frac{b_{2i}}{r_i}\alpha_i x u_i + \frac{b_{2i}^2}{r_i^2}\alpha_i^2 x^2,$$

and

$$u_i^2 + 2\frac{b_{2i}}{r_i}\alpha_i x u_i = \left(u_i + \frac{b_{2i}}{r_i}\alpha_i x\right)^2 - \frac{b_{2i}^2}{r_i^2}\alpha_i^2 x^2, \qquad (2.15)$$

*from which is it concluded that the optimal control input for the decision-
makers is given by*

$$u_i^* = -\frac{b_{2i}}{r_i}\alpha_i x, \ \forall i \in \mathcal{N}.$$

Replacing (2.15) in (2.14) yields

$$L_i(x, u_1, \ldots, u_n) - f_i(0, x(0)) = \frac{1}{2}(q_i(T) - \alpha_i(T))x(T)^2$$

$$+ \frac{1}{2} \int_0^T x^2 \left(q_i + \dot{\alpha}_i + 2b_1\alpha_i - \frac{b_{2i}^2}{r_i}\alpha_i^2 - 2\alpha_i \sum_{j \in \mathcal{N} \backslash \{i\}} \frac{b_{2j}^2}{r_j}\alpha_j \right) dt$$

$$+ \frac{1}{2} \int_0^T r_i \left(u_i + \frac{b_{2i}}{r_i}\alpha_i x \right)^2 dt.$$

Finally, notice that by matching $L_i(x, u_1, \ldots, u_n)$ with $f_i(0, x(0))$ we are obtaining the optimal cost functional. Thus, the announced result is obtained by minimizing all the terms in the right-hand side latter expression, completing the proof.

From the latter equation in the proof of Proposition 3, we can check the following. On the left-side hand, the optimal cost functional is equal to the value the ansatz gets at initial time, and on the right-hand side, the terminal boundary condition can be seen from the first line, the Riccati α_i differential equation can be seen from the second line, and the optimal control input u_i^* can be seen from the third line.

Another important remark is on the non-negativeness of the variables α_i, for all time t and all decision-maker $i \in \mathcal{N}$. This claim arises from the fact that the cost functionals $L_i(x, u_1, \ldots, u_n)$ is quadratic, convex and non-negative provided that the weight parameters q and r satisfy the established conditions introduced at the problem statement. Therefore, the ansatz for the optimal cost should be non-negative justifying the condition over α_i, for all $i \in \mathcal{N}$.

By following the same procedure that we performed with the optimal control problem scenario in Problem 2, and for the same reasons we exposed in the control case, we are also interested in evaluating the optimal game solution for the long term.

Let us consider now the infinite-horizon differential game cost functional as follows:

$$L_i^\infty(x, u_1, \ldots, u_n) = \frac{1}{2} \int_0^\infty \left(q_i x^2 + r_i u_i^2 \right) dt, \qquad (2.16)$$

for all $i \in \mathcal{N}$. Once again, the cost functional in (2.16) does not have a terminal-time cost component because of the time horizon. The differential game with this new infinite-time horizon cost functional is stated next in Problem 4.

Problem 4 (Infinite-Horizon Non-Cooperative Differential Game)
The infinite-horizon non-cooperative linear-quadratic differential game prob-

lem is given by

$$
\begin{cases}
\underset{u_i \in \mathcal{U}_i}{\text{minimize }} L_i^\infty (x, u_1, \ldots, u_n), \\[2ex]
\text{subject to} \\[2ex]
\dot{x} = b_1 x + \sum_{j \in \mathcal{N}} b_{2j} u_j, \\[1ex]
x(0) := x_0,
\end{cases}
$$

where the weight parameters in the cost functional are $q_i \in \mathbb{R}_{\geq 0}$, $r_i \in \mathbb{R}_{>0}$, and the system parameters are $b_0, b_1, b_{2j} \in \mathbb{R}$, for all $j \in \mathcal{N}$. Recall that these parameters can also be considered as function of time.

We present the semi-explicit solution for the infinite-horizon game problem in Proposition 4. We remind the reader about the fact that, different from the finite-time horizon case, in which a differential equation should be solved to compute the optimal control inputs, in the infinite-horizon case requires to solve a coupled algebraic equation.

Proposition 4 *The feedback-strategies infinite-horizon linear-quadratic differential Nash equilibrium for the non-cooperative problem 4 is given by:*

$$ u_i^* = -k_i x, $$

where

$$ k_i = \frac{b_{2i}}{r_i} \alpha_i^\infty, $$

and

$$ 0 = -q_i - 2b_1 \alpha_i^\infty + \frac{b_{2i}^2}{r_i} [\alpha_i^\infty]^2 + 2\alpha_i^\infty \sum_{j \in \mathcal{N} \setminus \{i\}} \frac{b_{2j}^2}{r_j} \alpha_j^\infty $$

is solved by α_i^∞, for all the decision-makers $i \in \mathcal{N}$, being the solution $[\alpha_1^\infty \ \cdots \ \alpha_n^\infty]^\top$ in the positive orthant.

Even though the infinite-time horizon solution does not require to compute the solution of Riccati differential equations, there is still the difficulty to simultaneously solve all the algebraic equations given that the coupling among all the decision-makers remains. A plausible approach in the solution of the algebraic equations consists of writing them in a unified form by means of a vector and matrix representation, i.e., using the variable $\alpha^\infty = [\alpha_1^\infty \ \cdots \ \alpha_n^\infty]^\top$.

Figure 2.6 presents the state-dependent feedback for the infinite-horizon linear-quadratic mean-field-free differential according to Proposition 4. In this

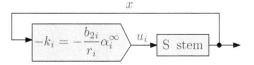

FIGURE 2.6
Feedback scheme for the infinite-horizon linear-quadratic mean-field-free differential game.

regard, it is not necessary to solve the ordinary differential equations presented in Proposition 3 but the value for the parameters α_i^∞, for all the decision-makers $i \in \mathcal{N}$. Next, we analyze conditions for the stability of the system state using the optimal control inputs for all the decision-makers.

Proof 4 (Proposition 4) *This proof of this Proposition is straight-forward and it is obtained from the proof of Proposition 3 and by computing the steady-state solution of the Riccati equations. Besides, replacing the optimal control inputs*

$$u_j^* = -b_{2j}r_j^{-1}\alpha_j^\infty x,$$

for all the decision-makers $j \in \mathcal{N}$, in the system dynamics shown in (2.11) yields

$$\dot{x} = \left(b_1 - \sum_{j\in\mathcal{N}} \frac{b_{2j}^2}{r_j}\alpha_j^\infty \right) x,$$

it is well known that under this expression, the state x vanishes as the multiplying factor is negative. Thus, we have the condition that

$$b_1 < \sum_{j\in\mathcal{N}} \frac{b_{2j}^2}{r_j}\alpha_j^\infty$$

to have exponential stability for the system, i.e., that $x(t) \to 0$ as $t \to \infty$.

The following sections will discuss game theoretical problems without considering mean-field terms, but incorporating a state and control-input-dependent diffusion term, i.e., we move from the analysis of deterministic differential games to, probably, the simplest stochastic differential game setup.

2.3 Stochastic Mean-Field-Free Differential Game

In mean-field-free games, the expected payoff functional is linear with respect to the probability measure of the (individual) state. Let us now consider a

Brownian motion describing the evolution of the system state as follows:

$$\mathrm{d}x = \underbrace{\left(b_1 x + \sum_{j \in \mathcal{N}} b_{2j} u_j \right) \mathrm{d}t}_{D_r(x,u)} + \underbrace{\sigma_0}_{D_f(x,u)} \mathrm{d}B, \qquad (2.17)$$

where $x \in \mathbb{R}$ corresponds to the system state and $u_i \in \mathbb{R}$ denotes the control input of the decision-maker $i \in \mathcal{N}$. Finally, $b_1, b_{2j}, \sigma_0 \in \mathbb{R}$, for all $j \in \mathcal{N}$ are the system state parameters that could also be considered as time-dependent functions, and B denotes a standard Brownian motion. In general, we consider a SDE of the form $\mathrm{d}x = D_r(x, u)\mathrm{d}t + D_f(x, u)\mathrm{d}B$, where the drift, denoted by $D_r(x, u)$, and the diffusion, denoted by $D_f(x, u)$, can be constant, time-dependent, state-dependent, and/or control-input-dependent. In this chapter, for instance, we address a constant diffusion term, but later on the book we will study other type of problems.

Each decision-maker behaves strategically intending to minimize its cost functional given by (2.12). Next, we present the stochastic differential game problem.

Problem 5 (Stochastic Differential Game Approach) *The stochastic linear-quadratic differential game problem is given by*

$$\begin{cases} \underset{u_i \in \mathcal{U}_i}{\text{minimize}} \ \mathbb{E}[L_i(x, u_1, \ldots, u_n)], \\[2mm] \text{subject to} \\[2mm] \mathrm{d}x = \left(b_1 x + \sum_{j \in \mathcal{N}} b_{2j} u_j \right) \mathrm{d}t + \sigma_0 \mathrm{d}B, \\[2mm] x(0) := x_0, \end{cases}$$

where the weight parameters in the cost functional are $q_i \in \mathbb{R}_{\geq 0}$, $r_i \in \mathbb{R}_{>0}$, and the system parameters are $b_0, b_1, b_{2j}, \sigma_0 \in \mathbb{R}$, for all $j \in \mathcal{N}$.

Proposition 5 establishes the solution corresponding to the differential game problem when a Brownian motion is incorporated in the dynamical system, which exhibits subtle differences with respect to the deterministic counterpart as we will discuss later on.

Proposition 5 *The feedback-strategies linear-quadratic stochastic differential Nash equilibrium for the Non-Cooperative Problem 5, and the optimal costs are given by:*

$$u_i^* = -\frac{b_{2i}}{r_i} \alpha_i x,$$

$$L_i(x, u_1^*, \ldots, u_n^*) = \frac{1}{2}\alpha_i(0)x_0^2 + \delta_i(0),$$

where α_i and δ_i solve the following differential equations:

$$\dot{\alpha}_i = -q_i - 2b_1\alpha_i + \frac{b_{2i}^2}{r_i}\alpha_i^2 + 2\alpha_i \sum_{j\in\mathcal{N}\backslash\{i\}} \frac{b_{2j}^2}{r_j}\alpha_j,$$

$$\dot{\delta}_i = -\alpha_i\frac{\sigma_0^2}{2},$$

with the following terminal boundary conditions:

$$\alpha_i(T) = q_i(T),$$
$$\delta_i(T) = 0.$$

We compare the solutions for the game-theoretical Problems 3 and 5 to see the effect of the Brownian motion. We will see in the coming discussion that there are several interesting similarities. The optimal control input for either the deterministic or the stochastic differential game coincide to each other (compare u_i^* in Proposition 3 and in Proposition 5).

Nevertheless, a new ordinary differential equation δ_i appears in the solution, which directly affects the optimal cost functional. The equivalence between the optimal control inputs emerge since δ_i does not appear in the optimal strategy, and the diffusion term is state-and-control-input independent.

Now, we present below the computation of the semi-explicit solution for the stochastic game problem. The reader will see that the main difference with respect to the computation we have studied so far in the book, relies on the integration formula.

Proof 5 (Proposition 5) *According to the structure of the cost functional and the system dynamics a guess functional (ansatz) is postulated for the optimal cost from time t up to T. We propose a quadratic guess functional of the following form:*

$$f_i(t, x) = \frac{1}{2}\alpha_i x^2 + \delta_i, \qquad (2.18)$$

for each decision-maker $i \in \mathrm{mathcal}N$, where α_i and δ_i are deterministic functions of time. Now, we compute the integration formula that is given by the Itô's formula, i.e.,

$$f_i(T, x(T)) - f_i(0, x(0)) = \int_0^T \left(\frac{1}{2}\dot{\alpha}_i x^2 + \dot{\delta}_i + \alpha_i x D_r(x, u) + \alpha_i \frac{\sigma_0^2}{2}\right) dt$$

$$+ \int_0^T \sigma_0\alpha_i x \, dB, \qquad (2.19)$$

where the drift is

$$D_r(x, u) = b_1 x + \sum_{j\in\mathcal{N}} b_{2j}u_j.$$

Then, replacing back the drift term in (2.19) yields

$$\mathbb{E}[f_i(T, x(T)) - f_i(0, x(0))] = \mathbb{E}\int_0^T \left(\frac{1}{2}\dot{\alpha}_i x^2 + \dot{\delta}_i + b_1\alpha_i x^2 + b_{2i}\alpha_i x u_i\right.$$

$$\left. + \alpha_i x \sum_{j\in\mathcal{N}\setminus\{i\}} b_{2j}u_j + \alpha_i\frac{\sigma_0^2}{2}\right)dt.$$

Now, we compute the gap $\mathbb{E}[L_i(x, u_1, \ldots, u_n) - f_i(0, x(0))]$. *Notice that* $f_i(0, x(0))$ *is the optimal cost from initial time* $t = 0$. *To this end, we use the latter expression coming from the Itô's formula and the cost functional in* (2.12), *obtaining the following expression:*

$$\mathbb{E}[L_i(x, u_1, \ldots, u_n) - f_i(0, x(0))] = \mathbb{E}[\frac{1}{2}(q_i - \alpha_i(T))x(T)^2] + \mathbb{E}[(0 - \delta_i(T))]$$

$$+ \frac{1}{2}\mathbb{E}\int_0^T \left(q_i x^2 + \dot{\alpha}_i x^2 + 2b_1\alpha_i x^2\right)dt$$

$$+ \mathbb{E}\int_0^T \left(\dot{\delta}_i + \alpha_i x \sum_{j\in\mathcal{N}\setminus\{i\}} b_{2j}u_j + \alpha_i\frac{\sigma_0^2}{2}\right)dt$$

$$+ \frac{1}{2}\mathbb{E}\int_0^T r_i\left(u_i^2 + 2\frac{b_{2i}}{r_i}\alpha_i x u_i\right)dt.$$

In this latter expression, we have isolated all those terms depending on the control input of the decision-maker $i \in \mathcal{N}$, *and we have gathered terms whose common factor is* x^2, *and those independent from both the square system state and the* i^{th} *control inputs. Then, we optimize in* u_i *by performing square completion for those terms depending on this control input. It yields*

$$u_i^2 + 2\frac{b_{2i}}{r_i}\alpha_i x u_i = \left(u_i + \frac{b_{2i}}{r_i}\alpha_i x\right)^2 - \frac{b_{2i}^2}{r_i^2}\alpha_i^2 x^2.$$

Replacing the completion of terms in the gap $\mathbb{E}[L_i(x, u_1, \ldots, u_n) - f_i(0, x(0))]$ *one arrives at the expression*

$$\mathbb{E}[L_i(x, u_1, \ldots, u_n) - f_i(0, x(0))] = \mathbb{E}[\frac{1}{2}(q_i - \alpha_i(T))x(T)^2] + \mathbb{E}[(0 - \delta_i(T))]$$

$$+ \frac{1}{2}\mathbb{E}\int_0^T \left(q_i + \dot{\alpha}_i + 2b_1\alpha_i - \frac{b_{2i}^2}{r_i}\alpha_i^2 - 2\alpha_i \sum_{j\in\mathcal{N}\setminus\{i\}} \frac{b_{2j}^2}{r_j}\alpha_j\right)x^2 dt$$

$$+ \mathbb{E}\int_0^T \left(\dot{\delta}_i + \alpha_i\frac{\sigma_0^2}{2}\right)dt + \frac{1}{2}\mathbb{E}\int_0^T r_i\left(u_i + \frac{b_{2i}}{r_i}\alpha_i x\right)^2 dt.$$

At this stage, there are four main left terms in the gap computation $\mathbb{E}[L_i(x, u_1, \ldots, u_n) - f_i(0, x(0))]$:

- *First, we have some terms at terminal time that will define the boundary conditions for the Riccati equations.*

- *Second, we have all those terms whose common factor is x^2.*

- *As third and fourth parts, we see the state and control-input-independent terms and those defining the optimal control input.*

   Finally, the announced result is obtained by minimizing the terms in the latter expression, completing the proof.

So far, we have only studied the continuous-time control and game problems. As debated during the introduction of the book in Chapter 1, the discrete-time direction is crucial for real implementation. This is because, although time is continuous, we can only get measurements about physical variables in a sampled manner. Motivated by the engineering implementation of control and game-theoretical solutions we will start the introduction of the discrete-time problems. We study next the same type of control and game problems that have been introduced above, e.g., the one decision-maker and multiple decision-makers mean-field-free game problems, but in discrete time. The solutions are equally computed by following the direct method and performing completion terms procedure , i.e., optimization over the decision variables given by the control inputs.

2.4   A Basic Discrete-Time Optimal Control Problem

Consider the following discrete-time system dynamics with a unique control input u_ℓ at time $\ell \in [k..k + N - 1]$:

$$x_{\ell+1} = b_1 x_\ell + b_2 u_\ell, \tag{2.20}$$

where $x \in \mathbb{R}$ corresponds to the system state and $u \in \mathbb{R}$ denotes the unique control input. $b_1, b_2 \in \mathbb{R}$ are system parameters and can be considered as functions of time. Here, the time window comprises the discrete steps from k up to $k + N$.

   The cost functional is given by

$$L(x, u) = q_{k+N} x_{k+N}^2 + \sum_{\ell=k}^{k+N-1} \left[q x_\ell^2 + r u_\ell^2 \right], \tag{2.21}$$

and $q \geq 0$ and $r > 0$ are the weight parameters that assign the importance of the control objectives in function of the system state and control input. The meaning of the cost functional is the same as in the previous optimization problems, i.e., the objectives consist of minimizing both the magnitude of the states and the used energy according to the prioritization defined by the weight parameters q and r. The difference control problem is stated next.

Problem 6 (Discrete-Time Mean-Field-Free Optimal Control)
The optimal control problem is given by

$$
\begin{cases}
\text{minimize } \limits_{u \in \mathcal{U}} L(x, u), \\
\\
\text{subject to} \\
\\
x_{\ell+1} = b_1 x_\ell + b_2 u_\ell, \quad \forall \ell \in [k..k+N-1], \\
x_k \triangleq x_0,
\end{cases} \tag{2.22}
$$

where the initial condition x_k is given, the weight parameters in the cost functional satisfy $q \geq 0$, $r > 0$, and the system parameters are $b_1, b_2 \in \mathbb{R}$.

Below, Proposition 6 presents the semi-explicit solution for this discrete-time control problem. Similarly as in the continuous-time approach, the optimal control input is of state-feedback form and it depends on an α equation that should be solved.

Proposition 6 *The optimal control input and optimal cost functional for Problem 6 are given by*

$$
u_\ell^* = -b_1 \frac{b_2}{r + \alpha_{\ell+1} b_2^2} \alpha_{\ell+1} x_\ell,
$$

$$
L(x, u^*) = \alpha_0 x_0^2,
$$

where α solves the following difference backward equation:

$$
\alpha_\ell = q - b_1^2 \frac{b_2^2}{r + \alpha_{\ell+1} b_2^2} \alpha_{\ell+1}^2 + b_1^2 \alpha_{\ell+1}, \tag{2.23}
$$

with the terminal condition

$$
\alpha_{k+N} = q_{k+N}.
$$

This simple and standard discrete-time control problem is solved using the solution for the multiple decision-maker problem as mentioned next. Then, the proof of the control case is obtained by following the same procedure as in the game problem.

Proof 6 (Proposition 6) *This proof is straight-forwardly obtained from the proof of Proposition 7 in the following section.*

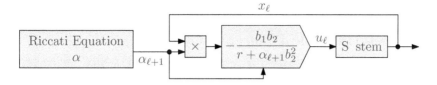

FIGURE 2.7
Feedback scheme for the linear-quadratic mean-field-free optimal control in discrete time.

Figure 2.7 shows the feedback scheme corresponding to the state-dependent law for the discrete-time mean-field-free optimal control according to the result displayed in Proposition 6. There is an important non-negativity condition over α term. In order to show the reasoning behind the positivity condition, recall that the guess functional is α-dependent (see later the guess used in the proof of Proposition 7). Further, the cost functional $L(x, u)$ is non-negative and convex according to the appropriate selection for the weight parameters. Consequently, α values should be non-negative throughout all the time interval. Remark 4 shows that this condition is satisfied for the evolution in (2.23) in Proposition 6.

Remark 4 *The Riccati equation in (2.23) has a non-negative solution. The expression in (2.23) can be re-written as follows:*

$$\alpha_\ell = q + b_1^2 \alpha_{\ell+1} \left(1 - \frac{\alpha_{\ell+1} b_2^2}{r + \alpha_{\ell+1} b_2^2} \right),$$

where it is concluded that

$$\alpha_{\ell+1} b_2^2 < r + \alpha_{\ell+1} b_2^2$$

since $r > 0$. Thus, it is claimed that

$$1 - \frac{\alpha_{\ell+1} b_2^2}{r + \alpha_{\ell+1} b_2^2} > 0.$$

Finally, given that there is a terminal boundary condition $\alpha_{k+N} = q_{k+N} > 0$. Then, we can see that $\alpha_\ell \geq 0$, for all ℓ within the time interval.

We have studied the setting when having a unique decision-maker for the discrete-time case. In the coming section, we will inspect the strategic interaction with multiple decision-makers, but also, we will enter into more details about the computation of the semi-explicit solution by applying the direct method. Extending the control problem considering an arbitrary number of decision-makers, one obtains a difference game, which is introduced next.

2.5 Deterministic Difference Games

Let $\mathcal{N} = \{1, \ldots, n\}$ be a set of $n \geq 2$, $n \in \mathbb{N}$, decision-makers. The decision-makers interact during a discrete time window $[k..k + N]$. The decision-maker $i \in \mathcal{N}$ selects a control action $u_{i,\ell}$ at time $\ell \in [k..k + N - 1]$. The system state dynamics under which all the decision-makers interact to each other is

$$x_{\ell+1} = b_1 x_\ell + \sum_{j \in \mathcal{N}} b_{2j} u_{j,\ell}, \tag{2.24}$$

where $x \in \mathbb{R}$ corresponds to the one dimensional system state, $u_i \in \mathbb{R}$ denotes the control input of the decision-maker $i \in \mathcal{N}$, and $b_1, b_{2j} \in \mathbb{R}$, for all $j \in \mathcal{N}$ are system parameters, which could be considered as time-dependent. The cost functional associated with each decision-maker $i \in \mathcal{N}$ is given by a quadratic form as follows:

$$L_i(x, u_1, \ldots, u_n) = q_{i,k+N} x_{k+N}^2 + \sum_{\ell=k}^{k+N-1} \left[q_i x_\ell^2 + r_i u_{i,\ell}^2 \right], \tag{2.25}$$

and with weight parameters satisfying the conditions $q_i \geq 0$ and $r_i > 0$, for all $i \in \mathcal{N}$. Once again, as we have been explaining throughout the book, these weights are tuned depending on the balance we want to assign to each one of the control objectives. This cost functional is simply the discrete-time analogy from the cost studied in (2.12). Using the dynamics presented in the difference equation in (2.24) and the cost functional in (2.25), a non-cooperative difference game is stated next.

Problem 7 (Non-Cooperative Difference Game) *The non-cooperative difference game problem for each $i \in \mathcal{N}$ is as follows:*

$$\begin{cases} \underset{u_i \in \mathcal{U}_i}{\text{minimize}} \ L_i(u_1, \ldots, u_n), \\[2ex] \text{subject to} \\[2ex] x_{\ell+1} = b_1 x_\ell + \sum_{j \in \mathcal{N}} b_{2j} u_{j,\ell}, \\[1ex] \forall \ell \in [k..k + N - 1], \end{cases} \tag{2.26}$$

where the initial system-state condition x_k is given, the weight parameters satisfy $q_i \geq 0$, $r_i > 0$, and the system parameters are $b_1, b_{2j} \in \mathbb{R}$, for all $j \in \mathcal{N}$, which could also be function of time.

Taking the statement of this game problem, we present again the definitions for both the best-response strategies and the Nash game solution concept in the frame of discrete time. The reader will notice that these concepts

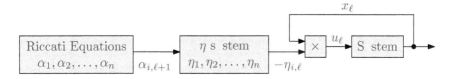

FIGURE 2.8
Feedback scheme for the linear-quadratic mean-field-free difference game.

are straight extension from the continuous-time setting. Each decision-maker plays a best-response strategy given the strategic selection made by the other decision-makers, and the game objective is to reach a Nash equilibrium.

Definition 6 *(Best-Response for the discrete-time game) For the non-cooperative approach, any control input sequence $\hat{u}_i^* := (u_{i,k}^*, \ldots, u_{i,k+N-1}^*)$, such that*

$$\hat{u}_i^* \in \arg\min_{u_i \in \mathcal{U}_i} [L_i(u_1, \ldots, u_n)],$$

is a best-response strategic sequence of the decision-maker $i \in \mathcal{N}$ against the strategies

$$\hat{u}_{-i} := (\hat{u}_1, \ldots, \hat{u}_{i-1}, \hat{u}_{i+1}, \ldots, \hat{u}_n)$$

selected from the other decision-makers $\mathcal{N} \setminus \{i\}$.

The set of Nash equilibria for the discrete-time game problem is defined by using the best-response. The equilibrium describes a situation in which no decision-maker is motivated to change their selected actions as defined next.

Definition 7 *(Nash equilibrium for the discrete-time game) Any feasible control input strategic sequence profile $(\hat{u}_i^*)_{i \in \mathcal{N}} := (\hat{u}_1^*, \ldots, \hat{u}_n^*)$ such that the $\hat{u}_i$ is a best response against $\hat{u}_{-i}^*$, for all $i \in \mathcal{N}$, is a Nash equilibrium.*

Proposition 7 exhibits the semi-explicit solution for the discrete-time game problem without considering mean-field terms yet. We discover several similarities in the information dependence structure such that all the decision-makers' solutions are coupled by means of the Riccati equations.

Proposition 7 *The optimal control input and cost functional, for each decision-maker $i \in \mathcal{N}$, of the difference game Problem 7 are given by*

$$u_{i,\ell}^* = -\eta_{i,\ell} x_\ell,$$
$$L_i(x, u_1^*, \ldots, u_n^*) = \alpha_{i,0} x_0^2,$$

where

$$\eta_{i,\ell} = -\frac{b_{2i}}{c_{i,\ell}} \alpha_{i,\ell+1} \sum_{j \in \mathcal{N} \setminus \{i\}} b_{2j} \eta_{j,\ell} + b_1 \frac{b_{2i}}{c_{i,\ell}} \alpha_{i,\ell+1}, \ \forall \ i \in \mathcal{N},$$

$$c_{i,\ell} = r_i + \alpha_{i,\ell+1} b_{2i}^2, \ \forall \ i \in \mathcal{N},$$

with the following difference equations:

$$\alpha_{i,\ell} = q_i - c_{i,\ell} \eta_{i,\ell}^2 + \alpha_{i,\ell+1} \left(b_1 - \sum_{j \in \mathcal{N} \backslash \{i\}} b_{2j} \eta_{j,\ell} \right)^2, \qquad (2.27)$$

for all $\ell \in [k..k + N - 1], i \in \mathcal{N}$, and with

$$\alpha_{i,k+N} = q_{i,k+N}$$

as a boundary terminal conditions.

Figure 2.8 shows the scheme for the decision-maker $i \in \mathcal{N}$ in the mean-field-free difference game according to Proposition 7. It can be seen that the Riccati equations $\alpha_{i,\ell}$ should be solved together with the value for $\eta = [\eta_1 \ \cdots \ \eta_n]^\top$.

This discrete-time game problem is solved by following the direct method so the reader can become familiar with the method before addressing other class of stochastic game-theoretical problems in discrete time. The proof of Proposition 7 is developed by following the method shown in Figure 1.18 in Chapter 1.

Proof 7 (Proposition 7) *Inspired from the structure of the cost functional and the system dynamics, we consider the following guess functional (ansatz) for the ith decision-maker :*

$$f_{i,\ell} = \alpha_{i,\ell} x_\ell^2, \ \forall \ \ell \in [k..k + N - 1], \qquad (2.28)$$

where $\alpha_{i,\ell}$ is a deterministic difference equation function of time. Now, making the analogy for the integration formula that takes place in the continuous-time case, we apply the telescopic sum decomposition given by

$$f_{i,k+N} = f_{i,k} + \sum_{\ell=k}^{k+N-1} (f_{i,\ell+1} - f_{i,\ell}),$$

in order to derive the difference between the guess functional and the cost functional. Hence, by replacing the term $f_{i,\ell+1}$ in terms of $\alpha_{i,k+\ell}$, one obtains the following:

$$f_{i,k+N} - f_{i,k} = \sum_{\ell=k}^{k+N-1} \left(\alpha_{i,\ell+1} x_{\ell+1}^2 - \alpha_{i,\ell} x_\ell^2 \right). \qquad (2.29)$$

Now, we compute separately the square system state for the time instance $\ell+1$ by using the system dynamics in (2.24). Then, we group independently those *terms chosen by the ith from those depending on the other decision-makers $j \in \mathcal{N} \setminus \{i\}$.*

$$x_{\ell+1}^2 = \left[b_1 x_\ell + \sum_{j \in \mathcal{N}} b_{2j} u_{j,\ell} \right]^2,$$

$$= b_1^2 x_\ell^2 + \left[\sum_{j \in \mathcal{N}} b_{2j} u_{j,\ell} \right]^2 + 2 b_1 x_\ell \sum_{j \in \mathcal{N}} b_{2j} u_{j,\ell},$$

$$= b_1^2 x_\ell^2 + b_{2i}^2 u_{i,\ell}^2 + 2 b_{2i} u_{i,\ell} \sum_{j \in \mathcal{N} \setminus \{i\}} b_{2j} u_{j,\ell} + \left[\sum_{j \in \mathcal{N} \setminus \{i\}} b_{2j} u_{j,\ell} \right]^2$$

$$+ 2 b_{2i} b_1 x_\ell u_{i,\ell} + 2 b_1 x_\ell \sum_{j \in \mathcal{N} \setminus \{i\}} b_{2j} u_{j,\ell}.$$

Replacing $x_{\ell+1}^2$ in (2.29) yields

$$f_{i,k+N} - f_{i,k} = \sum_{\ell=k}^{k+N-1} \bigg(\alpha_{i,\ell+1} b_1^2 x_\ell^2 + \alpha_{i,\ell+1} b_{2i}^2 u_{i,\ell}^2 + 2 b_{2i} \alpha_{i,\ell+1} u_{i,\ell} \sum_{j \in \mathcal{N} \setminus \{i\}} b_{2j} u_{j,\ell}$$

$$+ \alpha_{i,\ell+1} \left[\sum_{j \in \mathcal{N} \setminus \{i\}} b_{2j} u_{j,\ell} \right]^2 + 2 b_{2i} \left(b_1 x_\ell \right) \alpha_{i,\ell+1} u_{i,\ell}$$

$$+ 2 \left(b_1 x_\ell \right) \alpha_{i,\ell+1} \sum_{j \in \mathcal{N} \setminus \{i\}} b_{2j} u_{j,\ell} - \alpha_{i,\ell} x_\ell^2 \bigg).$$

Given that $f_{i,k}$ corresponds to the optimal cost for the problem from time instant k up to $k + N$, then we pursue to find the optimal cost inputs that match the cost functional $L_i(x, u_1, \ldots, u_n)$ with $f_{i,k}$. Please notice that this coming procedure is exactly the one we were computing for the continuous-time game problem. Now, we compute the gap $L_i(x, u_1, \ldots, u_n) - f_{i,k}$ by using the latter expression and the cost functional in (2.25) as follows:

$$L_i(x, u_1, \ldots, u_n) - f_{i,k} = q_i x_{k+N}^2 - \alpha_{i,k+N} x_{k+N}^2 + \sum_{\ell=k}^{k+N-1} \left(q_i x_\ell^2 + r_i u_{i,\ell}^2 \right)$$

$$+ \sum_{\ell=k}^{k+N-1} \left(\alpha_{i,\ell+1} b_1^2 x_\ell^2 + \alpha_{i,\ell+1} \left[\sum_{j \in \mathcal{N} \setminus \{i\}} b_{2j} u_{j,\ell} \right]^2 \right)$$

$$+ \sum_{\ell=k}^{k+N-1} \left(2 b_1 x_\ell \alpha_{i,\ell+1} \sum_{j \in \mathcal{N} \setminus \{i\}} b_{2j} u_{j,\ell} - \alpha_{i,\ell} x_\ell^2 \right)$$

$$+ \sum_{\ell=k}^{k+N-1} \left(\alpha_{i,\ell+1} b_{2i}^2 u_{i,\ell}^2 \right)$$

$$+ 2b_{2i}\alpha_{i,\ell+1}u_{i,\ell}\sum_{j\in\mathcal{N}\backslash\{i\}}b_{2j}u_{j,\ell} + 2b_{2i}b_1 x_\ell \alpha_{i,\ell+1}u_{i,\ell}\Bigg).$$

$$(2.30)$$

To optimize the cost functional, let us consider only the terms from (2.30) involving the strategic inputs $u_{i,\ell}$, i.e.,

$$\sum_{\ell=k}^{k+N-1}\left([r_i + \alpha_{i,\ell+1}b_{2i}^2]u_{i,\ell}^2 + 2b_{2i}\alpha_{i,\ell+1}\left(\sum_{j\in\mathcal{N}\backslash\{i\}}b_{2j}u_{j,\ell} + b_1 x_\ell\right)u_{i,\ell}\right)$$

$$= \sum_{\ell=k}^{k+N-1}[r_i + \alpha_{i,\ell+1}b_{2i}^2]\left(u_{i,\ell}^2 + 2\frac{b_{2i}\alpha_{i,\ell+1}}{r_i + \alpha_{i,\ell+1}b_{2i}^2}\left(\sum_{j\in\mathcal{N}\backslash\{i\}}b_{2j}\frac{u_{j,\ell}}{x_\ell} + b_1\right)x_\ell u_{i,\ell}\right).$$

For the sake of simplicity in the notation, we introduce some auxiliary variables to compact the latter terms depending on the optimal strategic selection of the ith decision-maker. These auxiliary variables are:

$$c_{i,\ell} = r_i + \alpha_{i,\ell+1}b_{2i}^2,$$

$$\eta_{i,\ell} = \frac{b_{2i}\alpha_{i,\ell+1}}{c_{i,\ell}}\left(\sum_{j\in\mathcal{N}\backslash\{i\}}b_{2j}\frac{u_{j,\ell}}{x_\ell} + b_1\right).$$

Thus, the terms depending on the strategies for the decision-maker $i \in \mathcal{N}$ is expressed as follows:

$$\sum_{\ell=k}^{k+N-1}c_{i,\ell}\left(u_{i,\ell}^2 + 2\eta_{i,\ell}x_\ell u_{i,\ell}\right).$$

Now, we find the control inputs that minimize the latter equation. To do this, we perform square completion procedure, i.e.,

$$u_{i,\ell}^2 + 2\eta_{i,\ell}x_\ell u_{i,\ell} = (u_{i,\ell} + \eta_{i,\ell}x_\ell)^2 - \eta_{i,\ell}^2 x_\ell^2. \qquad (2.31)$$

From (2.31) is deduced that

$$u_{j,\ell}^* = -\eta_{j,\ell}x_\ell, \ \forall j \in \mathcal{N},$$

with the coupling η_i terms for all the decision-makers given by

$$\eta_{i,\ell} = -\frac{b_{2i}\alpha_{i,\ell+1}}{c_{i,\ell}}\sum_{j\in\mathcal{N}\backslash\{i\}}b_{2j}\eta_{j,\ell} + b_1\frac{b_{2i}\alpha_{i,\ell+1}}{c_{i,\ell}}.$$

Hence, after the square completion and replacing in (2.30) for the gap $L_i(x,u_1,\ldots,u_n) - f_{i,k}$ yields

$$L_i(x,u_1,\ldots,u_n) - f_{i,k} = q_i x_{k+N}^2 - \alpha_{i,k+N}x_{k+N}^2$$

$$
+ \sum_{\ell=k}^{k+N-1} \left(q_i x_\ell^2 + \alpha_{i,\ell+1} b_1^2 x_\ell^2 + \alpha_{i,\ell+1} \left[\sum_{j \in \mathcal{N} \backslash \{i\}} b_{2j} \eta_{j,\ell} \right]^2 x_\ell^2 \right)
$$

$$
+ \sum_{\ell=k}^{k+N-1} \left(- 2 b_1 \alpha_{i,\ell+1} \sum_{j \in \mathcal{N} \backslash \{i\}} b_{2j} \eta_{j,\ell} x_\ell^2 - \alpha_{i,\ell} x_\ell^2 \right)
$$

$$
+ \sum_{\ell=k}^{k+N-1} c_{i,\ell} (u_{i,\ell} + \eta_{i,\ell} x_\ell)^2 - c_{i,\ell} \eta_{i,\ell}^2 x_\ell^2.
$$

By organizing terms one obtains that

$$
L_i(x, u_1, \ldots, u_n) - f_{i,k} = (q_i - \alpha_{i,k+N}) x_{k+N}^2
$$

$$
+ \sum_{\ell=k}^{k+N-1} \left(q_i + \alpha_{i,\ell+1} b_1^2 + \alpha_{i,\ell+1} \left[\sum_{j \in \mathcal{N} \backslash \{i\}} b_{2j} \eta_{j,\ell} \right]^2 \right.
$$

$$
\left. - 2 b_1 \alpha_{i,\ell+1} \sum_{j \in \mathcal{N} \backslash \{i\}} b_{2j} \eta_{j,\ell} - \alpha_{i,\ell} - c_{i,\ell} \eta_{i,\ell}^2 \right) x_\ell^2
$$

$$
+ \sum_{\ell=k}^{k+N-1} c_{i,\ell} (u_{i,\ell} + \eta_{i,\ell} x_\ell)^2,
$$

This last expression we have three main groups described next:

- *First line concerns the terminal time terms, which defines the boundary condition.*

- *Second and third line are composed of terms with common factor x_ℓ^2.*

- *Third line shows those terms defining the optimal control input.*

   Thus, the minimization of terms completes the proof.

The coming section pretends to present the same game problem addressed in this section by incorporating a noise, i.e., we continue our study by analyzing a stochastic difference game.

2.6 Stochastic Mean-Field-Free Difference Game

Let us preserve the same game setup from Section 2.5, with the same cost functional for the decision-makers as in (2.25). On the contrary, we take into consideration new system dynamics. Here, we consider the following stochastic system state dynamics:

$$x_{\ell+1} = b_1 x_\ell + \sum_{j \in \mathcal{N}} b_{2j} u_{j,\ell} + \sigma_0 w_{i,\ell}, \tag{2.32}$$

where $x \in \mathbb{R}$ corresponds to the system state, $u_i \in \mathbb{R}$ denotes the control input of the decision-maker $i \in \mathcal{N}$, and $b_1, b_{2j}, \sigma_0 \in \mathbb{R}$, for all $j \in \mathcal{N}$ are system parameters. We keep clarifying every time that these system parameters can be function of time, but we omit this option in order to ease notation. As mentioned above, the cost functionals for decision-makers is as in (2.25). Then, the new difference game problem is formally stated below.

Problem 8 (Stochastic Difference Game Approach) *The Stochastic Differential Game Problem for each $i \in \mathcal{N}$ is given by*

$$\begin{cases} \underset{u_i \in \mathcal{U}_i}{\text{minimize}} \ \mathbb{E}\left[L_i\left(u_1, \ldots, u_n\right)\right], \\ \\ \text{subject to} \\ \\ x_{\ell+1} = b_1 x_\ell + \sum_{j \in \mathcal{N}} b_{2j} u_{j,\ell} + s_0 w_{i,\ell}, \ \forall \ell \in [k..k + N - 1], \end{cases} \tag{2.33}$$

where the initial state x_k is given, the weight parameters in the cost are $q_i \geq 0$, $r_i > 0$, and the system parameters are $b_1, b_{2j}, \sigma_0 \in \mathbb{R}$, for all $j \in \mathcal{N}$.

Similarly, as it occurred with the continuous-time differential game, the optimal cost functional in discrete-time changes once stochastic dynamics are considered. Proposition 8 presents the semi-explicit solution corresponding to Problem 8.

Proposition 8 *The explicit solution comprising the optimal control input and optimal cost for the Problem 8 are*

$$u_{i,\ell}^* = -\eta_{i,\ell} x_\ell,$$

$$L_i(x, u_1^*, \ldots, u_n^*) = \alpha_{i,0} x_0^2 + \delta_{i,0},$$

where the auxiliary variables $\eta_{i,\ell}$ and $c_{i,\ell}$, for all $i \in \mathcal{N}$, are as follows:

$$\eta_{i,\ell} = -\alpha_{i,\ell+1} \frac{b_{2i}}{c_{i,\ell}} \sum_{j \in \mathcal{N} \setminus \{i\}} b_{2j} \eta_{j,\ell} + b_1 \frac{b_{2i}}{c_{i,\ell}} \alpha_{i,\ell+1}, \ \forall \ i \in \mathcal{N},$$

$$c_{i,\ell} = \alpha_{i,\ell+1} b_{2i}^2 + r_i, \ \forall \ i \in \mathcal{N},$$

and with the following difference equations:

$$\alpha_{i,\ell} = q_i - c_{i,\ell} \eta_{i,\ell}^2 + \alpha_{i,\ell+1} \left(b_1 - \sum_{j \in \mathcal{N} \setminus \{i\}} b_{2j} \eta_{j,\ell} \right)^2, \tag{2.34a}$$

$$\delta_{i,\ell} = \alpha_{i,\ell+1}\sigma_0^2\mathbb{E}[w_{i,\ell}^2] + \delta_{i,\ell+1}, \tag{2.34b}$$

for all $\ell \in [k..k + N - 1], i \in \mathcal{N}$ with the boundary conditions

$$\alpha_{i,k+N} = q_{i,k+N},$$
$$\delta_{i,k+N} = 0.$$

We make the same observations that were made for the continuous-time game problems. The optimal control inputs for both the discrete-time deterministic and stochastic difference games coincide due to the fact that the noise parameter σ_0 is independent of the system state and control inputs. Despite this coincidence, the optimal costs are different since a difference equation δ appears in the stochastic approach. The reader may notice that, from the result in Proposition 8, one can easily obtain the solution presented in Proposition 7, simply by making the parameter $\sigma_0 = 0$. In this regard, the more general and complexer problem we can solve, more variety of scenarios can be considered. Such situation will become more evident once we start to study the mean-field-type game problems in the coming sections of the book.

To finish this chapter, we show the computation of the semi-explicit solution for the stochastic difference game in discrete time by following each step of the direct method. After this, it is expected that the reader is now quite familiar to this simple method that is going to be used in the solution of more challenging problems adding mean-field terms, and optionally, other kinds of stochastic processes.

Proof 8 (Proposition 8) *As a first step, it is necessary to postulate an ansatz for the optimal cost functional. Inspired from the structure of the cost functional and the system dynamics, we consider the following guess functional, which is similar to the one considered for the deterministic case:*

$$f_{i,\ell} = \alpha_{i,\ell}x_\ell^2 + \delta_{i,\ell}, \ \forall \ \ell \in [k..k + N - 1], \tag{2.35}$$

where $\alpha_{i,\ell}$ and $\delta_{i,\ell}$ are deterministic difference equation functions of time. Now, we apply the telescopic sum decomposition given by

$$f_{i,k+N} = f_{i,k} + \sum_{\ell=k}^{k+N-1} (f_{i,\ell+1} - f_{i,\ell}).$$

Hence, by replacing the guess functional in the telescopic sum one arrives at the following expression using terms depending on α_i and δ_i:

$$f_{i,k+N} - f_{i,k} = \sum_{\ell=k}^{k+N-1} \left(\alpha_{i,\ell+1}x_{\ell+1}^2 + \delta_{i,\ell+1} - \alpha_{i,\ell}x_\ell^2 - \delta_{i,\ell}\right). \tag{2.36}$$

Again, it is necessary to compute $x_{\ell+1}^2$ in terms of time instant ℓ. The reader may note that this square term is different from the one computed in the *deterministic case due to the appearance of the noise and the parameter σ_0, i.e.,*

$$x_{\ell+1}^2 = \left[b_1 x_\ell + \sum_{j\in\mathcal{N}} b_{2j} u_{j,\ell} + \sigma_0 w_{i,\ell} \right]^2,$$

$$= b_1^2 x_\ell^2 + \left[\sum_{j\in\mathcal{N}} b_{2j} u_{j,\ell} \right]^2 + \sigma_0^2 w_{i,\ell}^2 + 2 b_1 x_\ell \sum_{j\in\mathcal{N}} b_{2j} u_{j,\ell}$$

$$+ 2 b_1 \sigma_0 x_\ell w_{i,\ell} + 2 \sum_{j\in\mathcal{N}} b_{2j} \sigma_0 u_{j,\ell} w_{i,\ell},$$

$$= b_1^2 x_\ell^2 + b_{2i}^2 u_{i,\ell}^2 + 2 b_{2i} u_{i,\ell} \sum_{j\in\mathcal{N}\setminus\{i\}} b_{2j} u_{j,\ell} + \left[\sum_{j\in\mathcal{N}\setminus\{i\}} b_{2j} u_{j,\ell} \right]^2 + \sigma_0^2 w_{i,\ell}^2$$

$$+ 2 b_1 x_\ell \sum_{j\in\mathcal{N}} b_{2j} u_{j,\ell} + 2 b_1 \sigma_0 x_\ell w_{i,\ell} + 2 \sum_{j\in\mathcal{N}} b_{2j} \sigma_0 u_{j,\ell} w_{i,\ell}.$$

In addition, we have separated those terms depending on the decisions made by $i \in \mathcal{N}$ from the decisions made by others $j \in \mathcal{N} \setminus \{i\}$. Replacing the square term $x_{\ell+1}^2$ back in the telescopic sum in (2.36) yields

$$f_{i,k+N} - f_{i,k} = \sum_{\ell=k}^{k+N-1} \left(\alpha_{i,\ell+1} b_1^2 x_\ell^2 + \alpha_{i,\ell+1} b_{2i}^2 u_{i,\ell}^2 \right.$$

$$+ 2\alpha_{i,\ell+1} b_{2i} u_{i,\ell} \sum_{j\in\mathcal{N}\setminus\{i\}} b_{2j} u_{j,\ell} + \alpha_{i,\ell+1} \left[\sum_{j\in\mathcal{N}\setminus\{i\}} b_{2j} u_{j,\ell} \right]^2$$

$$+ \alpha_{i,\ell+1} \sigma_0^2 w_{i,\ell}^2$$

$$+ 2\alpha_{i,\ell+1} b_1 x_\ell \sum_{j\in\mathcal{N}} b_{2j} u_{j,\ell} + 2\alpha_{i,\ell+1} b_1 \sigma_0 x_\ell w_{i,\ell}$$

$$\left. + 2\alpha_{i,\ell+1} \sum_{j\in\mathcal{N}} b_{2j} \sigma_0 u_{j,\ell} w_{i,\ell} + \delta_{i,\ell+1} - \alpha_{i,\ell} x_\ell^2 - \delta_{i,\ell} \right).$$

Now, we compute the gap $\mathbb{E}[L_i(x, u_1, \ldots, u_n) - f_{i,k}]$. Recall that it is desire to match the cost when applying the optimal control inputs with the optimal cost from time instant k up to $k + N$ given by $f_{i,k}$. Then,

$$L_i(x, u_1, \ldots, u_n) - f_{i,k} = \mathbb{E}(q_i - \alpha_{i,k+N}) x_{k+N}^2 + \mathbb{E}[0 - \delta_{i,k+N}]$$

$$+ \mathbb{E} \sum_{\ell=k}^{k+N-1} \left(\alpha_{i,\ell+1} b_1^2 + q_i - \alpha_{i,\ell} \right) x_\ell^2$$

$$+ \mathbb{E} \sum_{\ell=k}^{k+N-1} \left(\alpha_{i,\ell+1} \left[\sum_{j\in\mathcal{N}\setminus\{i\}} b_{2j} u_{j,\ell} \right]^2 + \alpha_{i,\ell+1} \sigma_0^2 w_{i,\ell}^2 \right.$$

$$+ 2\alpha_{i,\ell+1} b_1 x_\ell \sum_{j \in \mathcal{N} \setminus \{i\}} b_{2j} u_{j,\ell} + \delta_{i,\ell+1} - \delta_{i,\ell} \Bigg)$$

$$+ \mathbb{E} \sum_{\ell=k}^{k+N-1} \left(\alpha_{i,\ell+1} b_{2i}^2 + r_i \right) \left(u_{i,\ell}^2 \right.$$

$$+ 2\alpha_{i,\ell+1} \frac{b_{2i}}{\alpha_{i,\ell+1} b_{2i}^2 + r_i} \left(\sum_{j \in \mathcal{N} \setminus \{i\}} b_{2j} \frac{u_{j,\ell}}{x_\ell} + b_1 \right) x_\ell u_{i,\ell} \Bigg).$$

Pursuing the goal to ease notation, let us consider the auxiliary variables presented next.

$$c_{i,\ell} = \alpha_{i,\ell+1} b_{2i}^2 + r_i,$$

$$\eta_{i,\ell} = \alpha_{i,\ell+1} \frac{b_{2i}}{c_{i,\ell}} \left(\sum_{j \in \mathcal{N} \setminus \{i\}} b_{2j} \frac{u_{j,\ell}}{x_\ell} + b_1 \right).$$

Therefore, the last term depending on the control input $u_{i,\ell}$ in the gap $\mathbb{E}[L_i(x, u_1, \ldots, u_n) - f_{i,k}]$ is

$$\mathbb{E} \sum_{\ell=k}^{k+N-1} c_{i,\ell} \left(u_{i,\ell}^2 + 2\eta_{i,\ell} x_\ell u_{i,\ell} \right).$$

With this term in hand, it is necessary to optimize to minimize the cost functional in $u_{i,\ell}$. Here, we do this procedure by following a simple square completion to identify the optimal control input, i.e.,

$$\left(u_{i,\ell}^2 + 2\eta_{i,\ell} x_\ell u_{i,\ell} \right) = \left(u_{i,\ell} + \eta_{i,\ell} x_\ell \right)^2 - \eta_{i,\ell}^2 x_\ell^2. \tag{2.37}$$

From (2.37) it is concluded that

$$u_{j,\ell} = -\eta_{j,\ell} x_\ell, \ \forall j \in \mathcal{N},$$

and with the coupling η_i term for all the decision-makers as follows:

$$\eta_{i,\ell} = -\alpha_{i,\ell+1} \frac{b_{2i}}{c_{i,\ell}} \sum_{j \in \mathcal{N} \setminus \{i\}} b_{2j} \frac{\eta_{j,\ell} x_\ell}{x_\ell} + b_1 \alpha_{i,\ell+1} \frac{b_{2i}}{c_{i,\ell}}.$$

Replacing the computed new terms in the gap $L_i(x, u_1, \ldots, u_n) - f_{i,k}$ yields

$$L_i(x, u_1, \ldots, u_n) - f_{i,k} = \mathbb{E}(q_i - \alpha_{i,k+N}) x_{k+N}^2 + \mathbb{E}[0 - \delta_{i,k+N}]$$

$$+ \mathbb{E} \sum_{\ell=k}^{k+N-1} \left(\alpha_{i,\ell+1} b_1^2 + q_i - \alpha_{i,\ell} - c_{i,\ell} \eta_{i,\ell}^2 + \alpha_{i,\ell+1} \left[\sum_{j \in \mathcal{N} \setminus \{i\}} b_{2j} \eta_{j,\ell} \right]^2 \right)$$

$$- 2\alpha_{i,\ell+1}b_1 \sum_{j \in \mathcal{N}\backslash\{i\}} b_{2j}\eta_{j,\ell} \Bigg) x_\ell^2$$

$$+ \mathbb{E} \sum_{\ell=k}^{k+N-1} \left(\alpha_{i,\ell+1}\sigma_0^2 w_{i,\ell}^2 + \delta_{i,\ell+1} - \delta_{i,\ell} \right)$$

$$+ \mathbb{E} \sum_{\ell=k}^{k+N-1} c_{i,\ell} \left(u_{i,\ell} + \eta_{i,\ell}x_\ell \right)^2 .$$

Let us divide this expression into four different parts:

- *The first part comprises the terms at terminal time from which we deduce the boundary conditions for the Riccati equations.*

- *Second group is composed of the second and third lines, which correspond to those terms with common factor x_ℓ^2. These terms establish the α_i equations, for all $i \in \mathcal{N}$.*

- *Next, as a third group, we have the terms independent from both the square of the system state and the control inputs. These terms establish the δ_i equations, for all $i \in \mathcal{N}$.*

- *Finally, as the fourth group, we have the terms defining the optimal control input.*

Then, the minimization of terms completes the proof.

2.7 Exercises

1. Consider the following deterministic scalar-valued system state dynamics:

$$\dot{x} = b_1 x + b_2 u, \tag{2.38}$$

and the quadratic cost functional

$$L(x, u) = \frac{1}{2}qx(T)^2 + \frac{1}{2}\int_0^T qx^2 \mathrm{d}t. \tag{2.39}$$

Solve the mean-field-free control problem $\min_u L(x, u)$ with cost function in (2.39) subject to (2.38) and identify the implications that the term $u^2(t)$ does not appear in the running cost of $L(x, u)$.

Hint: Considering upper and lower bound constraints for the control inputs, the problem results in a bang-bang control law.

2. Consider the system dynamics with Brownian motion and control-input-dependent diffusion term

$$dx = b_1 x dt + u\left(b_2 dt + dB\right), \qquad (2.40)$$

Solve the mean-field-free control problem $\min_u \ L(x, u)$ with cost function in (2.39) subject to (2.40) and identify the implications that the term $u^2(t)$ does not appear in the running cost of $L(x, u)$.

Hint: Notice that even though there is no square term for u in the cost functional, it is possible to perform square completion now.

3. Consider the nonlinear system dynamics with Brownian motion and state-dependent diffusion term

$$dx = (b_1 x \ln(x) + b_2 u)\, dt + x \sigma dB, \qquad (2.41)$$

with $x_0 > 1$, and the cost functional

$$L(x, u) = q \ln(x(T)) + \int_0^T q \ln(x) dt. \qquad (2.42)$$

Solve the mean-field-free control problem $\min_u \ L(x, u)$ with cost function in (2.42) subject to (2.41).

Hint: An appropriate guess functional is given by

$$f(t, x) = \alpha \ln(x) + \delta.$$

4. Consider the nonlinear system dynamics with nonlinear state-dependent diffusion term

$$dx = x\left(-\frac{1}{2} + b_1 \ln(x) + b_2 u\right) dt + x\sqrt{\ln(x)}dB, \qquad (2.43)$$

with $x_0 > 1$, and the cost function

$$L(x, u) = q \ln^2(x(T)) + \int_0^T q \ln^2(x) + ru^2 dt. \qquad (2.44)$$

Solve the mean-field-free control problem $\min_u \ L(x, u)$ with cost function in (2.44) subject to (2.43).

Hint: An appropriate guess functional is given by

$$f(t, x) = \alpha \ln^2(x).$$

5. Now, let us consider n decision-makers, i.e., $\mathcal{N}$ is not singleton, and

with the system dynamics including a nonlinear state-dependent diffusion term

$$dx = \left(b_1 x \ln(x) + \sum_{j \in \mathcal{N}} b_{2j} u_j \right) dt + x\sigma dB, \qquad (2.45)$$

and the cost function of the decision maker $i \in \mathcal{N}$ is given by

$$L_i(x, u_1, \ldots, u_n) = q_i \ln(x(T)) + \int_0^T q_i \ln(x) dt. \qquad (2.46)$$

Solve the mean-field-free differential game problem $\min_u L_i(x, u_1, \ldots, u_n)$ with cost function in (2.46) subject to (2.45).

Hint: An appropriate guess functional is given by

$$f_i(t, x) = \alpha_i \ln(x) + \delta_i.$$

6. Let us consider n decision makers, i.e., $\mathcal{N}$ is not singleton, and the system dynamics

$$dx = x \left(-\frac{1}{2} + b_1 \ln(x) + \sum_{j \in \mathcal{N}} b_{2j} u_j \right) dt + x\sqrt{\ln(x)} dB, \qquad (2.47)$$

with $x_0 > 1$, and the cost function of the decision maker $i \in \mathcal{N}$ is

$$L_i(x, u_1, \ldots, u_n) = q_i \ln^2(x(T)) + \int_0^T q_i \ln^2(x) + r_i u_i^2 dt. \qquad (2.48)$$

Solve the mean-field-free differential game problem $\min_u L_i(x, u_1, \ldots, u_n)$ with cost function in (2.48) subject to (2.47).

Hint: An appropriate guess functional is given by

$$f_i(t, x) = \alpha_i \ln^2(x).$$

7. Propose a new non-linear mean-field-free control or game problem that could be addressed by using the direct method. Provide the guess functional and present the respective semi-explicit solution.

3

Mean-Field Games

In this chapter, we present a linear-quadratic mean-field game. This is a game in which the strategic selection of a single decision-maker is negligible.

Different from the game problems we have studied so far in this book, this time let us consider a quite large population of n decision-makers, i.e., $n \to \infty$. Notice that, due to the large number of decision-makers, the decisions made by an individual is completely imperceptible, i.e., a single player does not have any impact over the overall behavior of the population. In other words, *"a single decision-maker is too small in comparison to the total number of interactive decision-makers."* When this situation occurs the game is classified as non-atomic. Thus, the mean-field games are a class of non-atomic games. There are several works studying this class of games with a large population of decision-makers, e.g., see [12, 45, 85].

We also refer the reader to explore the problem settings for the so-called aggregative games in which each decision-maker basically interacts to the aggregative decision of all the other participants, but not with each other individually.

We recall the discussion presented in the work reported in [27] about key differences between the mean-field game with infinite number of decision-makers in the context of non-atomic games (in the mean-field games the large number of players make their choices despicable, i.e., negligible decision-makers), and mean-field-type approach with finite number of decision-makers in the context of atomic games. The consideration of large number of decision-makers has been analyzed under different scenarios. As it has been pointed out in [27], in the context of competitive market with large number of agents, the article [87], published in 1951, captures the assumption made in mean-field games with large number of agents:

"each of the participants is of the opinion that his own transactions do not influence the prevailing prices"

Abraham Wald
[87, page 380]

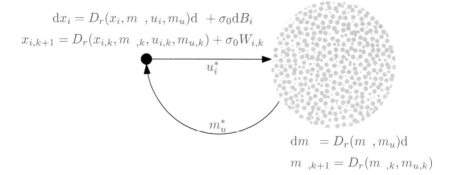

FIGURE 3.1
General Mean-Field Game Philosophy in either continuous or discrete time.
Each player (black node) strategically plays against a population mass of
infinite players (dark gray nodes).

Another comment on the impact over the population with mean-field term
was given in [88], published in 1953:

"When the number of participants becomes large, some hope
emerges that of the influence of every particular participant will
become negligible."

John von Neumman
[88, page 13]

Each decision-maker plays against the whole mass of players (aggregated
collection of decision-makers), which is described by a mean-field term for both
the system states denoted by m_x, and the strategies or control inputs denoted
by m_u as shown in Figure 3.1. Next, we present the set-up corresponding to
a simple mean-field game and we solve it by means of the direct method.

It is worth to mention that the study of mean-field games is not limited
to this simple settings we present in this chapter. However, we provide this
discussion to enrich later on the exposition of the mean-field-type games and to
show the reader about the suitability of applying the direct method even in the
context of non-atomic games assuming a large and homogeneous population
of players.

3.1 A Continuous-Time Deterministic Mean-Field Game

The interaction consists of infinitely many players (also known as decision-makers). Each player has an associated system state, which is affected by all the others' decisions through the mean-field terms denoted by m_x and m_u as follows:

$$\dot{x}_i = b_1 x_i + \bar{b}_1 m_x + b_2 u_i + \bar{b}_2 m_u, \tag{3.1a}$$

$$x_i(0) := x_{0,i}, \tag{3.1b}$$

$$m_x(0) := m_{x,0}, \tag{3.1c}$$

where $x_i \in \mathbb{R}$ corresponds to the system state, and $u_i \in \mathbb{R}$ denotes the control input of the decision-maker $i \in \mathcal{N}$. Hence, $b_1, \bar{b}_1, b_2, \bar{b}_2 \in \mathbb{R}$, for all $j \in \mathcal{N}$ are the system state parameters. Here, it is quite important to highlight that all these parameters are the same for all the decision-makers as in Remark 5. In other words, the population of the players is homogeneous. Having said that, we also mention that these system parameters can be considered as function of time if desired without this implying extra difficulty.

Remark 5 *Notice that the parameters* $b_1, \bar{b}_1, b_2, \bar{b}_2 \in \mathbb{R}$ *are identical for all* $j \in \mathcal{N}$ *since all the decision-makers are assumed to be homogeneous.*

We advise the reader to pay special attention on the necessity of considering Remark 5, which will be discussed later on when solving in a semi-explicit way the mean-field game problems. In addition, here we assume that there are as many system states as decision-makers. This situation is somehow different from the settings we have been considering in the previous chapters where we had a unique system state. Nevertheless, there is a tight relationship between the two approaches by considering as a unified state the vector of all the system states, i.e., $x = [x_1 \quad \dots x_n]^\top$. We refer the reader to the matrix-valued game problems in order to go deeper into this type of structures (see, e.g., Part IV in this book).

The mean-field terms m_x and m_u are directly associated with the system states and control inputs, respectively. Such terms are given by

$$m_x = \lim_{n \to \infty} \frac{1}{n} \sum_{j=1}^{n} x_j, \tag{3.2a}$$

$$m_u = \lim_{n \to \infty} \frac{1}{n} \sum_{j=1}^{n} u_j. \tag{3.2b}$$

The decision-makers' system states are coupled by means of the mean-field terms m_x and m_u. The system dynamics for x_i depends on m_x and m_u, and the latter mean-field terms depend on all the other homogeneous decision-makers. This means that there are coupling among all the decision-makers. Each decision-maker has an associated cost function as follows:

$$L_i(x, u_1, \ldots, u_n) = \frac{1}{2}q(T)x_i(T)^2 + \frac{1}{2}\bar{q}(T)m_x^2(T)$$

$$+ \frac{1}{2}\int_0^T \left(qx_i^2 + \bar{q}m_x^2 + ru_i^2 + \bar{r}m_u^2\right) \mathrm{d}t, \qquad (3.3)$$

which incorporates the mean-field terms m_x and m_u. Each decision-maker pursues to minimize its own state and applied energy. Besides, decision-makers also minimize the mean-field terms which are dependent on all other decision-makers. In addition, recall that x_i is coupled to all other players $\mathcal{N} \setminus \{i\}$ through m_x and m_u.

In comparison to the game problems studied in previous chapters, here we have more weight parameters in the cost functional and they determine the prioritization, balance or importance assigned to each control objective. In this case, there are four objectives attached to x_i, u_i, and their mean-field-related terms m_x, and m_u.

Remark 6 *Notice that the parameters $q, \bar{q}, r, \bar{r} \in \mathbb{R}$ are identical for all $j \in \mathcal{N}$ since all the decision-makers are assumed to be homogeneous. Indeed, these weight parameters can be considered as function of time.*

Next, we introduce the simplest mean-field deterministic game problem with homogeneous population of decision-makers.

Problem 9 (Deterministic Mean-Field Game Problem) *The linear-quadratic deterministic mean-field game problem is given by*

$$\begin{cases} \underset{u_i \in \mathcal{U}_i}{\text{minimize}} \ L_i(x, u_1, \ldots, u_n), \\[2mm] \text{subject to} \\[2mm] \dot{x}_i = b_1 x_i + \bar{b}_1 m_x + b_2 u_i + \bar{b}_2 m_u, \\ x_i(0) := x_{0i}, \end{cases}$$

where the weights in the cost functional are $q, q + \bar{q} \in \mathbb{R}_{\geq 0}$, $r, r + \bar{r} \in \mathbb{R}_{>0}$, and the system parameters $b_1, \bar{b}_1, b_2, \bar{b}_2 \in \mathbb{R}$, for all decision-makers $j \in \mathcal{N}$.

The semi-explicit solution for the mean-field deterministic problem is presented in Proposition 9, which defines not only an optimal solution for each decision-maker but also provides optimality for the mean-field control input term.

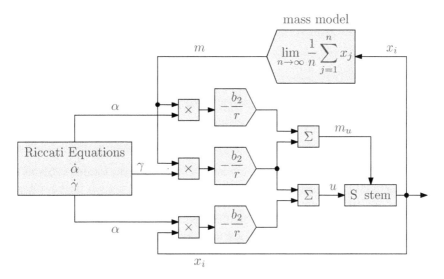

FIGURE 3.2
Feedback scheme for the linear-quadratic mean-field game in continuous time.

Proposition 9 *The feedback-strategies linear-quadratic mean-field Nash equilibrium for the Deterministic Mean-Field Game Problem 9, and the optimal costs are given by:*

$$u_i^* = -b_2 \frac{\alpha x_i + \gamma m_x}{r},$$

$$m_u^* = -\frac{b_2}{r}(\alpha + \gamma)m_x,$$

$$L_i(x, u_1^*, \dots, u_n^*) = \frac{1}{2}\alpha(0)x_{0i}^2 + \frac{1}{2}\beta(0)m_x(0)^2 + \gamma(0)m_x(0)x_{0i},$$

where α, β, and γ solve the following differential equations:

$$\dot{\alpha} = -q - 2b_1\alpha + \frac{b_2^2}{r}\alpha^2,$$

$$\dot{\beta} = -\bar{q} - 2\bar{b}_1\gamma - 2\beta\left(b_1 + \bar{b}_1 - \frac{b_2}{r}(b_2 + \bar{b}_2)(\alpha + \gamma)\right) + \frac{b_2^2}{r}\gamma^2 - \bar{r}\frac{b_2^2}{r^2}(\alpha + \gamma)^2$$

$$+ 2\bar{b}_2\gamma\frac{b_2}{r}(\alpha + \gamma),$$

$$\dot{\gamma} = -\bar{b}_1\alpha - b_1\gamma - \gamma\left(b_1 + \bar{b}_1 - \frac{b_2}{r}(b_2 + \bar{b}_2)(\alpha + \gamma)\right) + \frac{b_2^2}{r}\alpha\gamma$$

$$+ \bar{b}_2\alpha\frac{b_2}{r}(\alpha + \gamma),$$

with the following terminal boundary conditions:

$$\alpha(T) = q(T),$$

$$\beta(T) = \bar{q}(T),$$
$$\gamma(T) = 0.$$

Finally, the following differential equation

$$\dot{m}_x = \left(b_1 + \bar{b}_1 - \frac{b_2}{r} \left(b_2 + \bar{b}_2 \right) (\alpha + \gamma) \right) m_x,$$

defines the evolution of the mean-field state.

Please pay attention to the fact that, in contrast to the game problems we had considered in Chapter 2, more differential equations appear in the solution for the mean-field game problem. Besides, we have the following Remark 7 making an annotation for the optimal control input. Later on, when studying the mean-field-type games, we will see an evident difference regarding Remark 7 and the structure of the optimal solution.

Remark 7 *Notice that, the optimal control input in Proposition 9 is independent of the parameters $\bar{b}_2$. The number $\bar{b}_2$ does not appear directly in the equilibrium strategy but it does influence the functions γ and β.*

Figure 3.2 presents the scheme corresponding to the optimal solution of the mean-field game according to Proposition 9. It can be seen that two ordinary differential equations, α and γ, should be solved in order to compute the optimal control inputs u and m_u. The term γ does not have any effect over the optimal control input, but allows representing the optimal cost functional.

Proof 9 (Proposition 9) *This proof is straight-forwardly obtained from the proof of Proposition 10, which is introduced in the following section.*

Prior to moving to the stochastic mean-field counterpart, there is an interesting fact highlighted in Remark 8 showing a relationship between a Riccati equation in the mean-field case and the optimal control case. We suggest the reader to keep in mind this Remark when studying the mean-field-type games.

Remark 8 *Notice that, the evolution of the ordinary differential equation $\dot{\alpha}$ in Proposition 9 coincides with the one obtained in Proposition 1 that corresponds to the Riccati equation in the solution of an optimal control problem.*

3.2   A Continuous-Time Stochastic Mean-Field Game

In the previous section, we have studied the deterministic mean-field games. In this section, we extend the settings by incorporating a noise given by a

Brownian motion. Each player $i \in \mathcal{N}$ has an associated system state, which is affected by all the others' decisions by means of the mean-field terms as follows:

$$\mathrm{d}x_i = \underbrace{\left(b_1 x_i + \bar{b}_1 m_x + b_2 u_i + \bar{b}_2 m_u\right)}_{D_r(x_i, m_x, u_i, m_u)} \mathrm{d}t$$

$$+ \underbrace{\sigma_0}_{D_f(x_i, m_x, u_i, m_u)} \mathrm{d}B_i, \qquad (3.4a)$$

$$x_i(0) := x_{0,i}, \qquad (3.4b)$$

$$m_x(0) := m_{x,0}, \qquad (3.4c)$$

where $x_i \in \mathbb{R}$ corresponds to the system state, and $u_i \in \mathbb{R}$ denotes the control input of the decision maker $i \in \mathcal{N}$. the system state parameters are $b_1, \bar{b}_1, b_2, \bar{b}_2, \sigma_0 \in \mathbb{R}$, for all $j \in \mathcal{N}$. Hence, B_i denotes a standard Brownian motion. There are some mean-field terms m_x and m_u related to the system states and control inputs, respectively. Such terms are given by (3.2a) and (3.2b). Each decision-maker has an associated cost functional as in (3.3). For a general case in 3.4, we have named the drift term as $D_r(x_i, m_x, u_i, m_u)$, and the diffusion term as $D_f(x_i, m_x, u_i, m_u)$. Nevertheless, in this simplified example we consider a constant parameter for the diffusion term. The new mean-field game problem is formally stated next.

Problem 10 (Stochastic Mean-Field Game) *The Linear-Quadratic Mean-Field Game Problem is given by*

$$\begin{cases} \underset{u_i \in \mathcal{U}_i}{\text{minimize}} \, \mathbb{E}[L_i(x, u_1, \ldots, u_n)], \\[2mm] \text{subject to} \\[2mm] \mathrm{d}x_i = \left(b_1 x_i + \bar{b}_1 m_x + b_2 u_i + \bar{b}_2 m_u\right) \mathrm{d}t + \sigma_0 \mathrm{d}B_i, \\ x_i(0) := x_{0i}, \end{cases}$$

where the cost functional weight parameters are $q, q + \bar{q} \in \mathbb{R}_{\geq 0}$, $r, r + \bar{r} \in \mathbb{R}_{>0}$, and the system parameters are $b_1, \bar{b}_1, b_2, \bar{b}_2, \sigma_0 \in \mathbb{R}$, for all $j \in \mathcal{N}$.

Proposition 10 presents the semi-explicit solution for the stochastic mean-field game problem, which later on we will compare with respect to the deterministic counterpart as we did in the previous chapters of the book. Thus, we can observe the effect of incorporating a state and control-input independent diffusion.

Proposition 10 *The feedback-strategies linear-quadratic mean-field Nash equilibrium for the Non-Cooperative Problem 10, and the optimal costs are* *given by:*

$$u_i^* = -b_2 \frac{\alpha x_i + \gamma m_x}{r},$$

$$m_u^* = -\frac{b_2}{r}(\alpha + \gamma)m_x,$$

$$L_i(x, u_1^*, \dots, u_n^*) = \frac{1}{2}\alpha(0)x_{0i}^2 + \frac{1}{2}\beta(0)m_x(0)^2 + \gamma(0)m_x(0)x_{0i} + \delta(0),$$

where α, β, γ, and δ solve the following differential equations:

$$\dot{\alpha} = -q - 2b_1\alpha + \frac{b_2^2}{r}\alpha^2,$$

$$\dot{\beta} = -\bar{q} - 2\bar{b}_1\gamma - 2\beta\left(b_1 + \bar{b}_1 - \frac{b_2}{r}(b_2 + \bar{b}_2)(\alpha + \gamma)\right) + \frac{b_2^2}{r}\gamma^2$$

$$- \bar{r}\frac{b_2^2}{r^2}(\alpha + \gamma)^2 + 2\bar{b}_2\gamma\frac{b_2}{r}(\alpha + \gamma),$$

$$\dot{\gamma} = -\bar{b}_1\alpha - b_1\gamma - \gamma\left(b_1 + \bar{b}_1 - \frac{b_2}{r}(b_2 + \bar{b}_2)(\alpha + \gamma)\right)$$

$$+ \frac{b_2^2}{r}\alpha\gamma + \bar{b}_2\alpha\frac{b_2}{r}(\alpha + \gamma),$$

$$\dot{\delta} = -\frac{1}{2}\alpha\sigma_0^2,$$

with the terminal boundary conditions as follows:

$$\alpha(T) = q(T),$$

$$\beta(T) = \bar{q}(T),$$

$$\gamma(T) = \delta(T) = 0.$$

Finally, the evolution of the mean state is defined by the following differential equation:

$$\dot{m}_x = \left(b_1 + \bar{b}_1 - \frac{b_2}{r}(b_2 + \bar{b}_2)(\alpha + \gamma)\right)m_x.$$

Remark 9 shows some important aspects comparing the solution for both the deterministic and stochastic mean-field game problems and the existing relationship or similarities with the optimal control problem case.

Remark 9 *The evolution of the ordinary differential equation $\dot{\alpha}$ in Proposition 10 coincides with the one obtained in Proposition 1 that corresponds to the Riccati equation in the solution of an optimal control problem. Moreover, notice that the unique difference between the optimal solution for the deterministic and stochastic mean-field game problems relies on the optimal cost functional because of δ.*

Next, we construct the semi-explicit solution step-by-step for the stochastic mean-field game problem by applying the direct method. Indeed, from this proof one can infer the computation for the solution corresponding to the deterministic case.

Proof 10 (Proposition 10) *Recall that the first step to apply the direct method consists of defining an appropriate guess functional for the optimal cost. We propose a quadratic guess functional of the following form based on the structure evidenced in the cost functional:*

$$f_i(t, x_i, m_x) = \frac{1}{2}\alpha x_i^2 + \frac{1}{2}\beta m_x^2 + \gamma m_x x_i + \delta, \qquad (3.5)$$

where α, β, γ, and δ are deterministic functions of time. Notice that given the assumption that all the decision-makers are homogeneous, i.e., the system and cost parameters are identical for all the players, then it is not necessary to make difference among the deterministic functions α, β, γ, and δ. Let us recall the Itô's formula, i.e.,

$$f_i(T, x_i, m_x) - f_i(0, x_{0i}, m_x(0))$$
$$= \int_0^T \left(\frac{\partial f_i(t, x_i, m_x)}{\partial t} + \frac{\partial f_i(t, x_i, m_x)}{\partial x_i} D_r(x_i, m_x, u_i, m_u) \right.$$
$$\left. + \frac{\partial^2 f_i(t, x_i, m_x)}{\partial x_i^2} \frac{D_f(x_i, m_x, u_i, m_u)^2}{2} \right) dt$$
$$+ \int_0^T D_f(x_i, m_x, u_i, m_u) \frac{\partial f_i(t, x_i, m_x)}{\partial x_i} dB_i, \qquad (3.6)$$

with D_r and D_f denoting the drift and the diffusion terms, respectively, i.e.,

$$D_r(x_i, m_x, u_i, m_u) = b_1 x_i + \bar{b}_1 m_x + b_2 u_i + \bar{b}_2 m_u,$$

and $D_f(x_i, m_x, u_i, m_u) = \sigma_0$ (state and control input independent). To apply the Itô's formula to the guess functional, we first compute separately the following terms:

$$\frac{\partial f_i(t, x_i, m_x)}{\partial t} - \frac{1}{2}x_i^2\dot{\alpha} + \frac{1}{2}m_x^2\dot{\beta} + \beta m_x \dot{m}_x + m_x x_i \dot{\gamma} + \gamma x_i \dot{m}_x + \dot{\delta}, \quad (3.7a)$$

$$\frac{\partial f_i(t, x_i, m_x)}{\partial x_i} = \alpha x_i + \gamma m_x, \qquad (3.7b)$$

$$\frac{\partial^2 f_i(t, x_i, m_x)}{\partial x^2} = \alpha. \qquad (3.7c)$$

Replacing (3.7) in (3.6) yields

$$f_i(T, x_i, m_x) - f_i(0, x_{0i}, m_x(0))$$

$$= \int_0^T \left(\frac{1}{2} x_i^2 \dot{\alpha} + \frac{1}{2} m_x^2 \dot{\beta} + \beta m_x \dot{m}_x + m_x x_i \dot{\gamma} + \gamma x_i \dot{m}_x + \dot{\delta} \right.$$

$$+ (\alpha x_i + \gamma m_x)(b_1 x_i + \bar{b}_1 m_x + b_2 u_i + \bar{b}_2 m_u) + \left. \alpha \frac{\sigma_0^2}{2} \right) dt$$

$$+ \int_0^T \sigma_0 \alpha x_i \mathrm{d}B_i.$$

Then, we compute the gap $\mathbb{E}[L_i(x, u_1, \ldots, u_n) - f_i(0, x_{0i}, m_x(0))]$ *as follows:*

$$\mathbb{E}[L_i(x, u_1, \ldots, u_n) - f_i(0, x_{0i}, m_x(0))] = \frac{1}{2} \mathbb{E}(q(T) - \alpha(T)) x_i(T)^2$$

$$+ \frac{1}{2} \mathbb{E}(\bar{q}(T) - \beta(T)) m_x^2(T) + \mathbb{E}(0 - \gamma(T)) m_x(T) + \mathbb{E}(0 - \delta(T))$$

$$+ \mathbb{E} \int_0^T \left(\frac{1}{2} x_i^2 \dot{\alpha} + \frac{1}{2} m_x^2 \dot{\beta} + \beta m_x \dot{m}_x + m_x x_i \dot{\gamma} + \gamma x_i \dot{m}_x + \dot{\delta} \right.$$

$$+ b_1 \alpha x_i^2 + \bar{b}_1 \alpha x_i m_x + b_2 \alpha x_i u_i + \bar{b}_2 \alpha x_i m_u$$

$$+ b_1 \gamma m_x x_i + \bar{b}_1 \gamma m_x^2 + b_2 \gamma m_x u_i + \bar{b}_2 \gamma m_x m_u + \left. \alpha \frac{\sigma_0^2}{2} \right) dt, \qquad (3.8)$$

and grouping terms yields

$$\mathbb{E}[L_i(x, u_1, \ldots, u_n) - f_i(0, x_{0i}, m_x(0))] = \frac{1}{2} \mathbb{E}(q(T) - \alpha(T)) x_i(T)^2$$

$$+ \frac{1}{2} \mathbb{E}(\bar{q}(T) - \beta(T)) m_x^2(T) + \mathbb{E}(0 - \gamma(T)) m_x(T) + \mathbb{E}(0 - \delta(T))$$

$$+ \frac{1}{2} \mathbb{E} \int_0^T x_i^2 \left(q + \dot{\alpha} + 2b_1 \alpha \right) dt + \frac{1}{2} \mathbb{E} \int_0^T m_x^2 \left(\bar{q} + \dot{\beta} + 2\bar{b}_1 \gamma \right) dt$$

$$+ \frac{1}{2} \mathbb{E} \int_0^T m_x x_i \left(2\dot{\gamma} + 2\bar{b}_1 \alpha + 2b_1 \gamma \right) dt$$

$$+ \frac{1}{2} \mathbb{E} \int_0^T \left(2\beta m_x \dot{m}_x + 2\gamma x_i \dot{m}_x + 2\dot{\delta} + \alpha \sigma_0^2 \right) dt$$

$$+ \frac{1}{2} \mathbb{E} \int_0^T r \left(u_i^2 + 2b_2 \frac{\alpha x_i + \gamma m_x}{r} u_i \right) dt$$

$$+ \frac{1}{2} \mathbb{E} \int_0^T \left(\bar{r} m_u^2 + 2\bar{b}_2 \alpha x_i m_u + 2\bar{b}_2 \gamma m_x m_u \right) dt. \qquad (3.9)$$

Now we optimize and perform a square completion procedure for the terms depending on the strategies u_i, *i.e.,*

$$\left(u_i + b_2 \frac{\alpha x_i + \gamma m_x}{r} \right)^2 = u_i^2 + 2b_2 \frac{\alpha x_i + \gamma m_x}{r} u_i + b_2^2 \frac{(\alpha x_i + \gamma m_x)^2}{r^2},$$

where

$$(\alpha x_i + \gamma m_x)^2 = \alpha^2 x_i^2 + 2\alpha\gamma m_x x_i + \gamma^2 m_x^2.$$

Thus, the terms depending on the control inputs u_i can be conveniently re-written in the form

$$\left(u_i + b_2\frac{\alpha x_i + \gamma m_x}{r}\right)^2 = u_i^2 + 2b_2\frac{\alpha x_i + \gamma m_x}{r}u_i + \frac{b_2^2}{r^2}\alpha^2 x_i^2$$
$$+ 2\frac{b_2^2}{r^2}\alpha\gamma m_x x_i + \frac{b_2^2}{r^2}\gamma^2 m_x^2,$$

obtaining that the control-input-dependent terms in the gap $\mathbb{E}[L_i(x, u_1, \ldots, u_n) - f_i(0, x_{0i}, m_x(0))]$ can be expressed as follows:

$$u_i^2 + 2b_2\frac{\alpha x_i + \gamma m_x}{r}u_i = \left(u_i + b_2\frac{\alpha x_i + \gamma m_x}{r}\right)^2 - \frac{b_2^2}{r^2}\alpha^2 x_i^2$$
$$- 2\frac{b_2^2}{r^2}\alpha\gamma m_x x_i - \frac{b_2^2}{r^2}\gamma^2 m_x^2. \tag{3.10}$$

From (3.10), it is concluded that

$$u_i^* = -b_2\frac{\alpha x_i + \gamma m_x}{r}$$

by making the first right-hand side term in (3.10) zero (minimizing the term). From this optimal control input u_i^*, we compute the optimal mean-field input m_u^* taking advantage of the homogeneity of the population in the strategic interaction, i.e.,

$$m_u^* = \lim_{n\to\infty}\left[-\frac{b_2}{r}\alpha\left[\frac{1}{n}\sum_{j\in\mathcal{N}}x_j\right] - \frac{b_2}{r}\gamma\left[\frac{1}{n}\sum_{j\in\mathcal{N}}m_x\right]\right],$$
$$= -\frac{b_2}{r}(\alpha + \gamma)m_x.$$

In addition, the term m_u^2 will be needed in the computation of the solution, see the last line in (3.9). This term is computed next depending on m_x^2, i.e.,

$$m_u^2 = \frac{b_2^2}{r^2}(\alpha + \gamma)^2 m_x^2.$$

Now, replacing the square completion back in (3.8) together with the mean-field control-input-dependent term m_u yields

$$\mathbb{E}[L_i(x, u_1, \ldots, u_n) - f_i(0, x_{0i}, m_x(0))] = \frac{1}{2}\mathbb{E}(q(T) - \alpha(T))x_i(T)^2$$
$$+ \frac{1}{2}\mathbb{E}(\bar{q}(T) - \beta(T))m_x^2(T) + \mathbb{E}(0 - \gamma(T))m_x(T) + \mathbb{E}(0 - \delta(T))$$

$$+ \frac{1}{2}\mathbb{E}\int_0^T x_i^2\left(q + \dot{\alpha} + 2b_1\alpha\right)dt + \frac{1}{2}\mathbb{E}\int_0^T m_x^2\left(\bar{q} + \dot{\beta} + 2\bar{b}_1\gamma\right)dt$$

$$+ \frac{1}{2}\mathbb{E}\int_0^T m_x x_i\left(2\dot{\gamma} + 2\bar{b}_1\alpha + 2b_1\gamma\right)dt$$

$$+ \frac{1}{2}\mathbb{E}\int_0^T\left(2\beta m_x \dot{m}_x + 2\gamma x_i \dot{m}_x + 2\dot{\delta} + \alpha\sigma_0^2\right)dt$$

$$+ \frac{1}{2}\mathbb{E}\int_0^T\left(r\left(u_i + b_2\frac{\alpha x_i + \gamma m_x}{r}\right)^2 - \frac{b_2^2}{r}\alpha^2 x_i^2\right.$$

$$\left. - 2\frac{b_2^2}{r}\alpha\gamma m_x x_i - \frac{b_2^2}{r}\gamma^2 m_x^2\right)dt$$

$$+ \frac{1}{2}\mathbb{E}\int_0^T\left(\bar{r}\frac{b_2^2}{r^2}(\alpha + \gamma)^2 m_x^2 - 2\bar{b}_2\alpha\frac{b_2}{r}(\alpha + \gamma)x_i m_x -\right.$$

$$\left. - 2\bar{b}_2\gamma\frac{b_2}{r}(\alpha + \gamma)m_x^2\right)dt. \tag{3.11}$$

Note that it is required to compute the evolution of the mean-field state $\dot{m}_x$ since it appears in the computation of the gap $\mathbb{E}[L_i(x, u_1, \dots, u_n) - f_i(0, x_{0i}, m_x(0))]$, see the fifth line in (3.11). Here, we make a short parenthesis to compute this term. From (3.4) we obtain that

$$dx_i = \left(b_1 x_i + \bar{b}_1 m_x - \frac{b_2^2}{r}(\alpha x_i + \gamma m_x) - \frac{\bar{b}_2 b_2}{r}(\alpha + \gamma)m_x\right)dt$$

$$+ \sigma_0 dB_i,$$

$$\lim_{n\to\infty} d\frac{1}{n}\sum_{i\in\mathcal{N}} x_i = \lim_{n\to\infty}\left(b_1\frac{1}{n}\sum_{i\in\mathcal{N}} x_i + \bar{b}_1\frac{1}{n}\sum_{i\in\mathcal{N}} m_x - \frac{b_2^2}{r}\frac{1}{n}\sum_{i\in\mathcal{N}}(\alpha x_i + \gamma m_x)\right.$$

$$\left. - \frac{\bar{b}_2 b_2}{r}\frac{1}{n}\sum_{i\in\mathcal{N}}(\alpha + \gamma)m_x\right)dt + \sigma_0\lim_{n\to\infty}\frac{1}{n}\sum_{i\in\mathcal{N}} dB_i.$$

Thus, it is concluded that

$$dm_x = \left(b_1 m_x + \bar{b}_1 m_x - \frac{b_2^2}{r}(\alpha + \gamma)m_x - \frac{\bar{b}_2 b_2}{r}(\alpha + \gamma)m_x\right)dt,$$

$$= \left(b_1 + \bar{b}_1 - \frac{b_2}{r}(b_2 + \bar{b}_2)(\alpha + \gamma)\right)m_x dt.$$

We highlight that the computation of the latter expression is only possible thanks to the fact that all the parameters are the same for all the decision-makers (homogeneity). Replacing $\dot{m}_x$ in (3.11) yields

$$\mathbb{E}[L_i(x, u_1, \dots, u_n) - f_i(0, x_{0i}, m_x(0))] = \frac{1}{2}\mathbb{E}(q(T) - \alpha(T))x_i(T)^2$$

$$+ \frac{1}{2}\mathbb{E}(\bar{q}(T) - \beta(T))m_x^2(T) + \mathbb{E}(0 - \gamma(T))m_x(T) + \mathbb{E}(0 - \delta(T))$$

$$+ \frac{1}{2}\mathbb{E}\int_0^T x_i^2 \left(q + \dot{\alpha} + 2b_1\alpha\right) dt + \frac{1}{2}\mathbb{E}\int_0^T m_x^2 \left(\bar{q} + \dot{\beta} + 2\bar{b}_1\gamma\right) dt$$

$$+ \frac{1}{2}\mathbb{E}\int_0^T m_x x_i \left(2\dot{\gamma} + 2\bar{b}_1\alpha + 2b_1\gamma\right) dt$$

$$+ \frac{1}{2}\mathbb{E}\int_0^T \left(2\beta \left(b_1 + \bar{b}_1 - \frac{b_2}{r}(b_2 + \bar{b}_2)(\alpha + \gamma)\right) m_x^2 \right.$$

$$+ 2\gamma \left(b_1 + \bar{b}_1 - \frac{b_2}{r}(b_2 + \bar{b}_2)(\alpha + \gamma)\right) m_x x_i + 2\dot{\delta} + \alpha\sigma_0^2 \right) dt$$

$$+ \frac{1}{2}\mathbb{E}\int_0^T \left(r \left(u_i + b_2\frac{\alpha x_i + \gamma m_x}{r}\right)^2 - \frac{b_2^2}{r}\alpha^2 x_i^2 \right.$$

$$- 2\frac{b_2^2}{r}\alpha\gamma m_x x_i - \frac{b_2^2}{r}\gamma^2 m_x^2 \right) dt$$

$$+ \frac{1}{2}\mathbb{E}\int_0^T \left(\bar{r}\frac{b_2^2}{r^2}(\alpha + \gamma)^2 m_x^2 - 2\bar{b}_2\alpha\frac{b_2}{r}(\alpha + \gamma)x_i m_x \right.$$

$$- 2\bar{b}_2\gamma\frac{b_2}{r}(\alpha + \gamma)m_x^2 \right) dt.$$

Organizing terms yields

$$\mathbb{E}[L_i(x, u_1, \ldots, u_n) - f_i(0, x_{0i}, m_x(0))] = \frac{1}{2}\mathbb{E}(q(T) - \alpha(T))x_i(T)^2$$

$$+ \frac{1}{2}\mathbb{E}(\bar{q}(T) - \beta(T))m_x^2(T) + \mathbb{E}(0 - \gamma(T))m_x(T) + \mathbb{E}(0 - \delta(T))$$

$$+ \frac{1}{2}\mathbb{E}\int_0^T x_i^2 \left(q + \dot{\alpha} + 2b_1\alpha - \frac{b_2^2}{r}\alpha^2\right) dt$$

$$+ \frac{1}{2}\mathbb{E}\int_0^T m_x^2 \left(\bar{q} + \dot{\beta} + 2\bar{b}_1\gamma + 2\beta \left(b_1 + \bar{b}_1 - \frac{b_2}{r}(b_2 + \bar{b}_2)(\alpha + \gamma)\right)\right.$$

$$\left. - \frac{b_2^2}{r}\gamma^2 + \bar{r}\frac{b_2^2}{r^2}(\alpha + \gamma)^2 - 2\bar{b}_2\gamma\frac{b_2}{r}(\alpha + \gamma)\right) dt$$

$$+ \frac{1}{2}\mathbb{E}\int_0^T m_x x_i \left(2\dot{\gamma} + 2\bar{b}_1\alpha + 2b_1\gamma\right.$$

$$+ 2\gamma \left(b_1 + \bar{b}_1 - \frac{b_2}{r}(b_2 + \bar{b}_2)(\alpha + \gamma)\right) - 2\frac{b_2^2}{r}\alpha\gamma - 2\bar{b}_2\alpha\frac{b_2}{r}(\alpha + \gamma)\right) dt$$

$$+ \frac{1}{2}\mathbb{E}\int_0^T \left(2\dot{\delta} + \alpha\sigma_0^2\right) dt$$

$$+ \frac{1}{2} \mathbb{E} \int_0^T r \left(u_i + b_2 \frac{\alpha x_i + \gamma m_x}{r} \right)^2 dt. \tag{3.12}$$

Finally, we perform the process identification. Let us divide (3.12) into six groups:

- *First, in the first and second line, we observe the terminal-time-dependent terms that define the boundary conditions for the differential equations.*

- *The second and third group correspond to the lines from three to five comprising those terms with the quadratic common factors x_i^2 and m_x^2. These groups define the differential equations for α and β.*

- *Then, as a fourth and fifth groups, we have the terms with common factor $m_x x_i$ and those independent from the state and mean-field state. These groups establish the differential equations γ and δ.*

- *Finally, the sixth group defines the optimal control input.*

Thus, the announced result is obtained by minimizing the terms in the latter expression, completing the proof.

We proceed to study the mean-field game problem for the discrete time in next section.

3.3   A Discrete-Time Deterministic Mean-Field Game

For the discrete-time case, the decision-makers interact during a time window $[k..k + N - 1]$. Decision-maker $i \in \mathcal{N}$ selects a control action $u_{i,\ell}$ at time instant $\ell \in [k..k + N - 1]$. The ith decision-maker has an associated system state, which is affected by all the others' decisions $\mathcal{N} \setminus \{i\}$ by means of the mean-field terms. This is analogically the same setting we considered in the continuous-time problem, i.e.,

$$x_{i,\ell+1} = b_1 x_{i,\ell} + \bar{b}_1 m_{x,\ell} + b_2 u_{i,\ell} + \bar{b}_2 m_{u,\ell}, \tag{3.13}$$

where $x_i \in \mathbb{R}$ corresponds to the system state, and $u_i \in \mathbb{R}$ denotes the control input of the decision-maker $i \in \mathcal{N}$. Hence, $b_1, \bar{b}_1, b_2, \bar{b}_2 \in \mathbb{R}$, for all $j \in \mathcal{N}$ are the system state parameters, which can be taken as function of time as $b_1, \bar{b}_1, b_2, \bar{b}_2 : [0, T] \to \mathbb{R}$. There are some mean-field terms m_x and m_u associated with the system states and control inputs, respectively. Such terms are

$$m_{x,k} = \lim_{n \to \infty} \frac{1}{n} \sum_{j=1}^{n} x_{j,k}, \tag{3.14a}$$

$$m_{u,k} = \lim_{n \to \infty} \frac{1}{n} \sum_{j=1}^{n} u_{j,k}. \tag{3.14b}$$

Notice that the decision-makers' system states are coupled by means of the mean-field terms m_x and m_u. Each decision-maker has a mean-field cost functional as follows:

$$L_i(x, u_1, \ldots, u_n) = q_{k+N} x_{k+N}^2 + \bar{q}_{k+N} m_{x,k+N}^2$$
$$+ \sum_{\ell=k}^{k+N-1} \left[q x_\ell^2 + \bar{q} m_{x,\ell}^2 + r u_{i,\ell}^2 + \bar{r} m_{u,\ell}^2 \right], \tag{3.15}$$

In the same way that it was established for the continuous-time mean-field game problem, the following Remark highlights an important assumption that is made in this discrete-time mean-field game problem.

Remark 10 *Once again, it is pointed out that the interactive population of decision-makers is homogeneous. Then, the parameters $b_1, \bar{b}_1, b_2, \bar{b}_2$ in the system dynamics (3.13) and $q, \bar{q}, r, \bar{r}$ in the cost functional (3.15) are identical for all the players.*

The assumption indicated in Remark 10 is crucial at the moment the evolution of the mean-field state is computed as we will see in the proof of the discrete-time mean-field game problem later on, which is stated immediately next.

Problem 11 (Deterministic Discrete-Time Mean-Field Game) *The deterministic discrete-time mean-field game problem for each $i \in \mathcal{N}$ is given by*

$$\begin{cases} \underset{u_i \in \mathcal{U}_i}{\text{minimize}} \ L_i \left(u_1, \ldots, u_n \right), \\ \\ \text{subject to} \\ \\ x_{i,\ell+1} = b_1 x_{i,\ell} + \bar{b}_1 m_{x,\ell} + b_2 u_{i,\ell} + \bar{b}_2 m_{u,\ell}, \\ \forall \ell \in [k..k + N - 1], \end{cases} \tag{3.16}$$

where the initial condition $x_{i,k}$ is given, the weight parameters are $q \geq 0$, $\bar{q} > 0$, and $\bar{r} > 0$, and with the system parameters $b_1, \bar{b}_1, b_2, \bar{b}_2 \in \mathbb{R}$, for all $j \in \mathcal{N}$.

The solution for Problem 11 is of the state and mean-field-state-feedback form as shown in Proposition 11.

Proposition 11 *The optimal control input, mean-field control input and optimal cost functional for Problem 11 are given by*

$$u_{i,\ell}^* = -\frac{b_2}{c_\ell} \left\{ \left[\frac{1}{2} \left(b_1 + \bar{b}_1 \right) \gamma_{\ell+1} + \alpha_{\ell+1} \bar{b}_1 \right] \right.$$

$$+ \frac{b_2}{c_\ell} \left[\frac{1}{2} \left(b_2 + \bar{b}_2 \right) \gamma_{\ell+1} + \alpha_{\ell+1} \bar{b}_2 \right] \xi \Bigg\} m_{x,\ell}$$

$$- \frac{b_2}{c_\ell} \alpha_{\ell+1} b_1 x_{i,\ell}, \tag{3.17a}$$

$$m_{u,\ell}^* = \frac{\dfrac{b_2}{c_\ell} \left[\left(\dfrac{1}{2} \gamma_{\ell+1} + \alpha_{\ell+1} \right) \left(\bar{b}_1 + b_1 \right) \right]}{1 - \dfrac{b_2}{c_\ell} \left[\dfrac{1}{2} \left(b_2 + \bar{b}_2 \right) \gamma_{\ell+1} + \alpha_{\ell+1} \bar{b}_2 \right]} m_{x,\ell}, \tag{3.17b}$$

$$L_i(x, u_1^*, \ldots, u_n^*) = \alpha_k x_{i,k}^2 + \beta_k m_{x,k}^2 + \gamma_k m_{x,k} x_{i,k}, \tag{3.17c}$$

where α, β, and γ solve the recursive following equations:

$$\alpha_\ell = q + \alpha_{\ell+1} b_1^2 - \frac{b_2^2}{c_\ell} \alpha_{\ell+1}^2 b_1^2,$$

$$\beta_\ell = \bar{q} + \alpha_{\ell+1} \bar{b}_1^2 + \beta_{\ell+1} \left(b_1 + \bar{b}_1 \right)^2 + \left(b_1 + \bar{b}_1 \right) \bar{b}_1 \gamma_{\ell+1}$$

$$+ \left\{ \frac{b_2^2 \bar{r}}{c_\ell^2} + \frac{b_2^2}{c_\ell^2} \left[\alpha_{\ell+1} \bar{b}_2^2 + \beta_{\ell+1} \left(b_2 + \bar{b}_2 \right)^2 + \left(b_2 + \bar{b}_2 \right) \bar{b}_2 \gamma_{\ell+1} \right] \right.$$

$$\left. - \frac{b_2^4}{c_\ell^3} \left[\frac{1}{2} \left(b_2 + \bar{b}_2 \right) \gamma_{\ell+1} + \alpha_{\ell+1} \bar{b}_2 \right]^2 \right\} \xi^2$$

$$+ \left\{ \frac{b_2}{c_\ell} \left[\left(b_2 + \bar{b}_2 \right) \bar{b}_1 \gamma_{\ell+1} + \left(b_1 + \bar{b}_1 \right) \bar{b}_2 \gamma_{\ell+1} \right. \right.$$

$$\left. + 2\alpha_{\ell+1} \bar{b}_1 \bar{b}_2 + 2\beta_{\ell+1} \left(b_1 + \bar{b}_1 \right) \left(b_2 + \bar{b}_2 \right) \right]$$

$$\left. - 2\frac{b_2^3}{c_\ell^2} \left[\frac{1}{2} \left(b_1 + \bar{b}_1 \right) \gamma_{\ell+1} + \alpha_{\ell+1} \bar{b}_1 \right] \left[\frac{1}{2} \left(b_2 + \bar{b}_2 \right) \gamma_{\ell+1} + \alpha_{\ell+1} \bar{b}_2 \right] \right\} \xi$$

$$- \frac{b_2^2}{c_\ell} \left[\frac{1}{2} \left(b_1 + \bar{b}_1 \right) \gamma_{\ell+1} + \alpha_{\ell+1} \bar{b}_1 \right]^2,$$

$$\gamma_\ell = 2\alpha_{\ell+1} b_1 \bar{b}_1 + \left(b_1 + \bar{b}_1 \right) b_1 \gamma_{\ell+1} + \left\{ \frac{b_2}{c_\ell} \left[\left(b_2 + \bar{b}_2 \right) b_1 \gamma_{\ell+1} + 2\alpha_{\ell+1} b_1 \bar{b}_2 \right] \right.$$

$$\left. - 2\frac{b_2^3}{c_\ell^2} \alpha_{\ell+1} b_1 \left[\frac{1}{2} \left(b_2 + \bar{b}_2 \right) \gamma_{\ell+1} + \alpha_{\ell+1} \bar{b}_2 \right] \right\} \xi$$

$$- 2\frac{b_2^2}{c_\ell} \left[\frac{1}{2} \left(b_1 + \bar{b}_1 \right) \gamma_{\ell+1} + \alpha_{\ell+1} \bar{b}_1 \right] \alpha_{\ell+1} b_1,$$

for all $\ell \in [k..k + N - 1]$, with the terminal conditions

$$\alpha_{k+N} = q_{k+N},$$

$$\beta_{k+N} = \bar{q}_{k+N},$$

$$\gamma_{k+N} = 0,$$

and with the auxiliary variables

$$c_\ell = \alpha_{\ell+1} b_2^2 + r,$$

$$\xi = \frac{\left(\frac{1}{2}\gamma_{\ell+1} + \alpha_{\ell+1}\right)\left(\bar{b}_1 + b_1\right)}{1 - \frac{b_2}{c_\ell}\left[\frac{1}{2}\left(b_2 + \bar{b}_2\right)\gamma_{\ell+1} + \alpha_{\ell+1}\bar{b}_2\right]}$$

that appear in the optimal control strategy $u^*_{i,\ell}$, and in the difference equations β_ℓ and γ_ℓ. Finally, the evolution of the mean state is explicitly given by the following difference equation:

$$m_{x,\ell+1} = \left\{ \left(b_1 + \bar{b}_1\right) + \left(b_2 + \bar{b}_2\right) \frac{\frac{b_2}{c_\ell}\left[\left(\frac{1}{2}\gamma_{\ell+1} + \alpha_{\ell+1}\right)\left(\bar{b}_1 + b_1\right)\right]}{1 - \frac{b_2}{c_\ell}\left[\frac{1}{2}\left(b_2 + \bar{b}_2\right)\gamma_{\ell+1} + \alpha_{\ell+1}\bar{b}_2\right]} \right\} m_{x,\ell}.$$

The previous result also shows a condition to avoid a singularity in the evolution of the mean-field state, i.e., that the term

$$\frac{b_2}{c_\ell}\left[\frac{1}{2}\left(b_2 + \bar{b}_2\right)\gamma_{\ell+1} + \alpha_{\ell+1}\bar{b}_2\right] \neq 1.$$

Additionally, note that having the system state parameters already defined, one can find conditions over such that the mean-field state dynamics vanishes as time increases.

Proof 11 (Proposition 11) *This proof is straight-forwardly obtained from the proof of Proposition 12 in the following section.*

Figure 3.3 presents the complete scheme of the system-state feedback for the optimal mean-field game in discrete time according to Proposition 11. Both the continuous-time and discrete-time mean-field game schemes can be compared, and it can be identified that the optimal feedback in discrete time is complexer than the one in continuous time regarding computation. Nevertheless, the discrete-time laws are more suitable for real implementation.

Remark 11 *The α_ℓ term in Proposition 11 evolution coincides with the one obtained in Proposition 6 establishing a relationship between this game-theoretical approach with the control case.*

As it has been exposed in the section devoted to present the continuous-time case, we also proceed to show the solution for the mean-field game problem in discrete time by incorporating stochasticity in the coming section below.

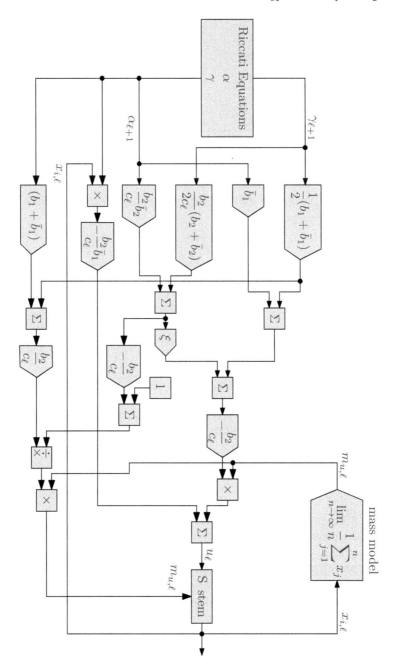

FIGURE 3.3
Feedback scheme for the linear-quadratic mean-field game in discrete time.

3.4 A Discrete-Time Stochastic Mean-Field Game

Now, let us consider the stochastic version of the mean-field problem. Each decision-maker has a new associated system state with stochasticity, which is affected by all the others' decisions, i.e.,

$$x_{i,\ell+1} = b_1 x_{i,\ell} + \bar{b}_1 m_{x,\ell} + b_2 u_{i,\ell} + \bar{b}_2 m_{u,\ell} + \sigma_0 w_{i,\ell}, \qquad (3.18)$$

where $x_i \in \mathbb{R}$ corresponds to the system state, and $u_i \in \mathbb{R}$ denotes the control input of the decision-maker $i \in \mathcal{N}$. Hence, $b_1, \bar{b}_1, b_2, \bar{b}_2, \sigma_0 \in \mathbb{R}$, for all $j \in \mathcal{N}$ are the system state parameters (recall these coefficients can be treated as function of time if desired). We still incorporate into the problem mean-field terms m_x and m_u associated with the system states and control inputs, respectively. Such terms are given by (3.14a) and (3.14b). Each decision-maker has the cost function in (3.15). The problem statement for the stochastic game is introduced next.

Problem 12 (Stochastic Discrete-Time Mean-Field Game) *The Stochastic Discrete-Time Mean-Field Game Problem for each decision-maker $i \in \mathcal{N}$ is*

$$
\begin{cases}
\underset{u_i \in \mathcal{U}_i}{\text{minimize}} \; \mathbb{E}\left[L_i\left(u_1, \ldots, u_n\right)\right], \\[2mm]
\text{subject to} \\[2mm]
x_{i,\ell+1} = b_1 x_{i,\ell} + \bar{b}_1 m_{x,\ell} + b_2 u_{i,\ell} + \bar{b}_2 m_{u,\ell} + \sigma_0 w_{i,\ell}, \\
\forall \ell \in [k..k+N-1],
\end{cases}
\qquad (3.19)
$$

where the initial state condition x_k is given, the weight parameters are $q \geq 0$, $\bar{q} \geq 0$, and $\bar{r} > 0$, and also the same for all the decision-makers. The system state parameters are $b_1, \bar{b}_1, b_2, \bar{b}_2, \sigma_0 \in \mathbb{R}$, for all $j \in \mathcal{N}$.

The solution for the mean-field game when the state and control-input-independent noise $\sigma_0 w_{i,\ell}$ is added to the system dynamics is shown in Proposition 12. Afterward, we make a remark that relates the result with the deterministic case and then we present the computation of the semi-explicit solution using the direct method.

Proposition 12 *The semi-explicit solution of Problem 12 is composed by the following optimal control input, optimal mean control, and optimal cost functional as follows:*

$$
u_{i,\ell}^* = -\frac{b_2}{c_\ell} \left\{ \left[\frac{1}{2}\left(b_1 + \bar{b}_1\right) \gamma_{\ell+1} + \alpha_{\ell+1} \bar{b}_1 \right] \right.
$$

$$+ \frac{b_2}{c_\ell} \left[\frac{1}{2} \left(b_2 + \bar{b}_2 \right) \gamma_{\ell+1} + \alpha_{\ell+1} \bar{b}_2 \right] \xi \bigg\} m_{x,\ell}$$

$$- \frac{b_2}{c_\ell} \alpha_{\ell+1} b_1 x_{i,\ell}, \tag{3.20a}$$

$$m^*_{u,\ell} = \frac{\dfrac{b_2}{c_\ell} \left[\left(\frac{1}{2} \gamma_{\ell+1} + \alpha_{\ell+1} \right) \left(\bar{b}_1 + b_1 \right) \right]}{1 - \dfrac{b_2}{c_\ell} \left[\frac{1}{2} \left(b_2 + \bar{b}_2 \right) \gamma_{\ell+1} + \alpha_{\ell+1} \bar{b}_2 \right]} m_{x,\ell}, \tag{3.20b}$$

$$L_i(x, u_1^*, \ldots, u_n^*) = \alpha_k x_{i,k}^2 + \beta_k m_{x,k}^2 + \gamma_k m_{x,k} x_{i,k} + \delta_k, \tag{3.20c}$$

where parameters are established such that

$$1 - \frac{b_2}{c_\ell} \left[\frac{1}{2} \left(b_2 + \bar{b}_2 \right) \gamma_{\ell+1} + \alpha_{\ell+1} \bar{b}_2 \right] \neq 0,$$

and the difference equations for α, β, γ, *and* δ *are:*

$$\alpha_\ell = q_i + \alpha_{\ell+1} b_1^2 - \frac{b_2^2}{c_\ell} \alpha_{\ell+1}^2 b_1^2,$$

$$\beta_\ell = \bar{q} + \alpha_{\ell+1} \bar{b}_1^2 + \beta_{\ell+1} \left(b_1 + \bar{b}_1 \right)^2 + \left(b_1 + \bar{b}_1 \right) \bar{b}_1 \gamma_{\ell+1}$$

$$+ \left\{ \frac{b_2^2 \bar{r}}{c_\ell^2} + \frac{b_2^2}{c_\ell^2} \left[\alpha_{\ell+1} \bar{b}_2^2 + \beta_{\ell+1} \left(b_2 + \bar{b}_2 \right)^2 + \left(b_2 + \bar{b}_2 \right) \bar{b}_2 \gamma_{\ell+1} \right] \right.$$

$$\left. - \frac{b_2^4}{c_\ell^3} \left[\frac{1}{2} \left(b_2 + \bar{b}_2 \right) \gamma_{\ell+1} + \alpha_{\ell+1} \bar{b}_2 \right]^2 \right\} \xi^2$$

$$+ \left\{ \frac{b_2}{c_\ell} \left[\left(b_2 + \bar{b}_2 \right) \bar{b}_1 \gamma_{\ell+1} + \left(b_1 + \bar{b}_1 \right) \bar{b}_2 \gamma_{\ell+1} \right.\right.$$

$$\left.\left. + 2\alpha_{\ell+1} \bar{b}_1 \bar{b}_2 + 2\beta_{\ell+1} \left(b_1 + \bar{b}_1 \right) \left(b_2 + \bar{b}_2 \right) \right] \right.$$

$$\left. - 2 \frac{b_2^3}{c_\ell^2} \left[\frac{1}{2} \left(b_1 + \bar{b}_1 \right) \gamma_{\ell+1} + \alpha_{\ell+1} \bar{b}_1 \right] \left[\frac{1}{2} \left(b_2 + \bar{b}_2 \right) \gamma_{\ell+1} + \alpha_{\ell+1} \bar{b}_2 \right] \right\} \xi$$

$$- \frac{b_2^2}{c_\ell} \left[\frac{1}{2} \left(b_1 + \bar{b}_1 \right) \gamma_{\ell+1} + \alpha_{\ell+1} \bar{b}_1 \right]^2,$$

$$\gamma_\ell = 2\alpha_{\ell+1} b_1 \bar{b}_1 + \left(b_1 + \bar{b}_1 \right) b_1 \gamma_{\ell+1} + \left\{ \frac{b_2}{c_\ell} \left[\left(b_2 + \bar{b}_2 \right) b_1 \gamma_{\ell+1} + 2\alpha_{\ell+1} b_1 \bar{b}_2 \right] \right.$$

$$\left. - 2 \frac{b_2^3}{c_\ell^2} \alpha_{\ell+1} b_1 \left[\frac{1}{2} \left(b_2 + \bar{b}_2 \right) \gamma_{\ell+1} + \alpha_{\ell+1} \bar{b}_2 \right] \right\} \xi$$

$$- 2 \frac{b_2^2}{c_\ell} \left[\frac{1}{2} \left(b_1 + \bar{b}_1 \right) \gamma_{\ell+1} + \alpha_{\ell+1} \bar{b}_1 \right] \alpha_{\ell+1} b_1,$$

$$\delta_\ell = \delta_{\ell+1} + \alpha_{\ell+1}\sigma_0^2 w_{i,\ell}^2,$$

for all $\ell \in [k..k + N - 1]$, and with the terminal boundary conditions

$$\alpha_{k+N} = q_{k+N},$$
$$\beta_{k+N} = \bar{q}_{k+N},$$
$$\gamma_{k+N} = \delta_{k+N} = 0.$$

In addition, the auxiliary variables are given by

$$c_\ell = \alpha_{\ell+1}b_2^2 + r,$$

$$\xi = \frac{\left(\frac{1}{2}\gamma_{\ell+1} + \alpha_{\ell+1}\right)(\bar{b}_1 + b_1)}{1 - \dfrac{b_2}{c_\ell}\left[\dfrac{1}{2}\left(b_2 + \bar{b}_2\right)\gamma_{\ell+1} + \alpha_{\ell+1}\bar{b}_2\right]}.$$

These variables appear in the optimal control strategy $u_{i,\ell}^*$ and the difference equations β_ℓ and γ_ℓ. Finally,

$$m_{x,\ell+1} = \left\{ (b_1 + \bar{b}_1) + (b_2 + \bar{b}_2)\frac{\dfrac{b_2}{c_\ell}\left[\left(\dfrac{1}{2}\gamma_{\ell+1} + \alpha_{\ell+1}\right)(\bar{b}_1 + b_1)\right]}{1 - \dfrac{b_2}{c_\ell}\left[\dfrac{1}{2}\left(b_2 + \bar{b}_2\right)\gamma_{\ell+1} + \alpha_{\ell+1}\bar{b}_2\right]} \right\} m_{x,\ell},$$

is the evolution of the mean state.

Remark 12 *Notice that, the α_ℓ in Proposition 12 evolution coincides with the one obtained in Proposition 6, which corresponds to a discrete-time control problem. Moreover, notice that the unique difference between the optimal solution for the discrete-time deterministic and stochastic mean-field game problems relies on the optimal cost functional because of δ.*

The proof for the stochastic mean-field solution, which is explained next, enables to directly deduce the solution for the deterministic counterpart.

Proof 12 (Proposition 12) *According to the structure of the cost functional and the system dynamics, let us consider the following guess functional (ansatz) for the optimal cost from time instant k up to $k + N$ for the ith decision-maker:*

$$f_{i,\ell} = \alpha_\ell x_{i,\ell}^2 + \beta_\ell m_{x,\ell}^2 + \gamma_\ell m_{x,\ell}x_{i,\ell} + \delta_\ell, \ \forall \ \ell \in [k..k + N - 1], \qquad (3.21)$$

where α, β, γ, and δ are deterministic difference equations function of time. Since the population is assumed to the homogeneous, then it is not necessary to distinguish among the Riccati equation given that all decision-makers act *also homogeneously. Now, we apply the telescopic sum decomposition to the guess functional given by*

$$f_{i,k+N} = f_{i,k} + \sum_{\ell=k}^{k+N-1} \left(f_{i,\ell+1} - f_{i,\ell}\right),$$

and we replace the right-hand side terms in function of the deterministic difference equations α, β, γ *and* δ, *i.e.,*

$$f_{i,k+N} - f_{i,k} = \sum_{\ell=k}^{k+N-1} \left(\alpha_{\ell+1}x_{i,\ell+1}^2 + \beta_{\ell+1}m_{x,\ell+1}^2 + \gamma_{\ell+1}m_{x,\ell+1}x_{i,\ell+1} + \delta_{\ell+1}\right.$$
$$\left. - \alpha_\ell x_{i,\ell}^2 - \beta_\ell m_{x,\ell}^2 - \gamma_\ell m_{x,\ell}x_{i,\ell} - \delta_\ell\right). \quad (3.22)$$

At this point, we observe that the terms $m_{x,\ell+1}$, $m_{x,\ell+1}^2$, and $x_{i,\ell+1}^2$, and the product $x_{i,\ell+1}m_{x,\ell+1}$ are required in the computation. Then, before continuing with the calculation of the telescopic sum $f_{i,k+N} - f_{i,k}$, we compute separately these four terms. First, from the system dynamics in (3.18) yields

$$\frac{1}{n}\sum_{i=1}^{n} x_{i,\ell+1} = b_1\frac{1}{n}\sum_{i=1}^{n} x_{i,\ell} + \bar{b}_1 m_{x,\ell} + b_2\frac{1}{n}\sum_{i=1}^{n} u_{i,\ell} + \bar{b}_2 m_{u,\ell} + \sigma_0\frac{1}{n}\sum_{i=1}^{n} w_{i,\ell},$$
$$m_{x,\ell+1} = \left(b_1 + \bar{b}_1\right) m_{x,\ell} + \left(b_2 + \bar{b}_2\right) m_{u,\ell}. \quad (3.23)$$

Second, using the expression $m_{x,\ell+1}$, *we compute its square as follows:*

$$m_{x,\ell+1}^2 = \left[\left(b_1 + \bar{b}_1\right) m_{x,\ell} + \left(b_2 + \bar{b}_2\right) m_{u,\ell}\right]^2,$$
$$= \left(b_1 + \bar{b}_1\right)^2 m_{x,\ell}^2 + \left(b_2 + \bar{b}_2\right)^2 m_{u,\ell}^2$$
$$+ 2\left(b_1 + \bar{b}_1\right)\left(b_2 + \bar{b}_2\right) m_{x,\ell}m_{u,\ell}. \quad (3.24)$$

Third, from (3.18) one obtains the evolution of $x_{i,\ell+1}^2$, *i.e.,*

$$x_{i,\ell+1}^2 = \left(b_1 x_{i,\ell} + \bar{b}_1 m_{x,\ell} + b_2 u_{i,\ell} + \bar{b}_2 m_{u,\ell} + \sigma_0 w_{i,\ell}\right)^2,$$
$$= b_1^2 x_{i,\ell}^2 + \bar{b}_1^2 m_{x,\ell}^2 + b_2^2 u_{i,\ell}^2 + \bar{b}_2^2 m_{u,\ell}^2 + \sigma_0^2 w_{i,\ell}^2$$
$$+ 2b_1\bar{b}_1 x_{i,\ell}m_{x,\ell} + 2b_1 b_2 x_{i,\ell}u_{i,\ell} + 2b_1\bar{b}_2 x_{i,\ell}m_{u,\ell} + 2b_1\sigma_0 x_{i,\ell}w_{i,\ell}$$
$$+ 2\bar{b}_1 b_2 m_{x,\ell}u_{i,\ell} + 2\bar{b}_1\bar{b}_2 m_{x,\ell}m_{u,\ell} + 2\bar{b}_1\sigma_0 m_{x,\ell}w_{i,\ell}$$
$$+ 2b_2\bar{b}_2 u_{i,\ell}m_{u,\ell} + 2b_2\sigma_0 u_{i,\ell}w_{i,\ell}$$
$$+ 2\bar{b}_2\sigma_0 m_{u,\ell}w_{i,\ell}.$$

And fourth

$$x_{i,\ell+1}m_{x,\ell+1} = \left[b_1 x_{i,\ell} + \bar{b}_1 m_{x,\ell} + b_2 u_{i,\ell} + \bar{b}_2 m_{u,\ell} + \sigma_0 w_{i,\ell}\right]$$
$$\left[\left(b_1 + \bar{b}_1\right) m_{x,\ell} + \left(b_2 + \bar{b}_2\right) m_{u,\ell}\right],$$

$$= (b_1 + \bar{b}_1) b_1 x_{i,\ell} m_{x,\ell} + (b_2 + \bar{b}_2) b_1 x_{i,\ell} m_{u,\ell}$$
$$+ (b_1 + \bar{b}_1) \bar{b}_1 m_{x,\ell}^2 + (b_2 + \bar{b}_2) \bar{b}_1 m_{x,\ell} m_{u,\ell}$$
$$+ (b_1 + \bar{b}_1) b_2 u_{i,\ell} m_{x,\ell} + (b_2 + \bar{b}_2) b_2 u_{i,\ell} m_{u,\ell}$$
$$+ (b_1 + \bar{b}_1) \bar{b}_2 m_{u,\ell} m_{x,\ell} + (b_2 + \bar{b}_2) \bar{b}_2 m_{u,\ell}^2$$
$$+ (b_1 + \bar{b}_1) \sigma_0 w_{i,\ell} m_{x,\ell} + (b_2 + \bar{b}_2) \sigma_0 w_{i,\ell} m_{u,\ell}.$$

Replacing terms in (3.22) yields

$$\mathbb{E}[f_{i,k+N} - f_{i,k}] = \mathbb{E} \sum_{\ell=k}^{k+N-1} \left(\alpha_{\ell+1} b_1^2 x_{i,\ell}^2 + \alpha_{\ell+1} \bar{b}_1^2 m_{x,\ell}^2 + \alpha_{\ell+1} b_2^2 u_{i,\ell}^2 \right.$$
$$+ \alpha_{\ell+1} \bar{b}_2^2 m_{u,\ell}^2 + \alpha_{\ell+1} \sigma_0^2 w_{i,\ell}^2 + 2\alpha_{\ell+1} b_1 \bar{b}_1 x_{i,\ell} m_{x,\ell} + 2\alpha_{\ell+1} b_1 b_2 x_{i,\ell} u_{i,\ell}$$
$$+ 2\alpha_{\ell+1} b_1 \bar{b}_2 x_{i,\ell} m_{u,\ell} + 2\alpha_{\ell+1} b_1 b_2 m_{x,\ell} u_{i,\ell} + 2\alpha_{\ell+1} \bar{b}_1 \bar{b}_2 m_{x,\ell} m_{u,\ell}$$
$$+ 2\alpha_{\ell+1} b_2 \bar{b}_2 u_{i,\ell} m_{u,\ell} + \beta_{\ell+1} (b_1 + \bar{b}_1)^2 m_{x,\ell}^2 + \beta_{\ell+1} (b_2 + \bar{b}_2)^2 m_{u,\ell}^2$$
$$+ 2\beta_{\ell+1} (b_1 + \bar{b}_1) (b_2 + \bar{b}_2) m_{x,\ell} m_{u,\ell} + \gamma_{\ell+1} m_{x,\ell+1} x_{i,\ell+1} + \delta_{\ell+1}$$
$$\left. - \alpha_\ell x_{i,\ell}^2 - \beta_\ell m_{x,\ell}^2 - \gamma_\ell m_{x,\ell} x_{i,\ell} - \delta_\ell \right),$$

and organizing terms by grouping them according to common factor and applying the expectation, one obtains

$$\mathbb{E}[f_{i,k+N} - f_{i,k}] = \mathbb{E} \sum_{\ell=k}^{k+N-1} \left(\left[\alpha_{\ell+1} b_1^2 - \alpha_\ell \right] x_{i,\ell}^2 \right.$$
$$+ \left[\alpha_{\ell+1} \bar{b}_1^2 + \beta_{\ell+1} (b_1 + \bar{b}_1)^2 - \beta_\ell + (b_1 + \bar{b}_1) \bar{b}_1 \gamma_{\ell+1} \right] m_{x,\ell}^2$$
$$+ \left[\alpha_{\ell+1} \bar{b}_2^2 + \beta_{\ell+1} (b_2 + \bar{b}_2)^2 + (b_2 + \bar{b}_2) \bar{b}_2 \gamma_{\ell+1} \right] m_{u,\ell}^2$$
$$+ \left[2\alpha_{\ell+1} b_1 \bar{b}_1 - \gamma_\ell + (b_1 + \bar{b}_1) b_1 \gamma_{\ell+1} \right] x_{i,\ell} m_{x,\ell}$$
$$+ \left[(b_2 + \bar{b}_2) b_1 \gamma_{\ell+1} + 2\alpha_{\ell+1} b_1 \bar{b}_2 \right] x_{i,\ell} m_{u,\ell}$$
$$+ \left[(b_2 + \bar{b}_2) \bar{b}_1 \gamma_{\ell+1} + (b_1 + \bar{b}_1) \bar{b}_2 \gamma_{\ell+1} + 2\alpha_{\ell+1} \bar{b}_1 \bar{b}_2 \right.$$
$$\left. + 2\beta_{\ell+1} (b_1 + \bar{b}_1) (b_2 + \bar{b}_2) \right] m_{x,\ell} m_{u,\ell}$$
$$+ \alpha_{\ell+1} b_2^2 u_{i,\ell}^2 + \left\{ \left[(b_1 + \bar{b}_1) b_2 \gamma_{\ell+1} + 2\alpha_{\ell+1} \bar{b}_1 b_2 \right] m_{x,\ell} \right.$$
$$+ \left[(b_2 + \bar{b}_2) b_2 \gamma_{\ell+1} + 2\alpha_{\ell+1} b_2 \bar{b}_2 \right] m_{u,\ell}$$
$$\left. + 2\alpha_{\ell+1} b_1 b_2 x_{i,\ell} \right\} u_{i,\ell} + \delta_{\ell+1} - \delta_\ell + \alpha_{\ell+1} \sigma_0^2 w_{i,\ell}^2 \bigg).$$

Continuing with the direct method, we compute the gap $\mathbb{E}[L_i(x, u_1, \ldots, u_n) -$

$f_{i,k}]$. *The cost matches with the optimal cost from time instant k up to $k+N$ when decision-makers play their the optimal strategies given by $f_{i,k}$.*

$$\mathbb{E}[L_i(x, u_1, \ldots, u_n) - f_{i,k}] = q_{k+N}x_{k+N}^2 + \bar{q}_{k+N}m_{x,k+N}^2 - \alpha_{k+N}x_{i,k+N}^2$$
$$- \beta_{k+N}m_{x,k+N}^2 - \gamma m_{x,k+N}x_{i,k+N} - \delta_{k+N}$$
$$+ \sum_{\ell=k}^{k+N-1}\left(q_i x_\ell^2 + \bar{q}m_{x,\ell}^2 + r u_{i,\ell}^2 + \bar{r}m_{u,\ell}^2 \right) + \mathbb{E}\sum_{\ell=k}^{k+N-1}\left(\left[\alpha_{\ell+1}b_1^2 - \alpha_\ell\right] x_{i,\ell}^2 \right)$$
$$+ \mathbb{E}\sum_{\ell=k}^{k+N-1}\left(\left[\alpha_{\ell+1}\bar{b}_1^2 + \beta_{\ell+1}\left(b_1 + \bar{b}_1\right)^2 - \beta_\ell + \left(b_1 + \bar{b}_1\right)\bar{b}_1\gamma_{\ell+1}\right] m_{x,\ell}^2 \right)$$
$$+ \mathbb{E}\sum_{\ell=k}^{k+N-1}\left(\left[\alpha_{\ell+1}\bar{b}_2^2 + \beta_{\ell+1}\left(b_2 + \bar{b}_2\right)^2 + \left(b_2 + \bar{b}_2\right)\bar{b}_2\gamma_{\ell+1}\right] m_{u,\ell}^2 \right)$$
$$+ \mathbb{E}\sum_{\ell=k}^{k+N-1}\left(\left[2\alpha_{\ell+1}b_1\bar{b}_1 - \gamma_\ell + \left(b_1 + \bar{b}_1\right)b_1\gamma_{\ell+1}\right] x_{i,\ell}m_{x,\ell} \right)$$
$$+ \mathbb{E}\sum_{\ell=k}^{k+N-1}\left(\left[\left(b_2 + \bar{b}_2\right)b_1\gamma_{\ell+1} + 2\alpha_{\ell+1}b_1\bar{b}_2\right] x_{i,\ell}m_{u,\ell} \right)$$
$$+ \mathbb{E}\sum_{\ell=k}^{k+N-1}\left(\Big[\left(b_2 + \bar{b}_2\right)\bar{b}_1\gamma_{\ell+1} + \left(b_1 + \bar{b}_1\right)\bar{b}_2\gamma_{\ell+1} + 2\alpha_{\ell+1}\bar{b}_1\bar{b}_2 \right.$$
$$\left. + 2\beta_{\ell+1}\left(b_1 + \bar{b}_1\right)\left(b_2 + \bar{b}_2\right)\Big] m_{x,\ell}m_{u,\ell} \right)$$
$$+ \mathbb{E}\sum_{\ell=k}^{k+N-1}\left(\alpha_{\ell+1}b_2^2 u_{i,\ell}^2 + \left\{ \left[\left(b_1 + \bar{b}_1\right)b_2\gamma_{\ell+1} + 2\alpha_{\ell+1}\bar{b}_1 b_2\right] m_{x,\ell} \right.\right.$$
$$\left.\left. + \left[\left(b_2 + \bar{b}_2\right)b_2\gamma_{\ell+1} + 2\alpha_{\ell+1}b_2\bar{b}_2\right] m_{u,\ell} + 2\alpha_{\ell+1}b_1 b_2 x_{i,\ell} \right\} u_{i,\ell} \right)$$
$$+ \mathbb{E}\sum_{\ell=k}^{k+N-1}\left(\delta_{\ell+1} - \delta_\ell + \alpha_{\ell+1}\sigma_0^2 w_{i,\ell}^2 \right). \tag{3.25}$$

We have extracted separately the terms depending on the control input, which are mainly coming from the quadratic form cost functional and the system dynamics. Then, we optimize the cost in $u_{i,\ell}$ by performing square completion. It yields

$$\mathbb{E}\sum_{\ell=k}^{k+N-1}\left(\alpha_{\ell+1}b_2^2 + r\right)\left(u_{i,\ell}^2 \right.$$
$$+ 2\frac{b_2}{\alpha_{\ell+1}b_2^2 + r}x_{i,\ell}\left\{ \left[\frac{1}{2}\left(b_1 + \bar{b}_1\right)\gamma_{\ell+1} + \alpha_{\ell+1}\bar{b}_1\right]\frac{m_{x,\ell}}{x_{i,\ell}} \right.$$

$$+ \left[\frac{1}{2} \left(b_2 + \bar{b}_2 \right) \gamma_{\ell+1} + \alpha_{\ell+1} \bar{b}_2 \right] \frac{m_{u,\ell}}{x_{i,\ell}} + \alpha_{\ell+1} b_1 \Bigg\} u_{i,\ell} \Bigg). \qquad (3.26)$$

In order to simplify the notation, let $c_\ell = \alpha_{\ell+1} b_2^2 + r$, *and*

$$\eta_{i,\ell} = \frac{b_2}{c_\ell} \Bigg\{ \left[\frac{1}{2} \left(b_1 + \bar{b}_1 \right) \gamma_{\ell+1} + \alpha_{\ell+1} \bar{b}_1 \right] \frac{m_{x,\ell}}{x_{i,\ell}}$$

$$+ \left[\frac{1}{2} \left(b_2 + \bar{b}_2 \right) \gamma_{\ell+1} + \alpha_{\ell+1} \bar{b}_2 \right] \frac{m_{u,\ell}}{x_{i,\ell}} + \alpha_{\ell+1} b_1 \Bigg\},$$

then, (3.26) *can be written as*

$$\mathbb{E} \sum_{\ell=k}^{k+N-1} c_\ell \left(u_{i,\ell}^2 + 2\eta_{i,\ell} x_{i,\ell} u_{i,\ell} \right).$$

It follows that

$$u_{i,\ell}^2 + 2\eta_{i,\ell} x_{i,\ell} u_{i,\ell} = \left(u_{i,\ell} + \eta_{i,\ell} x_{i,\ell} \right)^2 - \eta_{i,\ell}^2 x_{i,\ell}^2, \qquad (3.27)$$

from which it is concluded that the optimal control input for all the homogeneous decision-makers $i \in \mathcal{N}$ *is* $u_{i,\ell}^* = -\eta_{i,\ell} x_{i,\ell}$.

Going back to the auxiliary term $\eta_{i,\ell}$, one concludes that the optimal control input is, at this stage, expressed in terms of the optimal mean control input and it is not of the state and mean-state feedback form. Then, knowing the solution for the optimal control inputs, then we compute the optimal mean control input as follows:

$$- \lim_{n \to \infty} \frac{1}{n} \sum_{i=1}^{n} \eta_{i,\ell} x_{i,\ell} = - \lim_{n \to \infty} \frac{1}{n} \sum_{i=1}^{n} \frac{b_2}{c_\ell} \left[\frac{1}{2} \left(b_1 + \bar{b}_1 \right) \gamma_{\ell+1} + \alpha_{\ell+1} \bar{b}_1 \right] m_{x,\ell}$$

$$- \lim_{n \to \infty} \frac{1}{n} \sum_{i=1}^{n} \frac{b_2}{c_\ell} \left[\frac{1}{2} \left(b_2 + \bar{b}_2 \right) \gamma_{\ell+1} + \alpha_{\ell+1} \bar{b}_2 \right] m_{u,\ell}$$

$$- \lim_{n \to \infty} \frac{1}{n} \sum_{i=1}^{n} \frac{b_2}{c_\ell} \alpha_{\ell+1} b_1 x_{i,\ell},$$

$$= - \frac{b_2}{c_\ell} \left[\left(\frac{1}{2} \gamma_{\ell+1} + \alpha_{\ell+1} \right) \left(\bar{b}_1 + b_1 \right) \right] m_{x,\ell}$$

$$- \frac{b_2}{c_\ell} \left[\frac{1}{2} \left(b_2 + \bar{b}_2 \right) \gamma_{\ell+1} + \alpha_{\ell+1} \bar{b}_2 \right] m_{u,\ell},$$

and it follows that

$$m_{u,\ell}^* = \frac{\dfrac{b_2}{c_\ell} \left[\left(\frac{1}{2} \gamma_{\ell+1} + \alpha_{\ell+1} \right) \left(\bar{b}_1 + b_1 \right) \right]}{1 - \dfrac{b_2}{c_\ell} \left[\frac{1}{2} \left(b_2 + \bar{b}_2 \right) \gamma_{\ell+1} + \alpha_{\ell+1} \bar{b}_2 \right]} m_{x,\ell}, \qquad (3.28)$$

and

$$u_{i,\ell}^* = -\frac{b_2}{c_\ell}\left\{ \left[\frac{1}{2}\left(b_1 + \bar{b}_1\right)\gamma_{\ell+1} + \alpha_{\ell+1}\bar{b}_1 \right] \right.$$

$$+ \frac{b_2}{c_\ell}\left[\frac{1}{2}\left(b_2 + \bar{b}_2\right)\gamma_{\ell+1} + \alpha_{\ell+1}\bar{b}_2 \right] \frac{\left[\left(\frac{1}{2}\gamma_{\ell+1} + \alpha_{\ell+1}\right)\left(\bar{b}_1 + b_1\right)\right]}{1 - \frac{b_2}{c_\ell}\left[\frac{1}{2}\left(b_2 + \bar{b}_2\right)\gamma_{\ell+1} + \alpha_{\ell+1}\bar{b}_2\right]} \left.\vphantom{\frac{1}{2}}\right\} m_{x,\ell}$$

$$- \frac{b_2}{c_\ell}\alpha_{\ell+1}b_1 x_{i,\ell}. \tag{3.29}$$

From (3.27), before replacing the square completion in the gap

$$\mathbb{E}[L_i(x, u_1, \ldots, u_n) - f_{i,k}],$$

it is necessary to compute the product of quadratic terms $\eta_{i,\ell}^2 x_{i,\ell}^2$, *i.e.,*

$$\eta_{i,\ell}^2 x_{i,\ell}^2 = \frac{b_2^2}{c_\ell^2}\left\{ \left[\frac{1}{2}\left(b_1 + \bar{b}_1\right)\gamma_{\ell+1} + \alpha_{\ell+1}\bar{b}_1 \right] \frac{m_{x,\ell}}{x_{i,\ell}} \right.$$

$$\left. + \left[\frac{1}{2}\left(b_2 + \bar{b}_2\right)\gamma_{\ell+1} + \alpha_{\ell+1}\bar{b}_2 \right] \frac{m_{u,\ell}}{x_{i,\ell}} + \alpha_{\ell+1}b_1 \right\}^2 x_{i,\ell}^2,$$

$$= \frac{b_2^2}{c_\ell^2}\left\{ \left[\frac{1}{2}\left(b_1 + \bar{b}_1\right)\gamma_{\ell+1} + \alpha_{\ell+1}\bar{b}_1 \right]^2 \frac{m_{x,\ell}^2}{x_{i,\ell}^2} \right.$$

$$+ \left[\frac{1}{2}\left(b_2 + \bar{b}_2\right)\gamma_{\ell+1} + \alpha_{\ell+1}\bar{b}_2 \right]^2 \frac{m_{u,\ell}^2}{x_{i,\ell}^2} + \alpha_{\ell+1}^2 b_1^2$$

$$+ 2\left[\frac{1}{2}\left(b_1 + \bar{b}_1\right)\gamma_{\ell+1} + \alpha_{\ell+1}\bar{b}_1 \right]\left[\frac{1}{2}\left(b_2 + \bar{b}_2\right)\gamma_{\ell+1} + \alpha_{\ell+1}\bar{b}_2 \right] \frac{m_{x,\ell}m_{u,\ell}}{x_{i,\ell}^2}$$

$$+ 2\left[\frac{1}{2}\left(b_1 + \bar{b}_1\right)\gamma_{\ell+1} + \alpha_{\ell+1}\bar{b}_1 \right]\alpha_{\ell+1}b_1\frac{m_{x,\ell}}{x_{i,\ell}}$$

$$\left. + 2\left[\frac{1}{2}\left(b_2 + \bar{b}_2\right)\gamma_{\ell+1} + \alpha_{\ell+1}\bar{b}_2 \right]\alpha_{\ell+1}b_1\frac{m_{u,\ell}}{x_{i,\ell}} \right\} x_{i,\ell}^2,$$

$$= \frac{b_2^2}{c_\ell^2}\left\{ \left[\frac{1}{2}\left(b_1 + \bar{b}_1\right)\gamma_{\ell+1} + \alpha_{\ell+1}\bar{b}_1 \right]^2 m_{x,\ell}^2 \right.$$

$$+ \left[\frac{1}{2}\left(b_2 + \bar{b}_2\right)\gamma_{\ell+1} + \alpha_{\ell+1}\bar{b}_2 \right]^2 m_{u,\ell}^2 + \alpha_{\ell+1}^2 b_1^2 x_{i,\ell}^2$$

$$+ 2\left[\frac{1}{2}\left(b_1 + \bar{b}_1\right)\gamma_{\ell+1} + \alpha_{\ell+1}\bar{b}_1 \right]\left[\frac{1}{2}\left(b_2 + \bar{b}_2\right)\gamma_{\ell+1} + \alpha_{\ell+1}\bar{b}_2 \right] m_{x,\ell}m_{u,\ell}$$

$$+ 2\left[\frac{1}{2}\left(b_1 + \bar{b}_1\right)\gamma_{\ell+1} + \alpha_{\ell+1}\bar{b}_1 \right]\alpha_{\ell+1}b_1 m_{x,\ell}x_{i,\ell}$$

$$\left. + 2\left[\frac{1}{2}\left(b_2 + \bar{b}_2\right)\gamma_{\ell+1} + \alpha_{\ell+1}\bar{b}_2 \right]\alpha_{\ell+1}b_1 m_{u,\ell}x_{i,\ell} \right\},$$

and replacing (3.28) *in the latter expression and organizing terms we have*

$$
\eta_{i,\ell}^2 x_{i,\ell}^2 = \frac{b_2^2}{c_\ell^2} \left[\frac{1}{2} \left(b_1 + \bar{b}_1 \right) \gamma_{\ell+1} + \alpha_{\ell+1} \bar{b}_1 \right]^2 m_{x,\ell}^2
$$

$$
+ \frac{b_2^4}{c_\ell^4} \left[\frac{1}{2} \left(b_2 + \bar{b}_2 \right) \gamma_{\ell+1} + \alpha_{\ell+1} \bar{b}_2 \right]^2 \times
$$

$$
\frac{\left[\left(\frac{1}{2} \gamma_{\ell+1} + \alpha_{\ell+1} \right) \left(\bar{b}_1 + b_1 \right) \right]^2}{\left(1 - \frac{b_2}{c_\ell} \left[\frac{1}{2} \left(b_2 + \bar{b}_2 \right) \gamma_{\ell+1} + \alpha_{\ell+1} \bar{b}_2 \right] \right)^2} m_{x,\ell}^2
$$

$$
+ \frac{b_2^2}{c_\ell^2} \alpha_{\ell+1}^2 b_1^2 x_{i,\ell}^2
$$

$$
+ 2 \frac{b_2^3}{c_\ell^3} \left[\frac{1}{2} \left(b_1 + \bar{b}_1 \right) \gamma_{\ell+1} + \alpha_{\ell+1} \bar{b}_1 \right] \left[\frac{1}{2} \left(b_2 + \bar{b}_2 \right) \gamma_{\ell+1} + \alpha_{\ell+1} \bar{b}_2 \right] \times
$$

$$
\frac{\left[\left(\frac{1}{2} \gamma_{\ell+1} + \alpha_{\ell+1} \right) \left(\bar{b}_1 + b_1 \right) \right]}{1 - \frac{b_2}{c_\ell} \left[\frac{1}{2} \left(b_2 + \bar{b}_2 \right) \gamma_{\ell+1} + \alpha_{\ell+1} \bar{b}_2 \right]} m_{x,\ell}^2
$$

$$
+ 2 \frac{b_2^2}{c_\ell^2} \left[\frac{1}{2} \left(b_1 + \bar{b}_1 \right) + \alpha_{\ell+1} \bar{b}_1 \right] \alpha_{\ell+1} b_1 m_{x,\ell} x_{i,\ell}
$$

$$
+ 2 \frac{b_2^3}{c_\ell^3} \left[\frac{1}{2} \left(b_2 + \bar{b}_2 \right) + \alpha_{\ell+1} \bar{b}_2 \right] \alpha_{\ell+1} b_1 \times
$$

$$
\frac{\left[\left(\frac{1}{2} + \alpha_{\ell+1} \right) \left(\bar{b}_1 + b_1 \right) \right]}{1 - \frac{b_2}{c_\ell} \left[\frac{1}{2} \left(b_2 + \bar{b}_2 \right) + \alpha_{\ell+1} \bar{b}_2 \right]} m_{x,\ell} x_{i,\ell}.
$$

At this point, the square completion (3.27) *can be incorporated in the gap* $\mathbb{E}[L_i(x, u_1, \ldots, u_n) - f_{i,k}]$ *in* (3.25) *by using the optimal terms* (3.28) *and* (3.29) *obtaining the following:*

$$
\mathbb{E}[L_i(x, u_1, \ldots, u_n) - f_{i,k}] = (q_{k+N} - \alpha_{k+N}) x_{i,k+N}^2 + (\bar{q}_{k+N} - \beta_{k+N}) m_{x,k+N}^2
$$

$$
+ (0 - \gamma_{k+N}) m_{x,k+N} x_{i,k+N} + (0 - \delta_{k+N})
$$

$$
+ \mathbb{E} \sum_{\ell=k}^{k+N-1} \left(q_i + \alpha_{\ell+1} b_1^2 - \alpha_\ell - \frac{b_2^2}{c_\ell} \alpha_{\ell+1}^2 b_1^2 \right) x_{i,\ell}^2
$$

$$
+ \mathbb{E} \sum_{\ell=k}^{k+N-1} \left(\bar{q} + \alpha_{\ell+1} \bar{b}_1^2 + \beta_{\ell+1} \left(b_1 + \bar{b}_1 \right)^2 - \beta_\ell + \left(b_1 + \bar{b}_1 \right) \bar{b}_1 \gamma_{\ell+1} \right.
$$

$$
+ \left\{ \frac{b_2^2 \bar{r}}{c_\ell^2} + \frac{b_2^2}{c_\ell^2} \left[\alpha_{\ell+1} \bar{b}_2^2 + \beta_{\ell+1} \left(b_2 + \bar{b}_2 \right)^2 + \left(b_2 + \bar{b}_2 \right) \bar{b}_2 \gamma_{\ell+1} \right] \right.
$$

$$
\left. - \frac{b_2^4}{c_\ell^3} \left[\frac{1}{2} \left(b_2 + \bar{b}_2 \right) \gamma_{\ell+1} + \alpha_{\ell+1} \bar{b}_2 \right]^2 \right\} \xi^2
$$

$$
+ \left\{ \frac{b_2}{c_\ell} \left[\left(b_2 + \bar{b}_2 \right) \bar{b}_1 \gamma_{\ell+1} + \left(b_1 + \bar{b}_1 \right) \bar{b}_2 \gamma_{\ell+1} \right. \right.
$$

$$+ 2\alpha_{\ell+1} \bar{b}_1 \bar{b}_2 + 2\beta_{\ell+1} \left(b_1 + \bar{b}_1\right)\left(b_2 + \bar{b}_2\right) \Bigg]$$

$$- 2\frac{b_2^3}{c_\ell^2} \left[\frac{1}{2}\left(b_1 + \bar{b}_1\right)\gamma_{\ell+1} + \alpha_{\ell+1}\bar{b}_1\right]\left[\frac{1}{2}\left(b_2 + \bar{b}_2\right)\gamma_{\ell+1} + \alpha_{\ell+1}\bar{b}_2\right] \Bigg\}\xi$$

$$- \frac{b_2^2}{c_\ell}\left[\frac{1}{2}\left(b_1 + \bar{b}_1\right)\gamma_{\ell+1} + \alpha_{\ell+1}\bar{b}_1\right]^2\Bigg)m_{x,\ell}^2$$

$$+ \mathbb{E}\sum_{\ell=k}^{k+N-1}\Bigg(\left[2\alpha_{\ell+1}b_1\bar{b}_1 - \gamma_\ell + \left(b_1 + \bar{b}_1\right)b_1\gamma_{\ell+1}\right]$$

$$+\Bigg\{\frac{b_2}{c_\ell}\left[\left(b_2 + \bar{b}_2\right)b_1\gamma_{\ell+1} + 2\alpha_{\ell+1}b_1\bar{b}_2\right]$$

$$- 2\frac{b_2^3}{c_\ell^2}\alpha_{\ell+1}b_1\left[\frac{1}{2}\left(b_2 + \bar{b}_2\right)\gamma_{\ell+1} + \alpha_{\ell+1}\bar{b}_2\right]\Bigg\}\xi$$

$$- 2\frac{b_2^2}{c_\ell}\left[\frac{1}{2}\left(b_1 + \bar{b}_1\right)\gamma_{\ell+1} + \alpha_{\ell+1}\bar{b}_1\right]\alpha_{\ell+1}b_1\Bigg)x_{i,\ell}m_{x,\ell}$$

$$+ \mathbb{E}\sum_{\ell=k}^{k+N-1}\left(\delta_{\ell+1} - \delta_\ell + \alpha_{\ell+1}\sigma_0^2 w_{i,\ell}^2\right) + \mathbb{E}\sum_{\ell=k}^{k+N-1} c_\ell\left(u_{i,\ell} + \eta_{i,\ell}x_{i,\ell}\right)^2.$$

The latter long equation is divided into six main groups:

- Those terms at the terminal time instant

- Those terms common factor of the quadratic state

- Those terms common factor of the square mean state

- Those terms common factor of the product $x_{i,\ell}m_{x,\ell}$

- Those terms independent from the state, mean state, control input, and mean control input

- Those terms defining the optimal control input

The announced results are obtained by minimizing terms in the last expression.

3.5 Exercises

1. Consider a mean-field game with infinite-time horizon, where with cost associated with each decision-maker is

$$L_i^\infty(u_1, \ldots, u_n) = \frac{1}{2}\int_0^\infty \left(qx_i^2 + \bar{q}m_x^2 + ru_i^2 + \bar{r}m_u^2\right)\mathrm{d}t.$$

The mean-field game problem is given by

$$\underset{u_i \in \mathcal{U}_i}{\text{minimize}}\, L_i^\infty(u_1, \ldots, u_n)$$

subject to

$$\dot{x}_i = b_1 x_i + \bar{b}_1 m_x + b_2 u_i + \bar{b}_2 m_u,$$
$$x_i(0) := x_{0i},$$

where the weight parameters are $q, q + \bar{q} \in \mathbb{R}_{\geq 0}$, $r, r + \bar{r} \in \mathbb{R}_{>0}$, and the system state parameters are $b_1, \bar{b}_1, b_2, \bar{b}_2 \in \mathbb{R}$, for all $j \in \mathcal{N}$.

Hint: This problem is solved by computing the steady-state of the ordinary differential equation obtained in the finite-horizon mean-field game problem.

2. Solve the stochastic version of the problem presented before, i.e., the finite-horizon stochastic mean-field game problem.

3. Consider a discrete-time mean-field game with infinite-time horizon, where the cost for each decision-maker is as follows:

$$L_i^\infty(u_1, \ldots, u_n) = \sum_{\ell=k}^{\infty} \left[q x_\ell^2 + \bar{q} m_{x,\ell}^2 + r u_{i,\ell}^2 + \bar{r} m_{u,\ell}^2 \right].$$

The discrete-time mean-field game problem is given by

$$\underset{u_i \in \mathcal{U}_i}{\text{minimize}}\, L_i^\infty(u_1, \ldots, u_n),$$

subject to

$$x_{i,\ell+1} = b_1 x_{i,\ell} + \bar{b}_1 m_{x,\ell} + b_2 u_{i,\ell} + \bar{b}_2 m_{u,\ell},$$
$$\forall \ell \in [k..k + N - 1],$$

where the initial state condition x_k is given, and the weight parameters are $q \geq 0$, $\bar{q} \geq 0$, and $\bar{r} > 0$.

Hint: This problem is solved by computing the steady-state of the difference equation obtained in the discrete-time finite-horizon mean-field game problem.

4. Solve the discrete-time stochastic version of the problem presented before, i.e., the finite-horizon discrete-time stochastic mean-field game problem.

5. Solve the following finite-time horizon mean-field game problem with state- control-input-dependent diffusion terms, i.e., with the cost function for the ith decision-maker

$$L_i(x, u_1, \ldots, u_n) = \frac{1}{2} q x_i(T)^2 + \frac{1}{2} \bar{q} m_x^2(T)$$

$$+ \frac{1}{2} \int_0^T \left(qx_i^2 + \bar{q}m_x^2 + ru_i^2 + \bar{r}m_u^2 \right) dt.$$

The statement of the problem is

$$
\begin{cases}
\underset{u_i \in \mathcal{U}_i}{\text{minimize}} \, \mathbb{E}[L_i(x, u_1, \ldots, u_n)], \\[2mm]
\text{subject to} \\[2mm]
dx_i = \left(b_1 x_i + \bar{b}_1 m_x + b_2 u_i + \bar{b}_2 m_u \right) dt \\
\qquad + \left(\sigma_1 x_i + \bar{\sigma}_1 m_x + \sigma_2 u_i + \bar{\sigma}_2 m_u \right) dB_i, \\
x_i(0) := x_{0i},
\end{cases}
$$

where the weight parameters are $q, q + \bar{q} \in \mathbb{R}_{\geq 0}$, $r, r + \bar{r} \in \mathbb{R}_{>0}$, and the system state parameters are $b_1, \bar{b}_1, b_2, \bar{b}_2, \sigma_1, \bar{\sigma}_1, \sigma_2, \bar{\sigma}_2 \in \mathbb{R}$, for all $j \in \mathcal{N}$. Finally, B_i denotes a standard Brownian motion.

Hint: This problem can be solved by applying the direct method and using the following guess functional:

$$f_i(t, x_i, m_x) = \frac{1}{2}\alpha x_i^2 + \frac{1}{2}\beta m_x^2 + \gamma m_x x_i + \delta.$$

6. Solve the discrete-time finite-horizon mean-field game problem involving both state- and control-input-dependent terms. Consider the following cost function:

$$
L_i(x, u_1, \ldots, u_n) = qx_{k+N}^2 + \bar{q}m_{x,k+N}^2
$$
$$
+ \sum_{\ell=k}^{k+N-1} \left[qx_\ell^2 + \bar{q}m_{x,\ell}^2 + ru_{i,\ell}^2 + \bar{r}m_{u,\ell}^2 \right],
$$

and

$$
\begin{cases}
\underset{u_i \in \mathcal{U}_i}{\text{minimize}} \, \mathbb{E}\left[L_i\left(u_1, \ldots, u_n\right)\right], \\[2mm]
\text{subject to} \\[2mm]
x_{i,\ell+1} = \left(b_1 x_{i,\ell} + \bar{b}_1 m_{x,\ell} + b_2 u_{i,\ell} + \bar{b}_2 m_{u,\ell} \right) \\
\qquad + \left(\sigma_1 x_{i,\ell} + \bar{\sigma}_1 m_{x,\ell} + \sigma_2 u_{i,\ell} + \bar{\sigma}_2 m_{u,\ell} \right) w_{i,\ell}, \\
\forall \ell \in [k..k + N - 1],
\end{cases}
$$

where the initial condition x_k is given, and the weight parameters are $q_i \geq 0$, $\bar{q} \geq 0$, and $\bar{r} > 0$.

Hint: This problem can be solved by applying the direct method and using the following guess functional:

$$f_{i,\ell} = \alpha_\ell x_{i,\ell}^2 + \beta_\ell m_{x,\ell}^2 + \gamma_\ell m_{x,\ell} x_{i,\ell} + \delta_\ell.$$

Part III

One-Dimensional Mean-Field-Type Games

4

Continuous-Time Mean-Field-Type Games

The previous chapters mainly helped to make the reader familiar with different deterministic and stochastic game problems that do not belong to the mean-field-type class. Indeed, these previous chapters also served as introduction to the direct method, which we will continue using along the book. Here in this chapter, we start working with the main topic and objective of this book, we start studying mean-field-type control and game theory from the simplest setting, and progressively, we will introduce new concepts and terms later on.

Mean-field-type games are a class of stochastic games where the distribution of either the system states or the control inputs/strategies (or both) are taken into consideration for the cost functional and for the evolution of the system states in the underlying SDE. Under this class of games, the expected payoff functional is not necessarily linear in the probability measure of the system state. Games with distribution-dependent quantity-of-interest are getting special importance since they capture not only the mean but also the variance and higher order terms (recall the discussion presented in Section 1.2.2, Chapter 1). Therefore, within the framework of mean-field-type games and control theory, it is possible to quantify, study and minimize risk terms, which become crucial in the field of risk-aware engineering (recall the engineering applications with uncertainties mentioned in Section 1.2.3, Chapter 1).

In this chapter, we focus our attention over a linear-quadratic mean-field-type game for arbitrary finite number of players/decision-makers. The players' control actions are scalars. The state is driven by a continuous-time dynamical system that involves both the mean and variance of the variables-of-interest, i.e., system states and control inputs as in the works reported in [71, 73]. We address both the non-cooperative and fully-cooperative game approach, which are solved semi-explicitly by means of the direct method. The discrete time analogue will be discussed later on in Chapters 11 and 12 for the scalar and even for the matrix-valued cases, respectively. Next, we begin our trip throughout the mean-field-type theory by introducing the set-up for a scalar game and solve it by means of the direct method.

DOI: 10.1201/9781003098607-4

4.1   Mean-Field-Type Game Set-up

Consider a set of $n \geq 2$, $n \in \mathbb{N}$, decision-makers denoted by $\mathcal{N} = \{1, \ldots, n\}$. These agents strategically interact to each other by means of the following dynamical system:

$$\mathrm{d}x = \left(b_0 + b_1 x + \bar{b}_1 \mathbb{E}[x] + \sum_{j \in \mathcal{N}} b_{2j} u_j + \sum_{j \in \mathcal{N}} \bar{b}_{2j} \mathbb{E}[u_j] \right) \mathrm{d}t + \sigma_0 \mathrm{d}B, \quad (4.1\mathrm{a})$$

$$x(0) := x_0, \quad (4.1\mathrm{b})$$

where x, and $\mathbb{E}[x] \in \mathbb{R}$ correspond to the system state and its expectation, respectively. Similarly, u_i, and $\mathbb{E}[u_i] \in \mathbb{R}$ denote the control input of the decision-maker $i \in \mathcal{N}$, and its expected value, respectively. Hence, the system parameters $b_0, b_1, \bar{b}_1, b_{2j}, \bar{b}_{2j}, \sigma_0$, for all $j \in \mathcal{N}$, are scalar real functions of time, and B denotes a standard Brownian motion. In general, the underlying SDE is in the form

$$\mathrm{d}x = D_r(x, \mathbb{E}[x], u, \mathbb{E}[u]) \mathrm{d}t + D_f(x, \mathbb{E}[x], u, \mathbb{E}[u]) \mathrm{d}B$$

being D_r the drift term and D_f the diffusion term of mean-field-type. We start studying a simple case with a constant diffusion coefficient σ_0. Later on, we will consider more general cases. Each decision-maker is interested in minimizing its own cost functional, which is as follows:

$$L_i(x, u_1, \ldots, u_n) = \frac{1}{2} q_i(T) x(T)^2 + \frac{1}{2} \bar{q}_i(T) \mathbb{E}[x(T)]^2$$

$$+ \frac{1}{2} \int_0^T \left(q_i x^2 + \bar{q}_i \mathbb{E}[x]^2 + r_i u_i^2 + \bar{r}_i \mathbb{E}[u_i]^2 \right) \mathrm{d}t. \quad (4.2)$$

where the weight parameters in the cost functional $q_i, \bar{q}_i, r_i, \bar{r}_i$ are time-varying scalar deterministic functions. Different from the cost functionals studied in the previous chapters in this book, with the cost functional in (4.2) the decision-makers minimize the expected system state and the expected energy that is applied to the system (4.1a). The expected terms $\mathbb{E}[x]$ and $\mathbb{E}[u_i]$ are known as mean-field terms. The most important aspect of the cost functional incorporating these mean-field terms has to do with the risk consideration. To observe this, notice that the cost function $L_i(x, u_1, \ldots, u_n)$ in (4.2) can be conveniently re-written as follows:

$$L_i(x, u_1, \ldots, u_n) = \frac{1}{2} q_i(T) \left(x(T) - \mathbb{E}[x(T)] + \mathbb{E}[x(T)] \right)^2 + \frac{1}{2} \bar{q}_i(T) \mathbb{E}[x(T)]^2$$

$$+ \frac{1}{2} \int_0^T \Big(q_i \left(x - \mathbb{E}[x] + \mathbb{E}[x] \right)^2 + \bar{q}_i \mathbb{E}[x]^2$$

$$+ r_i \left(u_i - \mathbb{E}[u_i] + \mathbb{E}[u_i] \right)^2 + \bar{r}_i \mathbb{E}[u_i]^2 \Big) \mathrm{d}t.$$

Thus, the expectation $\mathbb{E}[L_i(x, u_1, \ldots, u_n)]$ can be written in terms of the subtraction $(x - \mathbb{E}[x])$ and the expected value $\mathbb{E}[x]$ as follows:

$$\mathbb{E}[L_i(x, u_1, \ldots, u_n)] = \frac{1}{2} q_i(T) \mathbb{E}[(x(T) - \mathbb{E}[x(T)])^2]$$
$$+ \frac{1}{2} \left(q_i(T) + \bar{q}_i(T) \right) \mathbb{E}[x(T)]^2$$
$$+ \frac{1}{2} \mathbb{E} \int_0^T \Big(q_i \left(x - \mathbb{E}[x] \right)^2 + (q_i + \bar{q}_i) \mathbb{E}[x]^2$$
$$+ r_i \left(u_i - \mathbb{E}[u_i] \right)^2 + (r_i + \bar{r}_i) \mathbb{E}[u_i]^2 \Big) \mathrm{d}t, \tag{4.3}$$

where it was taken into account that

$$\mathbb{E}[\mathbb{E}[x] \ (x - \mathbb{E}[x])] = 0,$$

and

$$\mathbb{E}[\mathbb{E}[u_j] \ (u_j - \mathbb{E}[u_j])] = 0,$$

for all $j \in \mathcal{N}$. Recall that the variance of a random variable x is defined as

$$\mathrm{var}[x] = \mathbb{E}[(x - \mathbb{E}[x])^2].$$

Hence, it yields that the cost functional that the decision-maker is pursuing to reduce is as follows:

$$\mathbb{E}[L_i(x, u_1, \ldots, u_n)] = \frac{1}{2} q_i(T) \mathrm{var}[x(T)] + \frac{1}{2} \left(q_i(T) + \bar{q}_i(T) \right) \mathbb{E}[x(T)]^2$$
$$+ \frac{1}{2} \int_0^T \Big(q_i \mathrm{var}[x] + (q_i + \bar{q}_i) \mathbb{E}[x]^2$$
$$+ r_i \mathrm{var}[u_i] + (r_i + \bar{r}_i) \mathbb{E}[u_i]^2 \Big) \mathrm{d}t, \tag{4.4}$$

showing that the proposed cost functional includes both the variance and mean for the states and the control inputs. Non-negativity of the cost requires that the coefficient functions $q_i, q_i + \bar{q}_i, r_i$, and $r_i + \bar{r}_i$ are non-negative.

Similarly, it is plausible to express the system dynamics in (4.1a) by using the same quadratic terms that appear in the cost functional. Thus, the construction of the solution can be computed more suitably. The SDE describing the evolution of the system state can be re-written as follows:

$$\mathrm{d}x = \Bigg(b_0 + b_1 \left(x - \mathbb{E}[x] \right) + \left(b_1 + \bar{b}_1 \right) \mathbb{E}[x] + \sum_{j \in \mathcal{N}} b_{2j} \left(u_j - \mathbb{E}[u_j] \right)$$
$$+ \sum_{j \in \mathcal{N}} \left(b_{2j} + \bar{b}_{2j} \right) \mathbb{E}[u_j] \Bigg) \mathrm{d}t + \sigma_0 \mathrm{d}B, \tag{4.5a}$$
$$\mathbb{E}[\dot{x}] = b_0 + \left(b_1 + \bar{b}_1 \right) \mathbb{E}[x] + \sum_{j \in \mathcal{N}} \left(b_{2j} + \bar{b}_{2j} \right) \mathbb{E}[u_j]. \tag{4.5b}$$

In (4.5b), we have also computed the evolution of the expected state $\mathbb{E}[x]$, i.e., the evolution of the mean-field term, which is given by a deterministic differential equation. Throughout the construction of the semi-explicit solution we will use this expression.

Next, we introduce the risk-neutral game problem where each player minimizes the expectation of its cost functional subject to the state dynamics. The interaction of the players (coupling among decision-makers) is through the state x and $\mathbb{E}[x]$ via the dynamics in (4.5).

Let $\mathcal{U}_i$, for all $i \in \mathcal{N}$, be the set of the control strategies that are square integrable and progressively measurable with respect to the Brownian motion. The non-cooperative mean-field-type game problem, which is equivalent to a risk-aware non-cooperative interactive decision-making problem is stated below.

Problem 13 (Non-Cooperative Game Problem) *The non-cooperative linear-quadratic mean-field-type game problem is given by*

$$
\begin{cases}
i \in \mathcal{N}, \\[4pt]
\underset{u_i \in \mathcal{U}_i}{\text{minimize}}\ \mathbb{E}[L_i(x, u_1, \ldots, u_n)], \\[8pt]
\text{subject to} \\[8pt]
\mathrm{d}x = \left(b_0 + b_1 x + \bar{b}_1 \mathbb{E}[x] + \sum_{j \in \mathcal{N}} b_{2j} u_j + \sum_{j \in \mathcal{N}} \bar{b}_{2j} \mathbb{E}[u_j] \right) \mathrm{d}t + \sigma_0 \mathrm{d}B, \\[8pt]
x(0) := x_0,
\end{cases}
$$

the initial state condition x_0 is generated from a distribution denoted by m_0 that has finite second moment, i.e., $\mathbb{E}[x_0^2] < +\infty$. The weight parameters are $q_i, q_i + \bar{q}_i \geq 0$, $r_i, r_i + \bar{r}_i > 0$, and the system parameters $b_0, b_1, \bar{b}_1, b_{2j}, \bar{b}_{2j}, \sigma_0 \in \mathbb{R}$, for all $j \in \mathcal{N}$, can be considered as deterministic functions of time.

As we studied in the differential games, each decision-maker $i \in \mathcal{N}$ does its best to minimize its own cost functional for given selected strategies by the other decision-makers $\mathcal{N} \setminus \{i\}$. Such best selection in the context of mean-field-type games is presented next in Definition 8.

Definition 8 (Mean-Field-Type Best-Response Strategy) *Any feasible strategy $u_i^* \in \mathcal{U}_i$ such that*

$$
u_i^* \in \arg \underset{u_i \in \mathcal{U}_i}{\text{minimize}}\ \mathbb{E}\left[L_i(x, u_1, \ldots, u_n)\right]
$$

is a Mean-field-type Best-Response strategy of the decision-maker $i \in \mathcal{N}$ against the strategies

$$
u_{-i} = \{u_1, \ldots, u_{i-1}, u_{i+1}, \ldots, u_n\}
$$

selected by the other decision makers $\mathcal{N} \setminus \{i\}$. *The set of Best-Response strate-gies is given by*

$$\mathrm{BR}_i : \prod_{j \in \mathcal{N} \setminus \{i\}} \mathcal{U}_j \to 2^{\mathcal{U}_i},$$

where $2^{\mathcal{U}_i}$ is the power set of all the possible subsets of $\mathcal{U}_i$.

The solution of the non-cooperative mean-field-type game Problem 13 is given by a Nash equilibrium. The set of Mean-field-type Nash equilibria is defined by using the mean-field-type best-response as shown next.

Definition 9 (Mean-Field-Type Nash equilibrium) *Any feasible strat-egy profile, denoted by*

$$[u_1^* \quad \cdots \quad u_n^*]^\top \in \prod_{j \in \mathcal{N}} \mathcal{U}_j,$$

such that the optimal $u_i^* \in \mathrm{BR}_i(u_{-i}^*)$, for all $i \in \mathcal{N}$, is a mean-field-type Nash equilibrium of the Linear-Quadratic Mean-Field-Type Game.

If all the decision-makers are selecting their best-response strategy, then they do not have incentives to modify their selection since they are doing their best pursuing to minimize the individual costs. We had discussed in previous chapters that this is the scenario in which none of the decision-makers would have a rational reason to unilaterally deviate its chosen strategies.

Figure 4.1 shows the general scheme corresponding to the non-cooperative mean-field-type game perturbed by a Brownian motion. Each players' strategic decision affects the evolution of a common state given by x and its expectation $\mathbb{E}[x]$. Moreover, each decision-maker has an individual objective consisting of the minimization of the cost function $L_i(x, u_1, \ldots, u_n)$, which involves both the mean and the variance of the system states and control inputs.

In the latter problem, the decision-makers do not help each other to min-imize their cost functionals, but they behave independently. The decision-makers behave selfishly since each one only cares about its own cost functional. In contrast, there is a scenario in which all the decision-makers cooperate to each other in order to pursue a common objective as presented in Problem 14.

Problem 14 (Cooperative Game Approach) *The cooperative* *linear-quadratic mean-field-type game problem is given by*

$$
\begin{cases}
L_0(x, u_1, \ldots, u_n) = \mathbb{E} \sum_{i \in \mathcal{N}} [L_i(x, u_1, \ldots, u_n)], \\[2mm]
\underset{\{u_i \in \mathcal{U}_i\}_{i \in \mathcal{N}}}{\text{minimize}} L_0(x, u_1, \ldots, u_n), \\[4mm]
\text{subject to} \\[4mm]
\mathrm{d}x = \left(b_0 + b_1 x + \bar{b}_1 \mathbb{E}[x] + \sum_{j \in \mathcal{N}} b_{2j} u_j + \sum_{j \in \mathcal{N}} \bar{b}_{2j} \mathbb{E}[u_j] \right) \mathrm{d}t + \sigma_0 \mathrm{d}B, \\[4mm]
x(0) := x_0,
\end{cases}
$$

where the weight parameters in the cost functional are

$$
q_0 = \left[\sum_{i \in \mathcal{N}} q_i \right] \geq 0,
$$

$$
q_0 + \bar{q}_0 = \left[\sum_{i \in \mathcal{N}} (q_i + \bar{q}_i) \right] \geq 0,
$$

and $r_j, r_j + \bar{r}_j > 0$. Again, the system parameters $b_0, b_1, \bar{b}_1, b_{2j}, \bar{b}_{2j}, \sigma_0 \in \mathbb{R}$, for all decision-makers $j \in \mathcal{N}$, can be considered as function of time or constant as desired.

Here, it is quite relevant to stand out that there is a tight relationship between the cooperative behavior with an automatic control problem. We refer the reader again to the introductory Section 1.3 in Chapter 1 where we presented the different game-theoretical solution concepts. One can also compare the illustration in Figure 1.9 with the illustrative example in Figure 1.16. This relation is stated in the following Remark. Therefore, the solution of the cooperative mean-field-type game problem corresponds to the solution of a mean-field-type control problem.

Remark 13 *Since all the decision-makers work jointly pursuing a common benefit, the aggregation of the strategic selections can be seen as a unique decision made by a unique player given by the coalition of decision-makers. In this regard, the cooperative game approach can be interpreted as a control problem.*

Remark 13 is also supported by comparing the semi-explicit solutions that emerge in the cooperative game and the stochastic control problem.

Figure 4.2 shows the general scheme of the fully-cooperative mean-field-type game, where all the decision-makers (players) optimize jointly a common cost functional composed of the sum of all the individual cost functions, i.e.,

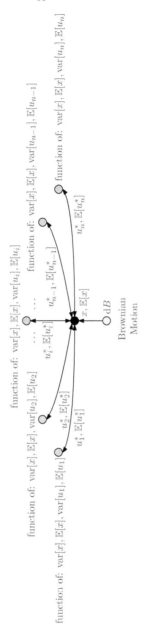

FIGURE 4.1
General scheme of the non-cooperative mean-field-type game. Dark gray nodes represent the n players, the black node represents the system state, and the light gray node represents the stochasticity affecting the system state.

$$L_0 = \mathbb{E} \sum_{i \in \mathcal{N}} L_i, \quad \text{(function of: var}[x], \; \mathbb{E}[x], \; \text{var}[u_i], \; \mathbb{E}[u_i])$$

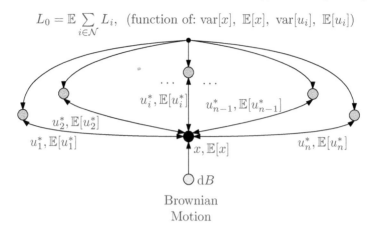

FIGURE 4.2
General scheme of the cooperative mean-field-type game. Dark gray nodes
represent the n players, the black node represents the system state, and the
light gray node represents the stochasticity affecting the system state.

$\sum_{i \in \mathcal{N}} L_i(x, u_1, \dots, u_n)$. In the following section, we focus on the computation of the mean-field-type game problems for the both scenarios, i.e., non-
cooperative and cooperative. As expected, we solve these problems by using
the so-called direct method.

4.2 Semi-explicit Solution of the Mean-Field-Type Game Problem

In this section, we present the semi-explicit solutions for both the non-
cooperative and cooperative mean-field-type game problems. First, we start
with the result for the non-cooperative scenario. Proposition 13 characterizes
the Nash equilibrium for the non-cooperative mean-field-type game.

Proposition 13 *The feedback-strategies linear-quadratic mean-field-type Nash
equilibrium for the non-cooperative problem 13, and the equilibrium costs are
given by:*

$$u_i^* = \mathbb{E}[u_i^*] - \frac{b_{2i}}{r_i} \alpha_i \left(x - \mathbb{E}[x] \right), \; \forall i \in \mathcal{N},$$

$$\mathbb{E}[u_i^*] = -\frac{(b_{2i} + \bar{b}_{2i})}{(r_i + \bar{r}_i)} \left[\beta_i \mathbb{E}[x] + \gamma_i \right], \; \forall i \in \mathcal{N},$$

$$L_i(x, u_1^*, \dots, u_n^*) = \frac{1}{2} \alpha_i(0) \left(x_0 - \mathbb{E}[x_0] \right)^2 + \frac{1}{2} \beta_i(0) \mathbb{E}[x_0]^2$$

$$+ \gamma_i(0)\mathbb{E}[x_0] + \delta_i(0), \quad \forall i \in \mathcal{N},$$

where α_i, β_i, γ_i, and δ_i solve the following differential equations:

$$\dot{\alpha}_i = -2b_1\alpha_i - q_i + 2\alpha_i \sum_{j \in \mathcal{N}\backslash\{i\}} \frac{b_{2j}^2}{r_j}\alpha_j + \frac{b_{2i}^2}{r_i}\alpha_i^2,$$

$$\dot{\beta}_i = -2\left(b_1 + \bar{b}_1\right)\beta_i - q_i - \bar{q}_i + 2\beta_i \sum_{j \in \mathcal{N}\backslash\{i\}} \frac{\left(b_{2j} + \bar{b}_{2j}\right)^2}{\left(r_j + \bar{r}_j\right)}\beta_j + \frac{\left(b_{2i} + \bar{b}_{2i}\right)^2}{\left(r_i + \bar{r}_i\right)}\beta_i^2,$$

$$\dot{\gamma}_i = -b_0\beta_i - \left(b_1 + \bar{b}_1\right)\gamma_i + \beta_i \sum_{j \in \mathcal{N}\backslash\{i\}} \frac{\left(b_{2j} + \bar{b}_{2j}\right)^2}{\left(r_j + \bar{r}_j\right)}\gamma_j$$

$$+ \gamma_i \sum_{j \in \mathcal{N}\backslash\{i\}} \frac{\left(b_{2j} + \bar{b}_{2j}\right)^2}{\left(r_j + \bar{r}_j\right)}\beta_j + \frac{\left(b_{2i} + \bar{b}_{2i}\right)^2}{\left(r_i + \bar{r}_i\right)}\beta_i\gamma_i,$$

$$\dot{\delta}_i = -b_0\gamma_i - \frac{1}{2}\alpha_i\sigma_0^2 + \gamma_i \sum_{j \in \mathcal{N}\backslash\{i\}} \frac{\left(b_{2j} + \bar{b}_{2j}\right)^2}{\left(r_j + \bar{r}_j\right)}\gamma_j + \frac{1}{2}\frac{\left(b_{2i} + \bar{b}_{2i}\right)^2}{\left(r_i + \bar{r}_i\right)}\gamma_i^2,$$

with the following boundary terminal conditions:

$$\alpha_i(T) = q_i(T),$$
$$\beta_i(T) = q_i(T) + \bar{q}_i(T),$$
$$\gamma_i(T) = \delta_i(T) = 0,$$

whenever these equations admit a solution with positive α_i and β_i.

The state-feedback optimal law for all the decision-makers presented in Proposition 13 is more sophisticated than the control input that is obtained for a mean-field-free differential game as studied in Chapter 2. Notice that the differences arises mainly because of the minimization of risks terms by means of the variance. Another interesting characteristic to realize about the solution is that the optimal control input is completely independent from the term δ, which is mainly affecting the computation of the optimal cost functional.

Remark 14 *Let us assume that $b_0 = 0$ in the SDE (4.5). Observe that under this situation the γ_j differential equation in Proposition 13 has a trivial solution given that the boundary conditions are $\gamma_j(T) = 0$, for all $j \in \mathcal{N}$. Therefore, the semi-explicit solution is simplified by making all $\gamma_j = 0$, for all $j \in \mathcal{N}$. This condition affects the optimal control input, the optimal cost, and the evolution of δ_i.*

Next, we present step by step the computation of the semi-explicit solution. Recall we do this procedure as indicated in the diagram shown in Figure 1.17 in Chapter 1.

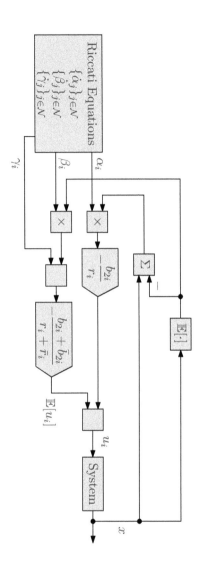

FIGURE 4.3
Feedback scheme for the linear-quadratic mean-field-type non-cooperative game problem.

Proof 13 (Proposition 13) *The reader is already familiar with the first step to compute the solution using the direct method. First, we require to postulate an appropriate guess functional for the optimal cost from time $t = 0$ to the terminal time T. This function is normally found by observing the structure of the problem. Thus, inspired by the quadratic form of the cost function, we propose a guess functional as follows*

$$f_i(t, x) = \frac{1}{2}\alpha_i \left(x - \mathbb{E}[x]\right)^2 + \frac{1}{2}\beta_i \mathbb{E}[x]^2 + \gamma_i \mathbb{E}[x] + \delta_i, \qquad (4.6)$$

where $\alpha_i, \beta_i, \gamma_i$, and δ_i are deterministic functions of time. Observe that, of course, the guess functional should incorporate the mean-field term $\mathbb{E}[x]$ given that it is expected that the optimal cost depends on it. Now, let us recall the Itô's formula, i.e.,

$$f_i(T, x) - f_i(0, x_0)$$

$$= \int_0^T \left(\frac{\partial f_i(t, x)}{\partial t} + \frac{\partial f_i(t, x)}{\partial x} D_r(x, u) + \frac{\partial^2 f_i(t, x)}{\partial x^2} \frac{D_f(x, u)^2}{2}\right) dt$$

$$+ \int_0^T D_f(x, u) \frac{\partial f_i(t, x)}{\partial x} dB, \qquad (4.7)$$

with $D_r(x, \mathbb{E}[x], u, \mathbb{E}[u])$ and $D_f(x, \mathbb{E}[x], u, \mathbb{E}[u])$ denoting the drift and the diffusion terms, respectively; which in a general case can be state and control-input-dependent terms involving the mean-field terms as well. Indeed, both the drift and diffusion can be function of time, but here we omit such dependence in the presentation of the integration formula.

$$D_r(x, \mathbb{E}[x], u, \mathbb{E}[u]) = b_0 + b_1 \left(x - \mathbb{E}[x]\right) + \left(b_1 + \bar{b}_1\right) \mathbb{E}[x]$$

$$+ \sum_{j \in \mathcal{N}} b_{2j} \left(u_j - \mathbb{E}[u_j]\right) + \sum_{j \in \mathcal{N}} \left(b_{2j} + \bar{b}_{2j}\right) \mathbb{E}[u_j],$$

and $D_f(x, \mathbb{E}[x], u, \mathbb{E}[u]) = \sigma_0$ (state and control-input independent).

Before continuing with the computation of the integration formula and for the sake of clarity in the reader learning process, we first calculate separately the following terms

$$\frac{\partial f_i(t, x)}{\partial t} = \frac{1}{2}\left(x - \mathbb{E}[x]\right)^2 \dot{\alpha}_i - \alpha_i \left(x - \mathbb{E}[x]\right) \mathbb{E}[\dot{x}] + \frac{1}{2}\mathbb{E}[x]^2 \dot{\beta}_i + \beta_i \mathbb{E}[x]\mathbb{E}[\dot{x}]$$

$$+ \mathbb{E}[x]\dot{\gamma}_i + \gamma_i \mathbb{E}[\dot{x}] + \dot{\delta}_i, \qquad (4.8a)$$

$$\frac{\partial f_i(t, x)}{\partial x} = \alpha_i \left(x - \mathbb{E}[x]\right), \qquad (4.8b)$$

$$\frac{\partial^2 f_i(t, x)}{\partial x^2} = \alpha_i. \qquad (4.8c)$$

Replacing (4.8a), (4.8b) and (4.8c) in (4.7) yields

$$f_i(T, x) - f_i(0, x_0) = \frac{1}{2}\int_0^T \left(\left(x - \mathbb{E}[x]\right)^2 \dot{\alpha}_i - 2\alpha_i \left(x - \mathbb{E}[x]\right) \mathbb{E}[\dot{x}] + \mathbb{E}[x]^2 \dot{\beta}_i\right)$$

$$+ 2\beta_i \mathbb{E}[x]\mathbb{E}[\dot{x}] + 2\mathbb{E}[x]\dot{\gamma}_i + 2\gamma_i \mathbb{E}[\dot{x}] + 2\dot{\delta}_i \Big) \mathrm{d}t$$

$$+ \int_0^T \alpha_i \left(x - \mathbb{E}[x]\right) \Big(b_0 + b_1 \left(x - \mathbb{E}[x]\right)$$

$$+ \left(b_1 + \bar{b}_1\right) \mathbb{E}[x] + \sum_{j \in \mathcal{N}} b_{2j} \left(u_j - \mathbb{E}[u_j]\right) + \sum_{j \in \mathcal{N}} \left(b_{2j} + \bar{b}_{2j}\right) \mathbb{E}[u_j] \Big) \mathrm{d}t$$

$$+ \frac{1}{2} \int_0^T \alpha_i \sigma_0^2 \mathrm{d}t + \int_0^T \sigma_0 \alpha_i \left(x - \mathbb{E}[x]\right) \mathrm{d}B. \tag{4.9}$$

In (4.9), we require the evolution of the mean-field state $\mathbb{E}[x]$, which was already introduced above in (4.5b). Using $\mathbb{E}[\dot{x}]$ and computing the expected value $\mathbb{E}[f_i(T, x) - f_i(0, x_0)]$ yields

$$\mathbb{E}[f_i(T, x) - f_i(0, x_0)] = \frac{1}{2} \mathbb{E} \int_0^T \Big[\left(x - \mathbb{E}[x]\right)^2 \dot{\alpha}_i + \mathbb{E}[x]^2 \dot{\beta}_i$$

$$+ 2\beta_i \mathbb{E}[x] \Big(b_0 + \left(b_1 + \bar{b}_1\right) \mathbb{E}[x] + \sum_{j \in \mathcal{N}} \left(b_{2j} + \bar{b}_{2j}\right) \mathbb{E}[u_j] \Big) + 2\mathbb{E}[x]\dot{\gamma}_i$$

$$+ 2\gamma_i \Big(b_0 + \left(b_1 + \bar{b}_1\right) \mathbb{E}[x] + \sum_{j \in \mathcal{N}} \left(b_{2j} + \bar{b}_{2j}\right) \mathbb{E}[u_j] \Big) + 2\dot{\delta}_i \Big] \mathrm{d}t$$

$$+ \mathbb{E} \int_0^T \alpha_i \left(x - \mathbb{E}[x]\right) \Big[b_1 \left(x - \mathbb{E}[x]\right) + \sum_{j \in \mathcal{N}} b_{2j} \left(u_j - \mathbb{E}[u_j]\right) \Big] \mathrm{d}t$$

$$+ \frac{1}{2} \mathbb{E} \int_0^T \alpha_i \sigma_0^2 \mathrm{d}t, \tag{4.10}$$

where it has been considered the fact that

$$\mathbb{E}\left[x - \mathbb{E}[x]\right] = 0,$$
$$\mathbb{E}\left[x - \mathbb{E}[x]\right] \mathbb{E}[x] = 0,$$
$$\mathbb{E}[\mathrm{d}B] = 0$$
$$\mathbb{E}\left[x - \mathbb{E}[x]\right] \mathbb{E}[u_j] = 0, \ \forall j \in \mathcal{N}.$$

Solving the parenthesis in (4.10), it follows that

$$\mathbb{E}[f_i(T, x) - f_i(0, x_0)] = \frac{1}{2} \mathbb{E} \int_0^T \Big[\left(x - \mathbb{E}[x]\right)^2 \dot{\alpha}_i + \mathbb{E}[x]^2 \dot{\beta}_i$$

$$+ 2 b_0 \beta_i \mathbb{E}[x] + 2 \left(b_1 + \bar{b}_1\right) \beta_i \mathbb{E}[x]^2 + 2 \sum_{j \in \mathcal{N}} \left(b_{2j} + \bar{b}_{2j}\right) \beta_i \mathbb{E}[x]\mathbb{E}[u_j] + 2\mathbb{E}[x]\dot{\gamma}_i$$

$$+ 2 b_0 \gamma_i + 2 \left(b_1 + \bar{b}_1\right) \gamma_i \mathbb{E}[x] + 2 \sum_{j \in \mathcal{N}} \left(b_{2j} + \bar{b}_{2j}\right) \gamma_i \mathbb{E}[u_j] + 2\dot{\delta}_i \Big] \mathrm{d}t$$

$$+ \mathbb{E} \int_0^T \Big[b_1 \alpha_i \left(x - \mathbb{E}[x]\right)^2 + \sum_{j \in \mathcal{N}} b_{2j} \left(u_j - \mathbb{E}[u_j]\right) \alpha_i \left(x - \mathbb{E}[x]\right) \Big] \mathrm{d}t$$

$$+ \frac{1}{2}\mathbb{E}\int_0^T \alpha_i \sigma_0^2 \mathrm{d}t,$$

and grouping terms according to common factors, putting aside all the terms depending on the decision variables for the ith decision-maker as $(u_i - \mathbb{E}[u_i])$ and $\mathbb{E}[u_i]$, and aside those terms depending on the other decision-makers as $(u_j - \mathbb{E}[u_j])$ and $\mathbb{E}[u_j]$, for all $j \in \mathcal{N} \setminus \{i\}$, yields

$$\mathbb{E}[f_i(T,x) - f_i(0,x_0)] = \frac{1}{2}\mathbb{E}\int_0^T (x - \mathbb{E}[x])^2 \left[\dot{\alpha}_i + 2b_1\alpha_i\right]\mathrm{d}t$$

$$+ \frac{1}{2}\mathbb{E}\int_0^T \mathbb{E}[x]^2 \left[\dot{\beta}_i + 2\left(b_1 + \bar{b}_1\right)\beta_i\right]\mathrm{d}t$$

$$+ \mathbb{E}\int_0^T \mathbb{E}[x]\left[b_0\beta_i + \dot{\gamma}_i + \left(b_1 + \bar{b}_1\right)\gamma_i\right]\mathrm{d}t$$

$$+ \mathbb{E}\int_0^T \left[\dot{\delta}_i + b_0\gamma_i + \frac{1}{2}\alpha_i\sigma_0^2\right]\mathrm{d}t$$

$$+ \mathbb{E}\int_0^T \sum_{j \in \mathcal{N}\setminus\{i\}} \left(b_{2j} + \bar{b}_{2j}\right)\left[\beta_i\mathbb{E}[x] + \gamma_i\right]\mathbb{E}[u_j]\mathrm{d}t$$

$$+ \mathbb{E}\int_0^T \sum_{j \in \mathcal{N}\setminus\{i\}} b_{2j}\left(u_j - \mathbb{E}[u_j]\right)\alpha_i\left(x - \mathbb{E}[x]\right)\mathrm{d}t$$

$$+ \mathbb{E}\int_0^T \left(b_{2i} + \bar{b}_{2i}\right)\left[\beta_i\mathbb{E}[x] + \gamma_i\right]\mathbb{E}[u_i]\mathrm{d}t$$

$$+ \mathbb{E}\int_0^T b_{2i}\left(u_i - \mathbb{E}[u_i]\right)\alpha_i\left(x - \mathbb{E}[x]\right)\mathrm{d}t.$$

Remember that $f_i(0,x_0)$ denotes the guess functional for the optimal cost for the time interval $[0,T]$. Then, it is desired to find the optimal control inputs such that it matches with the actual cost $L_i(x,u_1,\ldots,u_n)$. To do this, the following term requires the computation of the gap $\mathbb{E}[L_i(x,u_1,\ldots,u_n) - f_i(0,x_0)]$ as follows:

$$\mathbb{E}[L_i(x,u_1,\ldots,u_n) - f_i(0,x_0)] = \frac{1}{2}q_i(T)\mathbb{E}[(x(T) - \mathbb{E}[x(T)])^2]$$

$$+ \frac{1}{2}\left(q_i(T) + \bar{q}_i(T)\right)\mathbb{E}[x(T)]^2 - \frac{1}{2}\alpha_i(T)\mathbb{E}\left[x(T) - \mathbb{E}[x(T)]\right]^2$$

$$- \frac{1}{2}\beta_i(T)\mathbb{E}[x(T)]^2 - \gamma_i(T)\mathbb{E}[x(T)] - \delta_i(T)$$

$$+ \frac{1}{2}\mathbb{E}\int_0^T (x - \mathbb{E}[x])^2 \left[\dot{\alpha}_i + 2b_1\alpha_i + q_i\right]\mathrm{d}t$$

$$+ \frac{1}{2}\mathbb{E}\int_0^T \mathbb{E}[x]^2 \left[\dot{\beta}_i + 2\left(b_1 + \bar{b}_1\right)\beta_i + q_i + \bar{q}_i\right]\mathrm{d}t$$

$$+ \mathbb{E} \int_0^T \mathbb{E}[x] \left[b_0 \beta_i + \dot{\gamma}_i + \left(b_1 + \bar{b}_1 \right) \gamma_i \right] dt$$

$$+ \mathbb{E} \int_0^T \left[\dot{\delta}_i + b_0 \gamma_i + \frac{1}{2} \alpha_i \sigma_0^2 \right] dt$$

$$+ \mathbb{E} \int_0^T \sum_{j \in \mathcal{N} \setminus \{i\}} \left(b_{2j} + \bar{b}_{2j} \right) \left[\beta_i \mathbb{E}[x] + \gamma_i \right] \mathbb{E}[u_j] dt$$

$$+ \mathbb{E} \int_0^T \sum_{j \in \mathcal{N} \setminus \{i\}} b_{2j} \left(u_j - \mathbb{E}[u_j] \right) \alpha_i \left(x - \mathbb{E}[x] \right) dt$$

$$+ \frac{1}{2} \mathbb{E} \int_0^T \left(r_i + \bar{r}_i \right) \left[\mathbb{E}[u_i]^2 + 2 \frac{\left(b_{2i} + \bar{b}_{2i} \right)}{\left(r_i + \bar{r}_i \right)} \left[\beta_i \mathbb{E}[x] + \gamma_i \right] \mathbb{E}[u_i] \right] dt$$

$$+ \frac{1}{2} \mathbb{E} \int_0^T r_i \left[\left(u_i - \mathbb{E}[u_i] \right)^2 + 2 \frac{b_{2i}}{r_i} \left(u_i - \mathbb{E}[u_i] \right) \alpha_i \left(x - \mathbb{E}[x] \right) \right] dt. \tag{4.11}$$

We observe in (4.11) that quadratic system state, mean-field system state, control input, and mean-field control input terms appear in the gap $\mathbb{E}[L_i(x, u_1, \ldots, u_n)] - f_i(0, x_0)]$ coming from the cost functional in (4.2). Now, we perform optimization over the decision variables of the ith decision-maker. To do this, we make a square completion for the terms involving both $\mathbb{E}[u_i]$ and $\left(u_i - \mathbb{E}[u_i] \right)^2$ as follows:

$$\left(\mathbb{E}[u_i] + \frac{\left(b_{2i} + \bar{b}_{2i} \right)}{\left(r_i + \bar{r}_i \right)} \left[\beta_i \mathbb{E}[x] + \gamma_i \right] \right)^2 = \mathbb{E}[u_i]^2$$

$$+ 2 \frac{\left(b_{2i} + \bar{b}_{2i} \right)}{\left(r_i + \bar{r}_i \right)} \left[\beta_i \mathbb{E}[x] + \gamma_i \right] \mathbb{E}[u_i]$$

$$+ \frac{\left(b_{2i} + \bar{b}_{2i} \right)^2}{\left(r_i + \bar{r}_i \right)^2} \left[\beta_i^2 \mathbb{E}[x]^2 + 2 \beta_i \gamma_i \mathbb{E}[x] + \gamma_i^2 \right], \tag{4.12}$$

and

$$\left(u_i - \mathbb{E}[u_i] + \frac{b_{2i}}{r_i} \alpha_i \left(x - \mathbb{E}[x] \right) \right)^2 = \left(u_i - \mathbb{E}[u_i] \right)^2$$

$$+ 2 \frac{b_{2i}}{r_i} \left(u_i - \mathbb{E}[u_i] \right) \alpha_i \left(x - \mathbb{E}[x] \right)$$

$$+ \frac{b_{2i}^2}{r_i^2} \alpha_i^2 \left(x - \mathbb{E}[x] \right)^2 . \tag{4.13}$$

At this point, one can already deduce the optimal control input for the ith decision-maker, which is given by the value such that the quadratic term is *minimized. For example,*

$$\left(\mathbb{E}[u_i] + \frac{(b_{2i} + \bar{b}_{2i})}{(r_i + \bar{r}_i)}\left[\beta_i\mathbb{E}[x] + \gamma_i\right]\right)^2$$

is minimized with

$$\mathbb{E}[u_i] = -\frac{(b_{2i} + \bar{b}_{2i})}{(r_i + \bar{r}_i)}\left[\beta_i\mathbb{E}[x] + \gamma_i\right].$$

We will see this deduction in the identification process at the last step of the proof. Thus, we continue by replacing the square completion in (4.12) *and* (4.13) *in* (4.11) *one obtains*

$$\mathbb{E}[L_i(x, u_1, \ldots, u_n) - f_i(0, x_0)] = \frac{1}{2}q_i(T)\mathbb{E}[(x(T) - \mathbb{E}[x(T)])^2]$$

$$+ \frac{1}{2}\left(q_i(T) + \bar{q}_i(T)\right)\mathbb{E}[x(T)]^2 - \frac{1}{2}\alpha_i(T)\mathbb{E}\left[x(T) - \mathbb{E}[x(T)]\right]^2$$

$$- \frac{1}{2}\beta_i(T)\mathbb{E}[x(T)]^2 - \gamma_i(T)\mathbb{E}[x(T)] - \delta_i(T)$$

$$+ \frac{1}{2}\mathbb{E}\int_0^T (x - \mathbb{E}[x])^2\left[\dot{\alpha}_i + 2b_1\alpha_i + q_i\right]dt$$

$$+ \frac{1}{2}\mathbb{E}\int_0^T \mathbb{E}[x]^2\left[\dot{\beta}_i + 2\left(b_1 + \bar{b}_1\right)\beta_i + q_i + \bar{q}_i\right]dt$$

$$+ \mathbb{E}\int_0^T \mathbb{E}[x]\left[b_0\beta_i + \dot{\gamma}_i + \left(b_1 + \bar{b}_1\right)\gamma_i\right]dt$$

$$+ \mathbb{E}\int_0^T \left[\dot{\delta}_i + b_0\gamma_i + \frac{1}{2}\alpha_i\sigma_0^2\right]dt$$

$$- \mathbb{E}\int_0^T \sum_{j\in\mathcal{N}\setminus\{i\}} \frac{(b_{2j} + \bar{b}_{2j})^2}{(r_j + \bar{r}_j)}\beta_i\beta_j\mathbb{E}[x]^2dt$$

$$- \mathbb{E}\int_0^T \sum_{j\in\mathcal{N}\setminus\{i\}} \frac{(b_{2j} + \bar{b}_{2j})^2}{(r_j + \bar{r}_j)}\left[\beta_i\gamma_j + \gamma_i\beta_j\right]\mathbb{E}[x]dt$$

$$- \mathbb{E}\int_0^T \sum_{j\in\mathcal{N}\setminus\{i\}} \frac{(b_{2j} + \bar{b}_{2j})^2}{(r_j + \bar{r}_j)}\gamma_i\gamma_j dt$$

$$- \mathbb{E}\int_0^T \sum_{j\in\mathcal{N}\setminus\{i\}} \frac{b_{2j}^2}{r_j}\alpha_j\alpha_i(x - \mathbb{E}[x])^2 dt$$

$$- \frac{1}{2}\mathbb{E}\int_0^T \frac{(b_{2i} + \bar{b}_{2i})^2}{(r_i + \bar{r}_i)}\beta_i^2\mathbb{E}[x]^2dt$$

$$- \frac{1}{2}\mathbb{E}\int_0^T \frac{(b_{2i} + \bar{b}_{2i})^2}{(r_i + \bar{r}_i)}2\beta_i\gamma_i\mathbb{E}[x]dt$$

$$-\frac{1}{2}\mathbb{E}\int_0^T \frac{\left(b_{2i} + \bar{b}_{2i}\right)^2}{(r_i + \bar{r}_i)}\gamma_i^2 \mathrm{d}t - \frac{1}{2}\mathbb{E}\int_0^T \frac{b_{2i}^2}{r_i}\alpha_i^2\left(x - \mathbb{E}[x]\right)^2 \mathrm{d}t$$

$$+\frac{1}{2}\mathbb{E}\int_0^T r_i\left(u_i - \mathbb{E}[u_i] + \frac{b_{2i}}{r_i}\alpha_i\left(x - \mathbb{E}[x]\right)\right)^2 \mathrm{d}t$$

$$+\frac{1}{2}\mathbb{E}\int_0^T (r_i + \bar{r}_i)\left(\mathbb{E}[u_i] + \frac{\left(b_{2i} + \bar{b}_{2i}\right)}{(r_i + \bar{r}_i)}\left[\beta_i\mathbb{E}[x] + \gamma_i\right]\right)^2 \mathrm{d}t.$$

$$(4.14)$$

Once again, we organize the whole expression by grouping terms according to the common factors depending on the state and mean-field state terms. Grouping terms in (4.14) one arrives at the following:

$$\mathbb{E}[L_i(x, u_1, \ldots, u_n)] - f_i(0, x_0)] = \frac{1}{2}q_i(T)\mathbb{E}[(x(T) - \mathbb{E}[x(T)])^2]$$

$$+\frac{1}{2}\left(q_i(T) + \bar{q}_i(T)\right)\mathbb{E}[x(T)]^2 - \frac{1}{2}\alpha_i(T)\mathbb{E}\left[x(T) - \mathbb{E}[x(T)]\right]^2$$

$$-\frac{1}{2}\beta_i(T)\mathbb{E}[x(T)]^2 - \gamma_i(T)\mathbb{E}[x(T)] - \delta_i(T)$$

$$+\frac{1}{2}\mathbb{E}\int_0^T (x - \mathbb{E}[x])^2\left[\dot{\alpha}_i + 2b_1\alpha_i + q_i\right.$$

$$-2\alpha_i\sum_{j\in\mathcal{N}\backslash\{i\}}\frac{b_{2j}^2}{r_j}\alpha_j - \frac{b_{2i}^2}{r_i}\alpha_i^2\bigg]\mathrm{d}t$$

$$+\frac{1}{2}\mathbb{E}\int_0^T \mathbb{E}[x]^2\left[\dot{\beta}_i + 2\left(b_1 + \bar{b}_1\right)\beta_i + q_i + \bar{q}_i\right.$$

$$-2\beta_i\sum_{j\in\mathcal{N}\backslash\{i\}}\frac{\left(b_{2j} + \bar{b}_{2j}\right)^2}{(r_j + \bar{r}_j)}\beta_j - \frac{\left(b_{2i} + \bar{b}_{2i}\right)^2}{(r_i + \bar{r}_i)}\beta_i^2\bigg]\mathrm{d}t$$

$$+\mathbb{E}\int_0^T \mathbb{E}[x]\left[b_0\beta_i + \dot{\gamma}_i + \left(b_1 + \bar{b}_1\right)\gamma_i\right.$$

$$-\beta_i\sum_{j\in\mathcal{N}\backslash\{i\}}\frac{\left(b_{2j} + \bar{b}_{2j}\right)^2}{(r_j + \bar{r}_j)}\gamma_j$$

$$-\gamma_i\sum_{j\in\mathcal{N}\backslash\{i\}}\frac{\left(b_{2j} + \bar{b}_{2j}\right)^2}{(r_j + \bar{r}_j)}\beta_j - \frac{\left(b_{2i} + \bar{b}_{2i}\right)^2}{(r_i + \bar{r}_i)}\beta_i\gamma_i\bigg]\mathrm{d}t$$

$$+\mathbb{E}\int_0^T \left[\dot{\delta}_i + b_0\gamma_i + \frac{1}{2}\alpha_i\sigma_0^2 - \gamma_i\sum_{j\in\mathcal{N}\backslash\{i\}}\frac{\left(b_{2j} + \bar{b}_{2j}\right)^2}{(r_j + \bar{r}_j)}\gamma_j\right.$$

$$-\frac{1}{2}\frac{\left(b_{2i} + \bar{b}_{2i}\right)^2}{(r_i + \bar{r}_i)}\gamma_i^2\bigg]\mathrm{d}t$$

$$+ \frac{1}{2}\mathbb{E}\int_0^T r_i\left(u_i - \mathbb{E}[u_i] + \frac{b_{2i}}{r_i}\alpha_i\left(x - \mathbb{E}[x]\right)\right)^2 dt$$

$$+ \frac{1}{2}\mathbb{E}\int_0^T (r_i + \bar{r}_i)\left(\mathbb{E}[u_i] + \frac{(b_{2i} + \bar{b}_{2i})}{(r_i + \bar{r}_i)}\left[\beta_i\mathbb{E}[x] + \gamma_i\right]\right)^2 dt.$$

Let us divide this long equation into seven main groups:

- *Those terms corresponding to the terminal time, which define the boundary condition for the differential equations α_i, β_i, γ_i and δ_i*

- *Those terms whose common factor is $(x - \mathbb{E}[x])^2$, which define the α_i equation*

- *Those terms whose common factor is $\mathbb{E}[x]^2$, which define the β_i equation*

- *Those terms whose common factor is $\mathbb{E}[x]$, which define the γ_i equation*

- *Those terms independent from the system state, mean-field state, control inputs and mean-field control inputs. These terms define the δ_i equation*

- *Those terms depending on the decision variables for the ith decision-maker, i.e., $u_i - \mathbb{E}[u_i]$*

- *Those terms in function of the mean-field control input $\mathbb{E}[u_i]$*

Finally, the announced result is obtained by minimizing the terms completing the proof.

On the other hand, the solution for the cooperative approach, which coincides with a mean-field-type control problem is presented next in Proposition 14. This implies that, after studying the coming section, the reader will also learn how to solve mean-field-type control problems. Indeed, it is possible to reduce the number of decision-makers to a unique one.

Proposition 14 *The optimal feedback-strategies of the cooperative problem 14, and the optimal cost are given by:*

$$u_i^* = \mathbb{E}[u_i^*] - \frac{b_{2i}}{r_i}\alpha_0\left(x - \mathbb{E}[x]\right), \ \forall i \in \mathcal{N},$$

$$\mathbb{E}[u_i^*] - -\frac{(b_{2i} + \bar{b}_{2i})}{(r_i + \bar{r}_i)}\left[\beta_0\mathbb{E}[x] + \gamma_0\right], \ \forall i \in \mathcal{N},$$

$$L_0(x, u_1^*, \ldots, u_n^*) = \frac{1}{2}\alpha_0(0)\left(x_0 - \mathbb{E}[x_0]\right)^2$$

$$+ \frac{1}{2}\beta_0(0)\mathbb{E}[x_0]^2 + \gamma_0(0)\mathbb{E}[x_0] + \delta_0(0), \ \forall i \in \mathcal{N},$$

where α_0, β_0, γ_0, and δ_0 solve the following differential equations:

$$\dot{\alpha}_0 = -2b_1\alpha_0 - q_0 + \sum_{i\in\mathcal{N}} \frac{b_{2i}^2}{r_i}\alpha_0^2,$$

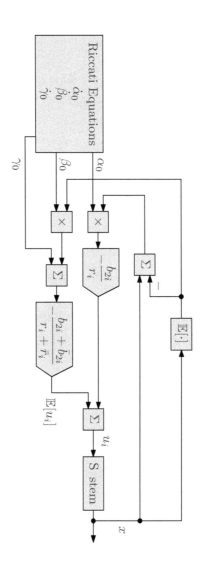

FIGURE 4.4
Feedback scheme for the linear-quadratic mean-field-type cooperative game
problem.

$$\dot{\beta}_0 = -2\left(b_1 + \bar{b}_1\right)\beta_0 - q_0 - \bar{q}_0 + \sum_{i \in \mathcal{N}} \frac{\left(b_{2i} + \bar{b}_{2i}\right)^2}{\left(r_i + \bar{r}_i\right)}\beta_0^2,$$

$$\dot{\gamma}_0 = -b_0\beta_0 - \left(b_1 + \bar{b}_1\right)\gamma_0 + \sum_{i \in \mathcal{N}} \frac{\left(b_{2i} + \bar{b}_{2i}\right)^2}{\left(r_i + \bar{r}_i\right)}\beta_0\gamma_0,$$

$$\dot{\delta}_0 = -b_0\gamma_0 - \frac{1}{2}\alpha_0\sigma_0^2 + \frac{1}{2}\sum_{i \in \mathcal{N}} \frac{\left(b_{2i} + \bar{b}_{2i}\right)^2}{\left(r_i + \bar{r}_i\right)}\gamma_0^2,$$

with terminal boundary conditions given by

$$\alpha_0(T) = q_0(T),$$
$$\beta_0(T) = q_0(T) + \bar{q}_0(T),$$
$$\gamma_0(T) = \delta_0(T) = 0,$$

whenever these equations admit a solution with positive α_0 and β_0.

Similar to the non-cooperative case, we observe in the semi-explicit solution that the optimal control inputs are independent from the term δ. However, the δ term allows computing the optimal cost. Next remark highlights conditions corresponding to the Riccati equations α_0 and β_0.

Remark 15 *Notice that*

$$\dot{\alpha}_0 = -2b_1\alpha_0 - q_0 + \sum_{i \in \mathcal{N}} \frac{b_{2i}^2}{r_i}\alpha_0^2,$$

with $\alpha_0(T) = q_0(T)$, admits a unique positive solution if

$$q_0(T), q_0(t) \geq 0, \quad and$$
$$\sum_{i \in \mathcal{N}} \frac{[b_{2i}(t)]^2}{r_i(t)} \geq \varepsilon > 0, \quad \forall t \in (0,T).$$

The same holds for β_0 with

$$q_0(T) + \bar{q}_0(T), q_0(t) + \bar{q}_0(t) \geq 0, \quad and$$
$$\sum_{i \in \mathcal{N}} \frac{\left(b_{2i}(t) + \bar{b}_{2i}(t)\right)^2}{\left(r_i(t) + r_i(t)\right)} > \varepsilon > 0,$$

for all time $t \in (0,T)$.

Figure 4.4 shows the feedback scheme for the cooperative mean-field-type game problem according to Proposition 14. It can be seen that the schemes for both the non-cooperative and cooperative scenarios are the same. However, the ordinary differential equations are different and the optimal trajectories differ from each other. We now compute the semi-explicit solution for the cooperative game problem.

Proof 14 (Proposition 14) *All the cost functionals for all the decision-makers have been summed up. Then, it is reasonable to solve the problem by using a similar guess functional as the one considered in the non-cooperative case. It turns out that the same structure for the ansatz for the optimal cost is suitable. Thus, inspired by the structure of the cost function, we propose a quadratic guess functional of the following form:*

$$f_0(t,x) = \frac{1}{2}\alpha_0 (x - \mathbb{E}[x])^2 + \frac{1}{2}\beta_0\mathbb{E}[x]^2 + \gamma_0\mathbb{E}[x] + \delta_0, \qquad (4.15)$$

where $\alpha_0, \beta_0, \gamma_0$, and δ_0 are deterministic functions of time. Following the same procedure as in the proof of Proposition 13, one obtains the following expression for the gap $\mathbb{E}[L_0(x,u_1,\ldots,u_n) - f_0(0,x_0)]$:

$$\mathbb{E}[L_0(x,u_1,\ldots,u_n) - f_0(0,x_0)] = \frac{1}{2}q_0(T)\mathbb{E}[(x(T) - \mathbb{E}[x(T)])^2]$$
$$+ \frac{1}{2}\left(q_0(T) + \bar{q}_0(T)\right)\mathbb{E}[x(T)]^2 - \frac{1}{2}\alpha_0(T)\mathbb{E}\left[x(T) - \mathbb{E}[x(T)]\right]^2$$
$$- \frac{1}{2}\beta_0(T)\mathbb{E}[x(T)]^2 - \gamma_0(T)\mathbb{E}[x(T)] - \delta_0(T)$$
$$+ \frac{1}{2}\mathbb{E}\int_0^T (x - \mathbb{E}[x])^2\left[\dot{\alpha}_0 + 2b_1\alpha_0 + q_i\right]dt$$
$$+ \frac{1}{2}\mathbb{E}\int_0^T \mathbb{E}[x]^2\left[\dot{\beta}_0 + 2\left(b_1 + \bar{b}_1\right)\beta_0 + q_0 + \bar{q}_0\right]dt$$
$$+ \mathbb{E}\int_0^T \mathbb{E}[x]\left[b_0\beta_0 + \dot{\gamma}_0 + \left(b_1 + \bar{b}_1\right)\gamma_0\right]dt + \mathbb{E}\int_0^T \left[\dot{\delta}_0 + b_0\gamma_0 + \frac{1}{2}\alpha_0\sigma_0^2\right]dt$$
$$+ \frac{1}{2}\mathbb{E}\int_0^T \sum_{j\in\mathcal{N}}(r_j + \bar{r}_j)\left[\mathbb{E}[u_j]^2 + 2\frac{(b_{2j} + \bar{b}_{2j})}{(r_j + \bar{r}_j)}\left[\beta_0\mathbb{E}[x] + \gamma_0\right]\mathbb{E}[u_j]\right]dt$$
$$+ \frac{1}{2}\mathbb{E}\int_0^T \sum_{j\in\mathcal{N}}r_j\left[(u_j - \mathbb{E}[u_j])^2 + 2\frac{b_{2j}}{r_j}(u_j - \mathbb{E}[u_j])\alpha_0(x - \mathbb{E}[x])\right]dt.$$
$$(4.16)$$

Note that we have simplified some details given that this procedure is quite similar to the one we did for the non-cooperative scenario. It follows that replacing the same square completion terms in (4.12) and (4.13) yields

$$\mathbb{E}[L_0(x,u_1,\ldots,u_n) - f_0(0,x_0)] = \frac{1}{2}q_0(T)\mathbb{E}[(x(T) - \mathbb{E}[x(T)])^2]$$
$$+ \frac{1}{2}\left(q_0(T) + \bar{q}_0(T)\right)\mathbb{E}[x(T)]^2 - \frac{1}{2}\alpha_0(T)\mathbb{E}\left[x(T) - \mathbb{E}[x(T)]\right]^2$$
$$- \frac{1}{2}\beta_0(T)\mathbb{E}[x(T)]^2 - \gamma_0(T)\mathbb{E}[x(T)] - \delta_0(T)$$
$$+ \frac{1}{2}\mathbb{E}\int_0^T (x - \mathbb{E}[x])^2\left[\dot{\alpha}_0 + 2b_1\alpha_0 + q_0\right]dt$$

$$+ \frac{1}{2}\mathbb{E}\int_0^T \mathbb{E}[x]^2 \Big[\dot{\beta}_0 + 2\left(b_1 + \bar{b}_1\right)\beta_0 + q_0 + \bar{q}_0\Big]dt$$

$$+ \mathbb{E}\int_0^T \mathbb{E}[x]\Big[b_0\beta_0 + \dot{\gamma}_0 + \left(b_1 + \bar{b}_1\right)\gamma_0\Big]dt$$

$$+ \mathbb{E}\int_0^T \Big[\dot{\delta}_0 + b_0\gamma_0 + \frac{1}{2}\alpha_0\sigma_0^2\Big]dt$$

$$+ \frac{1}{2}\mathbb{E}\int_0^T \sum_{j\in\mathcal{N}}(r_j + \bar{r}_j)\left(\mathbb{E}[u_j] + \frac{(b_{2j} + \bar{b}_{2j})}{(r_j + \bar{r}_j)}\Big[\beta_0\mathbb{E}[x] + \gamma_0\Big]\right)^2 dt$$

$$- \frac{1}{2}\mathbb{E}\int_0^T \sum_{j\in\mathcal{N}}\frac{(b_{2j} + \bar{b}_{2j})^2}{(r_j + \bar{r}_j)}\Big[\beta_0^2\mathbb{E}[x]^2 + 2\beta_0\gamma_0\mathbb{E}[x] + \gamma_0^2\Big]dt$$

$$+ \frac{1}{2}\mathbb{E}\int_0^T \sum_{j\in\mathcal{N}}r_j\left(u_j - \mathbb{E}[u_j] + \frac{b_{2j}}{r_j}\alpha_0\left(x - \mathbb{E}[x]\right)\right)^2 dt$$

$$- \frac{1}{2}\mathbb{E}\int_0^T \sum_{j\in\mathcal{N}}\frac{b_{2j}^2}{r_j}\alpha_0^2\left(x - \mathbb{E}[x]\right)^2 dt.$$

Grouping terms yields

$$\mathbb{E}[L_0(x, u_1, \ldots, u_n) - f_0(0, x_0)] = \frac{1}{2}q_0(T)\mathbb{E}[(x(T) - \mathbb{E}[x(T)])^2]$$

$$+ \frac{1}{2}\left(q_0(T) + \bar{q}_0(T)\right)\mathbb{E}[x(T)]^2 - \frac{1}{2}\alpha_0(T)\mathbb{E}\left[x(T) - \mathbb{E}[x(T)]\right]^2$$

$$- \frac{1}{2}\beta_0(T)\mathbb{E}[x(T)]^2 - \gamma_0(T)\mathbb{E}[x(T)] - \delta_0(T)$$

$$+ \frac{1}{2}\mathbb{E}\int_0^T \left(x - \mathbb{E}[x]\right)^2\Big[\dot{\alpha}_0 + 2b_1\alpha_0 + q_0 - \sum_{j\in\mathcal{N}}\frac{b_{2j}^2}{r_j}\alpha_0^2\Big]dt$$

$$+ \frac{1}{2}\mathbb{E}\int_0^T \mathbb{E}[x]^2\Bigg[\dot{\beta}_0 + 2\left(b_1 + \bar{b}_1\right)\beta_0 + q_0 + \bar{q}_0$$

$$- \sum_{j\in\mathcal{N}}\frac{(b_{2j} + \bar{b}_{2j})^2}{(r_j + \bar{r}_j)}\beta_0^2\Bigg]dt$$

$$+ \mathbb{E}\int_0^T \mathbb{E}[x]\Bigg[b_0\beta_0 + \dot{\gamma}_0 + \left(b_1 + \bar{b}_1\right)\gamma_0$$

$$- \sum_{j\in\mathcal{N}}\frac{(b_{2j} + \bar{b}_{2j})^2}{(r_j + \bar{r}_j)}\beta_0\gamma_0\Bigg]dt$$

$$+ \mathbb{E}\int_0^T \Big[\dot{\delta}_0 + b_0\gamma_0 + \frac{1}{2}\alpha_0\sigma_0^2 - \frac{1}{2}\sum_{j\in\mathcal{N}}\frac{(b_{2j} + \bar{b}_{2j})^2}{(r_j + \bar{r}_j)}\gamma_0^2\Big]dt$$

$$+ \frac{1}{2}\mathbb{E}\int_0^T \sum_{j\in\mathcal{N}}(r_j + \bar{r}_j)\left(\mathbb{E}[u_j] + \frac{(b_{2j}+\bar{b}_{2j})}{(r_j+\bar{r}_j)}\Big[\beta_0\mathbb{E}[x]+\gamma_0\Big]\right)^2 dt$$

$$+ \frac{1}{2}\mathbb{E}\int_0^T \sum_{j\in\mathcal{N}}r_j\left(u_j - \mathbb{E}[u_j] + \frac{b_{2j}}{r_j}\alpha_0(x - \mathbb{E}[x])\right)^2 dt.$$

The latter equation is then divided into seven groups:

- *Those terms at the terminal time T, which define the boundary conditions for the differential equations $\alpha_0, \beta_0, \gamma_0$, and δ_0*

- *Those terms with common factor $(x - \mathbb{E}[x])^2$. These terms will define the α_0 differential equation*

- *Those terms with common factor $\mathbb{E}[x]^2$. These terms will define the β_0 differential equation*

- *Those terms with common factor $\mathbb{E}[x]$. These terms will define the γ_0 differential equation*

- *Those terms independent from the state, mean-field state, control input, and mean-field control input terms. These terms will define the δ_0 differential equation*

- *And those two final groups comprising terms depending on the decision variables of the decision-makers.*

Finally, the announced result is obtained by minimizing the terms completing the proof.

In the following section, we show some numerical examples in order to illustrate the performance of the previously presented results and semi-explicit solutions. The simulations have been computed following the same approach as in [89] using a toolbox available together with this book.

4.3   Numerical Examples

We present some examples and numerical results for both non-cooperative and fully-cooperative mean-field-type games.

Example 1: Non-cooperative Scenario

Let us consider a three-player mean-field-type game with terminal time $T = 5\,\text{s}$ involving the following system state dynamics:

$$dx = \left(2 - \frac{3}{2}x - \frac{4}{3}\mathbb{E}[x] - 2u_1 + 4u_2 - 3u_3 + 3\mathbb{E}[u_1] - 5\mathbb{E}[u_2] + 2\mathbb{E}[u_3]\right)dt$$

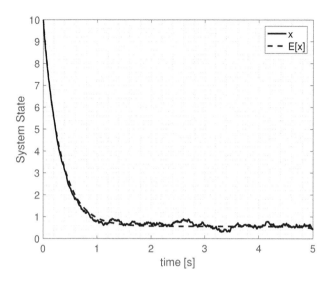

FIGURE 4.5
Evolution of the system state and its expectation for the scalar-value non-cooperative scenario.

$+\, 5\mathrm{d}B,$

and the following cost functions for each player:

$$L_1(x, u_1, u_2, u_3) = x(5)^2 + \frac{3}{4}\mathbb{E}[x(5)]^2$$

$$+ \frac{1}{2}\int_0^5 \left(x^2 + 3\mathbb{E}[x]^2 + 3u_1^2 + 4\mathbb{E}[u_1]^2\right) \mathrm{d}t,$$

$$L_2(x, u_1, u_2, u_3) = \frac{3}{2}x(5)^2 + \mathbb{E}[x(5)]^2$$

$$+ \frac{1}{2}\int_0^5 \left(2x^2 + 4\mathbb{E}[x]^2 + 2u_2^2 + 3\mathbb{E}[u_2]^2\right) \mathrm{d}t,$$

$$L_3(x, u_1, u_2, u_3) = 2x(5)^2 + \frac{5}{4}\mathbb{E}[x(5)]^2$$

$$+ \frac{1}{2}\int_0^5 \left(3x^2 + \mathbb{E}[x]^2 + 4u_3^2 + 5\mathbb{E}[u_3]^2\right) \mathrm{d}t.$$

Finally, consider the following initial conditions $x_0 = \mathbb{E}[x_0] = 10$. Applying Proposition 13, we obtain the results shown next.

Figure 4.5 presents the evolution of the system state x and its expectation $\mathbb{E}[x]$. It can be seen that the state is minimized according to the considered cost functions $L_1(x, u_1, u_2, u_3), \ldots, L_3(x, u_1, u_2, u_3)$.

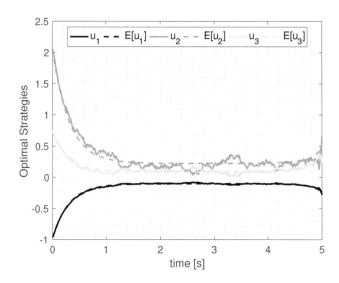

FIGURE 4.6
Evolution of the players' strategies and their expectation for the scalar-value non-cooperative scenario.

Figure 4.6 shows the evolution of the optimal players' strategies given by

$$u_1^* = -\frac{1}{7}\left[\beta_1\mathbb{E}[x] + \gamma_1\right] + \frac{2}{3}\alpha_1\left(x - \mathbb{E}[x]\right),$$

$$u_2^* = \frac{1}{5}\left[\beta_2\mathbb{E}[x] + \gamma_2\right] - 2\alpha_2\left(x - \mathbb{E}[x]\right),$$

$$u_3^* = \frac{1}{9}\left[\beta_3\mathbb{E}[x] + \gamma_3\right] + \frac{3}{4}\alpha_3\left(x - \mathbb{E}[x]\right).$$

We observe that, $\arg\max_{i \in \mathcal{N}}(u_i^2)$ is given by the second player due to the weights r_2 and $\bar{r}_2$ assigns less prioritization in $L_2(x, u_1, u_2, u_3)$ than $r_1, \bar{r}_1, r_3$, and $\bar{r}_3$ in $L_1(x, u_1, u_2, u_3)$ and $L_3(x, u_1, u_2, u_3)$, respectively.

Figure 4.7 presents the evolution of the Riccati equations $\alpha_1, \ldots, \alpha_3$ given by

$$\dot{\alpha}_1 = 3\alpha_1 - 1 + 2\alpha_1\left(8\alpha_2 + \frac{9}{4}\alpha_3\right) + \frac{4}{3}\alpha_1^2,$$

$$\dot{\alpha}_2 = 3\alpha_2 - 2 + 2\alpha_2\left(\frac{4}{3}\alpha_1 + \frac{9}{4}\alpha_3\right) + 8\alpha_2^2,$$

$$\dot{\alpha}_3 = 3\alpha_3 - 3 + 2\alpha_3\left(\frac{4}{3}\alpha_1 + 8\alpha_2^2\right) + \frac{9}{4}\alpha_3^2,$$

and it can be verified that the terminal boundary conditions are satisfied, i.e.,

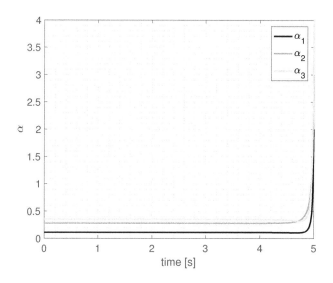

FIGURE 4.7
Evolution of the Riccati equations $\alpha_1, \ldots, \alpha_3$ for the scalar-value non-cooperative scenario.

$\alpha_1(5) = 2$, $\alpha_2(5) = 3$, and $\alpha_3(5) = 4$. On the other hand, Figure 4.8 presents the Riccati equations β_1, β_2, and β_3, which are as follows:

$$\dot{\beta}_1 = \frac{17}{3}\beta_1 - q_1 - \bar{q}_1 + 2\beta_1\left(\frac{1}{5}\beta_2 + \frac{1}{9}\beta_3\right) + \frac{1}{7}\beta_i^2,$$

$$\dot{\beta}_2 = \frac{17}{3}\beta_2 - q_2 - \bar{q}_2 + 2\beta_2\left(\frac{1}{7}\beta_1 + \frac{1}{9}\beta_3\right) + \frac{1}{5}\beta_i^2,$$

$$\dot{\beta}_3 = \frac{17}{3}\beta_3 - q_3 - \bar{q}_3 + 2\beta_3\left(\frac{1}{7}\beta_1 + \frac{1}{5}\beta_2\right) + \frac{1}{9}\beta_i^2.$$

Notice that the boundary conditions $\beta_1(5) = \frac{7}{2}$, $\beta_2(5) = 5$, and $\beta_3(5) = \frac{13}{2}$ are also satisfied. Figure 4.9 shows the evolution of the Riccati equations γ_1, γ_2, and γ_3, which are as follows:

$$\dot{\gamma}_1 = -2\beta_1 + \frac{17}{6}\gamma_1 + \beta_1\left(\frac{1}{5}\gamma_2 + \frac{1}{9}\gamma_3\right) + \gamma_1\left(\frac{1}{5}\beta_2 + \frac{1}{9}\beta_3\right) + \frac{1}{7}\beta_1\gamma_1,$$

$$\dot{\gamma}_2 = -2\beta_2 + \frac{17}{6}\gamma_2 + \beta_2\left(\frac{1}{7}\gamma_1 + \frac{1}{9}\gamma_3\right) + \gamma_2\left(\frac{1}{7}\beta_1 + \frac{1}{9}\beta_3\right) + \frac{1}{5}\beta_2\gamma_2,$$

$$\dot{\gamma}_3 = -2\beta_3 + \frac{17}{6}\gamma_3 + \beta_3\left(\frac{1}{7}\gamma_1 + \frac{1}{5}\gamma_2\right) + \gamma_3\left(\frac{1}{7}\beta_1 + \frac{1}{5}\beta_2\right) + \frac{1}{9}\beta_3\gamma_3,$$

and with the boundary conditions $\gamma_1(5) = \gamma_2(5) = \gamma_3(5) = 0$.

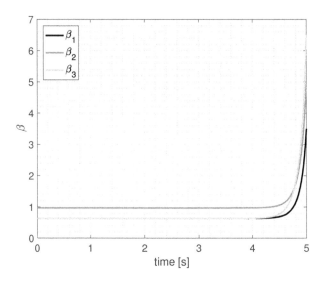

FIGURE 4.8
Evolution of the Riccati equations $\beta_1, \ldots, \beta_3$ for the scalar-value non-cooperative scenario.

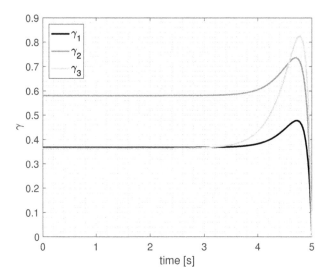

FIGURE 4.9
Evolution of the Riccati equations $\gamma_1, \ldots, \gamma_3$ for the scalar-value non-cooperative scenario.

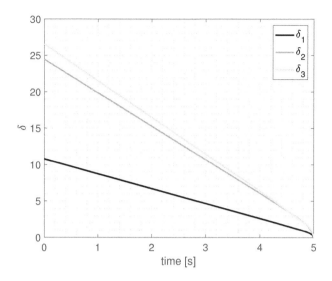

FIGURE 4.10
Evolution of the Riccati equations $\delta_1, \ldots, \delta_3$ for the scalar-value non-cooperative scenario.

Hence, the last Riccati equations δ_1, δ_2, and δ_3 are presented in Figure 4.10. These differential equations are as follows:

$$\dot{\delta}_1 = -2\gamma_1 - \frac{25}{2}\alpha_1 + \gamma_1\left(\frac{1}{5}\gamma_2 + \frac{1}{9}\gamma_3\right) + \frac{1}{14}\gamma_1^2,$$

$$\dot{\delta}_2 = -2\gamma_2 - \frac{25}{2}\alpha_2 + \gamma_2\left(\frac{1}{7}\gamma_1 + \frac{1}{9}\gamma_3\right) + \frac{1}{10}\gamma_2^2,$$

$$\dot{\delta}_3 = -2\gamma_3 - \frac{25}{2}\alpha_3 + \gamma_3\left(\frac{1}{7}\gamma_1 + \frac{1}{5}\gamma_2\right) + \frac{1}{18}\gamma_3^2,$$

and with the boundary conditions $\delta_1(5) = \delta_2(5) = \delta_3(5) = 0$.
We finally present the optimal costs which are given by

$$L_1(x, u_1^*, u_2^*, u_3^*) = L_0(x, u_1^*, u_2^*, u_3^*) = 50\beta_1(0) + 10\gamma_1(0) + \delta_1(0),$$

$$L_2(x, u_1^*, u_2^*, u_3^*) = L_0(x, u_1^*, u_2^*, u_3^*) = 50\beta_2(0) + 10\gamma_2(0) + \delta_2(0),$$

$$L_3(x, u_1^*, u_2^*, u_3^*) = L_0(x, u_1^*, u_2^*, u_3^*) = 50\beta_3(0) + 10\gamma_3(0) + \delta_3(0),$$

and shown in Figure 4.11. In addition, it can be seen from Figure 4.11 that all the players have different costs due to the associated weights for each player and how differently they influence over the system state.

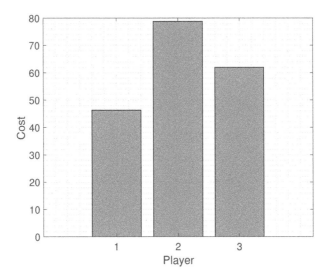

FIGURE 4.11
Optimal cost function for each player for the scalar-value non-cooperative scenario.

Example 2: Fully-cooperative Scenario

Now, we consider the case in which all the players select their strategies and optimize their costs jointly. Let us consider the following system state:

$$
\mathrm{d}x = \left(\frac{1}{2} - 2x + \mathbb{E}[x] - u_1 - 3u_2 + 3u_3 + 2\mathbb{E}[u_1] + 4\mathbb{E}[u_2] - 5\mathbb{E}[u_3] \right) \mathrm{d}t
$$
$$
+ 5\mathrm{d}B,
$$

with initial conditions $x_0 = \mathbb{E}[x_0] = 6$, and whose optimal evolution under the fully-cooperative mean-field-type game problem is presented in Figure 4.12. The cost functional is

$$
L_0(x, u_1, u_2, u_3)
$$
$$
= x(4)^2 + \frac{3}{2}\mathbb{E}[x(4)]^2 + \frac{1}{2}\int_0^4 \left(\frac{5}{2}x^2 + 3\mathbb{E}[x]^2 + u_1^2 + 2\mathbb{E}[u_1]^2 \right) \mathrm{d}t
$$
$$
+ 2x(4)^2 + \frac{5}{2}\mathbb{E}[x(4)]^2 + \frac{1}{2}\int_0^4 \left(\frac{9}{2}x^2 + 5\mathbb{E}[x]^2 + 2u_2^2 + 4\mathbb{E}[u_2]^2 \right) \mathrm{d}t
$$
$$
+ 3x(4)^2 + \frac{7}{2}\mathbb{E}[x(4)]^2 + \frac{1}{2}\int_0^4 \left(\frac{13}{2}x^2 + 7\mathbb{E}[x]^2 + 3u_3^2 + 6\mathbb{E}[u_3]^2 \right) \mathrm{d}t.
$$

The optimal strategies for the players are presented in Figure 4.13, which

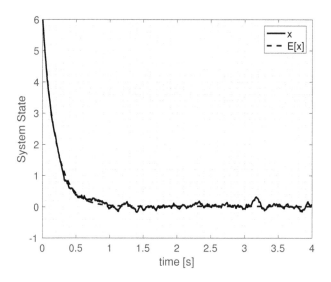

FIGURE 4.12
Evolution of the system state and its expectation for the scalar-value fully-cooperative scenario.

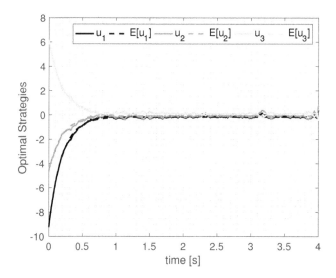

FIGURE 4.13
Evolution of the players' strategies and their expectation for the scalar-value fully-cooperative scenario.

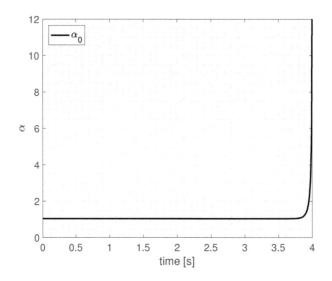

FIGURE 4.14
Evolution of the Riccati equation α_0 for the scalar-value fully-cooperative scenario.

are given by

$$u_1^* = -\frac{1}{3}\left[\beta_0\mathbb{E}[x] + \gamma_0\right] + \alpha_0\left(x - \mathbb{E}[x]\right),$$

$$u_2^* = -\frac{1}{6}\left[\beta_0\mathbb{E}[x] + \gamma_0\right] + \frac{3}{2}\alpha_0\left(x - \mathbb{E}[x]\right),$$

$$u_3^* = \frac{2}{9}\left[\beta_0\mathbb{E}[x] + \gamma_0\right] - \alpha_0\left(x - \mathbb{E}[x]\right),$$

Regarding the Riccati equations, Figure 4.14 shows the evolution of α_0, which is as follows:

$$\dot{\alpha}_0 = 4\alpha_0 - 12 + \left(1 + \frac{9}{2} + 3\right)\alpha_0^2.$$

Besides, it can be verified that the terminal boundary condition is satisfied, i.e., $\alpha_0(4) = 2 + 4 + 6 = 12$. Figure 4.15 shows the evolution of the Riccati equation β_0, which is as follows:

$$\dot{\beta}_0 = 2\beta_0 - 27 + \left(\frac{1}{3} + \frac{1}{6} + \frac{4}{9}\right)\beta_0^2,$$

satisfying its terminal boundary condition $\beta_0(4) = (2+4+6)+(3+5+7) = 27$. Now, the Riccati equations γ_0 and δ_0 are given by

$$\dot{\gamma}_0 = -\frac{1}{2}\beta_0 + \gamma_0 + \left(\frac{1}{3} + \frac{1}{6} + \frac{4}{9}\right)\beta_0\gamma_0,$$

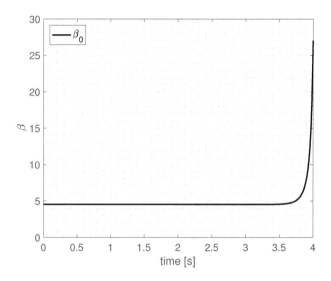

FIGURE 4.15
Evolution of the Riccati equation β_0 for the scalar-value fully-cooperative scenario.

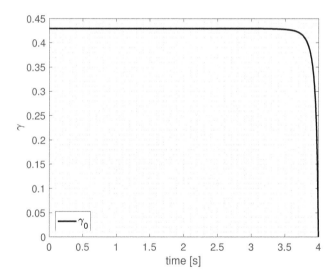

FIGURE 4.16
Evolution of the Riccati equation γ_0 for the scalar-value fully-cooperative scenario.

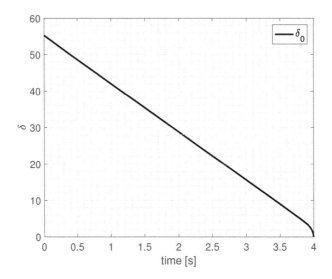

FIGURE 4.17
Evolution of the Riccati equation δ_0 for the scalar-value fully-cooperative scenario.

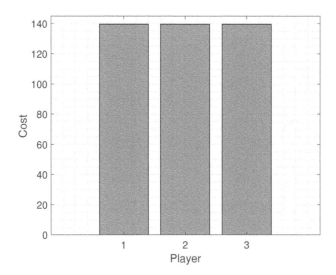

FIGURE 4.18
Cost function associated with each player for the scalar-value fully-cooperative scenario.

$$\dot{\delta}_0 = -\frac{1}{2}\gamma_0 - \frac{25}{2}\alpha_0 + \frac{1}{2}\left(\frac{1}{3} + \frac{1}{6} + \frac{4}{9}\right)\gamma_0^2,$$

and their evolution can be seen in Figures 4.16 and 4.17, respectively; where it can be checked that the boundary condition $\gamma_0(4) = \delta_0(4) = 0$ are satisfied.

Finally, Figure 4.18 shows that all the costs associated with the players are equally since the computed solution corresponds to the fully-cooperative scenario, i.e.,

$$L_i(x, u_1^*, u_2^*, u_3^*) = 18\beta_0(0) + 6\gamma_0(0) + \delta_i(0), \ \forall i \in \{1, 2, 3\}.$$

The model considered in this chapter can be extended by including $x, \mathbb{E}[x], u_j$, and $\mathbb{E}[u_j]$, for all $j \in \mathcal{N}$ in the diffusion as presented in Chapter 6.

4.4 Exercises

1. Solve the following linear-quadratic mean-field-type control problem. Consider the system dynamics

$$dx = \left(b_0 + b_1 x + \bar{b}_1 \mathbb{E}[x] + b_2 u + \bar{b}_2 \mathbb{E}[u]\right) dt + \sigma_0 dB, \quad (4.17)$$
$$x(0) := x_0,$$

and the cost functional

$$L(x, u) = \frac{1}{2}q(T)x(T)^2 + \frac{1}{2}\bar{q}(T)\mathbb{E}[x(T)]^2$$
$$+ \frac{1}{2}\int_0^T \left(qx^2 + \bar{q}\mathbb{E}[x]^2 + ru^2 + \bar{r}\mathbb{E}[u]^2\right) dt. \quad (4.18)$$

Solve the problem $\min_u L(x, u)$ with cost function in (4.18) subject to (4.17).

Hint: An appropriate guess functional for the optimal cost is given by

$$f(t, x) = \frac{1}{2}\alpha\left(x - \mathbb{E}[x]\right)^2 + \frac{1}{2}\beta\mathbb{E}[x]^2 + \gamma\mathbb{E}[x] + \delta.$$

Also recall that the mean-field-type control problem is similar to the cooperative mean-field-type game problem given that there is a unique cost functional to minimize. In contrast, for the control problem involves a unique decision-maker.

2. Solve the following infinite-time horizon linear-quadratic mean-field-type control problem. Consider the cost functional

$$L^{\infty}(u) = \frac{1}{2} \int_0^{\infty} \left(qx^2 + \bar{q}\mathbb{E}[x]^2 + ru^2 + \bar{r}\mathbb{E}[u]^2 \right) dt. \qquad (4.19)$$

Solve the problem $\min_u \; L^{\infty}(x, u)$ with cost function in (4.19) subject to (4.17). Also solve the infinite-time horizon problem for the cooperative mean-field-type game problem in order to compare with the control counterpart. Are these two solutions still similar?.

3. Solve the following linear-quadratic state-dependent diffusion mean-field-type control problem. Consider the system dynamics

$$dx = \left(b_2 u + \bar{b}_2 \mathbb{E}[u] \right) dt + \left(\sigma_1 x + \bar{\sigma}_1 \mathbb{E}[x] \right) dB, \qquad (4.20)$$
$$x(0) := x_0.$$

Solve the problem $\min_u \; L(x, u)$ with cost function in (4.18) subject to (4.20).

Hint: An appropriate guess functional for the optimal cost is given by

$$f(t, x) = \frac{1}{2}\alpha \left(x - \mathbb{E}[x] \right)^2 + \frac{1}{2}\beta\mathbb{E}[x]^2 + \gamma\mathbb{E}[x] + \delta.$$

5

Co-opetitive Mean-Field-Type Games

Chapter 1 introduced several solution concepts in the framework of game theory or interactive decision-making. So far, up to the immediately previous Chapter 4, we have only analyzed the non-cooperative setting whose solution is given by a Nash equilibrium, and the cooperative setting that is tightly related to a control problem. Here, we will learn about a new game-theoretical solution concept that is more flexible than the non-cooperative and cooperative scenarios in the sense it can consider a larger number of strategic possibilities.

Game theory has studied the interaction of decision-makers under different behavioral models. The most general example about this is evidenced in the two main branches of game theory, i.e., cooperative and non-cooperative (also known as coalitional) games, where power indexes such as the Shapley or Banzhaf values are studied to determine the impact, relevance and/or importance of a player over the output of the game. This versatility of the game theory allows the study of not only cooperation and competition, but also allows the analysis of concepts such as power, selfishness, altruism, self-abnegation, reputation, among many others. It is important to highlight that, when considering cooperative or non-cooperative games, it is assumed that all decision-makers follow the same behavioral model, i.e., all decision-makers either cooperate or compete. Nevertheless, a general framework would allow decision-makers perform heterogeneous behaviors, i.e., some decision-makers cooperating, some deciding in a selfish way, and others performing with partial cooperation and/or altruism.

In this chapter, we discuss the co-opetitive mean-field-type games, i.e., strategic interactions in which different levels of cooperation and competition can take place at the same time. Thus, this approach allows studying cooperation, partial cooperation, selfishness, altruism, self-abnegation, among others. We show that either the non-cooperative or the fully-cooperative solutions presented in the previous Chapter 4 can be retrieved by the appropriate selection of the co-opetitive parameters. In this regard, the study of co-opetitive games becomes a more general approach.

Full cooperation is performed when the interests and efforts from a group of decision-makers pursue to achieve a common objective as studied in [90]. On the other hand, competition refers to the case in which efforts concentrate in performing individually in a selfish manner. The combination of both behaviors is known as co-opetition [91–93], where all cases such as cooperation,

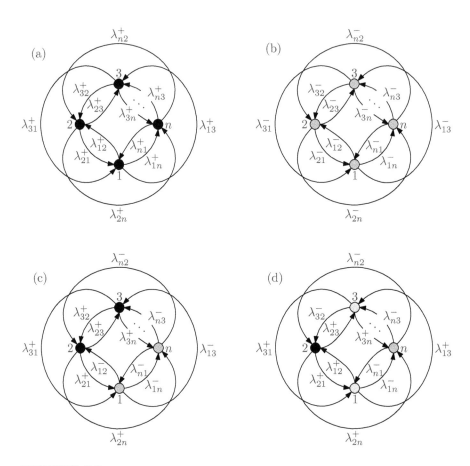

FIGURE 5.1
Co-opetition scheme λ_{ij}^+ (λ_{ij}^-) means $\lambda_{ij} > 0$, ($\lambda_{ij} < 0$). Black players are altruistic, dark gray players are spiteful, and light gray players cooperate and compete simultaneously. (a) Fully-altruistic scenario, (b) Fully-adversarial scenario, (c) Altruistic and adversarial players, and (d) Mixture of behaviors.

partial cooperation, selfishness, altruism, self-abnegation, etc can be considered by means of individual co-opetitive parameters. This possibility might enlarge the number of engineering system that can be modeled in a more realistic way.

Once it is possible to consider this heterogeneous behavior, the arising question concerns about how to design incentives to make cooperation occur. Indeed, how to create a model that allows co-opetitive parameters evolve along the time. For instance, in [75] authors study the case when the co-opetitive parameters change along the time. This question belongs to the mechanism design, which is a relevant approach in game-theory applied to engineering design.

Figure 5.1 shows the multiple possible co-opetitive relationships. Figure 5.1(a) shows a case in which all the players cooperate in the minimization of the other players' costs, i.e., all the co-opetitive parameters are positive. In contrast, Figure 5.1(b) presents a scenario in which all the players behave in a spiteful manner since they select the strategies seeking to increase the costs of the other players (deciding maliciously from the perspective of other players). Besides, Figure 5.1 corresponds to a scenario where some players cooperate whereas other players act against the interests of the others, i.e., a mixture between the situations presented in Figure 5.1(a) and 5.1(b). Finally, Figure 5.1(d) presents a case in which some players are partially cooperative and spiteful simultaneously with others (heterogeneous behaviors for a same decision-maker).

5.1 Co-opetitive Mean-Field-Type Game Set-up

Let us consider a set of $n \geq 2$ $(n \in \mathbb{N})$ decision-makers denoted by $\mathcal{N} = \{1, \ldots, n\}$. These agents (decision-makers) strategically interact to each other by means of the following SDE describing the evolution of a system state:

$$dx = \left(b_0 + b_1 x + \bar{b}_1 \mathbb{E}[x] + \sum_{j \in \mathcal{N}} b_{2j} u_j + \sum_{j \in \mathcal{N}} \bar{b}_{2j} \mathbb{E}[u_j] \right) dt + \sigma_0 dB, \quad (5.1a)$$

$$x(0) := x_0, \quad (5.1b)$$

where x, and $\mathbb{E}[x] \in \mathbb{R}$ correspond to the system state and its expectation, respectively. Similarly, u_i, and $\mathbb{E}[u_i] \in \mathbb{R}$ denote the control input of the decision-maker $i \in \mathcal{N}$, and its expected value, respectively. Hence, $b_0, b_1, \bar{b}_1, b_{2j}, \bar{b}_{2j}, \sigma_0 \in \mathbb{R}$, for all $j \in \mathcal{N}$ are the system state parameters, which are time-dependent functions; and the standard Brownian motion is denoted by B.

These decision-makers can interact to each other either by cooperating or competing (or following a mixture of cooperation and competition). Furthermore, there is a co-opetitive interaction in the game since decision-makers can interact with both partial/full selfishness or partial/full altruism. Such feature is determined by means of co-opetitive parameters given by

$$\Lambda_i = [\lambda_{i1} \quad \cdots \quad \lambda_{in}]^\top \in \mathbb{R}^n,$$

for all $i \in \mathcal{N}$. The co-opetitive parameters determine the ith decision-maker's levels of selfishness and altruism with respect to other decision-makers. More precisely, the term λ_{ij} determines the co-opetition of decision-maker $i \in \mathcal{N}$ with respect to the decision-maker $j \in \mathcal{N}$, i.e.,

- Partial altruism toward j: if $\lambda_{ij} > 0$ the ith decision-maker is considering positively j's payoff in this own behavior.

- Partial non-cooperation toward j : If $\lambda_{ij} = 0$ the ith decision-maker is not considering j's payoff in this own behavior.

- Partial Spite toward j: if $\lambda_{ij} < 0$ the ith decision-maker is considering negatively j's payoff in this own behavior.

We define a constant co-opetitive power denoted by $\pi_i \in \mathbb{R}_{>0}$ such that

$$\sum_{j \in \mathcal{N}} \lambda_{ij} = \pi_i, \ \forall i \in \mathcal{N},$$

allowing to consider heterogeneous co-opetitive powers throughout the set of decision-makers. In this regard, the co-opetitive power allows to consider a large variety of powers and how influential players are, e.g., strategic interactions where there are more powerful players than others. We could capture then, heterogeneous behaviors with heterogeneous powers. A player with higher co-opetitive power can have more influence over the game than those with small co-opetitive power. Besides, let

$$\Lambda_{i(-i)} = [\lambda_{ij}]_{j \in \mathcal{N} \setminus \{i\}} \in \mathbb{R}^{n-1}$$

be the vector of all the co-opetitive parameters excluding the ith parameter λ_{ii}, i.e.,

$$\Lambda_{i(-i)} = [\lambda_{i1} \quad \cdots \quad \lambda_{i(i-1)}(t), \lambda_{i(i+1)} \quad \lambda_{in}]^\top,$$

for all $i \in \mathcal{N}$.

Hence, let $L_i^{\text{C}}(x, u_1, \ldots, u_n, \Lambda_i)$ denote the co-opetitive cost with the co-opetitive parameters Λ_i, i.e.,

$$\begin{aligned} L_i^{\text{C}}(x, u_1, \ldots, u_n, \Lambda_i) &= L_i^{\text{S}}(x, u_1, \ldots, u_n, \lambda_{ii}) \\ &\quad + L_i^{\text{A}}(x, u_1, \ldots, u_n, \Lambda_{i(-i)}), \ \forall i \in \mathcal{N}, \end{aligned} \quad (5.2)$$

where $L_i^S(x, u_1, \ldots, u_n, \Lambda_i)$ denotes the *selfish* component of co-opetition, i.e.,

$$L_i^S(x, u_1, \ldots, u_n, \lambda_{ii}) = \frac{1}{2} \lambda_{ii} \left(q_i x^2(T) + \bar{q}_i \mathbb{E}[x(T)]^2 \right)$$

$$+ \frac{1}{2} \int_0^T \lambda_{ii} \left(q_i x^2 + \bar{q}_i \mathbb{E}[x]^2 + r_i u_i^2 + \bar{r}_i \mathbb{E}[u_i]^2 \right) dt, \ \forall i \in \mathcal{N}, \qquad (5.3)$$

and $L_i^A(x, u_1, \ldots, u_n, \Lambda_{i(-i)})$ denotes the *altruistic/self-reinforcing* (depending in the value of Λ_i) component of co-opetition, i.e.,

$$L_i^A(x, u_1, \ldots, u_n, \Lambda_{i(-i)}) = \frac{1}{2} \sum_{j \in \mathcal{N} \backslash \{i\}} \lambda_{ij} \left(q_j x^2(T) + \bar{q}_j \mathbb{E}[x(T)]^2 \right)$$

$$+ \frac{1}{2} \int_0^T \sum_{j \in \mathcal{N} \backslash \{i\}} \lambda_{ij} \left(q_j x^2 + \bar{q}_j \mathbb{E}[x]^2 + r_j u_j^2 + \bar{r}_j \mathbb{E}[u_j]^2 \right) dt, \ \forall i \in \mathcal{N}. \quad (5.4)$$

Once the co-opetitive cost function has been defined by means of selfish and altruism/self-reinforcing components, the co-opetitive linear-quadratic mean-field-type game problem is introduced next.

Problem 15 (Co-opetitive Mean-Field-Type Problem) *The linear-quadratic mean-field-type game problem with heterogeneous behavior for non-cooperation and cooperation (with co-opetitive parameters) is as follows:*

$$\begin{cases} \underset{u_i \in \mathcal{U}_i}{\text{minimize}} \, \mathbb{E}\left[L_i^C(x, u_1, \ldots, u_n, \Lambda_i) \right], \\[2mm] \text{subject to} \\[2mm] dx = \left(b_0 + b_1 x + \bar{b}_1 \mathbb{E}[x] + \sum_{j \in \mathcal{N}} b_{2j} u_j + \sum_{j \in \mathcal{N}} \bar{b}_{2j} \mathbb{E}[u_j] \right) dt + \sigma_0 dB, \\[1mm] x(0) = x_0, \\[1mm] \sum_{j \in \mathcal{N}} \lambda_{ij} = \pi_i, \ \forall i \in \mathcal{N}, \end{cases}$$

where the weight parameters in the cost functional are $q_i, q_i + \bar{q}_i \geq 0$, $r_i, r_i + \bar{r}_i > 0$, and the co-opetition powers are given.

With this new problem set-up, we recall the best-response strategy and Nash equilibrium for this new class of games known as co-opetitive games next.

Definition 10 (Co-opetitive Mean-Field-Type Best Response) *Any feasible strategy $u_i^* \in \mathcal{U}_i$ such that*

$$u_i^* \in \arg \inf_{u_i \in \mathcal{U}_i} \mathbb{E}\left[L_i^C(x, u_1, \ldots, u_n, \Lambda_i) \right]$$

is a co-opetitive mean-field-type Best-Response strategy of the decision maker $i \in \mathcal{N}$ against the strategies u_{-i} selected from the other decision makers $\mathcal{N} \setminus \{i\}$. The set of Best-Response strategies is given by

$$\text{BR}_i : \prod_{j \in \mathcal{N} \setminus \{i\}} \mathcal{U}_j \to 2^{\mathcal{U}_i},$$

where $2^{\mathcal{U}_i}$ is the power set of all the possible subsets of $\mathcal{U}_i$.

Consequently, the set of co-opetitive mean-field-type Nash equilibria is defined by using the co-opetitive mean-field-type best-response below.

Definition 11 (Co-opetitive Mean-field-type Nash equilibrium) *Any feasible strategy profile*

$$[u_1^* \quad \cdots \quad u_n^*]^\top \in \prod_{j \in \mathcal{N}} \mathcal{U}_j$$

such that the optimal $u_i^* \in \text{BR}_i(u_{-i}^*)$ is a mean-field-type Nash equilibrium of the co-opetitive linear-quadratic mean-field-type game.

Then as well, one can define the co-opetitive differential mean-field-free game problem by considering the evolution of the system state deterministic and excluding the mean-field terms. This problem becomes a particular case whose solution can be directly obtained from the solution of the co-opetitive mean-field-type game problem that is developed next. Based on this, we do not present the co-opetitive differential game problems in this book given that they can straight-forwardly solved following the direct method.

5.2 Semi-explicit Solution of the Co-opetitive Mean-Field-Type Game Problem

The semi-explicit solution for a co-opetitive scenario in the context of mean-field-type games is presented in Proposition 15. This result is suitable to study a risk-aware strategic interaction where there are heterogeneous behaviors that could also be considered time-varying.

Proposition 15 *The feedback-strategies linear-quadratic mean-field-type Nash equilibrium for the co-opetitive Problem 15, and the optimal costs are given by:*

$$u_i^* = \mathbb{E}[u_i] - \frac{b_{2i}}{\lambda_{ii} r_i} \alpha_i (x - \mathbb{E}[x]),$$

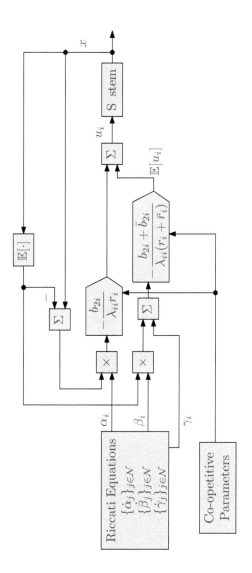

FIGURE 5.2
Feedback scheme for the linear-quadratic mean-field-type co-opetitive game problem.

$$\mathbb{E}[u_i^*] = -\frac{(b_{2i} + \bar{b}_{2i})}{\lambda_{ii}(r_i + \bar{r}_i)}\left[\beta_i \mathbb{E}[x] + \gamma_i\right]$$

$$L_i^C(x, u_1^*, \dots, u_n^*) = \frac{1}{2}\alpha_i(0)(x_0 - \mathbb{E}[x_0])^2 + \frac{1}{2}\beta_i(0)\mathbb{E}[x_0]^2$$
$$+ \gamma_i(0)\mathbb{E}[x_0] + \delta_i(0),$$

where α_i, β_i, γ_i, and δ_i solve the following differential equations:

$$\dot{\alpha}_i = -2b_1\alpha_i - \sum_{j\in\mathcal{N}}\lambda_{ij}q_j + 2\alpha_i\sum_{j\in\mathcal{N}\setminus\{i\}}\frac{b_{2j}^2}{\lambda_{jj}r_j}\alpha_j - \sum_{j\in\mathcal{N}\setminus\{i\}}\lambda_{ij}\frac{b_{2j}^2}{\lambda_{jj}^2 r_j}\alpha_j^2$$

$$+ \frac{b_{2i}^2}{\lambda_{ii}r_i}\alpha_i^2,$$

$$\dot{\beta}_i = -2\left(b_1 + \bar{b}_1\right)\beta_i - \sum_{j\in\mathcal{N}}\lambda_{ij}(q_j + \bar{q}_j) + 2\beta_i\sum_{j\in\mathcal{N}\setminus\{i\}}\frac{(b_{2j} + \bar{b}_{2j})^2}{\lambda_{jj}(r_j + \bar{r}_j)}\beta_j$$

$$- \sum_{j\in\mathcal{N}\setminus\{i\}}\lambda_{ij}\frac{(b_{2j} + \bar{b}_{2j})^2}{\lambda_{jj}^2(r_j + \bar{r}_j)}\beta_j^2 + \frac{(b_{2i} + \bar{b}_{2i})^2}{\lambda_{ii}(r_i + \bar{r}_i)}\beta_i^2,$$

$$\dot{\gamma}_i = -b_0\beta_i - \left(b_1 + \bar{b}_1\right)\gamma_i + \sum_{j\in\mathcal{N}\setminus\{i\}}\frac{(b_{2j} + \bar{b}_{2j})^2}{\lambda_{jj}(r_j + \bar{r}_j)}(\beta_i\gamma_j + \gamma_i\beta_j)$$

$$- \sum_{j\in\mathcal{N}\setminus\{i\}}\lambda_{ij}\frac{(b_{2j} + \bar{b}_{2j})^2}{\lambda_{jj}^2(r_j + \bar{r}_j)}\beta_j\gamma_j + \frac{(b_{2i} + \bar{b}_{2i})^2}{\lambda_{ii}(r_i + \bar{r}_i)}\beta_i\gamma_i,$$

$$\dot{\delta}_i = -b_0\gamma_i - \frac{1}{2}\alpha_i\sigma_0^2 + \gamma_i\sum_{j\in\mathcal{N}\setminus\{i\}}\frac{(b_{2j} + \bar{b}_{2j})^2}{\lambda_{jj}(r_j + \bar{r}_j)}\gamma_j$$

$$- \frac{1}{2}\sum_{j\in\mathcal{N}\setminus\{i\}}\lambda_{ij}\frac{(b_{2j} + \bar{b}_{2j})^2}{\lambda_{jj}^2(r_j + \bar{r}_j)}\gamma_j^2 + \frac{1}{2}\frac{(b_{2i} + \bar{b}_{2i})^2}{\lambda_{ii}(r_i + \bar{r}_i)}\gamma_i^2,$$

with terminal boundary conditions

$$\alpha_i(T) = \sum_{j\in\mathcal{N}}\lambda_{ij}q_j(T),$$

$$\beta_i(T) = \sum_{j\in\mathcal{N}}\lambda_{ij}(q_j(T) + \bar{q}_j(T)),$$

$$\gamma_i(T) = \delta_i(T) = 0,$$

whenever these equations admit a solution with positive α_i and β_i.

Figure 5.2 presents the feedback scheme for the co-opetitive mean-field-type game problem according to Proposition 15. Notice that the co-opetitive parameters directly affect/modify the gains in the computation of the optimal control inputs.

Remark 16 *The semi-explicit solution presented in Proposition 15 shows an interesting situation when a decision-maker is indifferent to its own cost functional, i.e., if we make decision-makers not to care about their benefits. This situation creates a singularity. Thus, the co-opetitive solutions require that* $\lambda_{jj} \neq 0$, *for all* $j \in \mathcal{N}$.

Next, we present the computation of the semi-explicit solution for the co-opetitive game by means of the direct method.

Proof 15 (Proposition 15) *Notice that, even though the game-theoretical solution concept is different from those treated in previous chapters, we still preserve the same structure for both the cost functional and system dynamics. Then, inspired by the structure of the cost function, we propose a quadratic guess functional of the following form:*

$$f_i(t, x) = \frac{1}{2}\alpha_i \left(x - \mathbb{E}[x]\right)^2 + \frac{1}{2}\beta_i \mathbb{E}[x]^2 + \gamma_i \mathbb{E}[x] + \delta_i, \tag{5.5}$$

where $\alpha_i, \beta_i, \gamma_i,$ *and* δ_i *are deterministic functions of time. Now, we recall the obtained expression in the proof of Proposition 13 after having applied the Itô's formula, i.e.,*

$$\mathbb{E}[f_i(T, x) - f_i(0, x_0)] = \frac{1}{2}\mathbb{E}\int_0^T (x - \mathbb{E}[x])^2 \left[\dot{\alpha}_i + 2b_1\alpha_i\right] dt$$

$$+ \frac{1}{2}\mathbb{E}\int_0^T \mathbb{E}[x]^2 \left[\dot{\beta}_i + 2\left(b_1 + \bar{b}_1\right)\beta_i\right] dt$$

$$+ \mathbb{E}\int_0^T \mathbb{E}[x]\left[b_0\beta_i + \dot{\gamma}_i + \left(b_1 + \bar{b}_1\right)\gamma_i\right] dt + \mathbb{E}\int_0^T \left[\dot{\delta}_i + b_0\gamma_i + \frac{1}{2}\alpha_i\sigma_0^2\right] dt$$

$$+ \mathbb{E}\int_0^T \sum_{j \in \mathcal{N}\setminus\{i\}} \left(b_{2j} + \bar{b}_{2j}\right)\left[\beta_i \mathbb{E}[x] + \gamma_i\right]\mathbb{E}[u_j] dt$$

$$+ \mathbb{E}\int_0^T \sum_{j \in \mathcal{N}\setminus\{i\}} b_{2j}\left(u_j - \mathbb{E}[u_j]\right)\alpha_i\left(x - \mathbb{E}[x]\right) dt$$

$$+ \mathbb{E}\int_0^T \left(b_{2i} + \bar{b}_{2i}\right)\left[\beta_i \mathbb{E}[x] + \gamma_i\right]\mathbb{E}[u_i] dt$$

$$+ \mathbb{E}\int_0^T b_{2i}\left(u_i - \mathbb{E}[u_i]\right)\alpha_i\left(x - \mathbb{E}[x]\right) dt.$$

We compute the gap $\mathbb{E}[L_i^C(x, u_1, \dots, u_n) - f_i(0, x_0)]$ *as follows:*

$$\mathbb{E}[L_i^C(x, u_1, \dots, u_n) - f_i(0, x_0)] \tag{5.6}$$

$$= \frac{1}{2}\lambda_{ii}\left(q_i \mathbb{E}\left[x(T) - \mathbb{E}[x(T)]\right]^2 + (q_i + \bar{q}_i)\mathbb{E}[x(T)]^2\right)$$

$$+ \frac{1}{2}\sum_{j \in \mathcal{N}\setminus\{i\}} \lambda_{ij}\left(q_j \mathbb{E}\left[x(T) - \mathbb{E}[x(T)]\right]^2 + (q_j + \bar{q}_j)\mathbb{E}[x(T)]^2\right)$$

$$- \frac{1}{2} \alpha_i(T) \mathbb{E}\left[x(T) - \mathbb{E}[x(T)]\right]^2 - \frac{1}{2} \beta_i(T) \mathbb{E}[x(T)]^2 - \gamma_i(T) \mathbb{E}[x(T)] - \delta_i(T)$$

$$+ \frac{1}{2} \mathbb{E} \int_0^T (x - \mathbb{E}[x])^2 \left[\dot{\alpha}_i + 2b_1 \alpha_i + \sum_{j \in \mathcal{N}} \lambda_{ij} q_j\right] dt$$

$$+ \frac{1}{2} \mathbb{E} \int_0^T \mathbb{E}[x]^2 \left[\dot{\beta}_i + 2\left(b_1 + \bar{b}_1\right) \beta_i + \sum_{j \in \mathcal{N}} \lambda_{ij} \left(q_j + \bar{q}_j\right)\right] dt$$

$$+ \mathbb{E} \int_0^T \mathbb{E}[x]\left[b_0 \beta_i + \dot{\gamma}_i + \left(b_1 + \bar{b}_1\right) \gamma_i\right] dt + \mathbb{E} \int_0^T \left[\dot{\delta}_i + b_0 \gamma_i + \frac{1}{2} \alpha_i \sigma_0^2\right] dt$$

$$+ \mathbb{E} \int_0^T \sum_{j \in \mathcal{N} \setminus \{i\}} \left(b_{2j} + \bar{b}_{2j}\right) \left[\beta_i \mathbb{E}[x] + \gamma_i\right] \mathbb{E}[u_j] dt$$

$$+ \mathbb{E} \int_0^T \sum_{j \in \mathcal{N} \setminus \{i\}} b_{2j} \left(u_j - \mathbb{E}[u_j]\right) \alpha_i \left(x - \mathbb{E}[x]\right) dt$$

$$+ \frac{1}{2} \mathbb{E} \int_0^T \sum_{j \in \mathcal{N} \setminus \{i\}} \lambda_{ij} r_j \left(u_j - \mathbb{E}[u_j]\right)^2 dt$$

$$+ \frac{1}{2} \mathbb{E} \int_0^T \sum_{j \in \mathcal{N} \setminus \{i\}} \lambda_{ij} \left(r_j + \bar{r}_j\right) \mathbb{E}[u_j]^2 dt$$

$$+ \frac{1}{2} \mathbb{E} \int_0^T \lambda_{ii} \left(r_i + \bar{r}_i\right) \left[\mathbb{E}[u_i]^2 + 2\frac{\left(b_{2i} + \bar{b}_{2i}\right)}{\lambda_{ii} \left(r_i + \bar{r}_i\right)} \left[\beta_i \mathbb{E}[x] + \gamma_i\right] \mathbb{E}[u_i]\right] dt$$

$$+ \frac{1}{2} \mathbb{E} \int_0^T \lambda_{ii} r_i \left[\left(u_i - \mathbb{E}[u_i]\right)^2 + 2\frac{b_{2i}}{\lambda_{ii} r_i} \left(u_i - \mathbb{E}[u_i]\right) \alpha_i \left(x - \mathbb{E}[x]\right)\right] dt. \quad (5.7)$$

Now, we perform square completion for the terms involving both $\mathbb{E}[u_i]$ *and* $(u_i - \mathbb{E}[u_i])$, *i.e.,*

$$\left(\mathbb{E}[u_i] + \frac{\left(b_{2i} + \bar{b}_{2i}\right)}{\lambda_{ii} \left(r_i + \bar{r}_i\right)} \left[\beta_i \mathbb{E}[x] + \gamma_i\right]\right)^2 = \mathbb{E}[u_i]^2$$

$$+ 2\frac{\left(b_{2i} + \bar{b}_{2i}\right)}{\lambda_{ii} \left(r_i + \bar{r}_i\right)} \left[\beta_i \mathbb{E}[x] + \gamma_i\right] \mathbb{E}[u_i]$$

$$+ \frac{\left(b_{2i} + \bar{b}_{2i}\right)^2}{\lambda_{ii}^2 \left(r_i + \bar{r}_i\right)^2} \left[\beta_i^2 \mathbb{E}[x]^2 + 2\beta_i \mathbb{E}[x]\gamma_i + \gamma_i^2\right], \quad (5.8)$$

and

$$\left(u_i - \mathbb{E}[u_i] + \frac{b_{2i}}{\lambda_{ii} r_i} \alpha_i \left(x - \mathbb{E}[x]\right)\right)^2 = \left(u_i - \mathbb{E}[u_i]\right)^2$$

$$+ 2\frac{b_{2i}}{\lambda_{ii} r_i} \left(u_i - \mathbb{E}[u_i]\right) \alpha_i \left(x - \mathbb{E}[x]\right) + \frac{b_{2i}^2}{\lambda_{ii}^2 r_i^2} \alpha_i^2 \left(x - \mathbb{E}[x]\right)^2. \quad (5.9)$$

Thus, replacing the square completion (5.8) and (5.9) in (5.7), one obtains after some algebraic procedure and grouping of terms that

$$
\mathbb{E}[L_i^C(x, u_1, \ldots, u_n) - f_i(0, x_0)]
$$

$$
= \frac{1}{2}\lambda_{ii}\left(q_i\mathbb{E}\left[x(T) - \mathbb{E}[x(T)]\right]^2 + (q_i + \bar{q}_i)\mathbb{E}[x(T)]^2\right)
$$

$$
+ \frac{1}{2}\sum_{j\in\mathcal{N}\backslash\{i\}}\lambda_{ij}\left(q_j\mathbb{E}\left[x(T) - \mathbb{E}[x(T)]\right]^2 + (q_j + \bar{q}_j)\mathbb{E}[x(T)]^2\right)
$$

$$
- \frac{1}{2}\alpha_i(T)\mathbb{E}\left[x(T) - \mathbb{E}[x(T)]\right]^2 - \frac{1}{2}\beta_i(T)\mathbb{E}[x(T)]^2
$$

$$
- \gamma_i(T)\mathbb{E}[x(T)] - \delta_i(T)
$$

$$
+ \frac{1}{2}\mathbb{E}\int_0^T (x - \mathbb{E}[x])^2 \left[\dot{\alpha}_i + 2b_1\alpha_i + \sum_{j\in\mathcal{N}}\lambda_{ij}q_j\right.
$$

$$
\left. - 2\alpha_i \sum_{j\in\mathcal{N}\backslash\{i\}}\frac{b_{2j}^2}{\lambda_{jj}r_j}\alpha_j + \sum_{j\in\mathcal{N}\backslash\{i\}}\lambda_{ij}\frac{b_{2j}^2}{\lambda_{jj}^2 r_j}\alpha_j^2 - \frac{b_{2i}^2}{\lambda_{ii}r_i}\alpha_i^2\right] dt
$$

$$
+ \frac{1}{2}\mathbb{E}\int_0^T \mathbb{E}[x]^2 \left[\dot{\beta}_i + 2\left(b_1 + \bar{b}_1\right)\beta_i + \sum_{j\in\mathcal{N}}\lambda_{ij}\left(q_j + \bar{q}_j\right)\right.
$$

$$
- 2\beta_i \sum_{j\in\mathcal{N}\backslash\{i\}}\frac{\left(b_{2j} + \bar{b}_{2j}\right)^2}{\lambda_{jj}\left(r_j + \bar{r}_j\right)}\beta_j + \sum_{j\in\mathcal{N}\backslash\{i\}}\lambda_{ij}\frac{\left(b_{2j} + \bar{b}_{2j}\right)^2}{\lambda_{jj}^2\left(r_j + \bar{r}_j\right)}\beta_j^2
$$

$$
\left. - \frac{\left(b_{2i} + \bar{b}_{2i}\right)^2}{\lambda_{ii}\left(r_i + \bar{r}_i\right)}\beta_i^2\right] dt
$$

$$
+ \mathbb{E}\int_0^T \mathbb{E}[x]\left[b_0\beta_i + \dot{\gamma}_i + \left(b_1 + \bar{b}_1\right)\gamma_i\right.
$$

$$
- \sum_{j\in\mathcal{N}\backslash\{i\}}\frac{\left(b_{2j} + \bar{b}_{2j}\right)^2}{\lambda_{jj}\left(r_j + \bar{r}_j\right)}\left(\beta_i\gamma_j + \gamma_i\beta_j\right)
$$

$$
\left. + \sum_{j\in\mathcal{N}\backslash\{i\}}\lambda_{ij}\frac{\left(b_{2j} + \bar{b}_{2j}\right)^2}{\lambda_{jj}^2\left(r_j + \bar{r}_j\right)}\beta_j\gamma_j - \frac{\left(b_{2i} + \bar{b}_{2i}\right)^2}{\lambda_{ii}\left(r_i + \bar{r}_i\right)}\beta_i\gamma_i\right] dt
$$

$$
+ \mathbb{E}\int_0^T \left[\dot{\delta}_i + b_0\gamma_i + \frac{1}{2}\alpha_i\sigma_0^2 - \gamma_i \sum_{j\in\mathcal{N}\backslash\{i\}}\frac{\left(b_{2j} + \bar{b}_{2j}\right)^2}{\lambda_{jj}\left(r_j + \bar{r}_j\right)}\gamma_j\right.
$$

$$
\left. + \frac{1}{2}\sum_{j\in\mathcal{N}\backslash\{i\}}\lambda_{ij}\frac{\left(b_{2j} + \bar{b}_{2j}\right)^2}{\lambda_{jj}^2\left(r_j + \bar{r}_j\right)}\gamma_j^2 - \frac{1}{2}\frac{\left(b_{2i} + \bar{b}_{2i}\right)^2}{\lambda_{ii}\left(r_i + \bar{r}_i\right)}\gamma_i^2\right] dt
$$

$$
+ \frac{1}{2}\mathbb{E}\int_0^T \lambda_{ii}\left(r_i + \bar{r}_i\right)\left(\mathbb{E}[u_i] + \frac{\left(b_{2i} + \bar{b}_{2i}\right)}{\lambda_{ii}\left(r_i + \bar{r}_i\right)}\left[\beta_i\mathbb{E}[x] + \gamma_i\right]\right)^2 dt
$$

$$+ \frac{1}{2}\mathbb{E}\int_0^T \lambda_{ii} r_i \left(u_i - \mathbb{E}[u_i] + \frac{b_{2i}}{\lambda_{ii} r_i}\alpha_i \left(x - \mathbb{E}[x]\right) \right)^2 dt.$$

We continue the same procedure by dividing this expression into several groups to minimize the cost function and make it match with the optimal cost in the interval $[0,T]$, i.e.,

- Those terms at terminal time T

- Those terms whose common factor is given by $(x - \mathbb{E}[x])^2$, $\mathbb{E}[x]^2$, and $\mathbb{E}[x]$

- Those terms independent from the state, mean-field state, and control inputs including their expectation

- And those terms depending on the decision variables

Finally, the announced result is obtained by minimizing the terms completing the proof.

We have mentioned at the beginning of this chapter that the result presented in Proposition 15 is more general than those introduced for non-cooperative and cooperative mean-field-type games in Chapter 4. In order to illustrate this claim, we show the connections between the co-opetitive result with other scenarios in the next section.

5.3 Connections between the Co-opetitive Solution with the Non-cooperative and Cooperative Solutions

5.3.1 Non-cooperative Relationship

We present the relationship between both the non-cooperative and cooperative solutions with the co-opetitive solution. To this end, let us recall the co-opetitive solution with selfish co-opetitive parameters, i.e., $\lambda_{ij} = 1$, if $i = j$, for all $i \in \mathcal{N}$, and $\lambda_{ij} = 0$, otherwise. Thus, recalling the differential equation $\dot{\alpha}_i$ for the co-opetitive solution we have

$$\dot{\alpha}_i = -2b_1\alpha_i - \sum_{j\in\mathcal{N}} \lambda_{ij}q_j + 2\alpha_i \sum_{j\in\mathcal{N}\setminus\{i\}} \frac{b_{2j}^2}{\lambda_{jj} r_j}\alpha_j$$

$$\underbrace{-\sum_{j\in\mathcal{N}\setminus\{i\}} \lambda_{ij}\frac{b_{2j}^2}{\lambda_{jj}^2 r_j}\alpha_j^2}_{=0} + \frac{b_{2i}^2}{\lambda_{ii} r_i}\alpha_i^2,$$

$$= -2b_1\alpha_i - q_i + 2\alpha_i \sum_{j\in\mathcal{N}\setminus\{i\}} \frac{b_{2j}^2}{r_j}\alpha_j + \frac{b_{2i}^2}{r_i}\alpha_i^2,$$

which corresponds to the differential equation presented in the non-cooperative mean-field-type game (see Proposition 13 in Chapter 4).

5.3.2 Cooperative Relationship

On the other hand, let us consider some co-opetitive parameters given by a cooperative scenario, i.e., $\lambda_{ij} = 1$, for all $i, j \in \mathcal{N}$. In this regard, notice that all the players have the same cost function and $\alpha_i = \alpha_0$, for all $i \in \mathcal{N}$. Thus, recalling the differential equation $\dot{\alpha}_i$ for the co-opetitive solution we have

$$
\begin{aligned}
\dot{\alpha}_i &= -2b_1\alpha_i - \sum_{j \in \mathcal{N}} \lambda_{ij} q_j + 2\alpha_i \sum_{j \in \mathcal{N}\setminus\{i\}} \frac{b_{2j}^2}{\lambda_{jj} r_j} \alpha_j \\
&\quad - \sum_{j \in \mathcal{N}\setminus\{i\}} \lambda_{ij} \frac{b_{2j}^2}{\lambda_{jj}^2 r_j} \alpha_j^2 + \frac{b_{2i}^2}{\lambda_{ii} r_i} \alpha_i^2, \\
&= -2b_1\alpha_i - \sum_{j \in \mathcal{N}} q_j + 2\alpha_i \sum_{j \in \mathcal{N}\setminus\{i\}} \frac{b_{2j}^2}{r_j} \alpha_j - \sum_{j \in \mathcal{N}\setminus\{i\}} \frac{b_{2j}^2}{r_j} \alpha_j^2 + \frac{b_{2i}^2}{r_i} \alpha_i^2, \\
&= -2b_1\alpha_0 - \sum_{j \in \mathcal{N}} q_j + 2 \sum_{j \in \mathcal{N}\setminus\{i\}} \frac{b_{2j}^2}{r_j} \alpha_0^2 - \sum_{j \in \mathcal{N}\setminus\{i\}} \frac{b_{2j}^2}{r_j} \alpha_0^2 + \frac{b_{2i}^2}{r_i} \alpha_0^2, \\
\dot{\alpha}_0 &= -2b_1\alpha_0 - q_0 + \sum_{j \in \mathcal{N}} \frac{b_{2j}^2}{r_j} \alpha_0^2,
\end{aligned}
$$

which corresponds to the differential equation presented in the cooperative mean-field-type game (see Proposition 14 in Chapter 4).

Remark 17 *The aforementioned equivalences between the co-opetitive and either the non-cooperative or cooperative solutions are obtained for all the differential equations $\alpha_i, \beta_i, \gamma_i, \delta_i$ for all $i \in \mathcal{N}$, as the co-opetitive parameters are selected in a fully cooperative or non-cooperative way. Therefore, the optimal control inputs are equivalent as well.*

We illustrate next some co-opetitive mean-field-type games by means of numerical simulations in order to observe the effect that the co-opetitive parameters have over the behavior of the optimal control inputs and the evolution of the system state.

5.4   Numerical Examples

Example 1: Partially Cooperation in a Co-opetitive Scenario

Consider a five-player co-opetitive mean-field-type game under the following system dynamics:

$$
dx = \left(1 + 2x + \mathbb{E}[x] - \frac{1}{2}u_1 + u_2 - \frac{3}{2}u_3 + 2u_4 - \frac{5}{2}u_5 \right.
$$

$$
\left. - \frac{1}{4}\mathbb{E}[u_1] + \frac{1}{2}\mathbb{E}[u_2] - \frac{3}{4}\mathbb{E}[u_3] + \mathbb{E}[u_4] - \frac{5}{4}\mathbb{E}[u_5] \right) dt + 3dB,
$$

and with the initial conditions $x_0 = \mathbb{E}[x_0] = 4$. First, we consider a partial cooperative scenario where all the players influences to each other as in Figure 5.1(a), and with the following co-opetitive parameters:

$$
\Lambda =
\begin{pmatrix}
1 & 0.1 & 0.2 & 0.3 & 0.4 \\
0.1 & 1 & 0.2 & 0.3 & 0.4 \\
0.1 & 0.2 & 1 & 0.3 & 0.4 \\
0.1 & 0.2 & 0.3 & 1 & 0.4 \\
0.1 & 0.2 & 0.3 & 0.4 & 1
\end{pmatrix}
$$

The cost function of the ith decision-maker with terminal time $T = \frac{3}{2}$ s is as follows:

$$
L_i^C(x, u_1, u_2, u_3, u_4, u_5, \Lambda) = \frac{1}{2} \left\{ \frac{1}{2}[\lambda_{i1}] \left(\left[x\left(\frac{3}{2}\right) \right]^2 + \mathbb{E}\left[x\left(\frac{3}{2}\right) \right]^2 \right) \right.
$$

$$
+ [\lambda_{i2}] \left(\left[x\left(\frac{3}{2}\right) \right]^2 + \mathbb{E}\left[x\left(\frac{3}{2}\right) \right]^2 \right) + \frac{3}{2}[\lambda_{i3}] \left(\left[x\left(\frac{3}{2}\right) \right]^2 + \mathbb{E}\left[x\left(\frac{3}{2}\right) \right]^2 \right)
$$

$$
+ 2[\lambda_{i4}] \left(\left[x\left(\frac{3}{2}\right) \right]^2 + \mathbb{E}\left[x\left(\frac{3}{2}\right) \right]^2 \right) + \frac{5}{2}[\lambda_{i5}] \left(\left[x\left(\frac{3}{2}\right) \right]^2 + \mathbb{E}\left[x\left(\frac{3}{2}\right) \right]^2 \right) \right\}
$$

$$
+ \frac{1}{2} \int_0^{\frac{3}{2}} \left\{ [\lambda_{i1}] \left(x^2 + \mathbb{E}[x]^2 + u_1^2 + \mathbb{E}[u_1]^2 \right) + 2[\lambda_{i2}] \left(x^2 + \mathbb{E}[x]^2 + u_2^2 + \mathbb{E}[u_2]^2 \right) \right.
$$

$$
+ 3[\lambda_{i3}] \left(x^2 + \mathbb{E}[x]^2 + u_3^2 + \mathbb{E}[u_3]^2 \right) + 4[\lambda_{i4}] \left(x^2 + \mathbb{E}[x]^2 + u_4^2 + \mathbb{E}[u_4]^2 \right)
$$

$$
\left. + 5[\lambda_{i5}] \left(x^2 + \mathbb{E}[x]^2 + u_5^2 + \mathbb{E}[u_5]^2 \right) \right\} dt,
$$

Figure 5.3 shows the evolution of the system state, when the following optimal strategies are applied:

$$
u_1^* = \frac{3}{8\lambda_{11}} \left[\beta_1 \mathbb{E}[x] + \gamma_1 \right] + \frac{1}{2\lambda_{11}} \alpha_1 (x - \mathbb{E}[x]),
$$

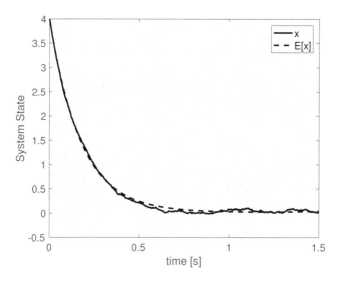

FIGURE 5.3
Evolution of the system state and its expectation.

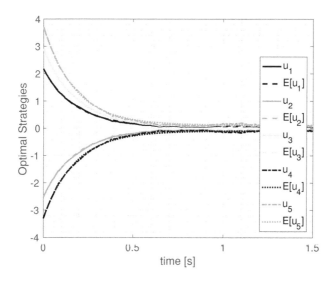

FIGURE 5.4
Evolution of the optimal control strategies for the partially cooperation in a co-opetitive scenario.

$$u_2^* = -\frac{3}{8\lambda_{22}}\left[\beta_2\mathbb{E}[x] + \gamma_2\right] - \frac{1}{2\lambda_{22}}\alpha_2\left(x - \mathbb{E}[x]\right),$$

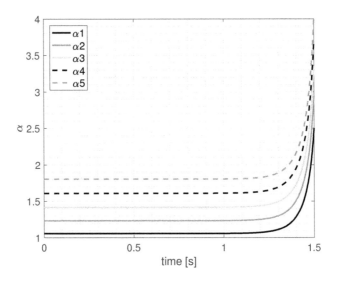

FIGURE 5.5
Evolution of the Riccati equations $\alpha_1, \ldots, \alpha_5$ for the partially cooperation in a co-opetitive scenario.

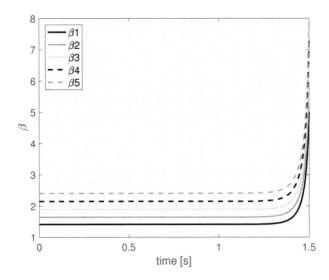

FIGURE 5.6
Evolution of the Riccati equations $\beta_1, \ldots, \beta_5$ for the partially cooperation in a co-opetitive scenario.

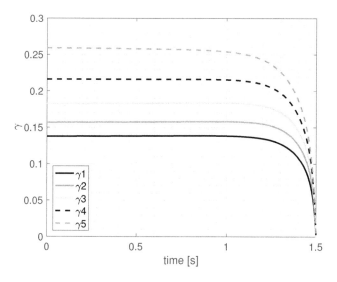

FIGURE 5.7
Evolution of the Riccati equations $\gamma_1, \ldots, \gamma_5$ for the partially cooperation in a co-opetitive scenario.

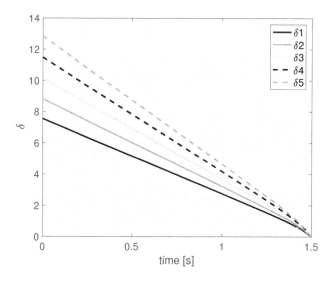

FIGURE 5.8
Evolution of the Riccati equations $\delta_1, \ldots, \delta_5$ for the partially cooperation in a co-opetitive scenario.

$$u_3^* = \frac{3}{8\lambda_{33}}\Big[\beta_3\mathbb{E}[x] + \gamma_3\Big] + \frac{1}{2\lambda_{33}}\alpha_3\,(x - \mathbb{E}[x])\,,$$

$$u_4^* = -\frac{3}{8\lambda_{44}}\Big[\beta_4\mathbb{E}[x] + \gamma_4\Big] - \frac{1}{2\lambda_{44}}\alpha_4\,(x - \mathbb{E}[x])\,,$$

$$u_5^* = \frac{3}{8\lambda_{55}}\Big[\beta_5\mathbb{E}[x] + \gamma_5\Big] + \frac{1}{2\lambda_{55}}\alpha_5\,(x - \mathbb{E}[x])\,,$$

which are presented in Figure 5.4.

The evolution of the Riccati equations $\alpha_1,\dots,\alpha_5$ is presented in Figure 5.5, which are as follows:

$$\dot{\alpha}_1 = -4\alpha_1 - \Big([\lambda_{11}] + 2[\lambda_{12}] + 3[\lambda_{13}] + 4[\lambda_{14}] + 5[\lambda_{15}]\Big)$$

$$+ 2\alpha_1\left(\frac{1}{2\lambda_{22}}\alpha_2 + \frac{3}{4\lambda_{33}}\alpha_3 + \frac{1}{\lambda_{44}}\alpha_4 + \frac{5}{4\lambda_{55}}\alpha_5\right)$$

$$- \left(\lambda_{12}\frac{1}{2\lambda_{22}^2}\alpha_2^2 + \lambda_{13}\frac{3}{4\lambda_{33}^2}\alpha_3^2 + \lambda_{14}\frac{1}{\lambda_{44}^2}\alpha_4^2 + \lambda_{15}\frac{5}{4\lambda_{55}^2}\alpha_5^2\right) + \frac{1}{4\lambda_{11}}\alpha_1^2,$$

$$\dot{\alpha}_2 = -4\alpha_2 - \Big([\lambda_{21}] + 2[\lambda_{22}] + 3[\lambda_{23}] + 4[\lambda_{24}] + 5[\lambda_{25}]\Big)$$

$$+ 2\alpha_2\left(\frac{1}{4\lambda_{11}}\alpha_1 + \frac{3}{4\lambda_{33}}\alpha_3 + \frac{1}{\lambda_{44}}\alpha_4 + \frac{5}{4\lambda_{55}}\alpha_5\right)$$

$$- \left(\lambda_{21}\frac{1}{4\lambda_{11}^2}\alpha_1^2 + \lambda_{23}\frac{3}{4\lambda_{33}^2}\alpha_3^2 + \lambda_{24}\frac{1}{\lambda_{44}^2}\alpha_4^2 + \lambda_{25}\frac{5}{4\lambda_{55}^2}\alpha_5^2\right) + \frac{1}{2\lambda_{22}}\alpha_2^2,$$

$$\dot{\alpha}_3 = -4\alpha_3 - \Big([\lambda_{31}] + 2[\lambda_{32}] + 3[\lambda_{33}] + 4[\lambda_{34}] + 5[\lambda_{35}]\Big)$$

$$+ 2\alpha_3\left(\frac{1}{4\lambda_{11}}\alpha_1 + \frac{1}{2\lambda_{22}}\alpha_2 + \frac{1}{\lambda_{44}}\alpha_4 + \frac{5}{4\lambda_{55}}\alpha_5\right)$$

$$- \left(\lambda_{31}\frac{1}{4\lambda_{11}^2}\alpha_1^2 + \lambda_{32}\frac{1}{2\lambda_{22}^2}\alpha_2^2 + \lambda_{34}\frac{1}{\lambda_{44}^2}\alpha_4^2 + \lambda_{35}\frac{5}{4\lambda_{55}^2}\alpha_5^2\right) + \frac{3}{4\lambda_{33}}\alpha_3^2,$$

$$\dot{\alpha}_4 = -4\alpha_4 - \Big([\lambda_{41}] + 2[\lambda_{42}] + 3[\lambda_{43}] + 4[\lambda_{44}] + 5[\lambda_{45}]\Big)$$

$$+ 2\alpha_4\left(\frac{1}{4\lambda_{11}}\alpha_1 + \frac{1}{2\lambda_{22}}\alpha_2 + \frac{3}{4\lambda_{33}}\alpha_3 + \frac{5}{4\lambda_{55}}\alpha_5\right)$$

$$- \left(\lambda_{41}\frac{1}{4\lambda_{11}^2}\alpha_1^2 + \lambda_{42}\frac{1}{2\lambda_{22}^2}\alpha_2^2 + \lambda_{43}\frac{3}{4\lambda_{33}^2}\alpha_3^2 + \lambda_{45}\frac{5}{4\lambda_{55}^2}\alpha_5^2\right) + \frac{1}{\lambda_{44}}\alpha_4^2,$$

$$\dot{\alpha}_5 = -4\alpha_5 - \Big([\lambda_{51}] + 2[\lambda_{52}] + 3[\lambda_{53}] + 4[\lambda_{54}] + 5[\lambda_{55}]\Big)$$

$$+ 2\alpha_5\left(\frac{1}{4\lambda_{11}}\alpha_1 + \frac{1}{2\lambda_{22}}\alpha_2 + \frac{3}{4\lambda_{33}}\alpha_3 + \frac{1}{\lambda_{44}}\alpha_4\right)$$

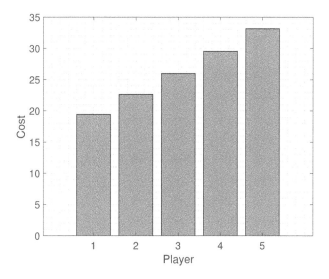

FIGURE 5.9
Optimal cost for the five players for the partially cooperation in a co-opetitive
scenario.

$$-\left(\lambda_{51}\frac{1}{4\lambda_{11}^2}\alpha_1^2 + \lambda_{52}\frac{1}{2\lambda_{22}^2}\alpha_2^2 + \lambda_{53}\frac{3}{4\lambda_{33}^2}\alpha_3^2 + \lambda_{54}\frac{1}{\lambda_{44}^2}\alpha_4^2\right) + \frac{5}{4\lambda_{55}}\alpha_5^2.$$

It can be verified in Figure 5.6 that the terminal boundary conditions are
satisfied, i.e.,

$$\alpha_1(1.5) = \left(\frac{1}{2}[\lambda_{11}] + [\lambda_{12}] + \frac{3}{2}[\lambda_{13}] + 2[\lambda_{14}] + \frac{5}{2}[\lambda_{15}]\right) = 2.5,$$

$$\alpha_2(1.5) = \left(\frac{1}{2}[\lambda_{21}] + [\lambda_{22}] + \frac{3}{2}[\lambda_{23}] + 2[\lambda_{24}] + \frac{5}{2}[\lambda_{25}]\right) = 2.95,$$

$$\alpha_3(1.5) = \left(\frac{1}{2}[\lambda_{31}] + [\lambda_{32}] + \frac{3}{2}[\lambda_{33}] + 2[\lambda_{34}] + \frac{5}{2}[\lambda_{35}]\right) = 3.35,$$

$$\alpha_4(1.5) = \left(\frac{1}{2}[\lambda_{41}] + [\lambda_{42}] + \frac{3}{2}[\lambda_{43}] + 2[\lambda_{44}] + \frac{5}{2}[\lambda_{45}]\right) = 3.7,$$

$$\alpha_5(1.5) = \left(\frac{1}{2}[\lambda_{51}] + [\lambda_{52}] + \frac{3}{2}[\lambda_{53}] + 2[\lambda_{54}] + \frac{5}{2}[\lambda_{55}]\right) = 4.$$

On the other hand, the Riccati equations corresponding to $\beta_1, \ldots, \beta_5$ are
presented in Figure 5.6, where the terminal boundary conditions

$$\beta_1(1.5) = ([\lambda_{11}] + 2[\lambda_{12}] + 3[\lambda_{13}] + 4[\lambda_{14}] + 5[\lambda_{15}]) = 5,$$
$$\beta_2(1.5) = ([\lambda_{21}] + 2[\lambda_{22}] + 3[\lambda_{23}] + 4[\lambda_{24}] + 5[\lambda_{25}]) = 5.9,$$

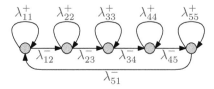

FIGURE 5.10
Co-opetitive parameters for the spiteful behavior in a co-opetitive scenario.

$$\beta_3(1.5) = ([\lambda_{31}] + 2[\lambda_{32}] + 3[\lambda_{33}] + 4[\lambda_{34}] + 5[\lambda_{35}]) = 6.7,$$
$$\beta_4(1.5) = ([\lambda_{41}] + 2[\lambda_{42}] + 3[\lambda_{43}] + 4[\lambda_{44}] + 5[\lambda_{45}]) = 7.4,$$
$$\beta_5(1.5) = ([\lambda_{51}] + 2[\lambda_{52}] + 3[\lambda_{53}] + 4[\lambda_{54}] + 5[\lambda_{55}]) = 8.$$

can be verified. Finally, Figures 5.7 and 5.8 present the evolution of the Riccati equations $\gamma_1, \ldots, \gamma_5$ and $\delta_1, \ldots, \delta_5$, whose terminal boundary conditions are null. The optimal costs for this co-opetitive mean-field-type game are given by

$$L_i^C(x, u_1^*, \ldots, u_5^*) = 8\beta_i(0) + 4\gamma_i(0) + \delta_i(0),$$

for all $i \in \mathcal{N}$, and they are presented in Figure 5.9.

Example 2: Partially Spiteful Behavior in a Co-opetitive Scenario

In contrast, we now consider a partially spiteful scenario corresponding to Figure 5.10, i.e., with the following co-opetitive parameters:

$$\Lambda = \begin{pmatrix} 1 & -0.1 & 0 & 0 & 0 \\ 0 & 1 & -0.1 & 0 & 0 \\ 0 & 0 & 1 & -0.1 & 0 \\ 0 & 0 & 0 & 1 & -0.1 \\ -0.1 & 0 & 0 & 0 & 1 \end{pmatrix}.$$

The cost functions corresponding to the five players with terminal time $T = 2\,\mathrm{s}$ are as follows:

$$L_1^C(x, u_1, u_2, \Lambda) = \frac{1}{2} \left\{ \frac{1}{2}[\lambda_{11}] \left(x^2(2) + \mathbb{E}[x(2)]^2\right) + [\lambda_{12}] \left(x^2(2) + \mathbb{E}[x(2)]^2\right) \right\}$$

$$+ \frac{1}{2} \int_0^2 \left\{ [\lambda_{11}] \left(5x^2 + 5\mathbb{E}[x]^2 + u_1^2 + \mathbb{E}[u_1]^2\right) \right.$$

$$\left. + [\lambda_{12}] \left(4x^2 + 4\mathbb{E}[x]^2 + 2u_2^2 + 2\mathbb{E}[u_2]^2\right) \right\} dt,$$

$$L_2^C(x, u_2, u_3, \Lambda) = \frac{1}{2} \left\{ [\lambda_{22}] \left(x^2(2) + \mathbb{E}[x(2)]^2\right) + \frac{3}{2}[\lambda_{23}] \left(x^2(2) + \mathbb{E}[x(2)]^2\right) \right\}$$

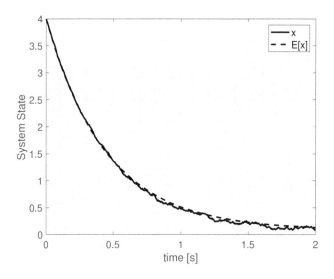

FIGURE 5.11
Evolution of the system state and its expectation for the spiteful behavior in a co-opetitive scenario.

$$+ \frac{1}{2} \int_0^2 \left\{ [\lambda_{22}] \left(4x^2 + 4\mathbb{E}[x]^2 + 2u_2^2 + 2\mathbb{E}[u_2]^2 \right) \right. $$

$$+ [\lambda_{23}] \left(3x^2 + 3\mathbb{E}[x]^2 + 3u_3^2 + 3\mathbb{E}[u_3]^2 \right) \Big\} dt,$$

$$L_3^C(x, u_3, u_4, \Lambda) = \frac{1}{2} \left\{ \frac{3}{2} [\lambda_{33}] \left(x^2(2) + \mathbb{E}[x(2)]^2 \right) + 2[\lambda_{34}] \left(x^2(2) + \mathbb{E}[x(2)]^2 \right) \right\}$$

$$+ \frac{1}{2} \int_0^2 \left\{ [\lambda_{33}] \left(3x^2 + 3\mathbb{E}[x]^2 + 3u_3^2 + 3\mathbb{E}[u_3]^2 \right) \right.$$

$$+ [\lambda_{34}] \left(2x^2 + 2\mathbb{E}[x]^2 + 4u_4^2 + 4\mathbb{E}[u_4]^2 \right) \Big\} dt,$$

$$L_4^C(x, u_4, u_5, \Lambda) = \frac{1}{2} \left\{ 2[\lambda_{44}] \left(x^2(2) + \mathbb{E}[x(2)]^2 \right) + \frac{5}{2} [\lambda_{45}] \left(x^2(2) + \mathbb{E}[x(2)]^2 \right) \right\}$$

$$+ \frac{1}{2} \int_0^2 \left\{ [\lambda_{44}] \left(2x^2 + 2\mathbb{E}[x]^2 + 4u_4^2 + 4\mathbb{E}[u_4]^2 \right) \right.$$

$$+ [\lambda_{45}] \left(x^2 + \mathbb{E}[x]^2 + 5u_5^2 + 5\mathbb{E}[u_5]^2 \right) \Big\} dt,$$

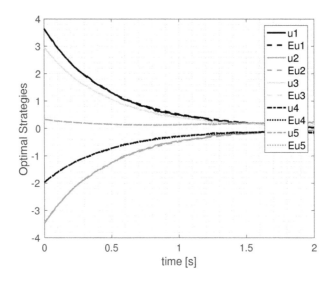

FIGURE 5.12
Evolution of the optimal strategies and their expectation for the spiteful behavior in a co-opetitive scenario.

$$L_5^C(x, u_1, u_5, \Lambda) = \frac{1}{2} \left\{ \frac{1}{2} [\lambda_{51}] \left(x^2(2) + \mathbb{E}[x(2)]^2 \right) + \frac{5}{2} [\lambda_{55}] \left(x^2(2) + \mathbb{E}[x(2)]^2 \right) \right\}$$

$$+ \frac{1}{2} \int_0^2 \left\{ [\lambda_{51}] \left(5x^2 + 5\mathbb{E}[x]^2 + u_1^2 + \mathbb{E}[u_1]^2 \right) \right.$$

$$\left. + [\lambda_{55}] \left(x^2 + \mathbb{E}[x]^2 + 5u_5^2 + 5\mathbb{E}[u_5]^2 \right) \right\} dt,$$

Figure 5.11 presents the evolution of the system state and its respective expectation, whereas the evolution of the state-dependent control inputs and their expectation is shown in Figure 5.12. The evolution of the Riccati equations, corresponding to this co-opetitive scenario involving spiteful behavior, are presented in Figures from 5.13 to 5.16. Notice that the terminal conditions

$$\alpha_1(2) = \left(\frac{1}{2} [\lambda_{11}] + [\lambda_{12}] \right) = 0.4,$$

$$\alpha_2(2) = \left([\lambda_{22}] + \frac{3}{2} [\lambda_{23}] \right) = 0.85,$$

$$\alpha_3(2) = \left(\frac{3}{2} [\lambda_{33}] + 2[\lambda_{34}] \right) = 1.3,$$

$$\alpha_4(2) = \left(2[\lambda_{44}] + \frac{5}{2} [\lambda_{45}] \right) = 1.75,$$

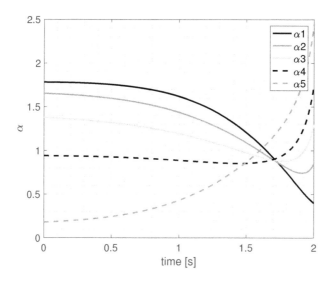

FIGURE 5.13
Evolution of the Riccati equations $\alpha_1, \ldots, \alpha_5$ for the spiteful behavior in a co-opetitive scenario.

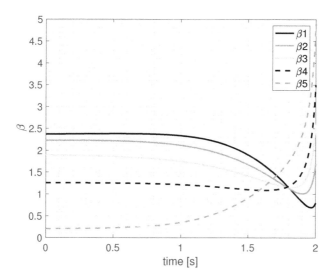

FIGURE 5.14
Evolution of the Riccati equations $\beta_1, \ldots, \beta_5$ for the spiteful behavior in a co-opetitive scenario.

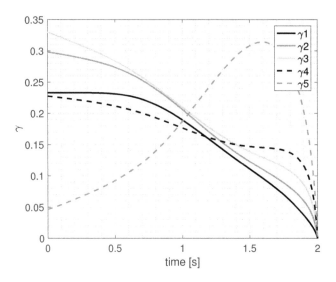

FIGURE 5.15
Evolution of the Riccati equations $\gamma_1, \ldots, \gamma_5$ for the spiteful behavior in a co-opetitive scenario.

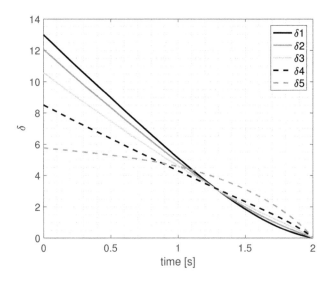

FIGURE 5.16
Evolution of the Riccati equations $\delta_1, \ldots, \delta_5$ for the spiteful behavior in a co-opetitive scenario.

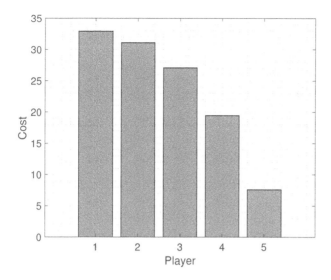

FIGURE 5.17
Optimal cost for the five players under the co-opetitive scenario with spiteful behavior.

$$\alpha_5(2) = \left(\frac{1}{2}[\lambda_{51}] + \frac{5}{2}[\lambda_{55}]\right) = 2.45.$$

and

$$\beta_1(2) = ([\lambda_{11}] + 2[\lambda_{12}]) = 0.8,$$
$$\beta_2(2) = ([2[\lambda_{22}] + 3[\lambda_{23}]) = 1.7,$$
$$\beta_3(2) = (3[\lambda_{33}] + 4[\lambda_{34}]) = 2.6,$$
$$\beta_4(2) = (4[\lambda_{44}] + 5[\lambda_{45}]) = 3.5,$$
$$\beta_5(2) = ([\lambda_{51}] + 5[\lambda_{55}]) = 4.9.$$

are satisfied for the differential equations $\alpha_1, \ldots, \alpha_5$ and $\beta_1, \ldots, \beta_5$, respectively. Moreover, the terminal values of $\gamma_1, \ldots, \gamma_5$ and $\delta_1, \ldots, \delta_5$ are zero fulfilling the terminal condition as well. Finally, Figure 5.17 shows the optimal costs associated with each player under this spiteful scenario, which is given by

$$L_i^C(x, u_1^*, \ldots, u_5^*) = 8\beta_i(0) + 4\gamma_i(0) + \delta_i(0).$$

Notice that, the higher cost is associated with the player with the smallest weight parameters, i.e., $q_1 = \bar{q}_1 = r_1 = \bar{r}_1 = 1$, for $t \in (0, T]$. In contrast, the player that most minimizes its cost is the one with the highest penalization parameters, i.e., $q_5 = \bar{q}_5 = r_5 = \bar{r}_5 = 5$, for $t \in (0, T]$. showing the interpretation of the results and how the decision-makers affect to each other.

In this chapter, we have presented a scenario that is more general than the fully-cooperative and non-cooperative cases. In fact, we have shown that the solutions presented in Chapter 4 can be retrieved from the co-opetitive solution computed in this chapter.

5.5   Exercises

1. Solve the co-opetitive mean-field-free game problem, i.e., for all $i \in \mathcal{N}$, consider the following cost

$$L_i^C(x, u_1, \ldots, u_n, \Lambda_i) = L_i^S(x, u_1, \ldots, u_n, \lambda_{ii})$$
$$+ L_i^A(x, u_1, \ldots, u_n, \Lambda_{i(-i)}), \ \forall i \in \mathcal{N}$$

with

$$L_i^S(x, u_1, \ldots, u_n, \lambda_{ii}) = \frac{1}{2}\lambda_{ii}q_i x^2(T)$$
$$+ \frac{1}{2}\int_0^T \lambda_{ii}\left(q_i x^2 + r_i u_i^2\right) dt,$$

$$L_i^A(x, u_1, \ldots, u_n, \Lambda_{i(-i)}) = \frac{1}{2}\sum_{j \in \mathcal{N}\setminus\{i\}}\lambda_{ij}q_j x^2(T)$$
$$+ \frac{1}{2}\int_0^T \sum_{j \in \mathcal{N}\setminus\{i\}}\lambda_{ij}\left(q_j x^2 + r_j u_j^2\right) dt,$$

and

$$\begin{cases} \underset{u_i \in \mathcal{U}_i}{\text{minimize}} \ \left[L_i^C(x, u_1, \ldots, u_n, \Lambda_i)\right], \\[2mm] \text{subject to} \\[2mm] \dot{x} = b_0 + b_1 x + \sum_{j \in \mathcal{N}} b_{2j} u_j, \\[2mm] x(0) = x_0, \\[2mm] \sum_{j \in \mathcal{N}} \lambda_{ij} = \pi_i, \ \forall i \in \mathcal{N}, \end{cases}$$

where the weight parameters in the cost functional are $q_i \geq 0$, $r_i > 0$, and the co-opetition powers are given.

2. Using the solution obtained in the previous problem, retrieved the solutions for the non-cooperative and cooperative mean-field-free game problems.

3. Write down the problem statement for an adversarial/robust two-decision-makers mean-field-type games. This is also known as a min-max mean-field-type game.

4. Re-write the adversarial/robust mean-field-type game problem as a co-opetitive mean-field-type game.

5. Retrieve the solution for the adversarial/robust mean-field-type game problem from the co-opetitive semi-explicit solution presented in this chapter.

6. After studying Chapters 11 and 12, solve a discrete-time co-opetitive mean-field-type game.

6

Mean-Field-Type Games with Jump-Diffusion and Regime Switching

The previous two chapters started the discussion on the mean-field-type game problems by using the simplest settings, i.e., with just a Brownian motion in the stochastic behavior of the system states, and considering a constant (or time-dependent) coefficient for the diffusion. The diffusion coefficient was not depending on the decisions that the players make neither on the evolution of the system state nor mean-field terms.

This chapter analyzes mean-field-type games beyond Brownian motion. Different from the previous chapters, here we consider that the diffusion component in the system dynamics depend on the state and control inputs. Besides, we consider jump-diffusion process and regime switching. Therefore, the problem addressed in this chapter is a more general mean-field-type game problem, which we solve in a semi-explicit manner by using the direct method. We present numerical results for these types of games.

We consider a linear-quadratic mean-field-type game problem under a jump-diffusion-regime switching state dynamics that include mean-field terms which are

- The conditional expectation of the state with respect to the filtration of the regime switching, in the drift, diffusion, and jump coefficient functions.

- The conditional expectation of the control-actions, in the drift, diffusion, and jump coefficient functions.

The state process is in one dimension. The instant cost function includes the conditional expectation of the state, control strategy, conditional variance of the state, covariance between the state and the control. We also include a switching regime, denoted by s, in the coefficients. Here, we are interested in finding mean-field-type 0-Nash equilibria, i.e., the exact solution for the mean-field-type game.

To solve the aforementioned problem in a semi-explicit way, we follow a direct method. In this more general case with other kinds of stochastic processes, the method starts by identifying a partial guess functional where the coefficients are regime-switching-dependent. Then, it uses Itô's formula, and its conditional expectation for jump-diffusion-regime switching processes, followed by a completion of squares for both control and expected value of the

DOI: 10.1201/9781003098607-6

control. Finally, the processes are identified by using an orthogonal decomposition technique and ordinary differential equations are derived in a semi-explicit way.

From the perspective of the engineering applications, the presented new game settings allow the consideration of a wide variety of uncertainties that can be modeled in different ways. For instance, we might have noisy uncertainties that can be captured and modeled by means of a Brownian motion, but also, we can take into consideration abrupt changes that are more suitably modeled by jump processes. To provide a possible concrete example, when considering renewable energies and power generation, one might have uncertainties associated with the behavior of the sun or the wind, which are modeled with the Brownian motion. On the other hand, one might model uncertainties that could create bigger changes such as an abrupt and unexpected climate change or the imposition of new laws by the government affecting abruptly the behavior of the power price in the market. In addition, the consideration of random coefficients is understood as having a Markov chain associated with the possible values the coefficients can take.

6.1   Mean-Field-Type Game Set-up

Consider the following state dynamics with jump-diffusion and regime switching on which $n \geq 2$ $(n \in \mathbb{N})$ decision-makers are strategically interacting, where $\mathcal{N} = \{1, \ldots, n\}$:

$$\mathrm{d}x = b\mathrm{d}t + \sigma\mathrm{d}B + \int_{\theta \in \Theta} \mu(\cdot, \theta)\tilde{N}(\mathrm{d}t, \mathrm{d}\theta), \qquad (6.1)$$

where

$$b = b_0 + b_1 x + \bar{b}_1 \mathbb{E}[x] + \sum_{j \in \mathcal{N}} b_{2j} u_j + \sum_{j \in \mathcal{N}} \bar{b}_{2j} \mathbb{E}[u_j],$$

$$\sigma = \sigma_0 + \sigma_1 x + \bar{\sigma}_1 \mathbb{E}[x] + \sum_{j \in \mathcal{N}} \sigma_{2j} u_j + \sum_{j \in \mathcal{N}} \bar{\sigma}_{2j} \mathbb{E}[u_j],$$

$$\mu(\cdot, \theta) = \mu_0 + \mu_1 x + \bar{\mu}_1 \mathbb{E}[x] + \sum_{j \in \mathcal{N}} \mu_{2j} u_j + \sum_{j \in \mathcal{N}} \bar{\mu}_{2j} \mathbb{E}[u_j],$$

with the coefficients $b_k(t, s); \sigma_k(t, s); \bar{b}_k(t, s); \bar{\sigma}_k(t, s) \in \mathbb{R}$ and $s \in \mathcal{S}$, is a regime switching, $\mathcal{S}$ is a non-empty and finite set of switching regimes. Moreover $\mu_k(t, s, \theta) \in \mathbb{R}$, $\theta \in \Theta$ where $\Theta \subset \mathbb{R}^d$ is the set of jump sizes, for all $k \in \{0, 1, 2\}$. The process B is a standard one-dimensional Brownian motion, $\tilde{N}(\mathrm{d}t, \mathrm{d}\theta) = N(\mathrm{d}t, \mathrm{d}\theta) - \nu(\mathrm{d}\theta)\mathrm{d}t$ is a compensated jump process with Radon measure ν over Θ. The switching process s switches among elements in $\mathcal{S}$ with transition rate $\tilde{q}_{ss'}$. The Q-matrix $Q = (\tilde{q}_{ss'}, \ (s, s') \in \mathcal{S}^2)$, satisfies

$$s \neq s', \tilde{q}_{ss'} > 0, \ \tilde{q}_{ss} = -\sum_{s' \neq s} \tilde{q}_{ss'}.$$

It is assumed that the processes $s(t), B(t), N(t, .), x_0$ are mutually independent.

The term $\mathbb{E}[x] = \mathbb{E}[x|\mathcal{F}^s]$ denotes the expected value of x given $s(t) = s$, and similarly, $\mathbb{E}[u] = \mathbb{E}[u|\mathcal{F}^s]$. These are called conditional state mean-
field and conditional control-action mean-field terms. The state dynamics are
therefore of conditional McKean-Vlasov type as the conditional probability
measure of the state appears in the drift, diffusion, and jump coefficients via
$(\mathbb{E}[x], \mathbb{E}[u])$. Below we denote a solution (if any) to (6.1) by $x := x^{u, \tau_0, s_0}$.

We associate a cost functional of mean-field type to the decision-maker
$i \in \mathcal{N}$ as

$$L_i(x, s, u_1, \ldots, u_n) = \frac{1}{2}q_i x^2(T) + \frac{1}{2}\bar{q}_i \mathbb{E}[x(T)]^2 + \varepsilon_{i3}(T)\mathbb{E}[x(T)]$$

$$+ \frac{1}{2}\int_0^T \left(q_i x^2 + \bar{q}_i \mathbb{E}[x]^2 + r_i u_i^2 + \bar{r}_i \mathbb{E}[u_i]^2\right) dt$$

$$+ \int_0^T \left(\varepsilon_{1i} x u_i + \bar{\varepsilon}_{1i}\mathbb{E}[x]\mathbb{E}[u_i] + \varepsilon_{2i}\mathbb{E}[u_i] + \varepsilon_{i3}\mathbb{E}[x]\right) dt,$$

where the weight parameters are switching-dependent, i.e., $q_i = q_i(t, s(t))$,
$\bar{q}_i = q_i(t, s(t))$; and $r_i = r_i(t, s(t))$, $\bar{r}_i = \bar{r}_i(t, s(t)) \in \mathbb{R}$, $\varepsilon_{ik} = \varepsilon_{ik}(t, s(t)) \in \mathbb{R}$.
All these coefficients are assumed to be integrable.

Figure 6.1 presents the general scheme of the non-cooperative mean-field-
type game with jumps and regime switching. It can be seen that the evolution
of the system state is not only dependent of the strategic selections from the
players, but also dependent of a Brownian motion, Poisson jumps, and regime
switching that affects the parameters in the dynamical system. Each player
has its own cost function aiming to minimize the variance of both the system
states and control inputs. Given the other decision-makers' strategies

$$u_{-i} = (u_1, \ldots, u_{i-1}, \ldots, u_{i+1}, u_n),$$

the statement of this mean-field-type game problem is presented next.

Problem 16 (Problem with jump-diffusion and regime switching)
The mean-field-type game problem with jump-diffusion process and random coefficients (regime-switching-dependent parameters) is as follows:

$$
\begin{cases}
\underset{u_i \in \mathcal{U}_i}{\text{minimize}} \ \mathbb{E}\left[L_i(x, s, u_1, \dots, u_n) | x(0) = x_0, s(0) = s_0 \right], \\[2mm]
\text{subject to} \\[2mm]
\mathrm{d}x = b\mathrm{d}t + \sigma \mathrm{d}B + \displaystyle\int_{\theta \in \Theta} \mu(\cdot, \theta)\tilde{N}(\mathrm{d}t, \mathrm{d}\theta), \\[2mm]
x(0) = x_0.
\end{cases}
\tag{6.2}
$$

the initial state condition x_0 is given. The weight parameters are $q_i, q_i + \bar{q}_i \geq 0$, $r_i, r_i + \bar{r}_i > 0$, and the system parameters of the functions b, σ, and μ, for all $j \in \mathcal{N}$ are time and switching dependent.

Let us briefly recall both the best-response and Nash equilibrium definitions that have been stated in Chapter 4 in the Definitions 8 and 9, respectively.

Definition 12 (Best Response) *A minimizer u_i^* to Problem (16) is called a best-response strategy of the ith decision-maker given the strategies u_{-i} of others. Hence, the set of best-response strategies of the ith decision-maker is denoted by $\mathrm{BR}_i(u_{-i})$.*

Definition 13 (Nash equilibrium) *the strategy profile $u^* \in \mathcal{U} = \prod_{j=1}^n \mathcal{U}_j$ is a Nash equilibrium if for all i, $u_i^* \in \mathrm{BR}_i(u_{-i}^*)$.*

We consider the following assumption:

$$
\begin{cases}
q_i(T), q_i(T) \geq 0, q_i(T) + \bar{q}_i(T) \geq 0, q_i + \bar{q}_i \geq 0, r_i > 0, r_i + \bar{r}_i > 0, \\[2mm]
\mathbb{E}[x_0^2] < +\infty, \ \displaystyle\int_0^T \int_{\theta \in \Theta} \phi_{ik}^2 \nu(\mathrm{d}\theta) < +\infty, \text{with } \phi_{ik} \in \{\mu_{ik}, \bar{\mu}_{ik}\}, \\[2mm]
|\mathcal{S}| < +\infty, b_k, \sigma_k^2, \bar{b}_k, \bar{\sigma}_k^2 \in L^1([0,T] \times \mathcal{S}, \mathbb{R}).
\end{cases}
\tag{6.3}
$$

Remark 18 *Under Assumption (8.4), the state dynamics has a unique solution for each $u \in \mathcal{U}$. Note that we do not impose boundedness conditions on the coefficients or Lipschitz conditions on the cost. The integrability condition above yields*

$$
\int_0^T [|b_0(t,s)| + |b_1(t,s)| + |\bar{b}_1(t,s)| + |b_{2j}(t,s)| + |\bar{b}_{2j}(t,s)|]\mathrm{d}t
$$

$$
+ \int_0^T [\sigma_0^2(t,s) + \sigma_1^2(t,s) + \bar{\sigma}_1^2(t,s) + \sigma_{2j}^2(t,s) + \bar{\sigma}_{2j}^2(t,s)]\mathrm{d}t
$$

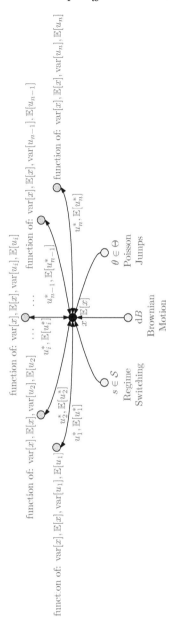

FIGURE 6.1
General scheme of the non-cooperative mean-field-type game with jumps and regime switching. Dark gray nodes represent the n players, the black node represents the system state, and the light gray node represents the stochasticity affecting the system state.

$$+ \int_0^T \int_{\theta \in \Theta} \Big[\mu_0^2(t,s,\theta) + \mu_1^2(t,s,\theta) + \bar{\mu}_1^2(t,s,\theta)$$

$$+ \mu_{2j}^2(t,s,\theta) + \bar{\mu}_{2j}^2(t,s,\theta) \Big] \nu(d\theta) dt < +\infty. \tag{6.4}$$

If both $r(t,s) \geq \tilde{\varepsilon} > 0$ *and* $r(t,s) + \bar{r}(t,s) \geq \hat{\varepsilon} > 0$ *for some* $\tilde{\varepsilon}$ *and* $\hat{\varepsilon}$, *then the cost function* $L_i(x,s,u_1,\ldots,u_n)$ *is coercive and continuous in* u_i. *Moreover, the set* U_i *is nonempty and convex. Therefore, Problem (16) is well defined, which implies existence of pure best-response strategies.*

In the next section, we compute the semi-explicit solution for this mean-field-type game with jump-diffusion process and regime switching (random coefficients).

We highlight that this proof is easy to follow but it is quite long since we construct a semi-explicit solution. Moreover, notice that we have state and control-input terms in both the diffusion and the jumps. These terms appear with the square in the Itô's formula, making the computation require extra auxiliary calculations along the proof. Hence, we will introduce all the details associated with the calculation. The procedure is the same we have been performing throughout the book.

6.2 Semi-explicit Solution of the Mean-Field-Type Game with Jump-Diffusion Process and Regime Switching

The mean-field-type game problem incorporates not only the Brownian motion, but also jumps and regime switching. Besides, all these stochastic processes are considered to be both state-and-control-input dependent. Then, its solution (see Theorem 1) is more general than the ones studied in Propositions 13 and 14.

Theorem 1 *The mean-field-type Nash equilibrium strategy and optimal cost are given by:*

$$u_i^* = \mathbb{E}[u_i]^* - \frac{\tilde{\tau}_i}{c_i}(x - \mathbb{E}[x]), \tag{6.5a}$$

$$\mathbb{E}[u_i]^* = -\frac{\bar{\tau}_i}{\bar{c}_i} = -\frac{\bar{\lambda}_{2i}\mathbb{E}[x]}{\bar{c}_i} - \frac{\bar{\lambda}_{i3}}{\bar{c}_i}, \tag{6.5b}$$

$$L_i^*(x,s,u_1^*,\ldots,u_n^*) = \frac{1}{2}\mathbb{E}[\alpha_i(0,s(0))(x_0 - \mathbb{E}[x_0])^2] + \frac{1}{2}\mathbb{E}[\beta_i(0,s(0))\mathbb{E}[x_0]^2]$$

$$+ \mathbb{E}[\gamma_i(0,s(0))\mathbb{E}[x_0]] + \mathbb{E}[\delta_i(0,s(0))], \tag{6.5c}$$

where α_i, β_i, γ_i, and δ_i solve the following differential equations:

$$\dot{\alpha}_i = -2\alpha_i b_1 - \alpha_i \left(\sigma_1^2 + \int_{\theta \in \Theta} \mu_1^2 \nu(d\theta) \right) - q_i - \sum_{s' \in \mathcal{S}} [\alpha_i(.,s') - \alpha_i(.,s)] \tilde{q}_{ss'}$$

$$- \alpha_i \left(\sum_{j \in \mathcal{N} \backslash \{i\}} \sigma_{2j} \frac{\tilde{\tau}_j}{c_j} \right)^2 - \int_{\theta \in \Theta} \alpha_i \left(\sum_{j \in \mathcal{N} \backslash \{i\}} \mu_{2j} \frac{\tilde{\tau}_j}{c_j} \right)^2 \nu(d\theta)$$

$$+ 2\alpha_i \sum_{j \in \mathcal{N} \backslash \{i\}} b_{2j} \frac{\tilde{\tau}_j}{c_j} + \frac{\tilde{\tau}_i^2}{c_i} + 2\alpha_i \sum_{j \in \mathcal{N} \backslash \{i\}} \left(\sigma_1 \sigma_{2j} + \int_{\theta \in \Theta} \mu_1 \mu_{2j} \nu(d\theta) \right) \frac{\tilde{\tau}_j}{c_j},$$

$$\dot{\beta}_i = -2\beta_i (b_1 + \bar{b}_1) - \alpha_i \left((\sigma_1 + \bar{\sigma}_1)^2 + \int_{\theta \in \Theta} (\mu_1 + \bar{\mu}_1)^2 \nu(d\theta) \right) - q_i - \bar{q}_i$$

$$- \sum_{s' \in \mathcal{S}} [\beta_i(.,s') - \beta_i(.,s)] \tilde{q}_{ss'} - \alpha_i \left(\sum_{j \in \mathcal{N} \backslash \{i\}} (\sigma_{2j} + \bar{\sigma}_{2j}) \frac{\bar{\lambda}_{2j}}{\bar{c}_j} \right)^2$$

$$- \alpha_i \int_{\theta \in \Theta} \left(\sum_{j \in \mathcal{N} \backslash \{i\}} (\mu_{2j} + \bar{\mu}_{2j}) \frac{\bar{\lambda}_{2j}}{\bar{c}_j} \right)^2 \nu(d\theta) + \frac{\bar{\lambda}_{2i}^2}{\bar{c}_i}$$

$$+ 2 \sum_{j \in \mathcal{N} \backslash \{i\}} \left[\alpha_i \Big((\sigma_1 + \bar{\sigma}_1)(\sigma_{2j} + \bar{\sigma}_{2j}) \right.$$

$$+ \int_{\theta \in \Theta} (\mu_1 + \bar{\mu}_1)(\mu_{2j} + \bar{\mu}_{2j}) \nu(d\theta) \Big) + \beta_i (b_{2j} + \bar{b}_{2j}) \Big] \frac{\bar{\lambda}_{2j}}{\bar{c}_j},$$

$$\dot{\gamma}_i = -\beta_i b_0 - \gamma_i (b_1 + \bar{b}_1) - \varepsilon_{i3} - \alpha_i \left(\sigma_0 (\sigma_1 + \bar{\sigma}_1) + \int_{\theta \in \Theta} \mu_0 (\mu_1 + \bar{\mu}_1) \nu(d\theta) \right)$$

$$- \sum_{s' \in \mathcal{S}} [\gamma_i(.,s') - \gamma_i(.,s)] \tilde{q}_{ss'}$$

$$- \alpha_i \left(\sum_{j \in \mathcal{N} \backslash \{i\}} (\sigma_{2j} + \bar{\sigma}_{2j}) \frac{\bar{\lambda}_{2j}}{\bar{c}_j} \right) \left(\sum_{j \in \mathcal{N} \backslash \{i\}} (\sigma_{2j} + \bar{\sigma}_{2j}) \frac{\bar{\lambda}_{j3}}{\bar{c}_j} \right)$$

$$- \alpha_i \int_{\theta \in \Theta} \left(\sum_{j \in \mathcal{N} \backslash \{i\}} (\mu_{2j} + \bar{\mu}_{2j}) \frac{\bar{\lambda}_{2j}}{\bar{c}_j} \right) \left(\sum_{j \in \mathcal{N} \backslash \{i\}} (\mu_{2j} + \bar{\mu}_{2j}) \frac{\bar{\lambda}_{j3}}{\bar{c}_j} \right) \nu(d\theta)$$

$$+ \sum_{j \in \mathcal{N} \backslash \{i\}} \left[\alpha_i \left(\sigma_0 (\sigma_{2j} + \bar{\sigma}_{2j}) + \int_{\theta \in \Theta} \mu_0 (\mu_{2j} + \bar{\mu}_{2j}) \nu(d\theta) \right) + \gamma_i (b_{2j} + \bar{b}_{2j}) \right] \frac{\bar{\lambda}_{2j}}{\bar{c}_j}$$

$$+ \sum_{j \in \mathcal{N} \backslash \{i\}} \left[\alpha_i \Big((\sigma_1 + \bar{\sigma}_1)(\sigma_{2j} + \bar{\sigma}_{2j}) \right.$$

$$+ \int_{\theta \in \Theta} (\mu_1 + \bar{\mu}_1)(\mu_{2j} + \bar{\mu}_{2j}) \nu(d\theta) \Big) + \beta_i (b_{2j} + \bar{b}_{2j}) \Big] \frac{\bar{\lambda}_{j3}}{\bar{c}_j} + \frac{\bar{\lambda}_{2i} \bar{\lambda}_{i3}}{\bar{c}_i},$$

$$\dot{\delta}_i = -\gamma_i b_0 - \frac{\alpha_i}{2}\left(\sigma_0^2 + \int_{\theta \in \Theta} \mu_0^2 \nu(\mathrm{d}\theta)\right) - \sum_{s' \in S}[\delta_i(.,s') - \delta_i(.,s)]\tilde{q}_{ss'}$$

$$- \alpha_i \frac{1}{2}\left(\sum_{j \in \mathcal{N}\setminus\{i\}}(\sigma_{2j} + \bar{\sigma}_{2j})\frac{\bar{\lambda}_{j3}}{\bar{c}_j}\right)^2$$

$$- \alpha_i \int_{\theta \in \Theta} \frac{1}{2}\left(\sum_{j \in \mathcal{N}\setminus\{i\}}(\mu_{2j} + \bar{\mu}_{2j})\frac{\bar{\lambda}_{j3}}{\bar{c}_j}\right)^2 \nu(\mathrm{d}\theta)$$

$$+ \sum_{j \in \mathcal{N}\setminus\{i\}}\left[\alpha_i\left(\sigma_0(\sigma_{2j} + \bar{\sigma}_{2j}) + \int_{\theta \in \Theta}\mu_0(\mu_{2j} + \bar{\mu}_{2j})\nu(\mathrm{d}\theta)\right) + \gamma_i(b_{2j} + \bar{b}_{2j})\right]\frac{\bar{\lambda}_{j3}}{\bar{c}_j}$$

$$+ \frac{\bar{\lambda}_{i3}^2}{2\bar{c}_i},$$

and satisfy the following terminal conditions:

$$\alpha_i(T, s(T)) = q_i(T),$$
$$\beta_i(T, s(T)) = q_i(T) + \bar{q}_i(T),$$
$$\gamma_i(T, s(T)) = \varepsilon_{i3}(T),$$
$$\delta_i(T, s(T)) = 0.$$

Moreover, the terms τ_i and $\bar{\tau}_i$ are obtained from $A\tilde{\tau} = v$, and $\bar{A}\bar{\tau} = \bar{v}$, being $A_{ii} = \bar{A}_{ii} = 1$,

$$A_{ij} = \frac{\alpha_i}{c_j}\left\{\sigma_{2i}\sigma_{2j} + \int_{\theta \in \Theta}\mu_{2i}\,\mu_{2j}\nu(\mathrm{d}\theta)\right\}, \ \forall j \neq i,$$

$$v_i = \varepsilon_{1i} + \alpha_i\left\{b_{2i} + \sigma_1\sigma_{2i} + \int_{\theta \in \Theta}\mu_1\mu_{2i}\nu(\mathrm{d}\theta)\right\},$$

$$\bar{A}_{ij} = \frac{\alpha_i}{\bar{c}_j}\left\{(\sigma_{2i} + \bar{\sigma}_{2i})(\sigma_{2j} + \bar{\sigma}_{2j}) + \int_{\theta \in \Theta}(\mu_{2i} + \bar{\mu}_{2i})(\mu_{2j} + \bar{\mu}_{2j})\nu(\mathrm{d}\theta)\right\}, \ \forall j \neq i,$$

$$\bar{v}_i = \varepsilon_{2i} + \alpha_i\left(\sigma_0(\sigma_{2i} + \bar{\sigma}_{2i}) + \int_{\theta \in \Theta}\mu_0(\mu_{2i} + \bar{\mu}_{2i})\nu(\mathrm{d}\theta)\right)$$

$$+ \gamma_i(b_{2i} + \bar{b}_{2i}) + \left\{\varepsilon_{1i} + \bar{\varepsilon}_{1i} + \alpha_i\left((\sigma_1 + \bar{\sigma}_1)(\sigma_{2i} + \bar{\sigma}_{2i})\right.\right.$$

$$\left.\left. + \int_{\theta \in \Theta}(\mu_1 + \bar{\mu}_1)(\mu_{2i} + \bar{\mu}_{2i})\nu(\mathrm{d}\theta)\right) + \beta_i(b_{2i} + \bar{b}_{2i})\right\}\mathbb{E}[x],$$

$$\bar{A}\bar{\tau} = \bar{A}\left(\bar{\lambda}_2\mathbb{E}[x] + \bar{\lambda}_3\right) = \bar{v},$$

$$c_i = r_i + \alpha_i[\sigma_{2i}^2 + \int_{\theta \in \Theta}\mu_{2i}^2\nu(\mathrm{d}\theta)],$$

$$\bar{c}_i = (r_i + \bar{r}_i) + \alpha_i\left\{(\sigma_{2i} + \bar{\sigma}_{2i})^2 + \int_{\theta \in \Theta}(\mu_{2i} + \bar{\mu}_{2i})^2\nu(\mathrm{d}\theta)\right\},$$

whenever these ordinary differential equations have a unique solution that does not blow up in $[0,T]$, and the matrices A and $\bar{A}$ are invertible.

Notice that the solution (6.5) is in state-and-mean-field feedback form. The reader may ask how to feedback the mean-field term $\mathbb{E}[x] = \mathbb{E}[x|\mathcal{F}^s]$. Remark 19 addresses this question.

Remark 19 *Taking the expectation in* (6.1), $\mathbb{E}[x|\mathcal{F}^s]$ *satisfies the following ordinary differential equation:*

$$\mathbb{E}[\dot{x}] = \left[b_0 - \sum_{j\in\mathcal{N}} (b_{2j} + \bar{b}_{2j}) \frac{\lambda_{j3}}{\bar{c}_j} \right] + \left[(b_1 + \bar{b}_1) - \sum_{j\in\mathcal{N}} (b_{2j} + \bar{b}_{2j}) \frac{\lambda_{2j}}{\bar{c}_j} \right] \mathbb{E}[x],$$

$$\mathbb{E}[x(0)] = \mathbb{E}[x_0].$$

Therefore, the mean-field term $\mathbb{E}[x|\mathcal{F}^s]$ *is given in a semi-explicit way:*

$$\mathbb{E}[x(t)] = \mathbb{E}[x(t)|\mathcal{F}^s_t],$$

$$= e^{\int_0^t \xi_1 dt'} \mathbb{E}[x_0] + e^{\int_0^t \xi_1 dt'} \int_0^t dt' e^{\int_0^{t'} \xi_1 dt''} \xi_0(t'),$$

where

$$\xi_1 = (b_1 + \bar{b}_1) - \sum_{j\in\mathcal{N}} (b_{2j} + \bar{b}_{2j}) \frac{\lambda_{2j}}{\bar{c}_j},$$

$$\xi_0 = b_0 - \sum_{j\in\mathcal{N}} (b_{2j} + \bar{b}_{2j}) \frac{\lambda_{j3}}{\bar{c}_j},$$

are function of β, λ.

As a preliminary step before computing the semi-explicit solution, we rewrite the problem, i.e., the cost functional and the system dynamics depending on the terms $(x - \mathbb{E}[x])$, $\mathbb{E}[x]$, $(u_i - \mathbb{E}[u_i])$, and $\mathbb{E}[u_i]$, for all $i \in \mathcal{N}$. This form allows us to identify the risk term (variance) appearing in the minimization problem. Consider the following problem, which is an equivalent reformulation of Problem 16:

Problem 17 *Let us reformulate Problem 16 in terms of risk, i.e.,*

$$\inf_{u_i} \mathbb{E}\left[\hat{L}_i(x, s, u_1, \ldots, u_n) | x(0) = x_0, s(0) = s_0 \right], \tag{6.6}$$

where

$$\hat{L}_i(x, s, u_1, \ldots, u_n) = \frac{1}{2} q_i(T)(x(T) - \mathbb{E}[x(T)])^2 + \frac{1}{2}(q_i(T) + \bar{q}_i(T))\mathbb{E}[x(T)]^2$$

$$+ \varepsilon_{i3}(T)\mathbb{E}[x(T)] + \frac{1}{2}\int_0^T \Big(q_i(x - \mathbb{E}[x])^2 + (q_i + \bar{q}_i)\mathbb{E}[x]^2$$

$$+ r_i(u_i - \mathbb{E}[u_i])^2 + (r_i + \bar{r}_i)\mathbb{E}[u_i]^2 \Big) dt$$

$$+ \int_0^T \Bigg(\varepsilon_{1i}(x - \mathbb{E}[x])(u_i - \mathbb{E}[u_i]) + (\varepsilon_{1i} + \bar{\varepsilon}_{1i})\mathbb{E}[x]\mathbb{E}[u_i]$$

$$+ \varepsilon_{2i}\mathbb{E}[u_i] + \varepsilon_{i3}\mathbb{E}[x] \Bigg) dt,$$

subject to

$$dx = \Bigg(b_0 + b_1(x - \mathbb{E}[x]) + (b_1 + \bar{b}_1)\mathbb{E}[x]$$

$$+ \sum_{j\in\mathcal{N}} b_{2j}(u_j - \mathbb{E}[u_j]) + \sum_{j\in\mathcal{N}}(b_{2j} + \bar{b}_{2j})\mathbb{E}[u_j] \Bigg) dt$$

$$+ \Bigg(\sigma_0 + \sigma_1(x - \mathbb{E}[x]) + (\sigma_1 + \bar{\sigma}_1)\mathbb{E}[x]$$

$$+ \sum_{j\in\mathcal{N}} \sigma_{2j}(u_j - \mathbb{E}[u_j]) + \sum_{j\in\mathcal{N}}(\sigma_{2j} + \bar{\sigma}_{2j})\mathbb{E}[u_j] \Bigg) dB$$

$$+ \int_{\theta\in\Theta} \Bigg(\mu_0 + \mu_1(x - \mathbb{E}[x]) + (\mu_1 + \bar{\mu}_1)\mathbb{E}[x] + \sum_{j\in\mathcal{N}} \mu_{2j}(u_j - \mathbb{E}[u_j])$$

$$+ \sum_{j\in\mathcal{N}}(\mu_{2j} + \bar{\mu}_{2j})\mathbb{E}[u_j] \Bigg) \tilde{N}(dt, d\theta), \qquad (6.7)$$

$$x(0) = x_0, \ s(0) = s_0.$$

Notice that the expectation of the cost $\hat{L}_i(x, s, u_1, \dots, u_n)$ shows the variance terms for the states and control inputs.

Proposition 16 shows that the solution of Problem 16 can be obtained by solving the reformulation of Problem 17.

Proposition 16 *Problem 16 is exactly the same as Problem 17.*

The equivalence of Problems 16 and 17 is related to what was discuss in Chapter 4 where we showed that the mean-field-type game problem incorporates risk terms. Such equivalence is also explained by means of the proof of Proposition 16.

Proof 16 (Proposition 16) *Use the decompositions $x = x - \mathbb{E}[x] + \mathbb{E}[x]$ and $u_j = u_j - \mathbb{E}[u_j] + \mathbb{E}[u_j]$ noticing that both $\mathbb{E}[\mathbb{E}[x](x - \mathbb{E}[x])] = 0$ and $\mathbb{E}[\mathbb{E}[u_j](u_j - \mathbb{E}[u_j])] = 0$, i.e., $x - \mathbb{E}[x] \perp \mathbb{E}[x]$ and $u_j - \mathbb{E}[u_j] \perp \mathbb{E}[u_j]$.*

Remark 20 *Notice that $\mathbb{E}[\hat{L}_i(x, s, u_1, \dots, u_n)|x(0) = x_0, s(0) = s_0]$ can be rewritten as*

$$\mathbb{E}[\hat{L}_i(x, s, u_1, \dots, u_n)|x(0) = x_0, s(0) = s_0]$$

$$= \frac{1}{2}\left[q_{iT} \operatorname{var}(x(T)) + (q_i(T) + \bar{q}_i(T))\mathbb{E}[x(T)]^2\right] + \varepsilon_{i3}(T)\mathbb{E}[x(T)]$$

$$+ \frac{1}{2}\int_0^T \left(q_i \operatorname{var}(x) + (q_i + \bar{q}_i)\mathbb{E}[x]^2 + r_i \operatorname{var}(u_i) + (r_i + \bar{r}_i)\mathbb{E}[u_i]^2\right) \mathrm{dt}$$

$$+ \int_0^T \left(\varepsilon_{1i}\operatorname{cov}(x, u_i) + (\varepsilon_{1i} + \bar{\varepsilon}_{1i})\mathbb{E}[x]\mathbb{E}[u_i] + \varepsilon_{2i}\mathbb{E}[u_i] + \varepsilon_{i3}\mathbb{E}[x]\right) \mathrm{dt},$$

where $\operatorname{cov}(x, u_j) = \mathbb{E}(x - \mathbb{E}[x])(u_j - \mathbb{E}[u_j])$ *is the conditional covariance and* $\operatorname{var}[y] = \mathbb{E}[y - \mathbb{E}[y]]^2$ *is the variance of* y, *which clearly indicates a weighted conditional variance-reduction problem depending on switching regime* s.

With the new convenient statement of the mean-field-type game problem with jump-diffusion process and regime switching (random coefficients), we construct the semi-explicit solution next.

Proof 17 (Theorem 1) *We construct a semi-explicit solution to Problems 16 and 17 via a direct method. Let the guess functional of decision-maker* $i \in \mathcal{N}$ *be*

$$f_i(t, x, s) = \frac{1}{2}\alpha_i(x - \mathbb{E}[x])^2 + \frac{1}{2}\beta_i\mathbb{E}[x]^2 + \gamma_i\mathbb{E}[x] + \delta_i,$$

where $\alpha_i = \alpha_i(t, s)$, $\beta_i = \beta_i(t, s)$, $\gamma_i = \gamma_i(t, s)$, $\delta_i = \delta_i(t, s)$, *and* $x = x^{u, x_0, s_0}(t)$ *solution to (6.7).*

It is crucial to emphasize that the Itô's formula we need to apply in this problem is different from the integration formula we have been using in the previous chapters. This is because the problem in this chapter incorporates new stochastic processes. The Itô's formula with jump-diffusion processes and switching is as follows:

$$f_i(T, x, s) - f_i(0, x_0, s_0)$$

$$= \int_0^T \left(\frac{\partial f_i(t, x, s)}{\partial t} + \frac{\partial f_i(t, x, s)}{\partial x}D_r(t, s, x, \mathbb{E}[x], u, \mathbb{E}[u])\right.$$

$$\left. + \frac{\partial^2 f_i(t, x, s)}{\partial x^2}\frac{D_f(t, s, x, \mathbb{E}[x], u, \mathbb{E}[u])^2}{2}\right) \mathrm{dt}$$

$$+ \int_0^T D_f(t, s, x, \mathbb{E}[x], u, \mathbb{E}[u])\frac{\partial f_i(t, x, s)}{\partial x}\mathrm{dB}$$

$$+ \int_0^T \int_{\theta \in \Theta} \left(f_i(t, x + \mu(t, \theta), s) - f_i(t, x, s) - \frac{\partial f_i(t, x, s)}{\partial x}\mu(t, \theta)\right) \nu(\mathrm{d}\theta)\mathrm{dt}$$

$$+ \int_0^T \int_{\theta \in \Theta} (f_i(t_-, x + \mu(t_-, \theta), s) - f_i(t_-, x, s)) \tilde{N}(\mathrm{dt}, \mathrm{d}\theta)$$

$$+ \int_0^T \sum_{s' \in \mathcal{S}} [f_i(., s') - f_i(., s)]\tilde{q}_{ss'}\mathrm{dt}. \tag{6.8}$$

Now, for the sake of clarity, we compute each term of the Itô's formula separately.

$$\frac{\partial f_i(t,x,s)}{\partial t} = \frac{1}{2}\left(x - \mathbb{E}[x]\right)^2 \dot{\alpha}_i - \alpha_i\left(x - \mathbb{E}[x]\right)\mathbb{E}[\dot{x}] + \frac{1}{2}\mathbb{E}[x]^2\dot{\beta}_i$$
$$+ \beta_i\mathbb{E}[x]\mathbb{E}[\dot{x}] + \mathbb{E}[x]\dot{\gamma}_i + \gamma_i\mathbb{E}[\dot{x}] + \dot{\delta}_i, \tag{6.9a}$$

$$\frac{\partial f_i(t,x,s)}{\partial x} = \alpha_i\left(x - \mathbb{E}[x]\right), \tag{6.9b}$$

$$\frac{\partial^2 f_i(t,x,s)}{\partial x^2} = \alpha_i, \tag{6.9c}$$

and the jump-related and switching-related terms are as follows:

$$f_i(t, x + \mu(t,\theta), s) - f_i(t,x,s) - \frac{\partial f_i(t,x,s)}{\partial x}\mu(t,\theta)$$
$$= \frac{1}{2}\alpha_i(x + \mu - \mathbb{E}[x])^2 - \frac{1}{2}\alpha_i(x - \mathbb{E}[x])^2 - \alpha_i\left(x - \mathbb{E}[x]\right)\mu(t,\theta),$$
$$= \frac{1}{2}\alpha_i(x - \mathbb{E}[x])^2 + \frac{1}{2}\alpha_i\mu^2 + \alpha_i(x - \mathbb{E}[x])\mu - \frac{1}{2}\alpha_i(x - \mathbb{E}[x])^2$$
$$- \alpha_i\left(x - \mathbb{E}[x]\right)\mu,$$
$$= \frac{1}{2}\alpha_i\mu^2, \tag{6.10}$$

and

$$\sum_{s'\in\mathcal{S}}[f_i(\cdot,s') - f_i(\cdot,s)]\tilde{q}_{ss'} = \frac{1}{2}\sum_{s'\in\mathcal{S}}\left(\alpha_i(\cdot,s') - \alpha_i(\cdot,s)\right)\left(x - \mathbb{E}[x]\right)^2$$
$$+ \frac{1}{2}\sum_{s'\in\mathcal{S}}\left(\beta_i(\cdot,s') - \beta_i(\cdot,s)\right)\mathbb{E}[x]^2$$
$$+ \sum_{s'\in\mathcal{S}}\left(\gamma_i(\cdot,s') - \gamma_i(\cdot,s)\right)\mathbb{E}[x]$$
$$+ \sum_{s'\in\mathcal{S}}\left(\delta_i(\cdot,s') - \delta_i(\cdot,s)\right). \tag{6.11}$$

Now, the terms in (6.9), (6.10), and (6.11) in the Itô's formula in (6.8), and also taking the expectation to the guess functional $f_i(t,x,s)$ yields

$$\mathrm{d}(\mathbb{E}[f_i(t,x,s)|s])$$
$$= \mathbb{E}\left[\frac{1}{2}\left\{\dot{\alpha}_i + 2\alpha_i b_1 + \sum_{s'\in\mathcal{S}}[\alpha_i(\cdot,s') - \alpha_i(\cdot,s)]\tilde{q}_{ss'}\right\}\left(x - \mathbb{E}[x]\right)^2\right.$$
$$+ \frac{1}{2}\left\{\dot{\beta}_i + 2\beta_i(b_1 + \bar{b}_1) + \sum_{s'\in\mathcal{S}}[\beta_i(\cdot,s') - \beta_i(\cdot,s)]\tilde{q}_{ss'}\right\}\mathbb{E}[x]^2$$
$$+ \left\{\dot{\gamma}_i + \beta_i b_0 + \gamma_i(b_1 + \bar{b}_1) + \sum_{s'\in\mathcal{S}}[\gamma_i(\cdot,s') - \gamma_i(\cdot,s)]\tilde{q}_{ss'}\right\}\mathbb{E}[x]$$

$$+ \left\{ \dot{\delta}_i + \gamma_i b_0 + \sum_{s' \in \mathcal{S}} [\delta_i(.,s') - \delta_i(.,s)] \tilde{q}_{ss'} \right\}$$

$$+ \frac{\alpha_i}{2} \left(\sigma^2 + \int_{\theta \in \Theta} \mu^2 \nu(\mathrm{d}\theta) \right) + \gamma_i \sum_{j \in \mathcal{N}} (b_{2j} + \bar{b}_{2j}) \mathbb{E}[u_j]$$

$$+ \alpha_i \sum_{j \in \mathcal{N}} b_{2j}(x - \mathbb{E}[x])(u_j - \mathbb{E}[u_j]) + \beta_i \sum_{j \in \mathcal{N}} (b_{2j} + \bar{b}_{2j}) \mathbb{E}[x] \mathbb{E}[u_j] \Bigg] \mathrm{d}t,$$

From the previous expression, we observe that it is necessary to compute the terms $\mathbb{E}[\alpha_i \sigma^2]$ and $\mathbb{E}\left[\int_{\theta \in \Theta} \alpha_i \mu^2 \nu(\mathrm{d}\theta) \right]$. These terms are long to compute. To avoid confusion, we calculate them separately next.

$$\mathbb{E}[\alpha_i \sigma^2] = \mathbb{E}\Bigg[\alpha_i \sigma_0^2 + \alpha_i \sigma_1^2 (x - \mathbb{E}[x])^2 + \alpha_i (\sigma_1 + \bar{\sigma}_1)^2 \mathbb{E}[x]^2$$

$$+ \alpha_i \left\{ \sum_{j \in \mathcal{N}} \sigma_{2j}(u_j - \mathbb{E}[u_j]) \right\}^2 + \alpha_i \left\{ \sum_{j \in \mathcal{N}} (\sigma_{2j} + \bar{\sigma}_{2j}) \mathbb{E}[u_j] \right\}^2$$

$$+ 2\alpha_i \sigma_0 (\sigma_1 + \bar{\sigma}_1) \mathbb{E}[x] + 2\alpha_i \sigma_0 \sum_{j \in \mathcal{N}} (\sigma_{2j} + \bar{\sigma}_{2j}) \mathbb{E}[u_j]$$

$$+ 2\alpha_i \sigma_1 \sum_{j \in \mathcal{N}} \sigma_{2j}(x - \mathbb{E}[x])(u_j - \mathbb{E}[u_j])$$

$$+ 2\alpha_i (\sigma_1 + \bar{\sigma}_1) \sum_{j \in \mathcal{N}} (\sigma_{2j} + \bar{\sigma}_{2j}) \mathbb{E}[x] \mathbb{E}[u_j] \Bigg],$$

and

$$\mathbb{E}\left[\int_{\theta \in \Theta} \alpha_i \mu^2 \nu(\mathrm{d}\theta) \right] = \mathbb{E} \int_{\theta \in \Theta} \nu(\mathrm{d}\theta) \Bigg[\alpha_i \mu_0^2 + \alpha_i \mu_1^2 (x - \mathbb{E}[x])^2$$

$$+ \alpha_i (\mu_1 + \bar{\mu}_1)^2 \mathbb{E}[x]^2 + \alpha_i \left\{ \sum_{j \in \mathcal{N}} \mu_{2j}(u_j - \mathbb{E}[u_j]) \right\}^2$$

$$+ \alpha_i \left\{ \sum_{j \in \mathcal{N}} (\mu_{2j} + \bar{\mu}_{2j}) \mathbb{E}[u_j] \right\}^2 + 2\alpha_i \mu_0 (\mu_1 + \bar{\mu}_1) \mathbb{E}[x]$$

$$+ 2\alpha_i \mu_0 \sum_{j \in \mathcal{N}} (\mu_{2j} + \bar{\mu}_{2j}) \mathbb{E}[u_j]$$

$$+ 2\alpha_i \mu_1 \sum_{j \in \mathcal{N}} \mu_{2j}(x - \mathbb{E}[x])(u_j - \mathbb{E}[u_j])$$

$$+ 2\alpha_i (\mu_1 + \bar{\mu}_1) \sum_{j \in \mathcal{N}} (\mu_{2j} + \bar{\mu}_{2j}) \mathbb{E}[x] \mathbb{E}[u_j] \Bigg],$$

Thus, we obtain that

$$\mathbb{E}\left[\frac{\alpha_i}{2} \left(\sigma^2 + \int_{\theta \in \Theta} \mu^2 \nu(\mathrm{d}\theta) \right) \right] = \frac{\alpha_i}{2} \left(\sigma_0^2 + \int_{\theta \in \Theta} \mu_0^2 \nu(\mathrm{d}\theta) \right)$$

$$+ \alpha_i \left(\sigma_0(\sigma_1 + \bar{\sigma}_1) + \int_{\theta \in \Theta} \mu_0(\mu_1 + \bar{\mu}_1)\nu(\mathrm{d}\theta) \right) \mathbb{E}[x]$$

$$+ \alpha_i \sum_{j \in \mathcal{N}} \left\{ \sigma_0(\sigma_{2j} + \bar{\sigma}_{2j}) + \int_{\theta \in \Theta} \mu_0(\mu_{2j} + \bar{\mu}_{2j})\nu(\mathrm{d}\theta) \right\} \mathbb{E}[u_j]$$

$$+ \frac{\alpha_i}{2} \left(\sigma_1^2 + \int_{\theta \in \Theta} \mu_1^2 \nu(\mathrm{d}\theta) \right) (x - \mathbb{E}[x])^2$$

$$+ \frac{\alpha_i}{2} \left((\sigma_1 + \bar{\sigma}_1)^2 + \int_{\theta \in \Theta} (\mu_1 + \bar{\mu}_1)^2 \nu(\mathrm{d}\theta) \right) \mathbb{E}[x]^2$$

$$+ \alpha_i \sum_{j \in \mathcal{N}} \left\{ (\sigma_1 + \bar{\sigma}_1)(\sigma_{2j} + \bar{\sigma}_{2j}) \right.$$

$$\left. + \int_{\theta \in \Theta} (\mu_1 + \bar{\mu}_1)(\mu_{2j} + \bar{\mu}_{2j})\nu(\mathrm{d}\theta) \right\} \mathbb{E}[x]\mathbb{E}[u_j]$$

$$+ \alpha_i \sum_{j \in \mathcal{N}} \left\{ \sigma_1 \sigma_{2j} + \int_{\theta \in \Theta} \mu_1 \mu_{2j} \nu(\mathrm{d}\theta) \right\} (x - \mathbb{E}[x])(u_j - \mathbb{E}[u_j])$$

$$+ \frac{\alpha_i}{2} \left\{ \sum_{j \in \mathcal{N}} \sigma_{2j}(u_j - \mathbb{E}[u_j]) \right\}^2$$

$$+ \int_{\theta \in \Theta} \frac{\alpha_i}{2} \left\{ \sum_{j \in \mathcal{N}} \mu_{2j}(u_j - \mathbb{E}[u_j]) \right\}^2 \nu(\mathrm{d}\theta)$$

$$+ \frac{\alpha_i}{2} \left\{ \sum_{j \in \mathcal{N}} (\sigma_{2j} + \bar{\sigma}_{2j})\mathbb{E}[u_j] \right\}^2$$

$$+ \int_{\theta \in \Theta} \frac{\alpha_i}{2} \left\{ \sum_{j \in \mathcal{N}} (\mu_{2j} + \bar{\mu}_{2j})\mathbb{E}[u_j] \right\}^2 \nu(\mathrm{d}\theta)$$

Thus, replacing the term $\mathbb{E}\left[\frac{\alpha_i}{2} \left(\sigma^2 + \int_{\theta \in \Theta} \mu^2 \nu(\mathrm{d}\theta) \right) \right]$, *the Itô's formula becomes*

$$\mathrm{d}(\mathbb{E}[f_i(t, x, s)|\mathcal{F}^s]) = \mathbb{E}\left[\frac{1}{2} \left\{ \dot{\alpha}_i + \alpha_i \left(\sigma_1^2 + \int_{\theta \in \Theta} \mu_1^2 \nu(\mathrm{d}\theta) \right) + 2\alpha_i b_1 \right. \right.$$

$$+ \sum_{s' \in \mathcal{S}} [\alpha_i(., s') - \alpha_i(., s)]\tilde{q}_{ss'} \right\} (x - \mathbb{E}[x])^2$$

$$+ \frac{1}{2} \left\{ \dot{\beta}_i + \alpha_i \left((\sigma_1 + \bar{\sigma}_1)^2 + \int_{\theta \in \Theta} (\mu_1 + \bar{\mu}_1)^2 \nu(\mathrm{d}\theta) \right) \right.$$

$$+ 2\beta_i(b_1 + \bar{b}_1) + \sum_{s' \in \mathcal{S}} [\beta_i(., s') - \beta_i(., s)]\tilde{q}_{ss'} \Big\} \mathbb{E}[x]^2$$

$$+ \Big\{ \dot{\gamma}_i + \beta_i b_0 + \alpha_i \Big(\sigma_0(\sigma_1 + \bar{\sigma}_1) + \int_{\theta \in \Theta} \mu_0(\mu_1 + \bar{\mu}_1)\nu(\mathrm{d}\theta) \Big)$$

$$+ \gamma_i(b_1 + \bar{b}_1) + \sum_{s' \in \mathcal{S}} [\gamma_i(., s') - \gamma_i(., s)]\tilde{q}_{ss'} \Big\} \mathbb{E}[x]$$

$$+ \Big\{ \dot{\delta}_i + \gamma_i b_0 + \frac{\alpha_i}{2} \Big(\sigma_0^2 + \int_{\theta \in \Theta} \mu_0^2 \nu(\mathrm{d}\theta) \Big)$$

$$+ \sum_{s' \in \mathcal{S}} [\delta_i(., s') - \delta_i(., s)]\tilde{q}_{ss'} \Big\}$$

$$+ \frac{\alpha_i}{2} \Big\{ \sum_{j \in \mathcal{N}} \sigma_{2j}(u_j - \mathbb{E}[u_j]) \Big\}^2$$

$$+ \int_{\theta \in \Theta} \frac{\alpha_i}{2} \Big\{ \sum_{j \in \mathcal{N}} \mu_{2j}(u_j - \mathbb{E}[u_j]) \Big\}^2 \nu(\mathrm{d}\theta)$$

$$+ \alpha_i \sum_{j \in \mathcal{N}} \Big\{ \sigma_1 \sigma_{2j} + \int_{\theta \in \Theta} \mu_1 \mu_{2j}\nu(\mathrm{d}\theta) \Big\} (x - \mathbb{E}[x])(u_j - \mathbb{E}[u_j])$$

$$+ \alpha_i \sum_{j \in \mathcal{N}} b_{2j}(x - \mathbb{E}[x])(u_j - \mathbb{E}[u_j])$$

$$+ \frac{\alpha_i}{2} \Big\{ \sum_{j \in \mathcal{N}} (\sigma_{2j} + \bar{\sigma}_{2j})\mathbb{E}[u_j] \Big\}^2$$

$$+ \int_{\theta \in \Theta} \frac{\alpha_i}{2} \Big\{ \sum_{j \in \mathcal{N}} (\mu_{2j} + \bar{\mu}_{2j})\mathbb{E}[u_j] \Big\}^2 \nu(\mathrm{d}\theta)$$

$$+ \gamma_i \sum_{j \in \mathcal{N}} (b_{2j} + \bar{b}_{2j})\mathbb{E}[u_j]$$

$$+ \alpha_i \sum_{j \in \mathcal{N}} \Big\{ \sigma_0(\sigma_{2j} + \bar{\sigma}_{2j}) + \int_{\theta \in \Theta} \mu_0(\mu_{2j} + \bar{\mu}_{2j})\nu(\mathrm{d}\theta) \Big\} \mathbb{E}[u_j]$$

$$+ \alpha_i \sum_{j \in \mathcal{N}} \Big\{ (\sigma_1 + \bar{\sigma}_1)(\sigma_{2j} + \bar{\sigma}_{2j})$$

$$+ \int_{\theta \in \Theta} (\mu_1 + \bar{\mu}_1)(\mu_{2j} + \bar{\mu}_{2j})\nu(\mathrm{d}\theta) \Big\} \mathbb{E}[x]\mathbb{E}[u_j]$$

$$+ \beta_i \sum_{j \in \mathcal{N}} (b_{2j} + \bar{b}_{2j})\mathbb{E}[x]\mathbb{E}[u_j] \Big] \mathrm{dt}.$$

Continuing with the direct method, recall that the guess functional $f_i(t,x,s)$ denotes the optimal cost from time t up to T. Therefore, we require to find the optimal control inputs such that the expected cost $\mathbb{E}[\hat{L}_i(x,s,u_1,\ldots,u_n)]$ matches with $\mathbb{E}[f_i(0,x_0,s_0)]$ given the filtration $\mathcal{F}^s$. Then, we compute the difference $\mathbb{E}[\hat{L}_i(x,s,u_1,\ldots,u_n) - f_i(0,x_0,s_0)|\mathcal{F}^s]$, i.e.,

$$\mathbb{E}[\hat{L}_i(x,s,u_1,\ldots,u_n) - f_i(0,x_0,s_0)|\mathcal{F}^s] = \frac{1}{2}\mathbb{E}[(q_{iT} - \alpha_i(T))(x(T) - \mathbb{E}[x(T)])^2$$

$$+ \frac{1}{2}\mathbb{E}[(q_{iT} + \bar{q}_{iT} - \beta_i(T))\mathbb{E}[x(T)]^2 + (\varepsilon_{i3}(T) - \gamma_{iT})\mathbb{E}[x(T)]$$

$$+ \mathbb{E}\int_0^T \left[\frac{1}{2}\left\{ \dot{\alpha}_i + 2\alpha_i b_1 + \alpha_i\left(\sigma_1^2 + \int_{\theta\in\Theta} \mu_1^2\nu(\mathrm{d}\theta)\right) + q_i \right.$$

$$\left. + \sum_{s'\in\mathcal{S}}[\alpha_i(.,s') - \alpha_i(.,s)]\tilde{q}_{ss'} \right\}(x - \mathbb{E}[x])^2$$

$$+ \frac{1}{2}\left\{ \dot{\beta}_i + \alpha_i\left((\sigma_1 + \bar{\sigma}_1)^2 + \int_{\theta\in\Theta}(\mu_1 + \bar{\mu}_1)^2\nu(\mathrm{d}\theta)\right) \right.$$

$$\left. + 2\beta_i(b_1 + \bar{b}_1) + q_i + \bar{q}_i + \sum_{s'\in\mathcal{S}}[\beta_i(.,s') - \beta_i(.,s)]\tilde{q}_{ss'} \right\}\mathbb{E}[x]^2$$

$$+ \left\{ \dot{\gamma}_i + \beta_i b_0 + \alpha_i\left(\sigma_0(\sigma_1 + \bar{\sigma}_1) + \int_{\theta\in\Theta}\mu_0(\mu_1 + \bar{\mu}_1)\nu(\mathrm{d}\theta)\right) \right.$$

$$\left. + \gamma_i(b_1 + \bar{b}_1) + \varepsilon_{i3} + \sum_{s'\in\mathcal{S}}[\gamma_i(.,s') - \gamma_i(.,s)]\tilde{q}_{ss'} \right\}\mathbb{E}[x]$$

$$+ \left\{ \dot{\delta}_i + \gamma_i b_0 + \frac{\alpha_i}{2}\left(\sigma_0^2 + \int_{\theta\in\Theta}\mu_0^2\nu(\mathrm{d}\theta)\right) \right.$$

$$\left. + \sum_{s'\in\mathcal{S}}[\delta_i(.,s') - \delta_i(.,s)]\tilde{q}_{ss'} \right\}$$

$$+ \frac{1}{2}r_i(u_i - \mathbb{E}[u_i])^2 + \frac{\alpha_i}{2}\left\{ \sum_{j\in\mathcal{N}}\sigma_{2j}(u_j - \mathbb{E}[u_j]) \right\}^2$$

$$+ \int_{\theta\in\Theta} \frac{\alpha_i}{2}\left\{ \sum_{j\in\mathcal{N}}\mu_{2j}(u_j - \mathbb{E}[u_j]) \right\}^2 \nu(\mathrm{d}\theta)$$

$$+ \varepsilon_{1i}(x - \mathbb{E}[x])(u_i - \mathbb{E}[u_i])$$

$$+ \alpha_i \sum_{j\in\mathcal{N}}\left\{ \sigma_1\sigma_{2j} + \int_{\theta\in\Theta}\mu_1\mu_{2j}\nu(\mathrm{d}\theta) \right\}(x - \mathbb{E}[x])(u_j - \mathbb{E}[u_j])$$

$$+ \alpha_i \sum_{j\in\mathcal{N}}b_{2j}(x - \mathbb{E}[x])(u_j - \mathbb{E}[u_j]) + \frac{1}{2}(r_i + \bar{r}_i)\mathbb{E}[u_i]^2$$

$$+ \frac{\alpha_i}{2} \left\{ \sum_{j \in \mathcal{N}} (\sigma_{2j} + \bar{\sigma}_{2j}) \mathbb{E}[u_j] \right\}^2$$

$$+ \int_{\theta \in \Theta} \frac{\alpha_i}{2} \left\{ \sum_{j \in \mathcal{N}} (\mu_{2j} + \bar{\mu}_{2j}) \mathbb{E}[u_j] \right\}^2 \nu(d\theta) + \varepsilon_{2i} \mathbb{E}[u_i]$$

$$+ \alpha_i \sum_{j \in \mathcal{N}} \left\{ \sigma_0(\sigma_{2j} + \bar{\sigma}_{2j}) + \int_{\theta \in \Theta} \mu_0(\mu_{2j} + \bar{\mu}_{2j}) \nu(d\theta) \right\} \mathbb{E}[u_j]$$

$$+ \gamma_i \sum_{j \in \mathcal{N}} (b_{2j} + \bar{b}_{2j}) \mathbb{E}[u_j] + \varepsilon_{1i} \mathbb{E}[x] \mathbb{E}[u_i]$$

$$+ \beta_i \sum_{j \in \mathcal{N}} (b_{2j} + \bar{b}_{2j}) \mathbb{E}[x] \mathbb{E}[u_j]$$

$$+ \alpha_i \sum_{j \in \mathcal{N}} \left\{ (\sigma_1 + \bar{\sigma}_1)(\sigma_{2j} + \bar{\sigma}_{2j}) \right.$$

$$\left. + \int_{\theta \in \Theta} (\mu_1 + \bar{\mu}_1)(\mu_{2j} + \bar{\mu}_{2j}) \nu(d\theta) \right\} \mathbb{E}[x] \mathbb{E}[u_j] \bigg] dt. \qquad (6.12)$$

We now group the terms depending on the control inputs from the previous term, *i.e.*,

$$\frac{1}{2} r_i (u_i - \mathbb{E}[u_i])^2 + \frac{\alpha_i}{2} \left\{ \sum_{j \in \mathcal{N}} \sigma_{2j} (u_j - \mathbb{E}[u_j]) \right\}^2$$

$$+ \int_{\theta \in \Theta} \frac{\alpha_i}{2} \left\{ \sum_{j \in \mathcal{N}} \mu_{2j} (u_j - \mathbb{E}[u_j]) \right\}^2 \nu(d\theta)$$

$$+ \varepsilon_{1i} (x - \mathbb{E}[x])(u_i - \mathbb{E}[u_i])$$

$$+ \alpha_i \sum_{j \in \mathcal{N}} \left\{ \sigma_1 \sigma_{2j} + \int_{\theta \in \Theta} \mu_1 \mu_{2j} \nu(d\theta) \right\} (x - \mathbb{E}[x])(u_j - \mathbb{E}[u_j])$$

$$+ \alpha_i \sum_{j \in \mathcal{N}} b_{2j} (x - \mathbb{E}[x])(u_j - \mathbb{E}[u_j])$$

$$= \frac{1}{2} r_i (u_i - \mathbb{E}[u_i])^2 + \frac{\alpha_i}{2} \{ \sigma_{2i} (u_i - \mathbb{E}[u_i]) \}^2 + \frac{\alpha_i}{2} \left\{ \sum_{j \in \mathcal{N} \setminus \{i\}} \sigma_{2j} (u_j - \mathbb{E}[u_j]) \right\}^2$$

$$+ \alpha_i \sigma_{2i} (u_i - \mathbb{E}[u_i]) \sum_{j \in \mathcal{N} \setminus \{i\}} \sigma_{2j} (u_j - \mathbb{E}[u_j]) + \int_{\theta \in \Theta} \frac{\alpha_i}{2} \{ \mu_{2i} (u_i - \mathbb{E}[u_i]) \}^2 \nu(d\theta)$$

$$+ \int_{\theta \in \Theta} \frac{\alpha_i}{2} \left\{ \sum_{j \in \mathcal{N} \setminus \{i\}} \mu_{2j} (u_j - \mathbb{E}[u_j]) \right\}^2 \nu(d\theta)$$

$$+ \int_{\theta \in \Theta} \alpha_i \{\mu_{2i}(u_i - \mathbb{E}[u_i])\} \left\{ \sum_{j \in \mathcal{N} \setminus \{i\}} \mu_{2j}(u_j - \mathbb{E}[u_j]) \right\} \nu(d\theta)$$

$$+ \varepsilon_{1i}(x - \mathbb{E}[x])(u_i - \mathbb{E}[u_i])$$

$$+ \alpha_i \left\{ \sigma_1 \sigma_{2i} + \int_{\theta \in \Theta} \mu_1 \mu_{2i} \nu(d\theta) \right\} (x - \mathbb{E}[x])(u_i - \mathbb{E}[u_i])$$

$$+ \alpha_i \sum_{j \in \mathcal{N} \setminus \{i\}} \left\{ \sigma_1 \sigma_{2j} + \int_{\theta \in \Theta} \mu_1 \mu_{2j} \nu(d\theta) \right\} (x - \mathbb{E}[x])(u_j - \mathbb{E}[u_j])$$

$$+ \alpha_i b_{2i}(x - \mathbb{E}[x])(u_i - \mathbb{E}[u_i]) + \alpha_i \sum_{j \in \mathcal{N} \setminus \{i\}} b_{2j}(x - \mathbb{E}[x])(u_j - \mathbb{E}[u_j]). \quad (6.13)$$

In order to simplify the latter expression, let us consider some auxiliary variables:

$$c_i = r_i + \alpha_i \sigma_{2i}^2 + \int_{\theta \in \Theta} \alpha_i \mu_{2i}^2 \nu(d\theta),$$

$$\tau_i = \alpha_i \sigma_{2i} \sum_{j \in \mathcal{N} \setminus \{i\}} \sigma_{2j}(u_j - \mathbb{E}[u_j])$$

$$+ \int_{\theta \in \Theta} \alpha_i \mu_{2i} \left\{ \sum_{j \in \mathcal{N} \setminus \{i\}} \mu_{2j}(u_j - \mathbb{E}[u_j]) \right\} \nu(d\theta)$$

$$+ \left\{ \varepsilon_{1i} + \alpha_i \left\{ b_{2i} + \sigma_1 \sigma_{2i} + \int_{\theta \in \Theta} \mu_1 \mu_{2i} \nu(d\theta) \right\} \right\} (x - \mathbb{E}[x]).$$

Using these auxiliary variables, we perform the square completion corresponding to the optimization over the decision variables of the ith decision-maker. The term in (6.13) is equivalent to

$$\frac{1}{2} c_i \left(u_i - \mathbb{E}[u_i] + \frac{\tau_i}{c_i} \right)^2 - \frac{\tau_i^2}{2c_i} + \frac{\alpha_i}{2} \left\{ \sum_{j \in \mathcal{N} \setminus \{i\}} \sigma_{2j}(u_j - \mathbb{E}[u_j]) \right\}^2$$

$$+ \int_{\theta \in \Theta} \frac{\alpha_i}{2} \left\{ \sum_{j \in \mathcal{N} \setminus \{i\}} \mu_{2j}(u_j - \mathbb{E}[u_j]) \right\}^2 \nu(d\theta)$$

$$+ \alpha_i \sum_{j \in \mathcal{N} \setminus \{i\}} \left\{ \sigma_1 \sigma_{2j} + \int_{\theta \in \Theta} \mu_1 \mu_{2j} \nu(d\theta) \right\} (x - \mathbb{E}[x])(u_j - \mathbb{E}[u_j])$$

$$+ \alpha_i \sum_{j \in \mathcal{N} \setminus \{i\}} b_{2j}(x - \mathbb{E}[x])(u_j - \mathbb{E}[u_j])$$

$$= \frac{1}{2} c_i \left(u_i - \mathbb{E}[u_i] + \frac{\tilde{\tau}_i}{c_i}(x - \mathbb{E}[x]) \right)^2$$

$$+ \left\{ \frac{\alpha_i}{2} \left\{ \sum_{j \in \mathcal{N} \setminus \{i\}} \sigma_{2j} \frac{\tilde{\tau}_j}{c_j} \right\} \right\}^2 + \int_{\theta \in \Theta} \frac{\alpha_i}{2} \left\{ \sum_{j \in \mathcal{N} \setminus \{i\}} \mu_{2j} \frac{\tilde{\tau}_j}{c_j} \right\}^2 \nu(\mathrm{d}\theta)$$

$$- \alpha_i \sum_{j \in \mathcal{N} \setminus \{i\}} b_{2j} \frac{\tilde{\tau}_j}{c_j}$$

$$- \alpha_i \sum_{j \in \mathcal{N} \setminus \{i\}} \left\{ \sigma_1 \sigma_{2j} + \int_{\theta \in \Theta} \mu_1 \mu_{2j} \nu(\mathrm{d}\theta) \right\} \frac{\tilde{\tau}_j}{c_j} - \frac{\tilde{\tau}_i^2}{2c_i} \right\} (x - \mathbb{E}[x])^2.$$

From the latter expression, one can deduce that optimal subtraction $u_i - \mathbb{E}[u_i]$, *for all* $i \in \mathcal{N}$, *which is*

$$u_i - \mathbb{E}[u_i] = -\frac{\tau_i}{c_i} = -\frac{\tilde{\tau}_i}{c_i}(x - \mathbb{E}[x]).$$

The auxiliary variable τ_i *depends on the terms* $u_j - \mathbb{E}[u_j]$, *for all* $j \in \mathcal{N} \setminus \{i\}$. *Using this optimal solution, one can now represent the auxiliary variable* τ_i *independent from the control inputs of all the other decision-makers. The vectorial function* τ *is obtained by solving the linear system of functions and is given by* $\tau = [A^{-1}v](x - \mathbb{E}[x])$, *where* $A_{ii} = 1$, *and*

$$A_{ij} = \frac{\alpha_i}{c_j} \left\{ \sigma_{2i} \sigma_{2j} + \int_{\theta \in \Theta} \mu_{2i} \, \mu_{2j} \nu(\mathrm{d}\theta) \right\}, \ \forall j \neq i,$$

$$v_i = \varepsilon_{1i} + \alpha_i \left\{ b_{2i} + \sigma_1 \sigma_{2i} + \int_{\theta \in \Theta} \mu_1 \mu_{2i} \nu(\mathrm{d}\theta) \right\},$$

whenever the matrix A *is invertible. Thus,* τ_i *can be written as* $\tau_i = \tilde{\tau}_i(x - \mathbb{E}[x])$ *with* $\tilde{\tau}_i := [A^{-1}v]_i$, *whenever the matrix* A *is invertible. Now, we follow a similar procedure for the optimization of the terms depending on* $\mathbb{E}[u_i]$, *i.e.,*

$$\frac{1}{2}(r_i + \bar{r}_i)\mathbb{E}[u_i]^2 + \alpha_i \left\{ \frac{1}{2}(\sigma_{2i} + \bar{\sigma}_{2i})^2 \mathbb{E}[u_i]^2 + \frac{1}{2} \left\{ \sum_{j \in \mathcal{N} \setminus \{i\}} (\sigma_{2j} + \bar{\sigma}_{2j})\mathbb{E}[u_j] \right\}^2 \right.$$

$$+ (\sigma_{2i} + \bar{\sigma}_{2i})\mathbb{E}[u_i] \sum_{j \in \mathcal{N} \setminus \{i\}} (\sigma_{2j} + \bar{\sigma}_{2j})\mathbb{E}[u_j] \right\}$$

$$+ \alpha_i \int_{\theta \in \Theta} \left\{ \frac{1}{2}(\mu_{2i} + \bar{\mu}_{2i})^2 \mathbb{E}[u_i]^2 + \frac{1}{2} \left\{ \sum_{j \in \mathcal{N} \setminus \{i\}} (\mu_{2j} + \bar{\mu}_{2j})\mathbb{E}[u_j] \right\}^2 \right.$$

$$+ (\mu_{2i} + \bar{\mu}_{2i})\mathbb{E}[u_i] \sum_{j \in \mathcal{N} \setminus \{i\}} (\mu_{2j} + \bar{\mu}_{2j})\mathbb{E}[u_j] \right\} \nu(\mathrm{d}\theta)$$

$$+ \left\{ \varepsilon_{2i} + \alpha_i \left(\sigma_0(\sigma_{2i} + \bar{\sigma}_{2i}) + \int_{\theta \in \Theta} \mu_0(\mu_{2i} + \bar{\mu}_{2i})\nu(\mathrm{d}\theta) \right) + \gamma_i(b_{2i} + \bar{b}_{2i}) \right\} \mathbb{E}[u_i]$$

$$+ \sum_{j \in \mathcal{N} \setminus \{i\}} \left\{ \alpha_i \left(\sigma_0(\sigma_{2j} + \bar{\sigma}_{2j}) + \int_{\theta \in \Theta} \mu_0(\mu_{2j} + \bar{\mu}_{2j})\nu(\mathrm{d}\theta) \right) + \gamma_i(b_{2j} + \bar{b}_{2j}) \right\} \mathbb{E}[u_j]$$

$$+ \left\{ \alpha_i \left((\sigma_1 + \bar{\sigma}_1)(\sigma_{2i} + \bar{\sigma}_{2i}) + \int_{\theta \in \Theta} (\mu_1 + \bar{\mu}_1)(\mu_{2i} + \bar{\mu}_{2i})\nu(\mathrm{d}\theta) \right) \right.$$

$$\left. + \varepsilon_{1i} + \bar{\varepsilon}_{1i} + \beta_i(b_{2i} + \bar{b}_{2i}) \right\} \mathbb{E}[x]\mathbb{E}[u_i]$$

$$+ \sum_{j \in \mathcal{N} \setminus \{i\}} \left\{ \beta_i(b_{2j} + \bar{b}_{2j}) + \alpha_i \left((\sigma_1 + \bar{\sigma}_1)(\sigma_{2j} + \bar{\sigma}_{2j}) \right. \right.$$

$$\left. \left. + \int_{\theta \in \Theta} (\mu_1 + \bar{\mu}_1)(\mu_{2j} + \bar{\mu}_{2j})\nu(\mathrm{d}\theta) \right) \right\} \mathbb{E}[x]\mathbb{E}[u_j]$$

$$= \frac{1}{2}(r_i + \bar{r}_i)\mathbb{E}[u_i]^2$$

$$+ \alpha_i \left[\frac{1}{2}(\sigma_{2i} + \bar{\sigma}_{2i})^2 \mathbb{E}[u_i]^2 + (\sigma_{2i} + \bar{\sigma}_{2i})\mathbb{E}[u_i] \sum_{j \in \mathcal{N} \setminus \{i\}} (\sigma_{2j} + \bar{\sigma}_{2j})\mathbb{E}[u_j] \right]$$

$$+ \alpha_i \int_{\theta \in \Theta} \left[\frac{1}{2}(\mu_{2i} + \bar{\mu}_{2i})^2 \mathbb{E}[u_i]^2 \right.$$

$$\left. + (\mu_{2i} + \bar{\mu}_{2i})\mathbb{E}[u_i] \sum_{j \in \mathcal{N} \setminus \{i\}} (\mu_{2j} + \bar{\mu}_{2j})\mathbb{E}[u_j] \right] \nu(\mathrm{d}\theta)$$

$$+ \left[\varepsilon_{2i} + \alpha_i \left(\sigma_0(\sigma_{2i} + \bar{\sigma}_{2i}) + \int_{\theta \in \Theta} \mu_0(\mu_{2i} + \bar{\mu}_{2i})\nu(\mathrm{d}\theta) \right) + \gamma_i(b_{2i} + \bar{b}_{2i}) \right] \mathbb{E}[u_i]$$

$$+ \left[\varepsilon_{1i} + \bar{\varepsilon}_{1i} + \alpha_i \left((\sigma_1 + \bar{\sigma}_1)(\sigma_{2i} + \bar{\sigma}_{2i}) + \int_{\theta \in \Theta} (\mu_1 + \bar{\mu}_1)(\mu_{2i} + \bar{\mu}_{2i})\nu(\mathrm{d}\theta) \right) \right.$$

$$\left. + \beta_i(b_{2i} + \bar{b}_{2i}) \right] \mathbb{E}[x]\mathbb{E}[u_i]$$

$$+ \alpha_i \left[\frac{1}{2} \left(\sum_{j \in \mathcal{N} \setminus \{i\}} (\sigma_{2j} + \bar{\sigma}_{2j})\mathbb{E}[u_j] \right)^2 \right.$$

$$\left. + \frac{1}{2} \int_{\theta \in \Theta} \left(\sum_{j \in \mathcal{N} \setminus \{i\}} (\mu_{2j} + \bar{\mu}_{2j})\mathbb{E}[u_j] \right)^2 \nu(\mathrm{d}\theta) \right]$$

$$+ \sum_{j \in \mathcal{N} \setminus \{i\}} \left[\alpha_i \left(\sigma_0(\sigma_{2j} + \bar{\sigma}_{2j}) + \int_{\theta \in \Theta} \mu_0(\mu_{2j} + \bar{\mu}_{2j})\nu(\mathrm{d}\theta) \right) + \gamma_i(b_{2j} + \bar{b}_{2j}) \right] \mathbb{E}[u_j]$$

$$+ \sum_{j \in \mathcal{N} \setminus \{i\}} \left[\alpha_i \left((\sigma_1 + \bar{\sigma}_1)(\sigma_{2j} + \bar{\sigma}_{2j}) \right. \right.$$

$$+ \int_{\theta \in \Theta} (\mu_1 + \bar{\mu}_1)(\mu_{2j} + \bar{\mu}_{2j})\nu(\mathrm{d}\theta) \bigg) + \beta_i(b_{2j} + \bar{b}_{2j}) \bigg] \mathbb{E}[x]\mathbb{E}[u_j] \qquad (6.14)$$

Let us consider the following auxiliary variables, which we introduce in order to compact the latter expression in a more friendly manner, i.e.,

$$\bar{c}_i = (r_i + \bar{r}_i) + \alpha_i \bigg[(\sigma_{2i} + \bar{\sigma}_{2i})^2 + \int_{\theta \in \Theta} (\mu_{2i} + \bar{\mu}_{2i})^2 \nu(\mathrm{d}\theta) \bigg]$$

$$\bar{\tau}_i = \alpha_i \bigg[(\sigma_{2i} + \bar{\sigma}_{2i}) \sum_{j \in \mathcal{N}\setminus\{i\}} (\sigma_{2j} + \bar{\sigma}_{2j})\mathbb{E}[u_j] \bigg] + \gamma_i(b_{2i} + \bar{b}_{2i})$$

$$+ \alpha_i \int_{\theta \in \Theta} \bigg[(\mu_{2i} + \bar{\mu}_{2i}) \sum_{j \in \mathcal{N}\setminus\{i\}} (\mu_{2j} + \bar{\mu}_{2j})\mathbb{E}[u_j] \bigg] \nu(\mathrm{d}\theta) + \varepsilon_{2i}$$

$$+ \alpha_i \bigg(\sigma_0(\sigma_{2i} + \bar{\sigma}_{2i}) + \int_{\theta \in \Theta} \mu_0(\mu_{2i} + \bar{\mu}_{2i})\nu(\mathrm{d}\theta) \bigg)$$

$$+ \bigg\{ \varepsilon_{1i} + \bar{\varepsilon}_{1i} + \beta_i(b_{2i} + \bar{b}_{2i}) + \alpha_i \bigg((\sigma_1 + \bar{\sigma}_1)(\sigma_{2i} + \bar{\sigma}_{2i})$$

$$+ \int_{\theta \in \Theta} (\mu_1 + \bar{\mu}_1)(\mu_{2i} + \bar{\mu}_{2i})\nu(\mathrm{d}\theta) \bigg) \bigg\} \mathbb{E}[x].$$

Using the auxiliary variables, the expression in (6.14) can be re-written as follows:

$$\frac{\bar{c}_i}{2} \bigg(\mathbb{E}[u_i] + \frac{\bar{\lambda}_{2i}\mathbb{E}[x] + \bar{\lambda}_{i3}}{\bar{c}_i} \bigg)^2 + \bigg\{ \alpha_i \frac{1}{2} \bigg(\sum_{j \in \mathcal{N}\setminus\{i\}} (\sigma_{2j} + \bar{\sigma}_{2j})\frac{\bar{\lambda}_{2j}}{\bar{c}_j} \bigg)^2$$

$$+ \alpha_i \int_{\theta \in \Theta} \frac{1}{2} \bigg(\sum_{j \in \mathcal{N}\setminus\{i\}} (\mu_{2j} + \bar{\mu}_{2j})\frac{\bar{\lambda}_{2j}}{\bar{c}_j} \bigg)^2 \nu(\mathrm{d}\theta) - \frac{\bar{\lambda}_{2i}^2}{2\bar{c}_i}$$

$$- \sum_{j \in \mathcal{N}\setminus\{i\}} \bigg[\beta_i(b_{2j} + \bar{b}_{2j}) + \alpha_i \bigg((\sigma_1 + \bar{\sigma}_1)(\sigma_{2j} + \bar{\sigma}_{2j})$$

$$+ \int_{\theta \in \Theta} (\mu_1 + \bar{\mu}_1)(\mu_{2j} + \bar{\mu}_{2j})\nu(\mathrm{d}\theta) \bigg) \bigg] \frac{\bar{\lambda}_{2j}}{\bar{c}_j} \bigg\} \mathbb{E}[x]^2$$

$$+ \bigg\{ \alpha_i \bigg(\sum_{j \in \mathcal{N}\setminus\{i\}} (\sigma_{2j} + \bar{\sigma}_{2j})\frac{\bar{\lambda}_{2j}}{\bar{c}_j} \bigg) \bigg(\sum_{j \in \mathcal{N}\setminus\{i\}} (\sigma_{2j} + \bar{\sigma}_{2j})\frac{\bar{\lambda}_{j3}}{\bar{c}_j} \bigg)$$

$$+ \alpha_i \int_{\theta \in \Theta} \bigg(\sum_{j \in \mathcal{N}\setminus\{i\}} (\mu_{2j} + \bar{\mu}_{2j})\frac{\bar{\lambda}_{2j}}{\bar{c}_j} \bigg) \bigg(\sum_{j \in \mathcal{N}\setminus\{i\}} (\mu_{2j} + \bar{\mu}_{2j})\frac{\bar{\lambda}_{j3}}{\bar{c}_j} \bigg) \nu(\mathrm{d}\theta)$$

$$- \sum_{j \in \mathcal{N}\setminus\{i\}} \bigg[\gamma_i(b_{2j} + \bar{b}_{2j}) + \alpha_i \bigg(\sigma_0(\sigma_{2j} + \bar{\sigma}_{2j}) + \int_{\theta \in \Theta} \mu_0(\mu_{2j} + \bar{\mu}_{2j})\nu(\mathrm{d}\theta) \bigg) \bigg] \frac{\bar{\lambda}_{2j}}{\bar{c}_j}$$

$$- \sum_{j \in \mathcal{N} \setminus \{i\}} \left[\beta_i (b_{2j} + \bar{b}_{2j}) + \alpha_i \left((\sigma_1 + \bar{\sigma}_1)(\sigma_{2j} + \bar{\sigma}_{2j}) \right. \right.$$

$$\left. \left. + \int_{\theta \in \Theta} (\mu_1 + \bar{\mu}_1)(\mu_{2j} + \bar{\mu}_{2j}) \nu(\mathrm{d}\theta) \right) \right] \frac{\bar{\lambda}_{j3}}{\bar{c}_j} - \frac{\bar{\lambda}_{2i} \bar{\lambda}_{i3}}{\bar{c}_i} \right\} \mathbb{E}[x]$$

$$+ \alpha_i \frac{1}{2} \left\{ \sum_{j \in \mathcal{N} \setminus \{i\}} (\sigma_{2j} + \bar{\sigma}_{2j}) \frac{\bar{\lambda}_{j3}}{\bar{c}_j} \right\}^2$$

$$+ \alpha_i \int_{\theta \in \Theta} \frac{1}{2} \left\{ \sum_{j \in \mathcal{N} \setminus \{i\}} (\mu_{2j} + \bar{\mu}_{2j}) \frac{\bar{\lambda}_{j3}}{\bar{c}_j} \right\}^2 \nu(\mathrm{d}\theta)$$

$$- \sum_{j \in \mathcal{N} \setminus \{i\}} \left[\gamma_i (b_{2j} + \bar{b}_{2j}) + \alpha_i \left(\sigma_0 (\sigma_{2j} + \bar{\sigma}_{2j}) + \int_{\theta \in \Theta} \mu_0 (\mu_{2j} + \bar{\mu}_{2j}) \nu(\mathrm{d}\theta) \right) \right] \frac{\bar{\lambda}_{j3}}{\bar{c}_j}$$

$$- \frac{\bar{\lambda}_{i3}^2}{2\bar{c}_i}.$$

Here, one can conclude the optimal mean-field control input for all the decision-makers $j \in \mathcal{N}$, i.e.,

$$\mathbb{E}[u_i] = -\frac{\bar{\lambda}_{2i} \mathbb{E}[x] + \bar{\lambda}_{i3}}{\bar{c}_i}, \ \forall i \in \mathcal{N}.$$

Therefore, the term $\bar{\tau}_i$, which has been presented depending on $\mathbb{E}[u_j]$, for all $j \in \mathcal{N} \setminus \{i\}$, can be written independent from the mean-field control inputs as follows:

$$\bar{\tau}_i = -\alpha_i \left[(\sigma_{2i} + \bar{\sigma}_{2i}) \sum_{j \in \mathcal{N} \setminus \{i\}} (\sigma_{2j} + \bar{\sigma}_{2j}) \frac{\bar{\tau}_j}{\bar{c}_j} \right] + \gamma_i (b_{2i} + \bar{b}_{2i})$$

$$- \alpha_i \int_{\theta \in \Theta} \left[(\mu_{2i} + \bar{\mu}_{2i}) \sum_{j \in \mathcal{N} \setminus \{i\}} (\mu_{2j} + \bar{\mu}_{2j}) \frac{\bar{\tau}_j}{\bar{c}_j} \right] \nu(\mathrm{d}\theta) + \varepsilon_{2i}$$

$$+ \alpha_i \left(\sigma_0 (\sigma_{2i} + \bar{\sigma}_{2i}) + \int_{\theta \in \Theta} \mu_0 (\mu_{2i} + \bar{\mu}_{2i}) \nu(\mathrm{d}\theta) \right)$$

$$+ \left\{ \varepsilon_{1i} + \bar{\varepsilon}_{1i} + \beta_i (b_{2i} + \bar{b}_{2i}) \right.$$

$$\left. + \alpha_i \left((\sigma_1 + \bar{\sigma}_1)(\sigma_{2i} + \bar{\sigma}_{2i}) + \int_{\theta \in \Theta} (\mu_1 + \bar{\mu}_1)(\mu_{2i} + \bar{\mu}_{2i}) \nu(\mathrm{d}\theta) \right) \right\} \mathbb{E}[x].$$

The vectorial function $\bar{\tau}$ is obtained by solving the linear system of functions and is given by $\bar{\tau} = [\bar{A}^{-1} \bar{v}]$, where $\bar{A}_{ii} = 1$, and

$$\bar{A}_{ij} = \frac{\alpha_i}{\bar{c}_j} \left[(\sigma_{2i} + \bar{\sigma}_{2i})(\sigma_{2j} + \bar{\sigma}_{2j}) + \int_{\theta \in \Theta} (\mu_{2i} + \bar{\mu}_{2i})(\mu_{2j} + \bar{\mu}_{2j}) \nu(\mathrm{d}\theta) \right],$$

$$\bar{v}_i = \varepsilon_{2i} + \alpha_i \left(\sigma_0(\sigma_{2i} + \bar{\sigma}_{2i}) + \int_{\theta \in \Theta} \mu_0(\mu_{2i} + \bar{\mu}_{2i})\nu(d\theta) \right) + \gamma_i(b_{2i} + \bar{b}_{2i})$$

$$+ \left[\varepsilon_{1i} + \bar{\varepsilon}_{1i} + \beta_i(b_{2i} + \bar{b}_{2i}) \right.$$

$$\left. + \alpha_i \left((\sigma_1 + \bar{\sigma}_1)(\sigma_{2i} + \bar{\sigma}_{2i}) + \int_{\theta \in \Theta} (\mu_1 + \bar{\mu}_1)(\mu_{2i} + \bar{\mu}_{2i})\nu(d\theta) \right) \right] \mathbb{E}[x],$$

Hence, $\bar{\tau}_i$ is expressed as $\bar{\tau} = \bar{A}^{-1}\bar{v} = \bar{\lambda}_2 \mathbb{E}[x] + \bar{\lambda}_3$, where $\bar{\tau}_i = \bar{\lambda}_{2i}\mathbb{E}[x] + \bar{\lambda}_{i3}$, and

$$\frac{\bar{\tau}_i^2}{2\bar{c}_i} = \frac{\bar{\lambda}_{2i}^2}{2\bar{c}_i}\mathbb{E}[x]^2 + \frac{\bar{\lambda}_{2i}\bar{\lambda}_{i3}}{\bar{c}_i}\mathbb{E}[x] + \frac{\bar{\lambda}_{i3}^2}{2\bar{c}_i}.$$

We replace the square completion in (6.12) to perform the identification process.

$$\mathbb{E}[\hat{L}_i(x, s, u_1, \ldots, u_n)|x(0) = x_0, s(0) = s_0] - \mathbb{E}[f(0, x_0, s_0)|s]$$

$$= \frac{1}{2}\mathbb{E}[(q_{iT} - \alpha_i(T))(x(T) - \mathbb{E}[x(T)])^2|s]$$

$$+ \frac{1}{2}\mathbb{E}[(q_{iT} + \bar{q}_{iT} - \beta_i(T))\mathbb{E}[x(T)]^2|s] + (\varepsilon_{i3}(T) - \gamma_{iT})\mathbb{E}[x(T)]$$

$$+ \mathbb{E}\int_0^T \left[\frac{1}{2}\left\{ \dot{\alpha}_i + 2\alpha_i b_1 + \alpha_i \left(\sigma_1^2 + \int_{\theta \in \Theta} \mu_1^2 \nu(d\theta) \right) + q_i \right. \right.$$

$$+ \sum_{s' \in \mathcal{S}} [\alpha_i(., s') - \alpha_i(., s)]\tilde{q}_{ss'}$$

$$+ \alpha_i \left(\sum_{j \in \mathcal{N}\backslash\{i\}} \sigma_{2j}\frac{\tilde{\tau}_j}{c_j} \right)^2 + \int_{\theta \in \Theta} \alpha_i \left(\sum_{j \in \mathcal{N}\backslash\{i\}} \mu_{2j}\frac{\tilde{\tau}_j}{c_j} \right)^2 \nu(d\theta)$$

$$- 2\alpha_i \sum_{j \in \mathcal{N}\backslash\{i\}} b_{2j}\frac{\tilde{\tau}_j}{c_j}$$

$$\left. - 2\alpha_i \sum_{j \in \mathcal{N}\backslash\{i\}} \left(\sigma_1\sigma_{2j} + \int_{\theta \in \Theta} \mu_1\mu_{2j}\nu(d\theta) \right)\frac{\tilde{\tau}_j}{c_j} - \frac{\tilde{\tau}_i^2}{c_i} \right\}(x - \mathbb{E}[x])^2$$

$$+ \frac{1}{2}\left\{ \dot{\beta}_i + \alpha_i \left((\sigma_1 + \bar{\sigma}_1)^2 + \int_{\theta \in \Theta} (\mu_1 + \bar{\mu}_1)^2 \nu(d\theta) \right) + q_i + \bar{q}_i \right.$$

$$+ \sum_{s' \in \mathcal{S}} [\beta_i(., s') - \beta_i(., s)]\tilde{q}_{ss'} + \alpha_i \left(\sum_{j \in \mathcal{N}\backslash\{i\}} (\sigma_{2j} + \bar{\sigma}_{2j})\frac{\bar{\lambda}_{2j}}{\bar{c}_j} \right)^2 + 2\beta_i(b_1 + \bar{b}_1)$$

$$+ \alpha_i \int_{\theta \in \Theta} \left(\sum_{j \in \mathcal{N}\backslash\{i\}} (\mu_{2j} + \bar{\mu}_{2j})\frac{\bar{\lambda}_{2j}}{\bar{c}_j} \right)^2 \nu(d\theta) - \frac{\bar{\lambda}_{2i}^2}{\bar{c}_i}$$

$$- 2\sum_{j \in \mathcal{N}\backslash\{i\}} \left[\beta_i(b_{2j} + \bar{b}_{2j}) + \alpha_i \left((\sigma_1 + \bar{\sigma}_1)(\sigma_{2j} + \bar{\sigma}_{2j}) \right. \right.$$

$$+ \int_{\theta \in \Theta} (\mu_1 + \bar{\mu}_1)(\mu_{2j} + \bar{\mu}_{2j})\nu(d\theta) \Big] \frac{\bar{\lambda}_{2j}}{\bar{c}_j} \Big\} \mathbb{E}[x]^2$$

$$+ \Big\{ \dot{\gamma}_i + \beta_i b_0 + \alpha_i \Big(\sigma_0(\sigma_1 + \bar{\sigma}_1) + \int_{\theta \in \Theta} \mu_0(\mu_1 + \bar{\mu}_1)\nu(d\theta) \Big)$$

$$+ \gamma_i(b_1 + \bar{b}_1) + \varepsilon_{i3} + \sum_{s' \in \mathcal{S}} [\gamma_i(., s') - \gamma_i(., s)]\tilde{q}_{ss'}$$

$$+ \alpha_i \Big(\sum_{j \in \mathcal{N} \setminus \{i\}} (\sigma_{2j} + \bar{\sigma}_{2j}) \frac{\bar{\lambda}_{2j}}{\bar{c}_j} \Big) \Big(\sum_{j \in \mathcal{N} \setminus \{i\}} (\sigma_{2j} + \bar{\sigma}_{2j}) \frac{\bar{\lambda}_{j3}}{\bar{c}_j} \Big)$$

$$+ \alpha_i \int_{\theta \in \Theta} \Big(\sum_{j \in \mathcal{N} \setminus \{i\}} (\mu_{2j} + \bar{\mu}_{2j}) \frac{\bar{\lambda}_{2j}}{\bar{c}_j} \Big) \Big(\sum_{j \in \mathcal{N} \setminus \{i\}} (\mu_{2j} + \bar{\mu}_{2j}) \frac{\bar{\lambda}_{j3}}{\bar{c}_j} \Big) \nu(d\theta)$$

$$- \sum_{j \in \mathcal{N} \setminus \{i\}} \Big[\alpha_i \Big(\sigma_0(\sigma_{2j} + \bar{\sigma}_{2j}) + \int_{\theta \in \Theta} \mu_0(\mu_{2j} + \bar{\mu}_{2j})\nu(d\theta) \Big) + \gamma_i(b_{2j} + \bar{b}_{2j}) \Big] \frac{\bar{\lambda}_{2j}}{\bar{c}_j}$$

$$- \sum_{j \in \mathcal{N} \setminus \{i\}} \Big[\alpha_i \Big((\sigma_1 + \bar{\sigma}_1)(\sigma_{2j} + \bar{\sigma}_{2j})$$

$$+ \int_{\theta \in \Theta} (\mu_1 + \bar{\mu}_1)(\mu_{2j} + \bar{\mu}_{2j})\nu(d\theta) \Big) + \beta_i(b_{2j} + \bar{b}_{2j}) \Big] \frac{\bar{\lambda}_{j3}}{\bar{c}_j} - \frac{\bar{\lambda}_{2i}\bar{\lambda}_{i3}}{\bar{c}_i} \Big\} \mathbb{E}[x]$$

$$+ \Big\{ \dot{\delta}_i + \gamma_i b_0 + \frac{\alpha_i}{2} \Big(\sigma_0^2 + \int_{\theta \in \Theta} \mu_0^2 \nu(d\theta) \Big) + \sum_{s' \in \mathcal{S}} [\delta_i(., s') - \delta_i(., s)]\tilde{q}_{ss'}$$

$$+ \alpha_i \frac{1}{2} \Big(\sum_{j \in \mathcal{N} \setminus \{i\}} (\sigma_{2j} + \bar{\sigma}_{2j}) \frac{\bar{\lambda}_{j3}}{\bar{c}_j} \Big)^2$$

$$+ \alpha_i \int_{\theta \in \Theta} \frac{1}{2} \Big(\sum_{j \in \mathcal{N} \setminus \{i\}} (\mu_{2j} + \bar{\mu}_{2j}) \frac{\bar{\lambda}_{j3}}{\bar{c}_j} \Big)^2 \nu(d\theta)$$

$$- \sum_{j \in \mathcal{N} \setminus \{i\}} \Big[\alpha_i \Big(\sigma_0(\sigma_{2j} + \bar{\sigma}_{2j})$$

$$+ \int_{\theta \in \Theta} \mu_0(\mu_{2j} + \bar{\mu}_{2j})\nu(d\theta) \Big) + \gamma_i(b_{2j} + \bar{b}_{2j}) \Big] \frac{\bar{\lambda}_{j3}}{\bar{c}_j} - \frac{\bar{\lambda}_{i3}^2}{2\bar{c}_i} \Big\}$$

$$+ \frac{1}{2} c_i \Big(u_i - \mathbb{E}[u_i] + \frac{\tilde{\tau}_i}{c_i}(x - \mathbb{E}[x]) \Big)^2 + \frac{\bar{c}_i}{2} \Big(\mathbb{E}[u_i] + \frac{\bar{\lambda}_{2i}\mathbb{E}[x] + \bar{\lambda}_{i3}}{\bar{c}_i} \Big)^2 \Big] dt.$$

We divide the previous expression into seven groups:

- *Those terms depending on the terminal time T. These terms define the boundary conditions for the differential equations α_i, β_i, γ_i and δ_i,*

- *Those terms with common factor $(x - \mathbb{E}[x])^2$ defining α_i equation,*

- *Those terms with common factor $\mathbb{E}[x]^2$ establishing the β_i equation,*

- *Those terms with $\mathbb{E}[x]$ common factor corresponding to γ_i equation,*

- *Those terms independent from the state, expected state, control inputs, and expected control inputs,*

- *Those terms defining the optimal solution for the optimal control inputs and their expectation.*

Finally, the minimization of the latest expression exhibits the announced results in the statement provided that c_i and $\bar{c}_i$ are positive completing the proof of our main result stated in Theorem 1.

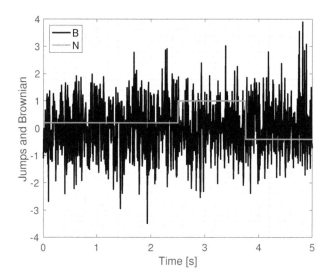

FIGURE 6.2
Brownian motion and two jumps.

6.3 Numerical Example

In this section we present a numerical examples to illustrate the performance of the mean-field-type game. Consider a two-player mean-field-type game under system state dynamics involving Brownian motion, Poisson jumps and regime switching $\mathcal{S} = \{s_1, s_2\}$ as in Figure 6.2, i.e.,

$$dx(t) = b(t, s(t))dt + \sigma(t, s(t))dB$$

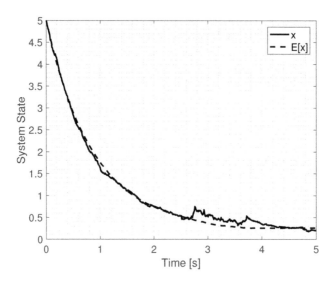

FIGURE 6.3

Evolution of the system state and its expectation for the scalar-value non-cooperative scenario with Brownian motion, Poisson jumps and regime switching.

$$+ \int_{\theta \in \Theta} \mu(t, s(t), \theta) \tilde{N}(dt, d\theta), \quad \text{with } \mathcal{S} = \{s_1, s_2\}.$$

For the first switching regime $s_1 \in \mathcal{S}$ we have the following drift, diffusion, and jump parameters:

$$b(s_1) = \frac{1}{2} - x + \frac{1}{10}\mathbb{E}[x] + \frac{1}{10}u_1 + \frac{1}{5}u_2 + \frac{1}{10}\mathbb{E}[u_1] + \frac{1}{5}\mathbb{E}[u_2],$$

$$\sigma(s_1) = 3 + 2x + 2\mathbb{E}[x] + 3u_1 + 3u_2 + \mathbb{E}[u_1] + 3\mathbb{E}[u_2],$$

$$\mu(s_1, \theta) = \frac{1}{10} + \frac{1}{10}x + \frac{1}{10}\mathbb{E}[x] + \frac{1}{10}u_1 + \frac{1}{5}u_2 + \frac{1}{10}\mathbb{E}[u_1] + \frac{1}{5}\mathbb{E}[u_2],$$

and for the second switching regime $s_2 \in \mathcal{S}$, the systems drift, diffusion, and jump parameters are:

$$b(s_2) = 1 - \frac{1}{2}x + \frac{1}{5}\mathbb{E}[x] + \frac{3}{10}u_1 + \frac{1}{10}u_2 + \frac{1}{5}\mathbb{E}[u_1] + \frac{2}{5}\mathbb{E}[u_2],$$

$$\sigma(s_2) = 4 + 3x + 3\mathbb{E}[x] + 4u_1 + \frac{33}{10}u_2 + 2\mathbb{E}[u_1] + \frac{35}{10}\mathbb{E}[u_2],$$

$$\mu(s_2, \theta) = \frac{1}{5} + \frac{1}{5}x + \frac{1}{5}\mathbb{E}[x] + \frac{3}{10}u_1 + \frac{1}{4}u_2 + \frac{1}{5}\mathbb{E}[u_1] + \frac{3}{10}\mathbb{E}[u_2].$$

The terminal time is considered to be $T = 5$ s. Regarding the cost functional, we have the following for the first switching regime $s_1 \in \mathcal{S}$, i.e.,

$$L_1(x, s_1, u_1, \dots, u_n) = 5x^2(5) + 5\mathbb{E}[x(5)]^2 + \mathbb{E}[x(5)]$$

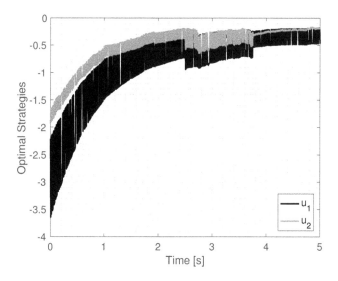

FIGURE 6.4
Evolution of the optimal strategies u_1^* and u_2^* for the scalar-value non-cooperative scenario with Brownian motion, Poisson jumps and regime switching.

$$+ \frac{1}{2} \int_0^5 \left(10x^2 + 10\mathbb{E}[x]^2 + u_1^2 + \mathbb{E}[u_1]^2\right) \mathrm{d}t$$

$$+ \int_0^5 \left(xu_1 + \mathbb{E}[x]\mathbb{E}[u_1] + \mathbb{E}[u_1] + \mathbb{E}[x]\right) \mathrm{d}t,$$

and

$$L_2(x, s_1, u_1, \ldots, u_n) = 10x^2(5) + 10\mathbb{E}[x(5)]^2 + 2\mathbb{E}[x(5)]$$

$$+ \frac{1}{2} \int_0^5 \left(20x^2 + 20\mathbb{E}[x]^2 + 2u_2^2 + 2\mathbb{E}[u_2]^2\right) \mathrm{d}t$$

$$+ \int_0^5 \left(2xu_2 + 2\mathbb{E}[x]\mathbb{E}[u_2] + 2\mathbb{E}[u_2] + 2\mathbb{E}[x]\right) \mathrm{d}t.$$

Now, the cost functions corresponding to the second regime switching $s_2 \in \mathcal{S}$ are as follows:

$$L_1(x, s_2, u_1, \ldots, u_n) = 6x^2(5) + \frac{15}{2}\mathbb{E}[x(5)]^2 + \frac{13}{10}\mathbb{E}[x(5)]$$

$$+ \frac{1}{2} \int_0^5 \left(12x^2 + 15\mathbb{E}[x]^2 + 3u_1^2 + 2\mathbb{E}[u_2]^2\right) \mathrm{d}t$$

$$+ \int_0^5 \left(\frac{11}{10}xu_1 + \frac{11}{10}\mathbb{E}[x]\mathbb{E}[u_1] + \frac{6}{5}\mathbb{E}[u_1] + \frac{13}{10}\mathbb{E}[x]\right) \mathrm{d}t,$$

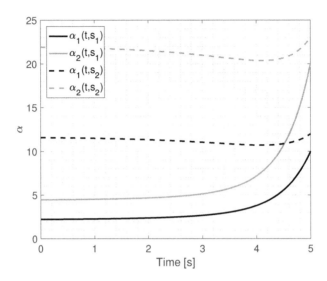

FIGURE 6.5
Evolution of the Riccati equations α_1 and α_2 for the scalar-value non-cooperative scenario with Brownian motion, Poisson jumps and regime switching.

and

$$
\begin{aligned}
L_2(x, s_2, u_1, \dots, u_n) =\ & \frac{23}{2}x^2(5) + \frac{25}{2}\mathbb{E}[x(5)]^2 + \frac{23}{10}\mathbb{E}[x(5)] \\
& + \frac{1}{2}\int_0^5 \left(23x^2 + 25\mathbb{E}[x]^2 + 4u_2^2 + 3\mathbb{E}[u_2]^2\right)\, \mathrm{d}t \\
& + \int_0^5 \left(\frac{21}{10}xu_2 + \frac{21}{10}\mathbb{E}[x]\mathbb{E}[u_2] + \frac{11}{5}\mathbb{E}[u_2] + \frac{23}{10}\mathbb{E}[x]\right)\, \mathrm{d}t.
\end{aligned}
$$

Figure 6.3 shows the optimal evolution of the system state and its expectation approaching to zero when the optimal strategies u_1^* and u_2^* are applied, which are presented in Figure 6.4. It can be seen in Figures 6.3 and 6.4 how the jumps presented in Figure 6.3 affect the evolution of the system states and the optimal strategies. These jumps are evident at times around 2.5 s and 3.7 s.

On the other hand, Figures 6.5 to 6.8 present the evolution of the differential equations to compute the optimal strategies. For instance, Figure 6.5 shows the evolution of $\alpha_1(t, s_1)$, $\alpha_1(t, s_2)$, $\alpha_2(t, s_1)$, and $\alpha_2(t, s_2)$; Figure 6.6 the evolution of $\beta_1(t, s_1)$, $\beta_1(t, s_2)$, $\beta_2(t, s_1)$, and $\beta_2(t, s_2)$; Figure 6.7 the evolution of $\gamma_1(t, s_1)$, $\gamma_1(t, s_2)$, $\gamma_2(t, s_1)$, and $\gamma_2(t, s_2)$; and Figure 6.8 the evolution of $\delta_1(t, s_1)$, $\delta_1(t, s_2)$, $\delta_2(t, s_1)$, and $\delta_2(t, s_2)$. Notice that the terminal boundary conditions are satisfied as stated in Theorem 1.

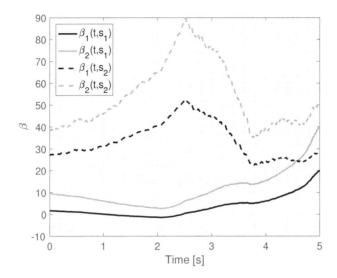

FIGURE 6.6
Evolution of the Riccati equations β_1 and β_2 for the scalar-value non-cooperative scenario with Brownian motion, Poisson jumps and regime switching.

In this chapter, we have presented a case where the obtained differential equations are not longer Riccati. This is an extended version of the cases studied in the previous chapters.

6.4 Exercises

1. Show that the solution of the Mean-Field-Type Games with Jump-Diffusion Process and Regime Switching presented in this chapter is more general than the solution discussed in Chapter 4 for the Non-Cooperative Mean-Field-Type Games considering constant coefficient diffusion without regime switching neither jumps.

   Hint: Show that the result presented in Proposition 13 can be directly retrieved from Theorem 1.

2. Solve a mean-field-type control problem with state-and control-input-dependent jump-diffusion processes and random coefficients,

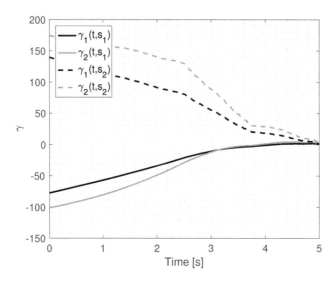

FIGURE 6.7
Evolution of the Riccati equations γ_1 and γ_2 for the scalar-value non-cooperative scenario with Brownian motion, Poisson jumps and switching.

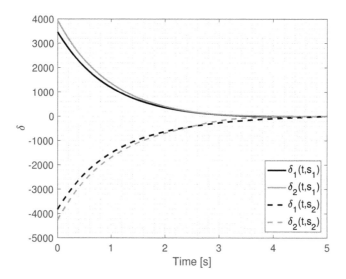

FIGURE 6.8
Evolution of the Riccati equations δ_1 and δ_2 for the scalar-value non-cooperative scenario with Brownian motion, Poisson jumps and switching.

i.e., consider the following dynamics:

$$dx = bdt + \sigma dB + \int_{\theta \in \Theta} \mu(\cdot, \theta) \tilde{N}(dt, d\theta),$$

where

$$b = b_0 + b_1 x + \bar{b}_1 \mathbb{E}[x] + b_2 u + \bar{b}_2 \mathbb{E}[u],$$
$$\sigma = \sigma_0 + \sigma_1 x + \bar{\sigma}_1 \mathbb{E}[x] + \sigma_2 u + \bar{\sigma}_2 \mathbb{E}[u],$$
$$\mu(\cdot, \theta) = \mu_0 + \mu_1 x + \bar{\mu}_1 \mathbb{E}[x] + \mu_2 u + \bar{\mu}_2 \mathbb{E}[u],$$

with the coefficients $b_k(t, s); \sigma_k(t, s); \bar{b}_k(t, s); \bar{\sigma}_k(t, s) \in \mathbb{R}$ and $s \in \mathcal{S}$, is a regime switching, $\mathcal{S}$ is a non-empty and finite set of switching regimes. Moreover $\mu_k(t, s, \theta) \in \mathbb{R}, \theta \in \Theta$ where $\Theta \subset \mathbb{R}^d$ is the set of jump sizes, for all $k \in \{0, 1, 2\}$. The cost functional given by

$$L(x, s, u) = \frac{1}{2} q(T) x^2(T) + \frac{1}{2} \bar{q}(T) \mathbb{E}[x(T)]^2 + \varepsilon_3(T) \mathbb{E}[x(T)]$$
$$+ \frac{1}{2} \int_0^T \left(q x^2 + \bar{q} \mathbb{E}[x]^2 + r u^2 + \bar{r} \mathbb{E}[u]^2 \right) dt$$
$$+ \int_0^T \left(\varepsilon_1 x u + \bar{\varepsilon}_1 \mathbb{E}[x] \mathbb{E}[u] + \varepsilon_2 \mathbb{E}[u] + \varepsilon_3 \mathbb{E}[x] \right) dt.$$

Hint: An appropriate guess functional is given by

$$f(t, x, s) = \frac{1}{2} \alpha \left(x - \mathbb{E}[x] \right)^2 + \frac{1}{2} \beta \mathbb{E}[x]^2 + \gamma \mathbb{E}[x] + \delta.$$

3. Extend the presented mean-field-type game with state- and control-input-dependent jump-diffusion processes and random coefficients for the co-opetitive scenario.

4. Show that both the non-cooperative and cooperative game solutions with state-and control-input-dependent jump-diffusion processes and random coefficients can be retrieved from the solution of the co-opetitive game problem (also considering the same stochastic processes). This is similar to what we made in Chapter 5 with the mean-field-type game problem with constant diffusion coefficient.

7

Mean-Field-Type Stackelberg Games

The idea of hierarchy dates back a least to 1934, when Stackelberg [94] introduced a game that models markets where some firms have stronger influence over others. Stackelberg games consist of only two players, which are known as a leader and a follower. The leader who moves first, decides an optimal strategy after anticipating the best response of the follower. Then, the follower eventually chooses the anticipated best response to optimize her cost or payoff. Therefore, this game is a game with two-level hierarchy. A dynamic LQ Stackelberg differential game was studied by Samaan and Cruz in [95]. A stochastic LQ Stackelberg differential game was investigated by Bagchi and Başar in [96]. Bensoussan et al. [97] derive a maximum principle for the leader's Stackelberg solution under the adapted closed-loop memoryless information structure.

When there are two or more players involved in this sequential strategic scheme, the Stackelberg game is called hierarchical game and it becomes more interesting and involved due to its multi-layer structure including various forms of information. The players act in sequential order such that each one of them is a leader for the previous and a follower of the next player in the hierarchy. For hierarchical mean-field-free differential games, see e.g. [98–102].

Only few works consider hierarchical structures in mean-field related games. Open-loop Stackelberg solutions are addressed in linear-quadratic setting in [103,104]; and in the context of large populations, mean-field Stackelberg games are investigated in [105–109]. Besides, the leader-follower configuration has been used in several problems and fields to illustrate and model a variety of hierarchical behaviors. For instance, in [110], a leader-follower stochastic differential game with asymmetric information is studied, and motivated by applications in finance, economics and management engineering. In [111], a large population leader-follower stochastic multi-agent system is analyzed with coupled cost functions and by using a mean-field LQG approach. Regarding control applications, [112] presents a tracking control design in a distributed manner in a multi-agent system configured in a leader-follower fashion, and it is shown that the setup can be used to model the power sharing problem in micro-grids. In [113], a security problem in networked control systems is studied by means of a Stackelberg approach, and in [114] a hierarchical control structure or sequential predictive control is designed for a large-scale water system. Other works have extended the two-hierarchical case for multiple layers, also studying the optimal order for the decision-makers throughout

DOI: 10.1201/9781003098607-7

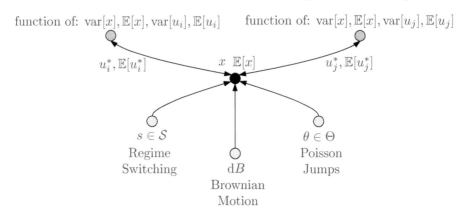

FIGURE 7.1

General scheme of the two-player Stackelberg mean-field-type game with jump-diffusion and regime switching.

the levels, and analyzing the most appropriate structure to reduce the social cost, i.e., hierarchical design [115], [116].

Hence, this topic is relevant and getting special importance because there is a large variety of engineering applications whose configuration is built in a hierarchical manner. Hence, we can also use and analyze this class of hierarchical game design in the context of risk-aware hierarchical optimal control techniques.

7.1   Mean-Field-Type Stackelberg Game Set-up

Consider the following state dynamics with jump-diffusion and regime switching involving only two decision-makers, i.e., $\{i, j\}$ (see Figure 7.1):

$$\mathrm{d}x = b\mathrm{d}t + \sigma \mathrm{d}B + \int_{\theta \in \Theta} \mu(\cdot, \theta)\tilde{N}(\mathrm{d}t, \mathrm{d}\theta), \qquad (7.1)$$

where the drift, diffusion, and jump components are as follows:

$$b = b_0 + b_1 x + \bar{b}_1 \mathbb{E}[x] + b_{2i} u_i + b_{2j} u_j + \bar{b}_{2i} \mathbb{E}[u_i] + \bar{b}_{2j} \mathbb{E}[u_j],$$
$$\sigma = \sigma_0 + \sigma_1 x + \bar{\sigma}_1 \mathbb{E}[x] + \sigma_{2i} u_i + \sigma_{2j} u_j + \bar{\sigma}_{2i} \mathbb{E}[u_i] + \bar{\sigma}_{2j} \mathbb{E}[u_j],$$
$$\mu(\cdot, \theta) = \mu_0 + \mu_1 x + \bar{\mu}_1 \mathbb{E}[x] + \mu_{2i} u_i + \mu_{2j} u_j + \bar{\mu}_{2i} \mathbb{E}[u_i] + \bar{\mu}_{2j} \mathbb{E}[u_j],$$

with time-and-switching-dependent coefficients :

$$b_k, \sigma_k, \bar{b}_k, \bar{\sigma}_k : [0, T] \times \mathcal{S} \to \mathbb{R},$$

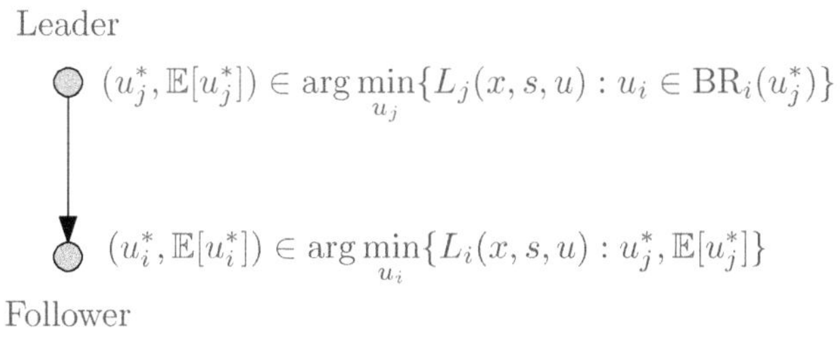

Leader

$$(u_j^*, \mathbb{E}[u_j^*]) \in \arg\min_{u_j}\{L_j(x, s, u) : u_i \in \mathrm{BR}_i(u_j^*)\}$$

$$(u_i^*, \mathbb{E}[u_i^*]) \in \arg\min_{u_i}\{L_i(x, s, u) : u_j^*, \mathbb{E}[u_j^*]\}$$

Follower

FIGURE 7.2
Hierarchical order in the Stackelberg mean-field-type game.

and $s \in \mathcal{S}$ is a regime switching, $\mathcal{S}$ is a non-empty and finite set of switching regimes. Hence,

$$\mu_k : [0, T] \times \mathcal{S} \times \Theta \to \mathbb{R},$$

with $\theta \in \Theta$ where $\Theta \subset \mathbb{R}^d$ is the set of jump sizes, for all $k \in \{0, 1, 2\}$. The process B is a standard one-dimensional Brownian motion, $\tilde{N}(dt, d\theta) = N(dt, d\theta) - \nu(d\theta)dt$ is a compensated jump process with Radon measure ν over Θ. The switching process s switches among elements in $\mathcal{S}$ with transition rate $\tilde{q}_{ss'}$. The $\tilde{Q}$-matrix $\tilde{Q} = (\tilde{q}_{ss'}, (s, s') \in \mathcal{S}^2)$, satisfies $s \neq s', \tilde{q}_{ss'} > 0$,

$$\tilde{q}_{ss} = -\sum_{s' \neq s} \tilde{q}_{ss'}.$$

In addition, it is assumed that the processes $s(t), B(t), N(t, .), x_0$ are mutually independent.

The term $\mathbb{E}[x] = \mathbb{E}[x|\mathcal{F}^s]$ denotes the expected value of x given the switching regime $s(t) = s$, and $\mathbb{E}[u] = \mathbb{E}[u|\mathcal{F}^s]$. These terms $\mathbb{E}[x]$ and $\mathbb{E}[u]$ are called conditional state mean-field, and conditional control-action mean-field terms, respectively. The state dynamics are therefore of conditional *McKean-Vlasov* type as the conditional probability measure of the state appears in the drift, diffusion, and jump coefficients via $(\mathbb{E}[x], \mathbb{E}[u_i], \mathbb{E}[u_j])$.

The reader can refer to the problem we studied in Chapter 6 noticing that the system dynamics considered there are equivalent to the SDE we study in this chapter. The main difference consists of the number of decision-makers in the strategic interaction.

In the decision maker set $\{i, j\}$, it is assumed that the decision-maker denoted by j is the leader and then decision-maker denoted by i is the followers. In this regard, the decision-maker i reacts against the strategic selection of the leader j. Figure 7.2 presents the hierarchical order for the leader-follower decision-makers.

We associate a cost functional of mean-field type of any decision-maker $k \in \{i, j\}$, either the leader or the follower as follows:

$$L_k(x, s, u_i, u_j) = \frac{1}{2}q_k(T)x^2(T) + \frac{1}{2}\bar{q}_k(T)\mathbb{E}[x(T)]^2$$

$$+ \frac{1}{2} \int_0^T \left(q_k x^2 + \bar{q}_k \mathbb{E}[x]^2 + r_k u_k^2 + \bar{r}_k \mathbb{E}[u_k]^2 \right.$$

$$+ \left. \varepsilon_{ij}(u_k - \mathbb{E}[u_k])(u_{k'} - \mathbb{E}[u_{k'}]) + \bar{\varepsilon}_{kk'}\mathbb{E}[u_k]\mathbb{E}[u_{k'}] \right) dt,$$

where k' denotes the other decision-maker different from $k \in \{i, j\}$ given by $\{i, j\} \setminus k$, and $q_k = q_k(t, s(t))$, $\bar{q}_k = q_k(t, s(t))$; and $r_k = r_k(t, s(t))$, $\bar{r}_k = \bar{r}_k(t, s(t)) \in \mathbb{R}$, $\varepsilon_{kk'} = \varepsilon_{kk'}(t, s(t)) \in \mathbb{R}$, $\bar{\varepsilon}_{kk'} = \bar{\varepsilon}_{kk'}(t, s(t)) \in \mathbb{R}$. All these coefficients are assumed to be integrable. Therefore, the cost function can be re-written for any $k \in \{i, j\}$ as

$$L_k(x, s, u_i, u_j) = \frac{1}{2} q_k (x(T) - \mathbb{E}[x(T)])^2 + \frac{1}{2} (q_k + \bar{q}_k)\mathbb{E}[x(T)]^2$$

$$+ \frac{1}{2} \int_0^T \left(q_k (x - \mathbb{E}[x])^2 + (q_k + \bar{q}_k)\mathbb{E}[x]^2 \right) dt$$

$$+ \frac{1}{2} \int_0^T \left(r_k (u_k - \mathbb{E}[u_k])^2 + (r_k + \bar{r}_k)\mathbb{E}[u_k]^2 \right) dt$$

$$+ \frac{1}{2} \int_0^T \left(\varepsilon_{kk'}(u_k - \mathbb{E}[u_k])(u_{k'} - \mathbb{E}[u_{k'}]) + \bar{\varepsilon}_{kk'}\mathbb{E}[u_k]\mathbb{E}[u_{k'}] \right) dt,$$

$$(7.2)$$

showing the risk-awareness in the two-level hierarchical game problem. Recall the equivalence between these two cost functional as it was discussed in Proposition 16. Furthermore, the cost functional in (7.2) also incorporates the covariance term $cov(u_k, u_{k'})$ creating a coupling between the control inputs for the leader and the follower.

It is clear that the strategic configuration of this game is different from the non-cooperative scenario where all the decision-makers make their strategic selection simultaneously. We recall that the solution for the non-cooperative game problem is given by a Nash equilibrium. In contrast, the solution of the two-level hierarchical game problem is given by a Stackelberg equilibrium, which is formally introduced next in Definition 14.

Definition 14 (Stackelberg solution) *A mean-field-type Stackelberg solution for a strategic interaction with decision-makers $\{i, j\}$, taking the decision-maker j as the leader and the decision-maker i as the follower, is given by (u_i^*, u_j^*) such that*

$$u_j^* \in \arg \min_{u_j \in \mathcal{U}_j} \{ L_j(x, s, u_i, u_j) : u_i \in \mathrm{BR}_i(u_j) \},$$

$$u_i^* \in \mathrm{BR}_i(u_j),$$

where $\mathrm{BR}_i(u_j)$ is the best response of the follower i against the strategic selection made by the leader j, i.e.,

$$u_i^* \in \arg \min_{u_i \in \mathcal{U}_i} \{ L_i(x, s, u_i, u_j) : u_j^* \}.$$

Now that the new settings for this game configuration have been establish, we continue by presenting the computation of the solution by using the direct method. In the following section, we present the semi-explicit Stackelberg solution for the aforementioned mean-field-type game problem.

7.2 Semi-explicit Solution of the Stackelberg Mean-Field-Type Game with Jump-Diffusion Process and Regime Switching

The semi-explicit solution for the two-level hierarchical mean-field-type game problem with a unique leader and follower, and with state-and-control-input-dependent jump-diffusion and regime switching processes (with random coefficients) is presented next in Theorem 2.

Theorem 2 *The mean-field-type Stackelberg solution with a leader decision-maker denoted by j and a follower decision-maker denoted by i is given by*

Leader j:

$$u_j^* = \mathbb{E}[u_j^*] - \frac{\bar{\phi}_j}{\bar{c}_j}\left(x - \mathbb{E}[x]\right),$$

$$\mathbb{E}[u_j^*] = -\frac{\lambda_j + \phi_j \mathbb{E}[x]}{c_j},$$

Follower i:

$$u_i^* = \mathbb{E}[u_i^*] + \left(\frac{\bar{\rho}_{ij}}{\bar{c}_i}\frac{\bar{\phi}_j}{\bar{c}_j} - \frac{\bar{\phi}_i}{\bar{c}_i}\right)\left(x - \mathbb{E}[x]\right),$$

$$\mathbb{E}[u_i^*] = -\frac{\lambda_i}{c_i} + \frac{\rho_{ij}}{c_i}\frac{\lambda_j}{c_j} + \left(\frac{\rho_{ij}}{c_i}\frac{\phi_j}{c_j} - \frac{\phi_i}{c_i}\right)\mathbb{E}[x],$$

where we have the following auxiliary parameters associated with the follower decision-maker i:

$$c_i = \left[\alpha_i\left((\sigma_{i2} + \bar{\sigma}_{i2})^2 + \int_{\theta\in\Theta}(\mu_{i2} + \bar{\mu}_{i2})^2\nu(\mathrm{d}\theta)\right) + r_i + \bar{r}_i\right],$$

$$\lambda_i = (b_{2i} + \bar{b}_{2i})\gamma_i + \alpha_i\left(\sigma_0(\sigma_{i2} + \bar{\sigma}_{i2}) + \int_{\theta\in\Theta}\mu_0(\mu_{i2} + \bar{\mu}_{i2})\nu(\mathrm{d}\theta)\right),$$

$$\phi_i = \left[(b_{2i} + \bar{b}_{2i})\beta_i\right.$$

$$\left. + \alpha_i\left((\sigma_1 + \bar{\sigma}_1)(\sigma_{i2} + \bar{\sigma}_{i2}) + \int_{\theta\in\Theta}(\mu_1 + \bar{\mu}_1)(\mu_{i2} + \bar{\mu}_{i2})\nu(\mathrm{d}\theta)\right)\right],$$

$$\rho_{ij} = \left[\alpha_i \left((\sigma_{i2} + \bar{\sigma}_{i2})(\sigma_{j2} + \bar{\sigma}_{j2}) + \int_{\theta \in \Theta} (\mu_{i2} + \bar{\mu}_{i2})(\mu_{j2} + \bar{\mu}_{j2})\nu(d\theta)\right) + \bar{\varepsilon}_{ij}\right],$$

$$\bar{c}_i = \left[\alpha_i \left(\sigma_{i2}^2 + \int_{\theta \in \Theta} \mu_{i2}^2 \nu(d\theta)\right) + r_i\right],$$

$$\bar{\phi}_i = \left[b_{2i}\alpha_i + \alpha_i \left(\sigma_1 \sigma_{i2} + \int_{\theta \in \Theta} \mu_1 \mu_{i2}\nu(d\theta)\right)\right],$$

$$\bar{\rho}_{ij} = \left[\alpha_i \left(\sigma_{i2}\sigma_{j2} + \int_{\theta \in \Theta} \mu_{i2}\mu_{j2}\nu(d\theta)\right) + \varepsilon_{ij}\right],$$

and the following auxiliary variables associated with the leader decision-maker j:

$$c_j = \left[\alpha_j \left((\sigma_{j2} + \bar{\sigma}_{j2})^2 + \int_{\theta \in \Theta} (\mu_{j2} + \bar{\mu}_{j2})^2 \nu(d\theta)\right) + r_j + \bar{r}_j \right.$$

$$- 2\alpha_j \left((\sigma_{j2} + \bar{\sigma}_{j2})(\sigma_{i2} + \bar{\sigma}_{i2}) + \int_{\theta \in \Theta} (\mu_{j2} + \bar{\mu}_{j2})(\mu_{i2} + \bar{\mu}_{i2})\nu(d\theta)\right.$$

$$- \frac{1}{4}\left((\sigma_{i2} + \bar{\sigma}_{i2})^2 + \int_{\theta \in \Theta} (\mu_{i2} + \bar{\mu}_{i2})^2 \nu(d\theta)\right)\frac{\rho_{ij}}{c_i}\right)\frac{\rho_{ij}}{c_i} - 2\bar{\varepsilon}_{ji}\frac{\rho_{ij}}{c_i}\right],$$

$$\lambda_j = (b_{2j} + \bar{b}_{2j})\gamma_j + \alpha_j \left(\sigma_0(\sigma_{j2} + \bar{\sigma}_{j2}) + \int_{\theta \in \Theta} \mu_0(\mu_{j2} + \bar{\mu}_{j2})\nu(d\theta)\right)$$

$$- \left[\alpha_j \left((\sigma_{j2} + \bar{\sigma}_{j2})(\sigma_{i2} + \bar{\sigma}_{i2})\right.\right.$$

$$\left.\left. + \int_{\theta \in \Theta} (\mu_{j2} + \bar{\mu}_{j2})(\mu_{i2} + \bar{\mu}_{i2})\nu(d\theta)\right) + \bar{\varepsilon}_{ji}\right]\frac{\lambda_i}{c_i}$$

$$+ \alpha_j \left((\sigma_{i2} + \bar{\sigma}_{i2})^2 + \int_{\theta \in \Theta} (\mu_{i2} + \bar{\mu}_{i2})^2 \nu(d\theta)\right)\frac{\lambda_i \rho_{ij}}{c_i^2}$$

$$- (b_{2i} + \bar{b}_{2i})\frac{\rho_{ij}}{c_i}\gamma_j - \alpha_j \left(\sigma_0(\sigma_{i2} + \bar{\sigma}_{i2}) + \int_{\theta \in \Theta} \mu_0(\mu_{i2} + \bar{\mu}_{i2})\nu(d\theta)\right)\frac{\rho_{ij}}{c_i},$$

$$\phi_j = \left[(b_{2j} + \bar{b}_{2j})\beta_j\right.$$

$$+ \alpha_j \left((\sigma_1 + \bar{\sigma}_1)(\sigma_{j2} + \bar{\sigma}_{j2}) + \int_{\theta \in \Theta} (\mu_1 + \bar{\mu}_1)(\mu_{j2} + \bar{\mu}_{j2})\nu(d\theta)\right)$$

$$- \alpha_j \left((\sigma_{j2} + \bar{\sigma}_{j2})(\sigma_{i2} + \bar{\sigma}_{i2}) + \int_{\theta \in \Theta} (\mu_{j2} + \bar{\mu}_{j2})(\mu_{i2} + \bar{\mu}_{i2})\nu(d\theta)\right)\frac{\phi_i}{c_i}$$

$$- \bar{\varepsilon}_{ji}\frac{\phi_i}{c_i} + \alpha_j \left((\sigma_{i2} + \bar{\sigma}_{i2})^2 + \int_{\theta \in \Theta} (\mu_{i2} + \bar{\mu}_{i2})^2 \nu(d\theta)\right)\frac{\phi_i \rho_{ij}}{c_i^2}$$

$$- (b_{2i} + \bar{b}_{2i})\frac{\rho_{ij}}{c_i}\beta_j$$

$$- \alpha_j \left((\sigma_1 + \bar{\sigma}_1)(\sigma_{i2} + \bar{\sigma}_{i2}) + \int_{\theta \in \Theta} (\mu_1 + \bar{\mu}_1)(\mu_{i2} + \bar{\mu}_{i2}) \nu(\mathrm{d}\theta) \right) \frac{\rho_{ij}}{c_i} \right],$$

$$\bar{c}_j = \left[\alpha_j \left[\left(\sigma_{j2}^2 + \int_{\theta \in \Theta} \mu_{j2}^2 \nu(\mathrm{d}\theta) \right) + \left(\sigma_{i2}^2 + \int_{\theta \in \Theta} \mu_{i2}^2 \nu(\mathrm{d}\theta) \right) \frac{\bar{\rho}_{ij}^2}{\bar{c}_i^2} \right] + r_j \right.$$

$$- \alpha_j \left(\sigma_{j2} \sigma_{i2} + \int_{\theta \in \Theta} \mu_{j2} \mu_{i2} \nu(\mathrm{d}\theta) \right) \frac{\bar{\rho}_{ij}}{\bar{c}_i} - \frac{\bar{\rho}_{ij}}{\bar{c}_i} \varepsilon_{ji} \right],$$

$$\bar{\phi}_j = \alpha_j \left[b_{2j} + \sigma_1 \sigma_{j2} + \int_{\theta \in \Theta} \mu_1 \mu_{j2} \nu(\mathrm{d}\theta) + \left(\sigma_{i2}^2 + \int_{\theta \in \Theta} \mu_{i2}^2 \nu(\mathrm{d}\theta) \right) \frac{\bar{\phi}_i \bar{\rho}_{ij}}{\bar{c}_i^2} \right.$$

$$- \left(b_{2i} + \left(\sigma_1 \sigma_{i2} + \int_{\theta \in \Theta} \mu_1 \mu_{i2} \nu(\mathrm{d}\theta) \right) \right) \frac{\bar{\rho}_{ij}}{\bar{c}_i}$$

$$- \alpha_j \left(\sigma_{j2} \sigma_{i2} + \int_{\theta \in \Theta} \mu_{j2} \mu_{i2} \nu(\mathrm{d}\theta) \right) \frac{\bar{\phi}_i}{\bar{c}_i} - \frac{\bar{\phi}_i}{\bar{c}_i} \varepsilon_{ji} \right].$$

Moreover, α_i, β_i, γ_i and δ_i for the follower decision-maker i, solve the following backward differential equations:

$$\dot{\alpha}_i = -q_i - 2b_1 \alpha_i - \sum_{s' \in \mathcal{S}} (\alpha_i(\cdot, s') - \alpha_i(\cdot, s)) - \alpha_i \left(\sigma_1^2 + \int_{\theta \in \Theta} \mu_1^2 \nu(\mathrm{d}\theta) \right)$$

$$- 2 \frac{\bar{\phi}_i \bar{\rho}_{ij}}{\bar{c}_i} \frac{\bar{\phi}_j}{\bar{c}_j} + \frac{\bar{\phi}_i^2}{\bar{c}_i} + \frac{\bar{\rho}_{ij}^2}{\bar{c}_i} \frac{\bar{\phi}_j^2}{\bar{c}_j^2} - \alpha_i \left(\sigma_{j2}^2 + \int_{\theta \in \Theta} \mu_{j2}^2 \nu(\mathrm{d}\theta) \right) \frac{\bar{\phi}_j^2}{\bar{c}_j^2}$$

$$+ 2 \left[b_{2j} \alpha_i + \alpha_i \left(\sigma_1 \sigma_{j2} + \int_{\theta \in \Theta} \mu_1 \mu_{j2} \nu(\mathrm{d}\theta) \right) \right] \frac{\bar{\phi}_j}{\bar{c}_j},$$

$$\dot{\beta}_i = -q_i - \bar{q}_i - 2(b_1 + \bar{b}_1) \beta_i - \sum_{s' \in \mathcal{S}} (\beta_i(\cdot, s') - \beta_i(\cdot, s))$$

$$- \alpha_i \left((\sigma_1 + \bar{\sigma}_1)^2 + \int_{\theta \in \Theta} (\mu_1 + \bar{\mu}_1)^2 \nu(\mathrm{d}\theta) \right) + \frac{\phi_i^2}{c_i}$$

$$+ \frac{\rho_{ij}^2}{c_i} \frac{\phi_j^2}{c_j^2} - 2 \frac{\phi_i \rho_{ij}}{c_i} \frac{\phi_j}{c_j} - \alpha_i \left((\sigma_{j2} + \bar{\sigma}_{j2})^2 + \int_{\theta \in \Theta} (\mu_{j2} + \bar{\mu}_{j2})^2 \nu(\mathrm{d}\theta) \right) \frac{\phi_j^2}{c_j^2}$$

$$+ 2 \left[(b_{2j} + \bar{b}_{2j}) \beta_i \right.$$

$$+ \alpha_i \left((\sigma_1 + \bar{\sigma}_1)(\sigma_{j2} + \bar{\sigma}_{j2}) + \int_{\theta \in \Theta} (\mu_1 + \bar{\mu}_1)(\mu_{j2} + \bar{\mu}_{j2}) \nu(\mathrm{d}\theta) \right) \right] \frac{\phi_j}{c_j},$$

$$\dot{\gamma}_i = -b_0 \beta_i - (b_1 + \bar{b}_1) \gamma_i - \sum_{s' \in \mathcal{S}} (\gamma_i(\cdot, s') - \gamma_i(\cdot, s))$$

$$- \alpha_i \left(\sigma_0(\sigma_1 + \bar{\sigma}_1) + \int_{\theta \in \Theta} \mu_0(\mu_1 + \bar{\mu}_1) \nu(\mathrm{d}\theta) \right) + \frac{\rho_{ij}^2}{c_i} \frac{\lambda_j \phi_j}{c_j^2} + \frac{\lambda_i \phi_i}{c_i}$$

$$-\frac{\lambda_i \rho_{ij}}{c_i}\frac{\phi_j}{c_j} - \frac{\phi_i \rho_{ij}}{c_i}\frac{\lambda_j}{c_j} - \alpha_i\left((\sigma_{j2} + \bar{\sigma}_{j2})^2 + \int_{\theta \in \Theta}(\mu_{j2} + \bar{\mu}_{j2})^2 \nu(d\theta)\right)\frac{\lambda_j \phi_j}{c_j^2}$$

$$+\left[(b_{2j} + \bar{b}_{2j})\beta_i\right.$$

$$\left. + \alpha_i\left((\sigma_1 + \bar{\sigma}_1)(\sigma_{j2} + \bar{\sigma}_{j2}) + \int_{\theta \in \Theta}(\mu_1 + \bar{\mu}_1)(\mu_{j2} + \bar{\mu}_{j2})\nu(d\theta)\right)\right]\frac{\lambda_j}{c_j}$$

$$+\left[(b_{2j} + \bar{b}_{2j})\gamma_i + \alpha_i\left(\sigma_0(\sigma_{j2} + \bar{\sigma}_{j2}) + \int_{\theta \in \Theta}\mu_0(\mu_{j2} + \bar{\mu}_{j2})\nu(d\theta)\right)\right]\frac{\phi_j}{c_j},$$

$$\dot{\delta}_i = -b_0 \gamma_i - \sum_{s' \in \mathcal{S}}(\delta_i(\cdot, s') - \delta_i(\cdot, s)) - \frac{1}{2}\alpha_i\left(\sigma_0^2 + \int_{\theta \in \Theta}\mu_0^2 \nu(d\theta)\right)$$

$$+\frac{1}{2}\frac{\lambda_i^2}{c_i} + \frac{1}{2}\frac{\rho_{ij}^2}{c_i}\frac{\lambda_j^2}{c_j^2} - \frac{\lambda_i \rho_{ij}}{c_i}\frac{\lambda_j}{c_j}$$

$$-\frac{1}{2}\alpha_i\left((\sigma_{j2} + \bar{\sigma}_{j2})^2 + \int_{\theta \in \Theta}(\mu_{j2} + \bar{\mu}_{j2})^2 \nu(d\theta)\right)\frac{\lambda_j^2}{c_j^2}$$

$$+\left[(b_{2j} + \bar{b}_{2j})\gamma_i + \alpha_i\left(\sigma_0(\sigma_{j2} + \bar{\sigma}_{j2}) + \int_{\theta \in \Theta}\mu_0(\mu_{j2} + \bar{\mu}_{j2})\nu(d\theta)\right)\right]\frac{\lambda_j}{c_j}$$

with the following terminal boundary condition:

$$\alpha_i(T) = q_i(T),$$
$$\beta_i(T) = q_i(T) + \bar{q}_i(T),$$
$$\gamma_i(T) = 0,$$
$$\delta_i(T) = 0,$$

and where α_j, β_j, γ_j, and δ_j of the leader decision-maker j solve the following backward differential equations:

$$\dot{\alpha}_j = -q_j - 2b_1 \alpha_j - \sum_{s' \in \mathcal{S}}(\alpha_j(\cdot, s') - \alpha_j(\cdot, s)) - \alpha_j\left(\sigma_1^2 + \int_{\theta \in \Theta}\mu_1^2 \nu(d\theta)\right)$$

$$- \alpha_j\left(\sigma_{i2}^2 + \int_{\theta \in \Theta}\mu_{i2}^2 \nu(d\theta)\right)\frac{\bar{\phi}_i^2}{\bar{c}_i^2}$$

$$+ 2\left[b_{2i}\alpha_j + \alpha_j\left(\sigma_1 \sigma_{i2} + \int_{\theta \in \Theta}\mu_1 \mu_{i2}\nu(d\theta)\right)\right]\frac{\bar{\phi}_i}{\bar{c}_i} + \frac{\bar{\phi}_j^2}{\bar{c}_j},$$

$$\dot{\beta}_j = -q_j - \bar{q}_j - 2(b_1 + \bar{b}_1)\beta_j - \sum_{s' \in \mathcal{S}}(\beta_j(\cdot, s') - \beta_j(\cdot, s))$$

$$- \alpha_j\left((\sigma_1 + \bar{\sigma}_1)^2 + \int_{\theta \in \Theta}(\mu_1 + \bar{\mu}_1)^2 \nu(d\theta)\right)$$

$$+ \left((\sigma_{i2} + \bar{\sigma}_{i2})^2 + \int_{\theta \in \Theta} (\mu_{i2} + \bar{\mu}_{i2})^2 \nu(\mathrm{d}\theta) \right) \frac{\phi_i^2}{c_i^2} + 2(b_{2i} + \bar{b}_{2i}) \frac{\phi_i}{c_i} \beta_j$$

$$+ 2\alpha_j \left((\sigma_1 + \bar{\sigma}_1)(\sigma_{i2} + \bar{\sigma}_{i2}) + \int_{\theta \in \Theta} (\mu_1 + \bar{\mu}_1)(\mu_{i2} + \bar{\mu}_{i2}) \nu(\mathrm{d}\theta) \right) \frac{\phi_i}{c_i} + \frac{\phi_j^2}{c_j},$$

$$\dot{\gamma}_j = -b_0 \beta_j - (b_1 + \bar{b}_1) \gamma_j - \sum_{s' \in \mathcal{S}} (\gamma_j(\cdot, s') - \gamma_j(\cdot, s))$$

$$- \alpha_j \left(\sigma_0(\sigma_1 + \bar{\sigma}_1) + \int_{\theta \in \Theta} \mu_0(\mu_1 + \bar{\mu}_1) \nu(\mathrm{d}\theta) \right)$$

$$- \alpha_j \left((\sigma_{i2} + \bar{\sigma}_{i2})^2 + \int_{\theta \in \Theta} (\mu_{i2} + \bar{\mu}_{i2})^2 \nu(\mathrm{d}\theta) \right) \frac{\lambda_i \phi_i}{c_i^2} + (b_{2i} + \bar{b}_{2i}) \frac{\lambda_i}{c_i} \beta_j$$

$$+ \alpha_j \left((\sigma_1 + \bar{\sigma}_1)(\sigma_{i2} + \bar{\sigma}_{i2}) + \int_{\theta \in \Theta} (\mu_1 + \bar{\mu}_1)(\mu_{i2} + \bar{\mu}_{i2}) \nu(\mathrm{d}\theta) \right) \frac{\lambda_i}{c_i}$$

$$+ (b_{2i} + \bar{b}_{2i}) \frac{\phi_i}{c_i} \gamma_j + \alpha_j \left(\sigma_0(\sigma_{i2} + \bar{\sigma}_{i2}) + \int_{\theta \in \Theta} \mu_0(\mu_{i2} + \bar{\mu}_{i2}) \nu(\mathrm{d}\theta) \right) \frac{\phi_i}{c_i}$$

$$+ \frac{\lambda_j \phi_j}{c_j},$$

$$\dot{\delta}_j = -b_0 \gamma_j - \sum_{s' \in \mathcal{S}} (\delta_j(\cdot, s') - \delta_j(\cdot, s))$$

$$- \frac{1}{2} \alpha_j \left(\sigma_0^2 + \int_{\theta \in \Theta} \mu_0^2 \nu(\mathrm{d}\theta) + \left((\sigma_{i2} + \bar{\sigma}_{i2})^2 + \int_{\theta \in \Theta} (\mu_{i2} + \bar{\mu}_{i2})^2 \nu(\mathrm{d}\theta) \right) \frac{\lambda_i^2}{c_i^2} \right)$$

$$+ (b_{2i} + \bar{b}_{2i}) \frac{\lambda_i}{c_i} \gamma_j + \alpha_j \left(\sigma_0(\sigma_{i2} + \bar{\sigma}_{i2}) + \int_{\theta \in \Theta} \mu_0(\mu_{i2} + \bar{\mu}_{i2}) \nu(\mathrm{d}\theta) \right) \frac{\lambda_i}{c_i}$$

$$+ \frac{\lambda_j^2}{2c_j},$$

being

$$\alpha_j(T) = q_j(T),$$
$$\beta_j(T) = q_j(T) + \bar{q}_j(T),$$
$$\gamma_j(T) = 0,$$
$$\delta_j(T) = 0,$$

the terminal boundary conditions.

From the previous result, we can observe that when solving the game-theoretical problems addressed in the previous chapters, the structure of the optimal control input was the same for all the decision-makers given that all of them played simultaneously. Here in contrast and generally, when considering multiple layers in the sequential strategic interaction, the control inputs corresponding to decision-makers acting at different layers do not coincide. However, there might be cases in which both the hierarchical equilibrium is the same as if all the decision-makers plays simultaneously as we will study later on in Section 7.3.

Below, we present the computation, step-by-step, of the semi-explicit solution applying the direct method.

Proof 18 (Theorem 2) *Even though we are now studying a new game-theoretical settings, notice that the structure of both the cost functional and the system dynamics is the same with respect to the problem we studied in Chapter 6. Therefore, it is expected we can use the same guess functional for the optimal cost from initial time t up to terminal time T given by $f_i(t, x, s)$. Then, let the guess functional of decision-maker i be*

$$f_i(t, x, s) = \frac{1}{2}\alpha_i(x - \mathbb{E}[x])^2 + \frac{1}{2}\beta_i\mathbb{E}[x]^2 + \gamma_i\mathbb{E}[x] + \delta_i,$$

where $\alpha_i = \alpha_i(t, s)$, $\beta_i = \beta_i(t, s)$, $\gamma_i = \gamma_i(t, s)$, and $\delta_i = \delta_i(t, s)$ are switching dependent, and $x = x^{u, x_0, s_0}(t)$. By applying Itô's formula with jump-diffusion processes and switching yields

$$f_i(T, x, s) - f_i(0, x_0, s_0)$$

$$= \int_0^T \left(\frac{\partial f_i(t, x, s)}{\partial t} + \frac{\partial f_i(t, x, s)}{\partial x} D_r(t, s, x, \mathbb{E}[x], u, \mathbb{E}[u]) \right.$$

$$+ \left. \frac{\partial^2 f_i(t, x, s)}{\partial x^2} \frac{D_f(t, s, x, \mathbb{E}[x], u, \mathbb{E}[u])^2}{2} \right) dt$$

$$+ \int_0^T D_f(t, s, x, \mathbb{E}[x], u, \mathbb{E}[u]) \frac{\partial f_i(t, x, s)}{\partial x} dB$$

$$+ \int_0^T \int_{\theta \in \Theta} \left(f_i(t, x + \mu(t, \theta), s) - f_i(t, x, s) - \frac{\partial f_i(t, x, s)}{\partial x}\mu(t, \theta), s \right) \nu(d\theta) dt$$

$$+ \int_0^T \int_{\theta \in \Theta} (f_i(t_-, x + \mu(t_-, \theta), s) - f_i(t_-, x)) \tilde{N}(dt, d\theta)$$

$$+ \int_0^T \sum_{s' \in \mathcal{S}} [f_i(., s') - f_i(., s)]\tilde{q}_{ss'} dt. \qquad (7.3)$$

For the sake of clarity, we compute first different terms appearing in (7.3) independently as follows:

$$\frac{\partial f_i(t, x)}{\partial t} = \frac{1}{2}(x - \mathbb{E}[x])^2 \dot{\alpha}_i - \alpha_i (x - \mathbb{E}[x]) \mathbb{E}[\dot{x}] + \frac{1}{2}\mathbb{E}[x]^2 \dot{\beta}_i$$

$$+ \beta_i\mathbb{E}[x]\mathbb{E}[\dot{x}] + \mathbb{E}[x]\dot{\gamma}_i + \gamma_i\mathbb{E}[\dot{x}] + \dot{\delta}_i, \qquad (7.4a)$$

$$\frac{\partial f_i(t, x)}{\partial x} = \alpha_i (x - \mathbb{E}[x]), \qquad (7.4b)$$

$$\frac{\partial^2 f_i(t, x)}{\partial x^2} = \alpha_i. \qquad (7.4c)$$

The jump-related terms are as follows:

$$f_i(t, x + \mu(t,\theta)) - f_i(t, x) - \frac{\partial f_i(t, x)}{\partial x}\mu(t, \theta)$$

$$= \frac{1}{2}\alpha_i(x + \mu - \mathbb{E}[x])^2 - \frac{1}{2}\alpha_i(x - \mathbb{E}[x])^2 - \alpha_i(x - \mathbb{E}[x])\mu(t, \theta),$$

$$= \frac{1}{2}\alpha_i(x - \mathbb{E}[x])^2 + \frac{1}{2}\alpha_i\mu^2 + \alpha_i(x - \mathbb{E}[x])\mu - \frac{1}{2}\alpha_i(x - \mathbb{E}[x])^2$$

$$- \alpha_i(x - \mathbb{E}[x])\mu,$$

$$= \frac{1}{2}\alpha_i\mu^2. \tag{7.5}$$

The switching-related terms are as follows:

$$\sum_{s' \in \mathcal{S}}[f_i(\cdot, s') - f_i(\cdot, s)]\tilde{q}_{ss'} = \frac{1}{2}\sum_{s' \in \mathcal{S}}(\alpha_i(\cdot, s') - \alpha_i(\cdot, s))(x - \mathbb{E}[x])^2$$

$$+ \frac{1}{2}\sum_{s' \in \mathcal{S}}(\beta_i(\cdot, s') - \beta_i(\cdot, s))\mathbb{E}[x]^2$$

$$+ \sum_{s' \in \mathcal{S}}(\gamma_i(\cdot, s') - \gamma_i(\cdot, s))\mathbb{E}[x]$$

$$+ \sum_{s' \in \mathcal{S}}(\delta_i(\cdot, s') - \delta_i(\cdot, s)). \tag{7.6}$$

The system dynamics can be re-written by the following expression:

$$dx = \Big(b_0 + b_1(x - \mathbb{E}[x]) + (b_1 + \bar{b}_1)\mathbb{E}[x] + b_{2i}(u_i - \mathbb{E}[u_i])$$

$$+ b_{2j}(u_j - \mathbb{E}[u_j]) + (b_{2i} + \bar{b}_{2i})\mathbb{E}[u_i] + (b_{2j} + \bar{b}_{2j})\mathbb{E}[u_j]\Big)dt$$

$$+ \Big(\sigma_0 + \sigma_1(x - \mathbb{E}[x]) + (\sigma_1 + \bar{\sigma}_1)\mathbb{E}[x] + \sigma_{i2}(u_i - \mathbb{E}[u_i])$$

$$+ \sigma_{j2}(u_j - \mathbb{E}[u_j]) + (\sigma_{i2} + \bar{\sigma}_{i2})\mathbb{E}[u_i] + (\sigma_{j2} + \bar{\sigma}_{j2})\mathbb{E}[u_j]\Big)dB$$

$$+ \int_{\theta \in \Theta}\Big(\mu_0 + \mu_1(x - \mathbb{E}[x]) + (\mu_1 + \bar{\mu}_1)\mathbb{E}[x] + \mu_{2i}(u_i - \mathbb{E}[u_i])$$

$$+ \mu_{2j}(u_j - \mathbb{E}[u_j]) + (\mu_{2i} + \bar{\mu}_{2i})\mathbb{E}[u_i] + (\mu_{j2} + \bar{\mu}_{j2})\mathbb{E}[u_j]\Big)\tilde{N}(dt, d\theta), \tag{7.7}$$

and the evolution of the mean state can be expressed as follows:

$$\mathbb{E}[\dot{x}] = b_0 + (b_1 + \bar{b}_1)\mathbb{E}[x] + (b_{2i} + \bar{b}_{2i})\mathbb{E}[u_i] + (b_{2j} + \bar{b}_{2j})\mathbb{E}[u_j]. \tag{7.8}$$

Replacing back the terms in (7.4)-(7.8) in the Itô's formula, one obtains

$$\mathbb{E}[f_i(T, s, x) - f_i(0, s_0, x_0)]$$

$$= \frac{1}{2}\mathbb{E}\int_0^T\Big[\dot{\alpha}_i + 2b_1\alpha_i + \sum_{s' \in \mathcal{S}}(\alpha_i(\cdot, s') - \alpha_i(\cdot, s))$$

$$+ \alpha_i \left(\sigma_1^2 + \int_{\theta \in \Theta} \mu_1^2 \nu(\mathrm{d}\theta) \right) \Bigg] (x - \mathbb{E}[x])^2 \, \mathrm{d}t$$

$$+ \frac{1}{2} \mathbb{E} \int_0^T \Bigg[\dot{\beta}_i + 2(b_1 + \bar{b}_1)\beta_i + \sum_{s' \in \mathcal{S}} (\beta_i(\cdot, s') - \beta_i(\cdot, s))$$

$$+ \alpha_i \left((\sigma_1 + \bar{\sigma}_1)^2 + \int_{\theta \in \Theta} (\mu_1 + \bar{\mu}_1)^2 \nu(\mathrm{d}\theta) \right) \Bigg] \mathbb{E}[x]^2 \mathrm{d}t$$

$$+ \mathbb{E} \int_0^T \Bigg[\dot{\gamma}_i + b_0 \beta_i + (b_1 + \bar{b}_1)\gamma_i + \sum_{s' \in \mathcal{S}} (\gamma_i(\cdot, s') - \gamma_i(\cdot, s))$$

$$+ \alpha_i \left(\sigma_0(\sigma_1 + \bar{\sigma}_1) + \int_{\theta \in \Theta} \mu_0(\mu_1 + \bar{\mu}_1)\nu(\mathrm{d}\theta) \right) \Bigg] \mathbb{E}[x]\mathrm{d}t$$

$$+ \mathbb{E} \int_0^T \Bigg[\dot{\delta}_i + b_0 \gamma_i + \sum_{s' \in \mathcal{S}} (\delta_i(\cdot, s') - \delta_i(\cdot, s)) + \frac{1}{2}\alpha_i \left(\sigma_0^2 + \int_{\theta \in \Theta} \mu_0^2 \nu(\mathrm{d}\theta) \right) \Bigg] \mathrm{d}t$$

$$+ \frac{1}{2} \mathbb{E} \int_0^T \alpha_i \left((\sigma_{i2} + \bar{\sigma}_{i2})^2 + \int_{\theta \in \Theta} (\mu_{i2} + \bar{\mu}_{i2})^2 \nu(\mathrm{d}\theta) \right) \mathbb{E}[u_i]^2 \mathrm{d}t$$

$$+ \frac{1}{2} \mathbb{E} \int_0^T 2 \Bigg\{ (b_{2i} + \bar{b}_{2i})\gamma_i + \alpha_i \left(\sigma_0(\sigma_{i2} + \bar{\sigma}_{i2}) + \int_{\theta \in \Theta} \mu_0(\mu_{i2} + \bar{\mu}_{i2})\nu(\mathrm{d}\theta) \right)$$

$$+ \Bigg[(b_{2i} + \bar{b}_{2i})\beta_i$$

$$+ \alpha_i \left((\sigma_1 + \bar{\sigma}_1)(\sigma_{i2} + \bar{\sigma}_{i2}) + \int_{\theta \in \Theta} (\mu_1 + \bar{\mu}_1)(\mu_{i2} + \bar{\mu}_{i2})\nu(\mathrm{d}\theta) \right) \Bigg] \mathbb{E}[x]$$

$$+ \alpha_i \Bigg((\sigma_{i2} + \bar{\sigma}_{i2})(\sigma_{j2} + \bar{\sigma}_{j2})$$

$$+ \int_{\theta \in \Theta} (\mu_{i2} + \bar{\mu}_{i2})(\mu_{j2} + \bar{\mu}_{j2})\nu(\mathrm{d}\theta) \Bigg) \mathbb{E}[u_j] \Bigg\} \mathbb{E}[u_i]\mathrm{d}t$$

$$+ \frac{1}{2} \mathbb{E} \int_0^T \alpha_i \left((\sigma_{j2} + \bar{\sigma}_{j2})^2 + \int_{\theta \in \Theta} (\mu_{j2} + \bar{\mu}_{j2})^2 \nu(\mathrm{d}\theta) \right) \mathbb{E}[u_j]^2 \mathrm{d}t$$

$$+ \frac{1}{2} \mathbb{E} \int_0^T 2 \Bigg\{ (b_{2j} + \bar{b}_{2j})\gamma_i + \alpha_i \left(\sigma_0(\sigma_{j2} + \bar{\sigma}_{j2}) + \int_{\theta \in \Theta} \mu_0(\mu_{j2} + \bar{\mu}_{j2})\nu(\mathrm{d}\theta) \right)$$

$$+ \Bigg[(b_{2j} + \bar{b}_{2j})\beta_i$$

$$+ \alpha_i \left((\sigma_1 + \bar{\sigma}_1)(\sigma_{j2} + \bar{\sigma}_{j2}) + \int_{\theta \in \Theta} (\mu_1 + \bar{\mu}_1)(\mu_{j2} + \bar{\mu}_{j2})\nu(\mathrm{d}\theta) \right) \Bigg] \mathbb{E}[x] \Bigg\} \mathbb{E}[u_j]\mathrm{d}t$$

$$+ \frac{1}{2}\mathbb{E}\int_0^T \alpha_i \left(\sigma_{i2}^2 + \int_{\theta \in \Theta} \mu_{i2}^2 \nu(\mathrm{d}\theta) \right) (u_i - \mathbb{E}[u_i])^2 \mathrm{d}t$$

$$+ \frac{1}{2}\mathbb{E}\int_0^T 2 \left\{ \left[b_{2i}\alpha_i + \alpha_i \left(\sigma_1 \sigma_{i2} + \int_{\theta \in \Theta} \mu_1 \mu_{i2} \nu(\mathrm{d}\theta) \right) \right] (x - \mathbb{E}[x]) \right.$$

$$+ \alpha_i \left(\sigma_{i2}\sigma_{j2} + \int_{\theta \in \Theta} \mu_{i2}\mu_{j2}\nu(\mathrm{d}\theta) \right) (u_j - \mathbb{E}[u_j]) \bigg\} (u_i - \mathbb{E}[u_i])\mathrm{d}t$$

$$+ \frac{1}{2}\mathbb{E}\int_0^T \alpha_i \left(\sigma_{j2}^2 + \int_{\theta \in \Theta} \mu_{j2}^2 \nu(\mathrm{d}\theta) \right) (u_j - \mathbb{E}[u_j])^2 \mathrm{d}t$$

$$+ \frac{1}{2}\mathbb{E}\int_0^T 2 \left\{ \left[b_{2j}\alpha_i + \alpha_i \left(\sigma_1 \sigma_{j2} + \int_{\theta \in \Theta} \mu_1 \mu_{j2} \nu(\mathrm{d}\theta) \right) \right] (x - \mathbb{E}[x]) \right\} \cdot$$

$$\cdot (u_j - \mathbb{E}[u_j])\mathrm{d}t,$$

Now we compute the gap $\mathbb{E}[L_i(x, s, u_i, u_j)|\mathcal{F}_t^s] - \mathbb{E}[f_i(0, s_0, x_0)]$. *It is desired to find the optimal control inputs such that the expected cost function* $\mathbb{E}[L_i(x, s, u_i, u_j)|\mathcal{F}_t^s]$ *matches with the ansatz for the optimal cost from time* $t = 0$ *up to terminal time* T, *which is given by* $\mathbb{E}[f_i(0, s_0, x_0)]$. *Performing the calculation one arrives at*

$$\mathbb{E}[L_i(x, s, u_i, u_j)|\mathcal{F}_t^s] - \mathbb{E}[f_i(0, s_0, x_0)] = \frac{1}{2}\mathbb{E}[(q_i - \alpha_i(T))(x(T) - \mathbb{E}[x(T)])^2]$$

$$+ \frac{1}{2}\mathbb{E}[(q_i + \bar{q}_i - \beta_i(T))\mathbb{E}[x(T)]^2] - \mathbb{E}[\gamma_i(T)\mathbb{E}[x(T)]] - \mathbb{E}[\delta_i(T)]$$

$$+ \frac{1}{2}\mathbb{E}\int_0^T \left[\dot{\alpha}_i + q_i + 2b_1\alpha_i + \sum_{s' \in \mathcal{S}} (\alpha_i(\cdot, s') - \alpha_i(\cdot, s)) \right.$$

$$+ \alpha_i \left(\sigma_1^2 + \int_{\theta \in \Theta} \mu_1^2 \nu(\mathrm{d}\theta) \right) \bigg] (x - \mathbb{E}[x])^2 \, \mathrm{d}t$$

$$+ \frac{1}{2}\mathbb{E}\int_0^T \left[\dot{\beta}_i + q_i + \bar{q}_i + 2(b_1 + \bar{b}_1)\beta_i + \sum_{s' \in \mathcal{S}} (\beta_i(\cdot, s') - \beta_i(\cdot, s)) \right.$$

$$+ \alpha_i \left((\sigma_1 + \bar{\sigma}_1)^2 + \int_{\theta \in \Theta} (\mu_1 + \bar{\mu}_1)^2 \nu(\mathrm{d}\theta) \right) \bigg] \mathbb{E}[x]^2 \mathrm{d}t$$

$$+ \mathbb{E}\int_0^T \left[\dot{\gamma}_i + b_0\beta_i + (b_1 + \bar{b}_1)\gamma_i + \sum_{s' \in \mathcal{S}} (\gamma_i(\cdot, s') - \gamma_i(\cdot, s)) \right.$$

$$+ \alpha_i \left(\sigma_0(\sigma_1 + \bar{\sigma}_1) + \int_{\theta \in \Theta} \mu_0(\mu_1 + \bar{\mu}_1)\nu(\mathrm{d}\theta) \right) \bigg] \mathbb{E}[x]\mathrm{d}t$$

$$+ \mathbb{E}\int_0^T \left[\dot{\delta}_i + b_0\gamma_i + \sum_{s' \in \mathcal{S}} (\delta_i(\cdot, s') - \delta_i(\cdot, s)) + \frac{1}{2}\alpha_i \left(\sigma_0^2 + \int_{\theta \in \Theta} \mu_0^2 \nu(\mathrm{d}\theta) \right) \right] \mathrm{d}t$$

$$+ \frac{1}{2}\mathbb{E}\int_0^T \left[\alpha_i \left((\sigma_{i2} + \bar{\sigma}_{i2})^2 + \int_{\theta \in \Theta} (\mu_{i2} + \bar{\mu}_{i2})^2 \nu(d\theta) \right) + r_i + \bar{r}_i \right] \mathbb{E}[u_i]^2 dt$$

$$+ \frac{1}{2}\mathbb{E}\int_0^T 2 \left\{ (b_{2i} + \bar{b}_{2i})\gamma_i + \alpha_i \left(\sigma_0(\sigma_{i2} + \bar{\sigma}_{i2}) + \int_{\theta \in \Theta} \mu_0(\mu_{i2} + \bar{\mu}_{i2})\nu(d\theta) \right) \right.$$

$$+ \left[(b_{2i} + \bar{b}_{2i})\beta_i \right.$$

$$+ \alpha_i \left((\sigma_1 + \bar{\sigma}_1)(\sigma_{i2} + \bar{\sigma}_{i2}) + \int_{\theta \in \Theta} (\mu_1 + \bar{\mu}_1)(\mu_{i2} + \bar{\mu}_{i2})\nu(d\theta) \right) \Bigg] \mathbb{E}[x]$$

$$+ \left[\alpha_i \left((\sigma_{i2} + \bar{\sigma}_{i2})(\sigma_{j2} + \bar{\sigma}_{j2}) \right. \right.$$

$$\left. \left. + \int_{\theta \in \Theta} (\mu_{i2} + \bar{\mu}_{i2})(\mu_{j2} + \bar{\mu}_{j2})\nu(d\theta) \right) + \bar{\varepsilon}_{ij} \right] \mathbb{E}[u_j] \right\} \mathbb{E}[u_i] dt$$

$$+ \frac{1}{2}\mathbb{E}\int_0^T \alpha_i \left((\sigma_{j2} + \bar{\sigma}_{j2})^2 + \int_{\theta \in \Theta} (\mu_{j2} + \bar{\mu}_{j2})^2 \nu(d\theta) \right) \mathbb{E}[u_j]^2 dt$$

$$+ \frac{1}{2}\mathbb{E}\int_0^T 2 \left\{ (b_{2j} + \bar{b}_{2j})\gamma_i + \alpha_i \left(\sigma_0(\sigma_{j2} + \bar{\sigma}_{j2}) + \int_{\theta \in \Theta} \mu_0(\mu_{j2} + \bar{\mu}_{j2})\nu(d\theta) \right) \right.$$

$$+ \left[(b_{2j} + \bar{b}_{2j})\beta_i + \alpha_i \left((\sigma_1 + \bar{\sigma}_1)(\sigma_{j2} + \bar{\sigma}_{j2}) \right. \right.$$

$$\left. \left. + \int_{\theta \in \Theta} (\mu_1 + \bar{\mu}_1)(\mu_{j2} + \bar{\mu}_{j2})\nu(d\theta) \right) \right] \mathbb{E}[x] \right\} \mathbb{E}[u_j] dt$$

$$+ \frac{1}{2}\mathbb{E}\int_0^T \left[\alpha_i \left(\sigma_{i2}^2 + \int_{\theta \in \Theta} \mu_{i2}^2 \nu(d\theta) \right) + r_i \right] (u_i - \mathbb{E}[u_i])^2 dt$$

$$+ \frac{1}{2}\mathbb{E}\int_0^T 2 \left\{ \left[b_{2i}\alpha_i + \alpha_i \left(\sigma_1\sigma_{i2} + \int_{\theta \in \Theta} \mu_1\mu_{i2}\nu(d\theta) \right) \right] (x - \mathbb{E}[x]) \right.$$

$$+ \left[\alpha_i \left(\sigma_{i2}\sigma_{j2} + \int_{\theta \in \Theta} \mu_{i2}\mu_{j2}\nu(d\theta) \right) + \varepsilon_{ij} \right] (u_j - \mathbb{E}[u_j]) \right\} (u_i - \mathbb{E}[u_i]) dt$$

$$+ \frac{1}{2}\mathbb{E}\int_0^T \alpha_i \left(\sigma_{j2}^2 + \int_{\theta \in \Theta} \mu_{j2}^2 \nu(d\theta) \right) (u_j - \mathbb{E}[u_j])^2 dt$$

$$+ \frac{1}{2}\mathbb{E}\int_0^T 2 \left\{ \left[b_{2j}\alpha_i + \alpha_i \left(\sigma_1\sigma_{j2} + \int_{\theta \in \Theta} \mu_1\mu_{j2}\nu(d\theta) \right) \right] (x - \mathbb{E}[x]) \right\} \cdot$$

$$\cdot (u_j - \mathbb{E}[u_j]) dt.$$

The next step in this procedure consists of performing optimization with respect to the decision-variables, in this case, corresponding to the follower. Given that the follower plays after the leader has made a decision, then the optimal cost, at this stage, is going to be expressed in function of the leader's action. To do this, let us first ease the notation for convenience. Let us consider the term

$$\eta_i = \frac{\lambda_i + \phi_i \mathbb{E}[x] + \rho_{ij} \mathbb{E}[u_j]}{c_i},$$

and using the auxiliary variables introduced in the statement of Theorem 2, we perform terms completion for the follower $\mathbb{E}[u_i]$. *We have the following term:*

$$\frac{1}{2} \mathbb{E} \int_0^T c_i \left(\mathbb{E}[u_i]^2 + 2 \, \eta_i \mathbb{E}[u_i] \right) dt$$

$$= \frac{1}{2} \mathbb{E} \int_0^T c_i \left(\mathbb{E}[u_i] + \eta_i \right)^2 dt - \frac{1}{2} \mathbb{E} \int_0^T c_i \eta_i^2 dt,$$

$$= \frac{1}{2} \mathbb{E} \int_0^T c_i \left(\mathbb{E}[u_i] + \eta_i \right)^2 dt - \frac{1}{2} \mathbb{E} \int_0^T \left(\frac{\lambda_i^2}{c_i} + \frac{\phi_i^2 \mathbb{E}[x]^2}{c_i} + \frac{\rho_{ij}^2 \mathbb{E}[u_j]^2}{c_i} \right.$$

$$\left. + \frac{2\lambda_i \phi_i \mathbb{E}[x]}{c_i} + \frac{2\lambda_i \rho_{ij} \mathbb{E}[u_j]}{c_i} + \frac{2\phi_i \rho_{ij} \mathbb{E}[x] \mathbb{E}[u_j]}{c_i} \right) dt.$$

From which it is concluded that the i-follower reacts to the signal of the j-leader $\mathbb{E}[u_j]$ *as follows:*

$$\mathbb{E}[u_i^*] = -\frac{\lambda_i}{c_i} - \frac{\phi_i}{c_i} \mathbb{E}[x] - \frac{\rho_{ij}}{c_i} \mathbb{E}[u_j],$$

Now, let us consider the following auxiliary variable:

$$\bar{\eta}_i = \frac{\bar{\phi}_i \left(x - \mathbb{E}[x] \right) + \bar{\rho}_{ij} (u_j - \mathbb{E}[u_j])}{\bar{c}_i}.$$

Then, we perform terms completion for the j—follower term $(u_i - \mathbb{E}[u_i])$. *We have the following term:*

$$\frac{1}{2} \mathbb{E} \int_0^T \bar{c}_i \left((u_i - \mathbb{E}[u_i])^2 + 2 \, \bar{\eta}_i (u_i - \mathbb{E}[u_i]) \right) dt$$

$$= \frac{1}{2} \mathbb{E} \int_0^T \bar{c}_i \left(u_i - \mathbb{E}[u_i] + \bar{\eta}_i \right)^2 dt - \frac{1}{2} \mathbb{E} \int_0^T \bar{c}_i \bar{\eta}_i^2 dt,$$

$$= \frac{1}{2} \mathbb{E} \int_0^T \bar{c}_i \left(u_i - \mathbb{E}[u_i] + \bar{\eta}_i \right)^2 dt - \frac{1}{2} \mathbb{E} \int_0^T \left(\frac{\bar{\phi}_i^2 \left(x - \mathbb{E}[x] \right)^2}{\bar{c}_i} \right.$$

$$\left. + \frac{\bar{\rho}_{ij}^2 (u_j - \mathbb{E}[u_j])^2}{\bar{c}_i} + 2 \frac{\bar{\phi}_i \bar{\rho}_{ij} \left(x - \mathbb{E}[x] \right) (u_j - \mathbb{E}[u_j])}{\bar{c}_i} \right) dt,$$

From which it is concluded that the i—follower reacts to the signal of the j−leader $u_j − \mathbb{E}[u_j]$ *as follows:*

$$u_i^* − \mathbb{E}[u_i^*] = −\frac{\bar{\phi}_i}{\bar{c}_i}(x − \mathbb{E}[x]) − \frac{\bar{\rho}_{ij}}{\bar{c}_i}(u_j − \mathbb{E}[u_j]),$$

Now, knowing how the follower i is going to response against the selection of the leader j, we compute the optimal solution for the leader. Therefore, following the same procedure as in the follower, we obtain that the gap $\mathbb{E}[L_j(x, s, u_i, u_j)|\mathcal{F}_t^s] − \mathbb{E}[f_j(0, s_0, x_0)]$. *In this expression, we have represented the follower's actions in terms of the leader's control inputs, i.e.,*

$$\mathbb{E}[L_j(x, s, u_i, u_j)|\mathcal{F}_t^s] − \mathbb{E}[f_j(0, s_0, x_0)] = \frac{1}{2}\mathbb{E}[(q_j − \alpha_j(T))(x(T) − \mathbb{E}[x(T)])^2]$$

$$+ \frac{1}{2}\mathbb{E}[(q_j + \bar{q}_j − \beta_j(T))\mathbb{E}[x(T)]^2] − \mathbb{E}[\gamma_j(T)\mathbb{E}[x(T)]] − \mathbb{E}[\delta_j(T)]$$

$$+ \frac{1}{2}\mathbb{E}\int_0^T \left[\dot{\alpha}_j + q_j + 2b_1\alpha_j + \sum_{s'\in\mathcal{S}}(\alpha_j(\cdot, s') − \alpha_j(\cdot, s))\right.$$

$$+ \alpha_j\left(\sigma_1^2 + \int_{\theta\in\Theta}\mu_1^2\nu(d\theta)\right) + \alpha_j\left(\sigma_{i2}^2 + \int_{\theta\in\Theta}\mu_{i2}^2\nu(d\theta)\right)\frac{\bar{\phi}_i^2}{\bar{c}_i^2}$$

$$\left. − 2\left[b_{2i}\alpha_j + \alpha_j\left(\sigma_1\sigma_{i2} + \int_{\theta\in\Theta}\mu_1\mu_{i2}\nu(d\theta)\right)\right]\frac{\bar{\phi}_i}{\bar{c}_i}\right](x − \mathbb{E}[x])^2 \, dt$$

$$+ \frac{1}{2}\mathbb{E}\int_0^T \left[\dot{\beta}_j + q_j + \bar{q}_j + 2(b_1 + \bar{b}_1)\beta_j + \sum_{s'\in\mathcal{S}}(\beta_j(\cdot, s') − \beta_j(\cdot, s))\right.$$

$$+ \alpha_j\left((\sigma_1 + \bar{\sigma}_1)^2 + \int_{\theta\in\Theta}(\mu_1 + \bar{\mu}_1)^2\nu(d\theta)\right.$$

$$+ \left((\sigma_{i2} + \bar{\sigma}_{i2})^2 + \int_{\theta\in\Theta}(\mu_{i2} + \bar{\mu}_{i2})^2\nu(d\theta)\right)\frac{\phi_i^2}{c_i^2}\right) − 2(b_{2i} + \bar{b}_{2i})\frac{\phi_i}{c_i}\beta_j$$

$$\left. − 2\alpha_j\left((\sigma_1 + \bar{\sigma}_1)(\sigma_{i2} + \bar{\sigma}_{i2}) + \int_{\theta\in\Theta}(\mu_1 + \bar{\mu}_1)(\mu_{i2} + \bar{\mu}_{i2})\nu(d\theta)\right)\frac{\phi_i}{c_i}\right]\mathbb{E}[x]^2 dt$$

$$+ \mathbb{E}\int_0^T \left[\dot{\gamma}_j + b_0\beta_j + (b_1 + \bar{b}_1)\gamma_j + \sum_{s'\in\mathcal{S}}(\gamma_j(\cdot, s') − \gamma_j(\cdot, s))\right.$$

$$+ \alpha_j\left(\sigma_0(\sigma_1 + \bar{\sigma}_1) + \int_{\theta\in\Theta}\mu_0(\mu_1 + \bar{\mu}_1)\nu(d\theta)\right)$$

$$+ \alpha_j\left((\sigma_{i2} + \bar{\sigma}_{i2})^2 + \int_{\theta\in\Theta}(\mu_{i2} + \bar{\mu}_{i2})^2\nu(d\theta)\right)\frac{\lambda_i\phi_i}{c_i^2} − (b_{2i} + \bar{b}_{2i})\frac{\lambda_i}{c_i}\beta_j$$

$$− \alpha_j\left((\sigma_1 + \bar{\sigma}_1)(\sigma_{i2} + \bar{\sigma}_{i2}) + \int_{\theta\in\Theta}(\mu_1 + \bar{\mu}_1)(\mu_{i2} + \bar{\mu}_{i2})\nu(d\theta)\right)\frac{\lambda_i}{c_i}$$

$$\left. − (b_{2i} + \bar{b}_{2i})\frac{\phi_i}{c_i}\gamma_j − \alpha_j\left(\sigma_0(\sigma_{i2} + \bar{\sigma}_{i2}) + \int_{\theta\in\Theta}\mu_0(\mu_{i2} + \bar{\mu}_{i2})\nu(d\theta)\right)\frac{\phi_i}{c_i}\right]\mathbb{E}[x]dt$$

$$+ \mathbb{E} \int_0^T \left[\dot{\delta}_j + b_0 \gamma_j + \sum_{s' \in \mathcal{S}} (\delta_j(\cdot, s') - \delta_j(\cdot, s)) \right.$$

$$+ \frac{1}{2} \alpha_j \left(\sigma_0^2 + \int_{\theta \in \Theta} \mu_0^2 \nu(\mathrm{d}\theta) + \left((\sigma_{i2} + \bar{\sigma}_{i2})^2 + \int_{\theta \in \Theta} (\mu_{i2} + \bar{\mu}_{i2})^2 \nu(\mathrm{d}\theta) \right) \frac{\lambda_i^2}{c_i^2} \right)$$

$$\left. - (b_{2i} + \bar{b}_{2i}) \frac{\lambda_i}{c_i} \gamma_j - \alpha_j \left(\sigma_0 (\sigma_{i2} + \bar{\sigma}_{i2}) + \int_{\theta \in \Theta} \mu_0 (\mu_{i2} + \bar{\mu}_{i2}) \nu(\mathrm{d}\theta) \right) \frac{\lambda_i}{c_i} \right] \mathrm{d}t$$

$$+ \frac{1}{2} \mathbb{E} \int_0^T \left[\alpha_j \left((\sigma_{j2} + \bar{\sigma}_{j2})^2 + \int_{\theta \in \Theta} (\mu_{j2} + \bar{\mu}_{j2})^2 \nu(\mathrm{d}\theta) \right) + r_j + \bar{r}_j \right.$$

$$- 2\alpha_j \left((\sigma_{j2} + \bar{\sigma}_{j2})(\sigma_{i2} + \bar{\sigma}_{i2}) + \int_{\theta \in \Theta} (\mu_{j2} + \bar{\mu}_{j2})(\mu_{i2} + \bar{\mu}_{i2}) \nu(\mathrm{d}\theta) \right.$$

$$\left. - \frac{1}{4} \left((\sigma_{i2} + \bar{\sigma}_{i2})^2 + \int_{\theta \in \Theta} (\mu_{i2} + \bar{\mu}_{i2})^2 \nu(\mathrm{d}\theta) \right) \frac{\rho_{ij}}{c_i} \right) \frac{\rho_{ij}}{c_i} - 2\bar{\varepsilon}_{ji} \frac{\rho_{ij}}{c_i} \right] \mathbb{E}[u_j]^2 \mathrm{d}t$$

$$+ \frac{1}{2} \mathbb{E} \int_0^T 2 \left\{ (b_{2j} + \bar{b}_{2j}) \gamma_j + \alpha_j \left(\sigma_0 (\sigma_{j2} + \bar{\sigma}_{j2}) + \int_{\theta \in \Theta} \mu_0 (\mu_{j2} + \bar{\mu}_{j2}) \nu(\mathrm{d}\theta) \right) \right.$$

$$- \left[\alpha_j \left((\sigma_{j2} + \bar{\sigma}_{j2})(\sigma_{i2} + \bar{\sigma}_{i2}) + \int_{\theta \in \Theta} (\mu_{j2} + \bar{\mu}_{j2})(\mu_{i2} + \bar{\mu}_{i2}) \nu(\mathrm{d}\theta) \right) + \bar{\varepsilon}_{ji} \right] \frac{\lambda_i}{c_i}$$

$$+ \alpha_j \left((\sigma_{i2} + \bar{\sigma}_{i2})^2 + \int_{\theta \in \Theta} (\mu_{i2} + \bar{\mu}_{i2})^2 \nu(\mathrm{d}\theta) \right) \frac{\lambda_i \rho_{ij}}{c_i^2}$$

$$- (b_{2i} + \bar{b}_{2i}) \frac{\rho_{ij}}{c_i} \gamma_j - \alpha_j \left(\sigma_0 (\sigma_{i2} + \bar{\sigma}_{i2}) + \int_{\theta \in \Theta} \mu_0 (\mu_{i2} + \bar{\mu}_{i2}) \nu(\mathrm{d}\theta) \right) \frac{\rho_{ij}}{c_i}$$

$$+ \left[(b_{2j} + \bar{b}_{2j}) \beta_j \right.$$

$$+ \alpha_j \left((\sigma_1 + \bar{\sigma}_1)(\sigma_{j2} + \bar{\sigma}_{j2}) + \int_{\theta \in \Theta} (\mu_1 + \bar{\mu}_1)(\mu_{j2} + \bar{\mu}_{j2}) \nu(\mathrm{d}\theta) \right)$$

$$- \alpha_j \left((\sigma_{j2} + \bar{\sigma}_{j2})(\sigma_{i2} + \bar{\sigma}_{i2}) + \int_{\theta \in \Theta} (\mu_{j2} + \bar{\mu}_{j2})(\mu_{i2} + \bar{\mu}_{i2}) \nu(\mathrm{d}\theta) \right) \frac{\phi_i}{c_i} - \bar{\varepsilon}_{ji} \frac{\phi_i}{c_i}$$

$$+ \alpha_j \left((\sigma_{i2} + \bar{\sigma}_{i2})^2 + \int_{\theta \in \Theta} (\mu_{i2} + \bar{\mu}_{i2})^2 \nu(\mathrm{d}\theta) \right) \frac{\phi_i \rho_{ij}}{c_i^2} - (b_{2i} + \bar{b}_{2i}) \frac{\rho_{ij}}{c_i} \beta_j$$

$$\left. \left. - \alpha_j \left((\sigma_1 + \bar{\sigma}_1)(\sigma_{i2} + \bar{\sigma}_{i2}) + \int_{\theta \in \Theta} (\mu_1 + \bar{\mu}_1)(\mu_{i2} + \bar{\mu}_{i2}) \nu(\mathrm{d}\theta) \right) \frac{\rho_{ij}}{c_i} \right] \mathbb{E}[x] \right\} \cdot$$

$$\cdot \mathbb{E}[u_j] \mathrm{d}t$$

$$+ \frac{1}{2} \mathbb{E} \int_0^T \left[\alpha_j \left[\left(\sigma_{j2}^2 + \int_{\theta \in \Theta} \mu_{j2}^2 \nu(\mathrm{d}\theta) \right) + \left(\sigma_{i2}^2 + \int_{\theta \in \Theta} \mu_{i2}^2 \nu(\mathrm{d}\theta) \right) \frac{\bar{\rho}_{ij}^2}{\bar{c}_i^2} \right] + r_j \right.$$

$$- \alpha_j \left(\sigma_{j2}\sigma_{i2} + \int_{\theta\in\Theta} \mu_{j2}\mu_{i2}\nu(d\theta) \right) \frac{\bar{\rho}_{ij}}{\bar{c}_i} - \frac{\bar{\rho}_{ij}}{\bar{c}_i}\varepsilon_{ji} \right] (u_j - \mathbb{E}[u_j])^2 dt$$

$$+ \frac{1}{2}\mathbb{E}\int_0^T 2 \left\{ \alpha_j \left[b_{2j} + \sigma_1\sigma_{j2} + \int_{\theta\in\Theta} \mu_1\mu_{j2}\nu(d\theta) \right. \right.$$

$$+ \left(\sigma_{i2}^2 + \int_{\theta\in\Theta} \mu_{i2}^2\nu(d\theta) \right) \frac{\bar{\phi}_i\bar{\rho}_{ij}}{\bar{c}_i^2}$$

$$- \left(b_{2i} + \left(\sigma_1\sigma_{i2} + \int_{\theta\in\Theta} \mu_1\mu_{i2}\nu(d\theta) \right) \right) \frac{\bar{\rho}_{ij}}{\bar{c}_i}$$

$$\left. - \alpha_j \left(\sigma_{j2}\sigma_{i2} + \int_{\theta\in\Theta} \mu_{j2}\mu_{i2}\nu(d\theta) \right) \frac{\bar{\phi}_i}{\bar{c}_i} - \frac{\bar{\phi}_i}{\bar{c}_i}\varepsilon_{ji} \right] (x - \mathbb{E}[x]) \right\}(u_j - \mathbb{E}[u_j])dt.$$

With this leader's cost-functional expression in hand, we continue by performing optimization with decision variables given by the leader's strategies. To simplify the notation, we introduce some auxiliary variables. Let us consider

$$\eta_j = \frac{\lambda_j + \phi_j\mathbb{E}[x]}{c_j}.$$

We perform terms completion for the leader $\mathbb{E}[u_j]$, we have

$$\frac{1}{2}\mathbb{E}\int_0^T c_j \left(\mathbb{E}[u_j]^2 + 2 \eta_j\mathbb{E}[u_j] \right) dt$$

$$= \frac{1}{2}\mathbb{E}\int_0^T c_j (\mathbb{E}[u_j] + \eta_j)^2 dt - \frac{1}{2}\mathbb{E}\int_0^T c_j\eta_j^2 dt,$$

$$= \frac{1}{2}\mathbb{E}\int_0^T c_j (\mathbb{E}[u_j] + \eta_j)^2 dt - \frac{1}{2}\mathbb{E}\int_0^T \frac{\lambda_j^2}{c_j} dt$$

$$- \frac{1}{2}\mathbb{E}\int_0^T \frac{\phi_j^2}{c_j}\mathbb{E}[x]^2 dt - \mathbb{E}\int_0^T \frac{\lambda_j\phi_j}{c_j}\mathbb{E}[x]dt,$$

from which it is concluded that

$$\mathbb{E}[u_j^*] = -\frac{\lambda_j}{c_j} - \frac{\phi_j}{c_j}\mathbb{E}[x].$$

On the other hand, considering the variable

$$\bar{\eta}_j = \frac{\bar{\phi}_j}{\bar{c}_j} (x - \mathbb{E}[x]),$$

we perform terms completion for the leader term $(u_j - \mathbb{E}[u_j])$. We have the following term:

$$\frac{1}{2}\mathbb{E}\int_0^T \bar{c}_j \left((u_j - \mathbb{E}[u_j])^2 + 2 \bar{\eta}_j(u_j - \mathbb{E}[u_j]) \right) dt$$

$$= \frac{1}{2}\mathbb{E}\int_0^T \bar{c}_j \left(u_j - \mathbb{E}[u_j] + \bar{\eta}_j\right)^2 dt - \frac{1}{2}\mathbb{E}\int_0^T \bar{c}_j \bar{\eta}_j^2 dt,$$

$$= \frac{1}{2}\mathbb{E}\int_0^T \bar{c}_j \left(u_j - \mathbb{E}[u_j] + \bar{\eta}_j\right)^2 dt - \frac{1}{2}\mathbb{E}\int_0^T \frac{\bar{\phi}_j^2}{\bar{c}_j} \left(x - \mathbb{E}[x]\right)^2 dt,$$

from which it is concluded that

$$u_j^* - \mathbb{E}[u_j^*] = -\frac{\bar{\phi}_j}{\bar{c}_j} \left(x - \mathbb{E}[x]\right).$$

At this point, it is quite important to highlight that the optimal control input for the leader does not depend on the follower's decisions, but the optimal solution is of state and mean-field state feedback form. Now, replacing back the optimal control inputs and the square-completion in $\mathbb{E}[L_j(x, s, u_i, u_j)|\mathcal{F}_t^s] - \mathbb{E}[f_j(0, s_0, x_0)]$ yields for the leader j that

$$\mathbb{E}[L_j(x, s, u_i, u_j)|\mathcal{F}_t^s] - \mathbb{E}[f_j(0, s_0, x_0)] = \frac{1}{2}\mathbb{E}[(q_j - \alpha_j(T))\,(x(T) - \mathbb{E}[x(T)])^2]$$

$$+ \frac{1}{2}\mathbb{E}[(q_j + \bar{q}_j - \beta_j(T))\mathbb{E}[x(T)]^2] - \mathbb{E}[\gamma_j(T)\mathbb{E}[x(T)]] - \mathbb{E}[\delta_j(T)]$$

$$+ \frac{1}{2}\mathbb{E}\int_0^T \left[\dot{\alpha}_j + q_j + 2b_1\alpha_j + \sum_{s'\in\mathcal{S}}(\alpha_j(\cdot, s') - \alpha_j(\cdot, s))\right.$$

$$+ \alpha_j\left(\sigma_1^2 + \int_{\theta\in\Theta}\mu_1^2\nu(\mathrm{d}\theta)\right) + \alpha_j\left(\sigma_{i2}^2 + \int_{\theta\in\Theta}\mu_{i2}^2\nu(\mathrm{d}\theta)\right)\frac{\bar{\phi}_i^2}{\bar{c}_i^2}$$

$$\left. - 2\left[b_{2i}\alpha_j + \alpha_j\left(\sigma_1\sigma_{i2} + \int_{\theta\in\Theta}\mu_1\mu_{i2}\nu(\mathrm{d}\theta)\right)\right]\frac{\bar{\phi}_i}{\bar{c}_i} - \frac{\bar{\phi}_j^2}{\bar{c}_j}\right](x - \mathbb{E}[x])^2\,dt$$

$$+ \frac{1}{2}\mathbb{E}\int_0^T \left[\dot{\beta}_j + q_j + \bar{q}_j + 2(b_1 + \bar{b}_1)\beta_j + \sum_{s'\in\mathcal{S}}(\beta_j(\cdot, s') - \beta_j(\cdot, s))\right.$$

$$+ \alpha_j\left((\sigma_1 + \bar{\sigma}_1)^2 + \int_{\theta\in\Theta}(\mu_1 + \bar{\mu}_1)^2\nu(\mathrm{d}\theta) + \left((\sigma_{i2} + \bar{\sigma}_{i2})^2\right.\right.$$

$$\left.\left. + \int_{\theta\in\Theta}(\mu_{i2} + \bar{\mu}_{i2})^2\nu(\mathrm{d}\theta)\right)\frac{\phi_i^2}{c_i^2}\right) - 2(b_{2i} + \bar{b}_{2i})\frac{\phi_i}{c_i}\beta_j - \frac{\phi_j^2}{c_j}$$

$$\left. - 2\alpha_j\left((\sigma_1 + \bar{\sigma}_1)(\sigma_{i2} + \bar{\sigma}_{i2}) + \int_{\theta\in\Theta}(\mu_1 + \bar{\mu}_1)(\mu_{i2} + \bar{\mu}_{i2})\nu(\mathrm{d}\theta)\right)\frac{\phi_i}{c_i}\right]\mathbb{E}[x]^2\,dt$$

$$+ \mathbb{E}\int_0^T \left[\dot{\gamma}_j + b_0\beta_j + (b_1 + \bar{b}_1)\gamma_j + \sum_{s'\in\mathcal{S}}(\gamma_j(\cdot, s') - \gamma_j(\cdot, s))\right.$$

$$\left. + \alpha_j\left(\sigma_0(\sigma_1 + \bar{\sigma}_1) + \int_{\theta\in\Theta}\mu_0(\mu_1 + \bar{\mu}_1)\nu(\mathrm{d}\theta)\right)\right.$$

$$+ \alpha_j \left((\sigma_{i2} + \bar{\sigma}_{i2})^2 + \int_{\theta \in \Theta} (\mu_{i2} + \bar{\mu}_{i2})^2 \nu(d\theta) \right) \frac{\lambda_i \phi_i}{c_i^2} - (b_{2i} + \bar{b}_{2i}) \frac{\lambda_i}{c_i} \beta_j$$

$$- \alpha_j \left((\sigma_1 + \bar{\sigma}_1)(\sigma_{i2} + \bar{\sigma}_{i2}) + \int_{\theta \in \Theta} (\mu_1 + \bar{\mu}_1)(\mu_{i2} + \bar{\mu}_{i2}) \nu(d\theta) \right) \frac{\lambda_i}{c_i}$$

$$- (b_{2i} + \bar{b}_{2i}) \frac{\phi_i}{c_i} \gamma_j - \alpha_j \left(\sigma_0 (\sigma_{i2} + \bar{\sigma}_{i2}) + \int_{\theta \in \Theta} \mu_0 (\mu_{i2} + \bar{\mu}_{i2}) \nu(d\theta) \right) \frac{\phi_i}{c_i}$$

$$\left. - \frac{\lambda_j \phi_j}{c_j} \right] \mathbb{E}[x] dt$$

$$+ \mathbb{E} \int_0^T \left[\dot{\delta}_j + b_0 \gamma_j + \sum_{s' \in \mathcal{S}} (\delta_j(\cdot, s') - \delta_j(\cdot, s)) \right.$$

$$+ \frac{1}{2} \alpha_j \left(\sigma_0^2 + \int_{\theta \in \Theta} \mu_0^2 \nu(d\theta) + \left((\sigma_{i2} + \bar{\sigma}_{i2})^2 + \int_{\theta \in \Theta} (\mu_{i2} + \bar{\mu}_{i2})^2 \nu(d\theta) \right) \frac{\lambda_i^2}{c_i^2} \right)$$

$$- (b_{2i} + \bar{b}_{2i}) \frac{\lambda_i}{c_i} \gamma_j - \alpha_j \left(\sigma_0 (\sigma_{i2} + \bar{\sigma}_{i2}) + \int_{\theta \in \Theta} \mu_0 (\mu_{i2} + \bar{\mu}_{i2}) \nu(d\theta) \right) \frac{\lambda_i}{c_i}$$

$$\left. - \frac{\lambda_j^2}{2 c_j} \right] dt$$

$$+ \frac{1}{2} \mathbb{E} \int_0^T c_j \left(\mathbb{E}[u_j] + \eta_j \right)^2 dt + \frac{1}{2} \mathbb{E} \int_0^T \bar{c}_j \left(u_j - \mathbb{E}[u_j] + \bar{\eta}_j \right)^2 dt,$$

The latter expression is divided into the following groups:

- *Those terms evaluated at the terminal time, which define the boundary conditions for the differential equations*

- *Those terms with common factors $(x - \mathbb{E}[x])^2$, $\mathbb{E}[x]^2$, and $\mathbb{E}[x]$*

- *Those terms independent from the system state, control inputs and mean-field terms*

- *And the terms depending on the control inputs associated with the leader decision-maker.*

   The announced results for the leader j are obtained from minimizing the terms in the latter expression.

   Regarding the follower, we replace the square completions for both the leader and follower in $\mathbb{E}[L_i(x, s, u_i, u_j)|\mathcal{F}_t^s] - \mathbb{E}[f_i(0, s_0, x_0)]$, obtaining the following:

$$\mathbb{E}[L_i(x, s, u_i, u_j)|\mathcal{F}_t^s] - \mathbb{E}[f_i(0, s_0, x_0)] = \frac{1}{2} \mathbb{E}[(q_i - \alpha_i(T)) (x(T) - \mathbb{E}[x(T)])^2]$$

$$+ \frac{1}{2} \mathbb{E}[(q_i + \bar{q}_i - \beta_i(T)) \mathbb{E}[x(T)]^2] - \mathbb{E}[\gamma_i(T) \mathbb{E}[x(T)]] - \mathbb{E}[\delta_i(T)]$$

$$
+ \frac{1}{2}\mathbb{E}\int_0^T \left[\dot{\alpha}_i + q_i + 2b_1\alpha_i + \sum_{s'\in\mathcal{S}} (\alpha_i(\cdot,s') - \alpha_i(\cdot,s)) \right.
$$

$$
+ \alpha_i\left(\sigma_1^2 + \int_{\theta\in\Theta}\mu_1^2\nu(\mathrm{d}\theta)\right) + 2\frac{\bar{\phi}_i\bar{\rho}_{ij}}{\bar{c}_i}\frac{\bar{\phi}_j}{\bar{c}_j}
$$

$$
- \frac{\bar{\phi}_i^2}{\bar{c}_i} - \frac{\bar{\rho}_{ij}^2}{\bar{c}_i}\frac{\bar{\phi}_j^2}{\bar{c}_j^2} + \alpha_i\left(\sigma_{j2}^2 + \int_{\theta\in\Theta}\mu_{j2}^2\nu(\mathrm{d}\theta)\right)\frac{\bar{\phi}_j^2}{\bar{c}_j^2}
$$

$$
- 2\left[b_{2j}\alpha_i + \alpha_i\left(\sigma_1\sigma_{j2} + \int_{\theta\in\Theta}\mu_1\mu_{j2}\nu(\mathrm{d}\theta)\right)\right]\frac{\bar{\phi}_j}{\bar{c}_j} \right] (x - \mathbb{E}[x])^2 \, \mathrm{d}t
$$

$$
+ \frac{1}{2}\mathbb{E}\int_0^T \left[\dot{\beta}_i + q_i + \bar{q}_i + 2(b_1 + \bar{b}_1)\beta_i + \sum_{s'\in\mathcal{S}} (\beta_i(\cdot,s') - \beta_i(\cdot,s)) \right.
$$

$$
+ \alpha_i\left((\sigma_1 + \bar{\sigma}_1)^2 + \int_{\theta\in\Theta}(\mu_1 + \bar{\mu}_1)^2\nu(\mathrm{d}\theta)\right) - \frac{\phi_i^2}{c_i} - \frac{\rho_{ij}^2}{c_i}\frac{\phi_j^2}{c_j^2}
$$

$$
+ 2\frac{\phi_i\rho_{ij}}{c_i}\frac{\phi_j}{c_j} + \alpha_i\left((\sigma_{j2} + \bar{\sigma}_{j2})^2 + \int_{\theta\in\Theta}(\mu_{j2} + \bar{\mu}_{j2})^2\nu(\mathrm{d}\theta)\right)\frac{\phi_j^2}{c_j^2}
$$

$$
- 2\left[(b_{2j} + \bar{b}_{2j})\beta_i \right.
$$

$$
\left. + \alpha_i\left((\sigma_1 + \bar{\sigma}_1)(\sigma_{j2} + \bar{\sigma}_{j2}) + \int_{\theta\in\Theta}(\mu_1 + \bar{\mu}_1)(\mu_{j2} + \bar{\mu}_{j2})\nu(\mathrm{d}\theta)\right)\right]\frac{\phi_j}{c_j} \right]\mathbb{E}[x]^2\mathrm{d}t
$$

$$
+ \mathbb{E}\int_0^T \left[\dot{\gamma}_i + b_0\beta_i + (b_1 + \bar{b}_1)\gamma_i + \sum_{s'\in\mathcal{S}} (\gamma_i(\cdot,s') - \gamma_i(\cdot,s)) \right.
$$

$$
+ \alpha_i\left(\sigma_0(\sigma_1 + \bar{\sigma}_1) + \int_{\theta\in\Theta}\mu_0(\mu_1 + \bar{\mu}_1)\nu(\mathrm{d}\theta)\right) - \frac{\rho_{ij}^2}{c_i}\frac{\lambda_j\phi_j}{c_j^2} - \frac{\lambda_i\phi_i}{c_i}
$$

$$
+ \frac{\lambda_i\rho_{ij}}{c_i}\frac{\phi_j}{c_j} + \frac{\phi_i\rho_{ij}}{c_i}\frac{\lambda_j}{c_j} + \alpha_i\left((\sigma_{j2} + \bar{\sigma}_{j2})^2 + \int_{\theta\in\Theta}(\mu_{j2} + \bar{\mu}_{j2})^2\nu(\mathrm{d}\theta)\right)\frac{\lambda_j\phi_j}{c_j^2}
$$

$$
- \left[(b_{2j} + \bar{b}_{2j})\beta_i \right.
$$

$$
\left. + \alpha_i\left((\sigma_1 + \bar{\sigma}_1)(\sigma_{j2} + \bar{\sigma}_{j2}) + \int_{\theta\in\Theta}(\mu_1 + \bar{\mu}_1)(\mu_{j2} + \bar{\mu}_{j2})\nu(\mathrm{d}\theta)\right)\right]\frac{\lambda_j}{c_j}
$$

$$
- \left[(b_{2j} + \bar{b}_{2j})\gamma_i \right.
$$

$$
\left. + \alpha_i\left(\sigma_0(\sigma_{j2} + \bar{\sigma}_{j2}) + \int_{\theta\in\Theta}\mu_0(\mu_{j2} + \bar{\mu}_{j2})\nu(\mathrm{d}\theta)\right)\right]\frac{\phi_j}{c_j} \right]\mathbb{E}[x]\mathrm{d}t
$$

$$+ \mathbb{E} \int_0^T \left[\dot{\delta}_i + b_0 \gamma_i + \sum_{s' \in \mathcal{S}} (\delta_i(\cdot, s') - \delta_i(\cdot, s)) + \frac{1}{2} \alpha_i \left(\sigma_0^2 + \int_{\theta \in \Theta} \mu_0^2 \nu(d\theta) \right) \right.$$

$$- \frac{1}{2} \frac{\lambda_i^2}{c_i} - \frac{1}{2} \frac{\rho_{ij}^2 \lambda_j^2}{c_i c_j^2} + \frac{\lambda_i \rho_{ij}}{c_i} \frac{\lambda_j}{c_j}$$

$$+ \frac{1}{2} \alpha_i \left((\sigma_{j2} + \bar{\sigma}_{j2})^2 + \int_{\theta \in \Theta} (\mu_{j2} + \bar{\mu}_{j2})^2 \nu(d\theta) \right) \frac{\lambda_j^2}{c_j^2}$$

$$\left. - \left[(b_{2j} + \bar{b}_{2j}) \gamma_i + \alpha_i \left(\sigma_0 (\sigma_{j2} + \bar{\sigma}_{j2}) + \int_{\theta \in \Theta} \mu_0 (\mu_{j2} + \bar{\mu}_{j2}) \nu(d\theta) \right) \right] \frac{\lambda_j}{c_j} \right] dt$$

$$+ \frac{1}{2} \mathbb{E} \int_0^T c_i \left(\mathbb{E}[u_i] + \eta_i \right)^2 dt + \frac{1}{2} \mathbb{E} \int_0^T \bar{c}_i \left(u_i - \mathbb{E}[u_i] + \bar{\eta}_i \right)^2 dt,$$

Notice that, we had already computed the leader's optimal solution in terms of the system state and the mean-field state. Then, the optimal control input for the follower is not longer expressed in terms of the leader's control input, but in terms of the state feedback form. The latter expression, which does not depend on the leader's actions, is grouped as follows:

- *Those terms evaluated at the terminal time T*

- *Those terms with common factors $(x - \mathbb{E}[x])^2$, $\mathbb{E}[x]^2$, and $\mathbb{E}[x]$*

- *Those terms independent from the system state, control inputs and mean-field terms*

- *And the terms depending on the control inputs associated with the follower decision-maker.*

The announced results for the follower i are obtained from minimizing the terms in the latter expression, completing the proof.

7.3   When Nash Solution Corresponds to Stackelberg Solution for Mean-Field-Type Games

According to the square completion performed in the proof of Theorem 2, one can identify an equivalence between the Nash equilibrium and the Stackelberg solution under certain conditions as stated in the following remark.

Remark 21 *The Stackelberg solution corresponds to the Nash equilibrium if the following conditions are met:*

$$\rho_{ij} = 0, \qquad\qquad \bar{\rho}_{ij} = 0,$$

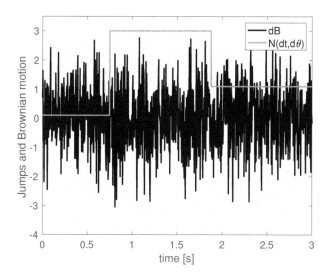

FIGURE 7.3
Brownian motion and two jumps.

$$\rho_{ji} = 0, \qquad\qquad \bar{\rho}_{ji} = 0,$$

where

$$\rho_{ij} = \left[\alpha_i \left((\sigma_{i2} + \bar{\sigma}_{i2})(\sigma_{j2} + \bar{\sigma}_{j2}) + \int_{\theta \in \Theta} (\mu_{i2} + \bar{\mu}_{i2})(\mu_{j2} + \bar{\mu}_{j2}) \nu(\mathrm{d}\theta) \right) + \bar{\varepsilon}_{ij} \right],$$

$$\bar{\rho}_{ij} = \left[\alpha_i \left(\sigma_{i2}\sigma_{j2} + \int_{\theta \in \Theta} \mu_{i2}\mu_{j2}\nu(\mathrm{d}\theta) \right) + \varepsilon_{ij} \right].$$

Thus, the optimal control inputs are not influential over each other obtaining an equivalence between Stackelberg and Nash solutions.

7.4 Numerical Example

Let us consider a singleton set of regime switching $\mathcal{S} = \{s\}$, and Brownian motion and jump-diffusion as presented in Figure 7.3. The following system dynamics with leader j and follower i:

$$\mathrm{d}x = \left(5 - 0.1x - 0.2\mathbb{E}[x] + 2u_i + u_j + \mathbb{E}[u_i] + 2\mathbb{E}[u_j] \right) \mathrm{d}t$$
$$+ \left(5 + x + 3\mathbb{E}[x] + 0.1u_i + 0.1u_j + 0.2\mathbb{E}[u_i] + 0.2\mathbb{E}[u_j] \right) \mathrm{d}B$$

$$+ \int_{\theta \in \Theta} \Big(3 + x + 2\mathbb{E}[x] + 0.2u_i + 0.4u_j + 0.3\mathbb{E}[u_i] + 0.2\mathbb{E}[u_j]\Big) \tilde{N}(\mathrm{d}t, \mathrm{d}\theta),$$

The cost functional for the leader is given by

$$L_j(x, s, u_i, u_j) = \frac{3}{2}(x(T) - \mathbb{E}[x(T)])^2 + \frac{5}{2}\mathbb{E}[x(T)]^2$$

$$+ \frac{1}{2} \int_0^T \Big(3(x - \mathbb{E}[x])^2 + 5\mathbb{E}[x]^2\Big)\, \mathrm{d}t$$

$$+ \frac{1}{2} \int_0^T \Big(7(u_j - \mathbb{E}[u_j])^2 + 14\mathbb{E}[u_j]^2\Big)\, \mathrm{d}t$$

$$+ \frac{1}{2} \int_0^T \Big(0.2(u_j - \mathbb{E}[u_j])(u_i - \mathbb{E}[u_i]) + 0.2\mathbb{E}[u_j]\mathbb{E}[u_i]\Big)\, \mathrm{d}t,$$

and for the follower

$$L_i(x, s, u_i, u_j) = (x(T) - \mathbb{E}[x(T)])^2 + \frac{5}{2}\mathbb{E}[x(T)]^2$$

$$+ \frac{1}{2} \int_0^T \Big(2(x - \mathbb{E}[x])^2 + 5\mathbb{E}[x]^2\Big)\, \mathrm{d}t$$

$$+ \frac{1}{2} \int_0^T \Big(5(u_i - \mathbb{E}[u_i])^2 + 10\mathbb{E}[u_i]^2\Big)\, \mathrm{d}t$$

$$+ \frac{1}{2} \int_0^T \Big(0.1(u_i - \mathbb{E}[u_i])(u_j - \mathbb{E}[u_j]) + 0.1\mathbb{E}[u_i]\mathbb{E}[u_j]\Big)\, \mathrm{d}t.$$

Figure 7.4 shows the evolution of the system state and its respective expectation. Figure 7.5 presents the evolution of the optimal control inputs for both the leader and follower decision-makers. Regarding the differential equations, Figures 7.6, 7.7, 7.8, and 7.9 exhibit the evolutions of the α, β, γ, and δ equations for both the leader and the follower. In those shown trajectories, we can verify that the terminal boundary conditions are satisfied according to the result presented in this chapter.

We have studied leader-follower mean-field-type games and have computed the semi-explicit Stackelberg solution involving two players. This Stackelberg solution concept can be extended to multiple leaders and followers, i.e., a team of leader and followers. Indeed, several layers or levels can be incorporated. When the number of levels is greater than two, then the game in known as hierarchical mean-field-type games. In case the reader is interested in studying this type of games, we recommend the works reported in [115], [116].

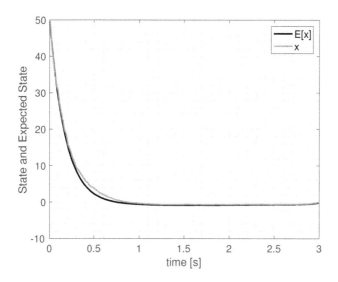

FIGURE 7.4

Evolution of the system state and its expectation for the scalar-value Stackelberg scenario with Brownian motion and Poisson jumps.

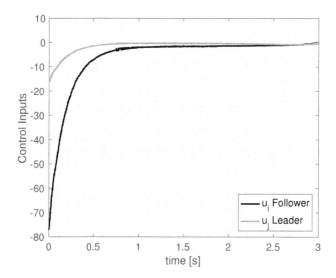

FIGURE 7.5

Evolution of the optimal control inputs u_i^* and u_j^* for the scalar-value Stackelberg scenario with Brownian motion and Poisson jumps.

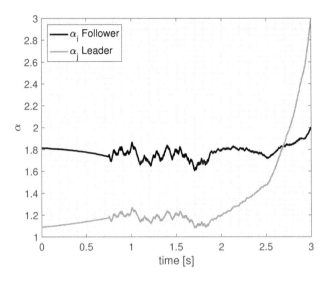

FIGURE 7.6
Evolution of the differential equations α_i and α_j for the scalar-value Stackelberg scenario with Brownian motion and Poisson jumps.

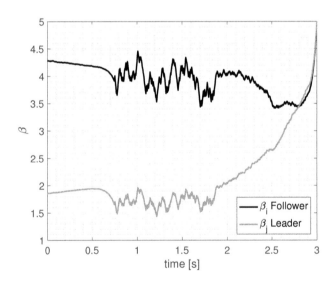

FIGURE 7.7
Evolution of the differential equations β_i and β_j for the scalar-value Stackelberg scenario with Brownian motion and Poisson jumps.

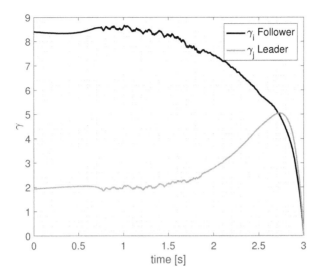

FIGURE 7.8
Evolution of the differential equations γ_i and γ_j for the scalar-value Stackelberg scenario with Brownian motion and Poisson jumps.

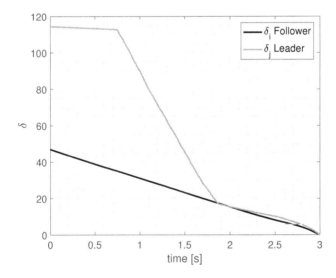

FIGURE 7.9
Evolution of the differential equations δ_i and δ_j for the scalar-value Stackelberg scenario with Brownian motion and Poisson jumps.

7.5   Exercises

1. Let us consider the following system dynamics with two decision-makers

$$dx = \left(b_0 + b_1 x + \bar{b}_1 \mathbb{E}[x] + b_{21} u_1 \right.$$

$$\left. + b_{22} u_2 + \bar{b}_{21} \mathbb{E}[u_1] + \bar{b}_{22} \mathbb{E}[u_2] \right) dt + \sigma_0 dB, \qquad (7.9)$$

$$x(0) := x_0.$$

Consider decision-maker 1 as the leader, decision-maker 2 as the follower, and the cost function for any $k \in \{1, 2\}$

$$L_k(x, u_1, u_2) = \frac{1}{2} q_k(T) x(T)^2 + \frac{1}{2} \bar{q}_k(T) \mathbb{E}[x(T)]^2 \qquad (7.10)$$

$$+ \frac{1}{2} \int_0^T \left(q_k x^2 + \bar{q}_k \mathbb{E}[x]^2 + r_k u_k^2 + \bar{r}_k \mathbb{E}[u_k]^2 \right) dt$$

$$+ \int_0^T \left(\epsilon_{kk'} (u_k - \mathbb{E}[u_k])(u_{k'} - \mathbb{E}[u_{k'}]) + \bar{\epsilon}_{kk'} \mathbb{E}[u_k] \mathbb{E}[u_{k'}] \right) dt,$$

where $k' = \{1, 2\} \setminus \{k\}$. Solve the Stackelberg problem with cost function in (7.10) subject to (7.9).

2. Establish the required condition over the Stackelberg problem in the previous exercise such that its solution coincides with the Nash equilibrium

3. Let us consider the following system dynamics with three decision-makers

$$dx = \left(b_0 + b_1 x + \bar{b}_1 \mathbb{E}[x] + b_{21} u_1 + b_{22} u_2 + b_{23} u_3 \right.$$

$$\left. + \bar{b}_{21} \mathbb{E}[u_1] + \bar{b}_{22} \mathbb{E}[u_2] + \bar{b}_{23} \mathbb{E}[u_3] \right) dt + \sigma_0 dB, \qquad (7.11)$$

$$x(0) := x_0.$$

Consider the cost functional for any $k \in \{1, 2, 3\}$ as follows:

$$L_k(x, u_1, \ldots, u_3) = \frac{1}{2} q_k(T) x(T)^2 + \frac{1}{2} \bar{q}_k(T) \mathbb{E}[x(T)]^2$$

$$+ \frac{1}{2} \int_0^T \left(q_k x^2 + \bar{q}_k \mathbb{E}[x]^2 + r_k u_k^2 + \bar{r}_k \mathbb{E}[u_k]^2 \right) dt,$$

$$(7.12)$$

Assume that decision-makers play in the following sequence: first decision-makers makes a choice 1, then decision-maker 2 reacts to the selection made by decision-maker 1, and then finally decision-maker 3 plays. Solve this Hierarchical mean-field-type game problem.

4. Does the solution of the previous three-player hierarchical mean-field-type game coincide with the Nash equilibrium?. Explain and justify your answer.

5. After studying Chapters 11 and 12, solve a discrete-time Stackelberg mean-field-type game problem.

8

Berge Equilibrium in Mean-Field-Type Games

One of the most important features and desirable outcomes in interacting decision-makers is the emergence of co-opetitive behaviors [75]. Such a behavior can be observed in a Berge solution in wide range of games. In time-independent games in strategic form, the Berge solution can be seen as a mutual support between the players. The Berge solution concept breaks the dilemma issue in the Prisoner's dilemma as well as in the forwarding dilemma in communication networks, which is the similar dilemma in the context of engineering, more precisely, in the context of telecommunications and data networks. The Berge solution enforces the emergence of mutual support and collaboration in these games. In the static duopoly game, there is wide range of parameters under which the refined Berge solution offers better outcome to both players than the Nash equilibrium outcome, i.e., the refined Berge solution Pareto-dominates the Nash equilibrium solution. In this regard, it is quite important to take into consideration Berge solution for the systems engineering design.

These aforementioned preliminary results on Berge solution motivates its investigation in other class of games including time-dependent games. Time-dependent cooperative solution concept such as bargaining, altruism and empathy have been examined in mean-field-type games (see [70] and the references therein). However, the Berge solution concept has not been considered in the context of mean-field-type game theory so far, which is the main topic addressed in this chapter.

The mean-field-type Nash equilibrium analysis were conducted in Chapter 6, and the Stackelberg solution were treated in Chapter 7. Different from the problems addressed in those previous chapters, here we are interested in analyzing a Berge solution of such type of mean-field-type problems. In order to find a Berge solution in a semi-explicit way, we follow a direct method.

Prior to presenting the Berge mean-field-type game problem statement, we first introduce the Berge solution concept by means of a two-decision-makers case. Afterward, we focus on the mean-field-type game problem and its semi-explicit solution.

DOI: 10.1201/9781003098607-8

8.1   On the Berge Solution Concept

The Berge solution concept was introduced in [117, page 20]. See also the works reported in from [118] to [126] for recent investigation of the Berge solution. If the decision-makers have chosen a strategy profile that forms a Berge solution, and a decision-maker, let us name it i, sticks to the chosen strategy but some of other players change their strategies, then i's objective function will not improve. This is a resilience to a deviation by other players or other teams. For two decision-makers, the strategy profile $u^* = (u_1^*, u_2^*)$ is a (constrained) Berge solution if the following holds:

$$L_1(x, u_1^*, u_2^*) = \inf_{u_2 \in U_2(u_1^*)} L_1(x, u_1^*, u_2),$$

$$L_2(x, u_1^*, u_2^*) = \inf_{u_1 \in U_1(u_2^*)} L_2(x, u_1, u_2^*),$$

where $U_i(u_j^*)$ is the set of constrained strategies of player i given the other's strategy u_j^*. For instance, $U_i(u_j^*)$ means that the feasible set for the decision-maker i depends on the selection made by the decision-maker j.

The Berge strategy yields the best objective outcome to the others' players who also play the Berge strategies. In (constrained) Berge solution, a deviation by one or more other players cannot improve the objective to a player who does not deviate. Note that, the (constrained) Berge solution concept is different from the classical (constrained) Cournot-Nash equilibrium solution concept which is given by

$$L_1(x, u_1^*, u_2^*) = \inf_{u_1 \in U_1(u_2^*)} L_1(x, u_1, u_2^*),$$

$$L_2(x, u_1^*, u_2^*) = \inf_{u_2 \in U_2(u_1^*)} L_2(x, u_1^*, u_2).$$

By exchanging (permuting) the objective functions one can derive a (constrained) Cournot-Nash equilibrium of a new (constrained) game with exchanged objectives. However, for three or more players, one quickly realize that the (constrained) Berge solution concept is a totally different concept and focuses on mutual support between the players.

8.2   Berge Mean-Field-Type Game Problem

Consider two players with the following state dynamics with jump-diffusion and regime switching:

$$dx = bdt + \sigma dB + \int_\Theta \mu(\cdot, \theta)\tilde{N}(dt, d\theta), \tag{8.1a}$$

$$x(0) = x_0, s(0) = s_0, \tag{8.1b}$$

where

$$b = b_0 + b_1 x + \bar{b}_1 \bar{x} + \sum_{\ell \in \{i,j\}} b_{\ell 2} u_\ell + \sum_{\ell \in \{i,j\}} \bar{b}_{\ell 2} \bar{u}_\ell,$$

$$\sigma = \sigma_0 + \sigma_1 x + \bar{\sigma}_1 \bar{x} + \sum_{\ell \in \{i,j\}} \sigma_{\ell 2} u_\ell + \sum_{\ell \in \{i,j\}} \bar{\sigma}_{\ell 2} \bar{u}_\ell,$$

$$\mu(\cdot, \theta) = \mu_0 + \mu_1 x + \bar{\mu}_1 \bar{x} + \sum_{\ell \in \{i,j\}} \mu_{\ell 2} u_\ell + \sum_{\ell \in \{i,j\}} \bar{\mu}_{\ell 2} \bar{u}_\ell,$$

with drift, diffusion, and jump terms as follows: $b_k(t, s)$; $\sigma_k(t, s)$; $\bar{b}_k(t, s)$; $\bar{\sigma}_k(t, s) \in \mathbb{R}$ and $s \in \mathcal{S}$, where $\mathcal{S}$ is the set of switching regimes. Moreover $\mu_k(t, s, \theta) \in \mathbb{R}$, $\theta \in \Theta$ where Θ is the set of jump sizes, for all $k \in \{0, 1, 2\}$. The process B is a standard one-dimensional Brownian motion, $\tilde{N}(dt, d\theta) = N(dt, d\theta) - \nu(d\theta)dt$ is a compensated jump process with Radon measure ν over Θ. The switching process switches between elements in $\mathcal{S}$ with transition rate $\tilde{q}_{ss'}$. It is assumed that processes $s(t), B(t), N(t, .)$ are mutually independent. This system dynamics are the same general dynamics considered in Chapters 6 and 7. Indeed, here we consider only two decision-makers in the strategic interaction as we also did in Chapter 7.

The term $\mathbb{E}[x] = \mathbb{E}[x|s]$ denotes the expected value of x given the switching regime $s(t) = s$, and similarly we have $\mathbb{E}[u] = \mathbb{E}[u|s]$. These terms $\mathbb{E}[x]$ and $\mathbb{E}[u]$ are called conditional state mean-field and conditional control-action mean-field terms. The state dynamics is therefore of conditional *McKean-Vlasov* type as the conditional probability measure of the state appears in the drift, diffusion, and jump coefficients via ($\mathbb{E}[x]$, $\mathbb{E}[u]$). Below we denote a solution (if any) to (8.1) by $x := x^{u, x_0, s_0}$.

We associate a cost functional of mean-field type of each decision-maker $\ell \in \{i, j\}$ as

$$L_\ell(x, s, u_i, u_j) = \frac{1}{2} q_\ell (x(T) - \mathbb{E}[x(T)])^2 + \frac{1}{2}(q_\ell + \bar{q}_\ell)\mathbb{E}[x(T)]^2$$

$$+ \frac{1}{2} \int_0^T \Big(q_\ell(x - \mathbb{E}[x])^2 + (q_\ell + \bar{q}_\ell)\mathbb{E}[x]^2$$

$$+ r_\ell(u_\ell - \mathbb{E}[u_\ell])^2 + (r_\ell + \bar{r}_\ell)\mathbb{E}[u_\ell]^2$$

$$+ \epsilon_{\ell\ell'}(u_\ell - \mathbb{E}[u_\ell])(u_{\ell'} - \mathbb{E}[u_{\ell'}]) + \bar{\epsilon}_{\ell\ell'}\mathbb{E}[u_\ell]\mathbb{E}[u_{\ell'}] \Big) dt,$$

where $\ell' \in \{i, j\} \setminus \{\ell\}$, and the weight parameters are $q_{\ell T} = q_\ell(T, s(T))$, $q_\ell = q_\ell(t, s(t))$, $\bar{q}_{\ell T} = \bar{q}_\ell(T, s(T))$, $\bar{q}_\ell = \bar{q}_\ell(t, s(t))$; and $r_\ell = r_\ell(t, s(t))$, $\bar{r}_\ell = \bar{r}_\ell(t, s(t)) \in \mathbb{R}$, $\epsilon_{\ell\ell'} = \epsilon_{\ell\ell'}(t, s(t)) \in \mathbb{R}$. The Berge mean-field-type game problem is stated next in Problem 18.

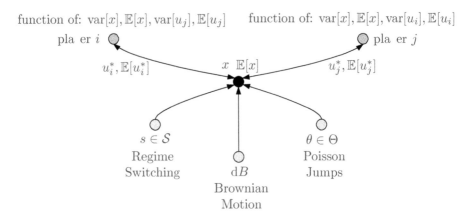

FIGURE 8.1
General scheme of the two-player Berge mean-field-type game with jump-diffusion and regime switching.

Problem 18 *Given u_i, define the Berge (support) response problem of j to i as*

$$
\begin{cases}
\text{minimize}_{u_j \in \mathcal{U}_j} \ \mathbb{E}\left[L_i(x, s, u_i, u_j) | x(0) = x_0, s(0) = s_0\right], \\[2mm]
\text{subject to} \\[2mm]
dx = bdt + \sigma dB + \displaystyle\int_\Theta \mu(\cdot, \theta)\tilde{N}(dt, d\theta), \\[1mm]
x(0) = x_0, s(0) = s_0.
\end{cases}
\tag{8.2}
$$

the initial state condition x_0 is given. The weight parameters are $q_i, q_i + \bar{q}_i \geq 0$, $r_i, r_i + \bar{r}_i > 0$, and the system parameters of the functions b, σ, and μ are time and switching dependent.

Next, we define the Berge support response and the Berge solution. Be aware that the introduction of these two concepts follow a similar exposition as we were doing with the best-response and Nash equilibrium for the non-cooperative game counterpart problem. Nevertheless, the game structure in this chapter is of mutual support.

Definition 15 (Berge Support Response) *A minimizer u_j^* to Problem (8.2) is called a Berge support response strategy of the jth decision-maker given the strategy u_i of others. Hence, the set of support response strategies of the jth decision-maker to i is denoted by $SR_j(u_i) \subset \mathcal{U}_j$.*

The Berge support response introduced in Definition 15 assists to define the Berge solution next in Definition 16.

Definition 16 *[Berge solution] A strategy $u^* \in \mathcal{U} = \mathcal{U}_i \times \mathcal{U}_j$ is a Berge solution if $u_i^* \in \mathrm{SR}_i(u_j^*)$ and $u_j^* \in \mathrm{SR}_j(u_i^*)$. A Berge solution solves the following system:*

$$\begin{cases} \mathbb{E}\left[L_i(x,s,u_i^*,u_j^*)|x(0)=x_0, s(0)=s_0\right] \\ \qquad = \inf_{u_j \in \mathcal{U}_j} \mathbb{E}\left[L_i(x,s,u_i^*,u_j)|x(0)=x_0, s(0)=s_0\right], \\ \mathbb{E}\left[L_j(x,s,u_i^*,u_j^*)|x(0)=x_0, s(0)=s_0\right] \\ \qquad = \inf_{u_i \in \mathcal{U}_i} \mathbb{E}\left[L_j(x,s,u_i,u_j^*)|x(0)=x_0, s(0)=s_0\right]. \end{cases} \tag{8.3}$$

Figure 8.1 presents the scheme for the Berge solutions.

We consider the following assumption:

$$\begin{cases} q_{iT}, q_i \geq 0, q_{iT} + \bar{q}_{iT} \geq 0, \\ q_i + \bar{q}_i \geq 0, r_i \geq 0, r_i + \bar{r}_i \geq 0, \\ \mathbb{E}[x_0^2] < +\infty, \quad \int_0^T \int_\Theta \phi_{ik}^2 \nu(\mathrm{d}\theta) < +\infty, \\ \text{with } \phi_{ik} \in \{\mu_{ik}, \bar{\mu}_{ik}\}, \text{ and} \\ |\mathcal{S}| < +\infty, b_k, \sigma_k^2, \bar{b}_k, \bar{\sigma}_k^2 \in L^1([0,T] \times \mathcal{S}, \mathbb{R}), \forall k \in \{0,1,2\}. \end{cases} \tag{8.4}$$

Remark 22 *Under Assumption (8.4), the state dynamics has solution for each $u \in \mathcal{U}$. We do not impose boundedness conditions on the coefficients or Lipschitz conditions on the cost. The integrability condition above yields*

$$\int_0^T \left\{ |b_0(t,s)| + |b_1(t,s)| + |\bar{b}_1(t,s)| + |b_{2j}(t,s)| + |\bar{b}_{2j}(t,s)| \right\} \mathrm{d}t$$

$$+ \int_0^T \left\{ \sigma_0^2(t,s) + \sigma_1^2(t,s) + \bar{\sigma}_1^2(t,s) + \sigma_{2j}^2(t,s) + \bar{\sigma}_{2j}^2(t,s) \right\} \mathrm{d}t$$

$$+ \int_0^T \left\{ \int_\Theta [\mu_0^2(t,s,\theta) + \mu_1^2(t,s,\theta) \right.$$

$$\left. + \bar{\mu}_1^2(t,s,\theta) + \mu_{2j}^2(t,s,\theta) + \bar{\mu}_{2j}^2(t,s,\theta)]\nu(\mathrm{d}\theta) \right\} \mathrm{d}t < +\infty. \tag{8.5}$$

If both

$$r(t,s) \geq \tilde{\varepsilon} > 0$$

and

$$r(t,s) + \bar{r}(t,s) \geq \hat{\varepsilon} > 0$$

for some $\tilde{\varepsilon}$ and $\hat{\varepsilon}$, then the cost function $L_i(x,s,u_i,u_j)$ is coercive and continuous in u_i. Moreover, the set $\mathcal{U}_i$ is nonempty and convex. Therefore, Problem 18 is well defined, which implies existence of pure best-response strategies. Here, we are interested in Berge support response strategies which is different than Nash best-response strategies. To do so, one looks for the coefficient of the control of decision-maker j.

In the coming section, we compute the Berge solution in a semi-explicit manner by following the direct method.

8.3   Semi-explicit Mean-Field-Type Berge Solution

We present the main result discussed in this chapter in Theorem 3, consisting on the semi-explicit solution for Problem 18.

Theorem 3 *The Berge support control strategy and the Berge solution cost are given by:*

$$u_\ell^* = -\lambda_{1\ell} - \lambda_{2\ell}\mathbb{E}[x] - \lambda_{3\ell}(x - \mathbb{E}[x]), \ \forall \ell \in \{i,j\},$$

$$\lambda_{1\ell} = \left(1 - \frac{\rho_{\ell k}}{c_\ell}\frac{\rho_{k\ell}}{c_k}\right)^{-1}\left[\frac{\lambda_\ell}{c_\ell} - \frac{\rho_{\ell k}}{c_\ell}\frac{\lambda_k}{c_k}\right],$$

$$\lambda_{2\ell} = \left(1 - \frac{\rho_{\ell k}}{c_\ell}\frac{\rho_{k\ell}}{c_k}\right)^{-1}\left(\frac{\phi_\ell}{c_\ell} - \frac{\rho_{\ell k}}{c_\ell}\frac{\phi_k}{c_k}\right)\mathbb{E}[x],$$

$$\lambda_{3\ell} = \left(1 - \frac{\bar{\rho}_{\ell k}}{\bar{c}_\ell}\frac{\bar{\rho}_{k\ell}}{\bar{c}_k}\right)^{-1}\left(\frac{\bar{\phi}_\ell}{\bar{c}_\ell} - \frac{\bar{\rho}_{\ell k}}{\bar{c}_\ell}\frac{\bar{\phi}_k}{\bar{c}_k}\right),$$

$$\forall \ell \in \{i,j\}, k \in \{i,j\} \setminus \{\ell\},$$

$$\mathbb{E}[L_\ell^*(x,s,u_i^*,u_j^*)] = \frac{1}{2}\mathbb{E}[\alpha_\ell(0,s(0))(x_0 - \bar{x}_0)^2] + \frac{1}{2}\mathbb{E}[\beta_\ell(0,s(0))\bar{x}_0^2]$$
$$+ \mathbb{E}[\gamma_\ell(0,s(0))\bar{x}_0] + \mathbb{E}[\delta_\ell(0,s(0))], \ \forall \ell \in \{i,j\}.$$

where α_ℓ, β_ℓ, γ_ℓ, and δ_ℓ solve the following differential equations:

$$\dot{\alpha}_\ell = -q_\ell - 2b_1\alpha_\ell - \sum_{s' \in \mathcal{S}}(\alpha_\ell(\cdot,s') - \alpha_\ell(\cdot,s)) - \alpha_\ell\left(\sigma_1^2 + \int_{\theta \in \Theta}\mu_1^2\nu(d\theta)\right)$$

$$+ \frac{\bar{\phi}_k^2}{\bar{c}_k} + \frac{\bar{\rho}_{k\ell}^2}{\bar{c}_k}\lambda_{3\ell}^2 + 2\frac{\bar{\phi}_k\bar{\rho}_{k\ell}}{\bar{c}_k}\lambda_{3\ell}$$

$$+ 2\lambda_{3\ell}\left[b_{2\ell}\alpha_\ell + \alpha_\ell\left(\sigma_1\sigma_{\ell 2} + \int_{\theta \in \Theta}\mu_1\mu_{\ell 2}\nu(d\theta)\right)\right]$$

$$- \lambda_{3\ell}^2\left[\alpha_\ell\left(\sigma_{\ell 2}^2 + \int_{\theta \in \Theta}\mu_{\ell 2}^2\nu(d\theta)\right) + r_\ell\right],$$

$$\dot{\beta}_\ell = -q_\ell - \bar{q}_\ell - 2(b_1 + \bar{b}_1)\beta_\ell - \sum_{s' \in \mathcal{S}}(\beta_\ell(\cdot,s') - \beta_\ell(\cdot,s))$$

$$- \alpha_\ell\left((\sigma_1 + \bar{\sigma}_1)^2 + \int_{\theta \in \Theta}(\mu_1 + \bar{\mu}_1)^2\nu(d\theta)\right)$$

$$- \left[\alpha_\ell\left((\sigma_{\ell 2} + \bar{\sigma}_{\ell 2})^2 + \int_{\theta \in \Theta}(\mu_{\ell 2} + \bar{\mu}_{\ell 2})^2\nu(d\theta)\right) + r_\ell + \bar{r}_\ell\right]\lambda_{2\ell}^2$$

$$+ 2\left[(b_{2\ell} + \bar{b}_{2\ell})\beta_\ell + \alpha_\ell\left((\sigma_1 + \bar{\sigma}_1)(\sigma_{\ell2} + \bar{\sigma}_{\ell2})\right.\right.$$

$$\left.\left.+ \int_{\theta\in\Theta}(\mu_1 + \bar{\mu}_1)(\mu_{\ell2} + \bar{\mu}_{\ell2})\nu(d\theta)\right)\right]\lambda_{2\ell}$$

$$+ \frac{\rho_{k\ell}^2}{c_k}\lambda_{2\ell}^2 + \frac{\phi_k^2}{c_k} - 2\frac{\phi_k\rho_{k\ell}}{c_k}\lambda_{2\ell},$$

$$\dot{\gamma}_\ell = -b_0\beta_\ell - (b_1 + \bar{b}_1)\gamma_\ell - \sum_{s'\in\mathcal{S}}(\gamma_\ell(\cdot, s') - \gamma_\ell(\cdot, s))$$

$$- \alpha_\ell\left(\sigma_0(\sigma_1 + \bar{\sigma}_1) + \int_{\theta\in\Theta}\mu_0(\mu_1 + \bar{\mu}_1)\nu(d\theta)\right)$$

$$- \left[\alpha_\ell\left((\sigma_{\ell2} + \bar{\sigma}_{\ell2})^2 + \int_{\theta\in\Theta}(\mu_{\ell2} + \bar{\mu}_{\ell2})^2\nu(d\theta)\right) + r_\ell + \bar{r}_\ell\right]\lambda_{1\ell}\lambda_{2\ell}$$

$$+ \left[(b_{2\ell} + \bar{b}_{2\ell})\beta_\ell + \alpha_\ell\left((\sigma_1 + \bar{\sigma}_1)(\sigma_{\ell2} + \bar{\sigma}_{\ell2})\right.\right.$$

$$\left.\left.+ \int_{\theta\in\Theta}(\mu_1 + \bar{\mu}_1)(\mu_{\ell2} + \bar{\mu}_{\ell2})\nu(d\theta)\right)\right]\lambda_{1\ell}$$

$$+ \alpha_\ell\left(\sigma_0(\sigma_{\ell2} + \bar{\sigma}_{\ell2}) + \int_{\theta\in\Theta}\mu_0(\mu_{\ell2} + \bar{\mu}_{\ell2})\nu(d\theta)\right)\lambda_{2\ell}$$

$$+ (b_{2\ell} + \bar{b}_{2\ell})\gamma_\ell\lambda_{2\ell} + \frac{1}{2}\frac{\rho_{k\ell}^2}{c_k}2\lambda_{1\ell}\lambda_{2\ell} + \frac{\lambda_k\phi_k}{c_k} - \frac{\lambda_k\rho_{k\ell}}{c_k}\lambda_{2\ell} - \frac{\phi_k\rho_{k\ell}}{c_k}\lambda_{1\ell},$$

$$\dot{\delta}_\ell = -b_0\gamma_\ell - \sum_{s'\in\mathcal{S}}(\delta_\ell(\cdot, s') - \delta_\ell(\cdot, s)) - \frac{1}{2}\alpha_\ell\left(\sigma_0^2 + \int_{\theta\in\Theta}\mu_0^2\nu(d\theta)\right)$$

$$- \frac{1}{2}\left[\alpha_\ell\left((\sigma_{\ell2} + \bar{\sigma}_{\ell2})^2 + \int_{\theta\in\Theta}(\mu_{\ell2} + \bar{\mu}_{\ell2})^2\nu(d\theta)\right) + r_\ell + \bar{r}_\ell\right]\lambda_{1\ell}^2$$

$$+ (b_{2\ell} + \bar{b}_{2\ell})\gamma_\ell\lambda_{1\ell} + \alpha_\ell\left(\sigma_0(\sigma_{\ell2} + \bar{\sigma}_{\ell2}) + \int_{\theta\in\Theta}\mu_0(\mu_{\ell2} + \bar{\mu}_{\ell2})\nu(d\theta)\right)\lambda_{1\ell}$$

$$+ \frac{1}{2}\frac{\lambda_k^2}{c_k} + \frac{1}{2}\frac{\rho_{k\ell}^2}{c_k}\lambda_{1\ell}^2 - \frac{\lambda_k\rho_{k\ell}}{c_k}\lambda_{1\ell}, \quad \forall \ell \in \{i, j\}, k \in \{i, j\} \setminus \{\ell\},$$

with terminal boundary conditions given by:

$$\alpha_i(T) = q_i(T),$$
$$\beta_i(T) = q_i(T) + \bar{q}_i(T),$$
$$\gamma_i(T) = \delta_i(T) = 0,$$

and c_ℓ, λ_ℓ, ϕ_ℓ, $\rho_{\ell k}$, $\bar{c}_\ell$, $\bar{\phi}_\ell$, and $\bar{\rho}_{\ell k}$ are the following auxiliary variables:

$$c_\ell = \alpha_k\left((\sigma_{\ell2} + \bar{\sigma}_{\ell2})^2 + \int_{\theta\in\Theta}(\mu_{\ell2} + \bar{\mu}_{\ell2})^2\nu(d\theta)\right),$$

$$\lambda_\ell = (b_{2\ell} + \bar{b}_{2\ell})\gamma_k + \alpha_k \left(\sigma_0(\sigma_{\ell 2} + \bar{\sigma}_{\ell 2}) + \int_{\theta \in \Theta} \mu_0(\mu_{\ell 2} + \bar{\mu}_{\ell 2})\nu(d\theta) \right),$$

$$\phi_\ell = (b_{2\ell} + \bar{b}_{2\ell})\beta_k$$

$$+ \alpha_k \left((\sigma_1 + \bar{\sigma}_1)(\sigma_{\ell 2} + \bar{\sigma}_{\ell 2}) + \int_{\theta \in \Theta} (\mu_1 + \bar{\mu}_1)(\mu_{\ell 2} + \bar{\mu}_{\ell 2})\nu(d\theta) \right),$$

$$\rho_{\ell k} = \alpha_k \left((\sigma_{k2} + \bar{\sigma}_{k2})(\sigma_{\ell 2} + \bar{\sigma}_{\ell 2}) + \int_{\theta \in \Theta} (\mu_{k2} + \bar{\mu}_{k2})(\mu_{\ell 2} + \bar{\mu}_{\ell 2})\nu(d\theta) \right) + \bar{\varepsilon}_{k\ell},$$

$$\bar{c}_\ell = \alpha_k \left(\sigma_{\ell 2}^2 + \int_{\theta \in \Theta} \mu_{\ell 2}^2 \nu(d\theta) \right),$$

$$\bar{\phi}_\ell = b_{2\ell}\alpha_k + \alpha_k \left(\sigma_1 \sigma_{\ell 2} + \int_{\theta \in \Theta} \mu_1 \mu_{\ell 2} \nu(d\theta) \right),$$

$$\bar{\rho}_{\ell k} = \alpha_k \left(\sigma_{k2}\sigma_{\ell 2} + \int_{\theta \in \Theta} \mu_{k2}\mu_{\ell 2}\nu(d\theta) \right) + \varepsilon_{k\ell}, \ \forall \ell \in \{i, j\}, k \in \{i, j\} \setminus \{\ell\},$$

whenever these ordinary differential equations have a solution that does not blow up in $[0, T]$, with

$$c_k \neq 0, \ \forall k \in \{i, j\},$$

$$\left(\frac{\rho_{\ell k}}{c_\ell} \frac{\rho_{k\ell}}{c_k} \right) \neq 1.$$

By observing the optimal control input in function of λ terms, and considering that these λ terms are function of the c terms, then we make the following remark below to avoid singularity in the Berge solution. Therefore, the diffusion function in the underlying SDE is crucial in the computation of the semi-explicit solution for Berge-like mean-field-type game problems.

Remark 23 *Notice that, c_ℓ and $\bar{c}_\ell$ terms show that contrary to the non-cooperative game problem whose solution is given by a Nash equilibrium (see Chapter 6), the Berge problem should incorporate control-input-dependent jump-diffusion terms, i.e, the terms*

$$(\sigma_{\ell 2} + \bar{\sigma}_{\ell 2})^2 + \int_{\theta \in \Theta} (\mu_{\ell 2} + \bar{\mu}_{\ell 2})^2 \nu(d\theta),$$

and

$$\sigma_{\ell 2}^2 + \int_{\theta \in \Theta} \mu_{\ell 2}^2 \nu(d\theta)$$

must be different than zero.

Below, we present the step-by-step computation of the semi-explicit solution.

Proof 19 (Theorem 3) *Once again, we have a problem whose cost functional and system dynamics are the same as in Chapters 6 and 7. We develop a proof of optimality using a direct method. Let the guess functional of* *decision-maker i be*

$$f_i(t, x, s) = \frac{1}{2}\alpha_i(x - \mathbb{E}[x])^2 + \frac{1}{2}\beta_i\mathbb{E}[x]^2 + \gamma_i\mathbb{E}[x] + \delta_i,$$

where $\alpha_i = \alpha_i(t, s)$, $\beta_i = \beta_i(t, s)$, $\gamma_i = \gamma_i(t, s)$, and $\delta_i = \delta_i(t, s)$ are time and regime-switching dependent, and $x = x^{u,x_0,s_0}(t)$. By applying Itô's formula with jump-diffusion processes and switching one obtains

$$\mathbb{E}[f_i(T, s, x) - f_i(0, s_0, x_0)] = \frac{1}{2}\mathbb{E}\int_0^T \left[\dot{\alpha}_i + 2b_1\alpha_i + \sum_{s' \in \mathcal{S}} (\alpha_i(\cdot, s') - \alpha_i(\cdot, s)) \right.$$

$$+ \alpha_i \left(\sigma_1^2 + \int_{\theta \in \Theta} \mu_1^2\nu(\mathrm{d}\theta) \right) \Bigg] (x - \mathbb{E}[x])^2 \, \mathrm{d}t$$

$$+ \frac{1}{2}\mathbb{E}\int_0^T \left[\dot{\beta}_i + 2(b_1 + \bar{b}_1)\beta_i + \sum_{s' \in \mathcal{S}} (\beta_i(\cdot, s') - \beta_i(\cdot, s)) \right.$$

$$+ \alpha_i \left((\sigma_1 + \bar{\sigma}_1)^2 + \int_{\theta \in \Theta} (\mu_1 + \bar{\mu}_1)^2\nu(\mathrm{d}\theta) \right) \Bigg] \mathbb{E}[x]^2\mathrm{d}t$$

$$+ \mathbb{E}\int_0^T \left[\dot{\gamma}_i + b_0\beta_i + (b_1 + \bar{b}_1)\gamma_i + \sum_{s' \in \mathcal{S}} (\gamma_i(\cdot, s') - \gamma_i(\cdot, s)) \right.$$

$$+ \alpha_i \left(\sigma_0(\sigma_1 + \bar{\sigma}_1) + \int_{\theta \in \Theta} \mu_0(\mu_1 + \bar{\mu}_1)\nu(\mathrm{d}\theta) \right) \Bigg] \mathbb{E}[x]\mathrm{d}t$$

$$+ \mathbb{E}\int_0^T \left[\dot{\delta}_i + b_0\gamma_i + \sum_{s' \in \mathcal{S}} (\delta_i(\cdot, s') - \delta_i(\cdot, s)) + \frac{1}{2}\alpha_i \left(\sigma_0^2 + \int_{\theta \in \Theta} \mu_0^2\nu(\mathrm{d}\theta) \right) \right] \mathrm{d}t$$

$$+ \frac{1}{2}\mathbb{E}\int_0^T \alpha_i \left((\sigma_{i2} + \bar{\sigma}_{i2})^2 + \int_{\theta \in \Theta} (\mu_{i2} + \bar{\mu}_{i2})^2\nu(\mathrm{d}\theta) \right) \mathbb{E}[u_i]^2\mathrm{d}t$$

$$+ \frac{1}{2}\mathbb{E}\int_0^T 2 \left\{ (b_{2i} + \bar{b}_{2i})\gamma_i + \alpha_i \left(\sigma_0(\sigma_{i2} + \bar{\sigma}_{i2}) + \int_{\theta \in \Theta} \mu_0(\mu_{i2} + \bar{\mu}_{i2})\nu(\mathrm{d}\theta) \right) \right.$$

$$+ \left[(b_{2i} + \bar{b}_{2i})\beta_i + \alpha_i \left((\sigma_1 + \bar{\sigma}_1)(\sigma_{i2} + \bar{\sigma}_{i2}) \right. \right.$$

$$+ \left. \left. \int_{\theta \in \Theta} (\mu_1 + \bar{\mu}_1)(\mu_{i2} + \bar{\mu}_{i2})\nu(\mathrm{d}\theta) \right) \right] \mathbb{E}[x] \Bigg\} \mathbb{E}[u_i]\mathrm{d}t$$

$$+ \frac{1}{2}\mathbb{E}\int_0^T \alpha_i \left((\sigma_{j2} + \bar{\sigma}_{j2})^2 + \int_{\theta \in \Theta} (\mu_{j2} + \bar{\mu}_{j2})^2\nu(\mathrm{d}\theta) \right) \mathbb{E}[u_j]^2\mathrm{d}t$$

$$+ \frac{1}{2}\mathbb{E}\int_0^T 2 \left\{ (b_{2j} + \bar{b}_{2j})\gamma_i + \alpha_i \left(\sigma_0(\sigma_{j2} + \bar{\sigma}_{j2}) + \int_{\theta \in \Theta} \mu_0(\mu_{j2} + \bar{\mu}_{j2})\nu(\mathrm{d}\theta) \right) \right.$$

$$+ \left[(b_{2j} + \bar{b}_{2j})\beta_i + \alpha_i \left((\sigma_1 + \bar{\sigma}_1)(\sigma_{j2} + \bar{\sigma}_{j2}) \right. \right.$$

$$+ \left. \left. \int_{\theta \in \Theta} (\mu_1 + \bar{\mu}_1)(\mu_{j2} + \bar{\mu}_{j2})\nu(\mathrm{d}\theta) \right) \right] \mathbb{E}[x]$$

$$+ \alpha_i \left((\sigma_{i2} + \bar{\sigma}_{i2})(\sigma_{j2} + \bar{\sigma}_{j2}) \right.$$

$$+ \left. \left. \int_{\theta \in \Theta} (\mu_{i2} + \bar{\mu}_{i2})(\mu_{j2} + \bar{\mu}_{j2})\nu(\mathrm{d}\theta) \right) \mathbb{E}[u_i] \right\} \mathbb{E}[u_j]\mathrm{d}t$$

$$+ \frac{1}{2}\mathbb{E}\int_0^T \alpha_i \left(\sigma_{i2}^2 + \int_{\theta \in \Theta} \mu_{i2}^2\nu(\mathrm{d}\theta) \right)(u_i - \mathbb{E}[u_i])^2\mathrm{d}t$$

$$+ \frac{1}{2}\mathbb{E}\int_0^T 2 \left\{ \left[b_{2i}\alpha_i \right. \right.$$

$$+ \alpha_i \left. \left(\sigma_1\sigma_{i2} + \int_{\theta \in \Theta} \mu_1\mu_{i2}\nu(\mathrm{d}\theta) \right) \right] (x - \mathbb{E}[x]) \right\}(u_i - \mathbb{E}[u_i])\mathrm{d}t$$

$$+ \frac{1}{2}\mathbb{E}\int_0^T \alpha_i \left(\sigma_{j2}^2 + \int_{\theta \in \Theta} \mu_{j2}^2\nu(\mathrm{d}\theta) \right)(u_j - \mathbb{E}[u_j])^2\mathrm{d}t$$

$$+ \frac{1}{2}\mathbb{E}\int_0^T 2 \left\{ \left[b_{2j}\alpha_i + \alpha_i \left(\sigma_1\sigma_{j2} + \int_{\theta \in \Theta} \mu_1\mu_{j2}\nu(\mathrm{d}\theta) \right) \right] (x - \mathbb{E}[x]) \right.$$

$$+ \alpha_i \left. \left(\sigma_{i2}\sigma_{j2} + \int_{\theta \in \Theta} \mu_{i2}\mu_{j2}\nu(\mathrm{d}\theta) \right)(u_i - \mathbb{E}[u_i]) \right\}(u_j - \mathbb{E}[u_j])\mathrm{d}t.$$

Now we compute the gap $\mathbb{E}[L_i(x, s, u_i, u_j)|\mathcal{F}_t^s] - \mathbb{E}[f_i(0, s_0, x_0)]$ *as follows:*

$$\mathbb{E}[L_i(x, s, u_i, u_j)|\mathcal{F}_t^s] - \mathbb{E}[f_i(0, s_0, x_0)] = \frac{1}{2}\mathbb{E}[(q_i - \alpha_i(T))(x(T) - \mathbb{E}[x(T)])^2]$$

$$+ \frac{1}{2}\mathbb{E}[(q_i + \bar{q}_i - \beta_i(T))\mathbb{E}[x(T)]^2] - \mathbb{E}[\gamma_i(T)\mathbb{E}[x(T)]] - \mathbb{E}[\delta_i(T)]$$

$$+ \frac{1}{2}\mathbb{E}\int_0^T \left[\dot{\alpha}_i + q_i + 2b_1\alpha_i + \sum_{s' \in \mathcal{S}} (\alpha_i(\cdot, s') - \alpha_i(\cdot, s)) \right.$$

$$+ \alpha_i \left. \left(\sigma_1^2 + \int_{\theta \in \Theta} \mu_1^2\nu(\mathrm{d}\theta) \right) \right](x - \mathbb{E}[x])^2 \,\mathrm{d}t$$

$$+ \frac{1}{2}\mathbb{E}\int_0^T \left[\dot{\beta}_i + q_i + \bar{q}_i + 2(b_1 + \bar{b}_1)\beta_i + \sum_{s' \in \mathcal{S}} (\beta_i(\cdot, s') - \beta_i(\cdot, s)) \right.$$

$$+ \alpha_i \left. \left((\sigma_1 + \bar{\sigma}_1)^2 + \int_{\theta \in \Theta} (\mu_1 + \bar{\mu}_1)^2\nu(\mathrm{d}\theta) \right) \right] \mathbb{E}[x]^2\mathrm{d}t$$

$$+ \mathbb{E} \int_0^T \left[\dot{\gamma}_i + b_0 \beta_i + (b_1 + \bar{b}_1)\gamma_i + \sum_{s' \in \mathcal{S}} (\gamma_i(\cdot, s') - \gamma_i(\cdot, s)) \right.$$

$$\left. + \alpha_i \left(\sigma_0(\sigma_1 + \bar{\sigma}_1) + \int_{\theta \in \Theta} \mu_0(\mu_1 + \bar{\mu}_1)\nu(d\theta) \right) \right] \mathbb{E}[x] dt$$

$$+ \mathbb{E} \int_0^T \left[\dot{\delta}_i + b_0 \gamma_i + \sum_{s' \in \mathcal{S}} (\delta_i(\cdot, s') - \delta_i(\cdot, s)) + \frac{1}{2}\alpha_i \left(\sigma_0^2 + \int_{\theta \in \Theta} \mu_0^2 \nu(d\theta) \right) \right] dt$$

$$+ \frac{1}{2}\mathbb{E} \int_0^T \left[\alpha_i \left((\sigma_{i2} + \bar{\sigma}_{i2})^2 + \int_{\theta \in \Theta} (\mu_{i2} + \bar{\mu}_{i2})^2 \nu(d\theta) \right) + r_i + \bar{r}_i \right] \mathbb{E}[u_i]^2 dt$$

$$+ \frac{1}{2}\mathbb{E} \int_0^T 2 \left\{ (b_{2i} + \bar{b}_{2i})\gamma_i + \alpha_i \left(\sigma_0(\sigma_{i2} + \bar{\sigma}_{i2}) + \int_{\theta \in \Theta} \mu_0(\mu_{i2} + \bar{\mu}_{i2})\nu(d\theta) \right) \right.$$

$$+ \left[(b_{2i} + \bar{b}_{2i})\beta_i \right.$$

$$\left. + \alpha_i \left((\sigma_1 + \bar{\sigma}_1)(\sigma_{i2} + \bar{\sigma}_{i2}) + \int_{\theta \in \Theta} (\mu_1 + \bar{\mu}_1)(\mu_{i2} + \bar{\mu}_{i2})\nu(d\theta) \right) \right] \mathbb{E}[x] \left. \right\} \mathbb{E}[u_i] dt$$

$$+ \frac{1}{2}\mathbb{E} \int_0^T \alpha_i \left((\sigma_{j2} + \bar{\sigma}_{j2})^2 + \int_{\theta \in \Theta} (\mu_{j2} + \bar{\mu}_{j2})^2 \nu(d\theta) \right) \mathbb{E}[u_j]^2 dt$$

$$+ \frac{1}{2}\mathbb{E} \int_0^T 2 \left\{ (b_{2j} + \bar{b}_{2j})\gamma_i + \alpha_i \left(\sigma_0(\sigma_{j2} + \bar{\sigma}_{j2}) + \int_{\theta \in \Theta} \mu_0(\mu_{j2} + \bar{\mu}_{j2})\nu(d\theta) \right) \right.$$

$$+ \left[(b_{2j} + \bar{b}_{2j})\beta_i \right.$$

$$\left. + \alpha_i \left((\sigma_1 + \bar{\sigma}_1)(\sigma_{j2} + \bar{\sigma}_{j2}) + \int_{\theta \in \Theta} (\mu_1 + \bar{\mu}_1)(\mu_{j2} + \bar{\mu}_{j2})\nu(d\theta) \right) \right] \mathbb{E}[x]$$

$$+ \left[\alpha_i \left((\sigma_{i2} + \bar{\sigma}_{i2})(\sigma_{j2} + \bar{\sigma}_{j2}) \right. \right.$$

$$\left. \left. + \int_{\theta \in \Theta} (\mu_{i2} + \bar{\mu}_{i2})(\mu_{j2} + \bar{\mu}_{j2})\nu(d\theta) \right) + \bar{\varepsilon}_{ij} \right] \mathbb{E}[u_i] \left. \right\} \mathbb{E}[u_j] dt$$

$$+ \frac{1}{2}\mathbb{E} \int_0^T \left[\alpha_i \left(\sigma_{i2}^2 + \int_{\theta \in \Theta} \mu_{i2}^2 \nu(d\theta) \right) + r_i \right] (u_i - \mathbb{E}[u_i])^2 dt$$

$$+ \frac{1}{2}\mathbb{E} \int_0^T 2 \left\{ \left[b_{2i}\alpha_i \right. \right.$$

$$\left. \left. + \alpha_i \left(\sigma_1 \sigma_{i2} + \int_{\theta \in \Theta} \mu_1 \mu_{i2}\nu(d\theta) \right) \right] (x - \mathbb{E}[x]) \right\} (u_i - \mathbb{E}[u_i]) dt$$

$$+ \frac{1}{2}\mathbb{E} \int_0^T \alpha_i \left(\sigma_{j2}^2 + \int_{\theta \in \Theta} \mu_{j2}^2 \nu(\mathrm{d}\theta) \right) (u_j - \mathbb{E}[u_j])^2 \mathrm{d}t$$

$$+ \frac{1}{2}\mathbb{E} \int_0^T 2 \left\{ \left[b_{2j}\alpha_i + \alpha_i \left(\sigma_1 \sigma_{j2} + \int_{\theta \in \Theta} \mu_1 \mu_{j2} \nu(\mathrm{d}\theta) \right) \right] (x - \mathbb{E}[x]) \right.$$

$$\left. + \left[\alpha_i \left(\sigma_{i2}\sigma_{j2} + \int_{\theta \in \Theta} \mu_{i2}\mu_{j2}\nu(\mathrm{d}\theta) \right) + \varepsilon_{ij} \right] (u_i - \mathbb{E}[u_i]) \right\} (u_j - \mathbb{E}[u_j]) \mathrm{d}t,$$

To find the Berge solution, we take the cost functional of the ith decision-maker and optimize it in the decision variables of the jth decision-maker. Therefore, we perform optimization with the control inputs $(u_j - \mathbb{E}[u_j])$ and $\mathbb{E}[u_j]$. Prior to performing the square completion procedure for these terms, we compact the notation by introducing the following auxiliary variables:

$$c_j = \alpha_i \left((\sigma_{j2} + \bar{\sigma}_{j2})^2 + \int_{\theta \in \Theta} (\mu_{j2} + \bar{\mu}_{j2})^2 \nu(\mathrm{d}\theta) \right),$$

$$\lambda_j = (b_{2j} + \bar{b}_{2j})\gamma_i + \alpha_i \left(\sigma_0 (\sigma_{j2} + \bar{\sigma}_{j2}) + \int_{\theta \in \Theta} \mu_0 (\mu_{j2} + \bar{\mu}_{j2}) \nu(\mathrm{d}\theta) \right),$$

$$\phi_j = (b_{2j} + \bar{b}_{2j})\beta_i$$
$$+ \alpha_i \left((\sigma_1 + \bar{\sigma}_1)(\sigma_{j2} + \bar{\sigma}_{j2}) + \int_{\theta \in \Theta} (\mu_1 + \bar{\mu}_1)(\mu_{j2} + \bar{\mu}_{j2}) \nu(\mathrm{d}\theta) \right),$$

$$\rho_{ji} = \alpha_i \left((\sigma_{i2} + \bar{\sigma}_{i2})(\sigma_{j2} + \bar{\sigma}_{j2}) + \int_{\theta \in \Theta} (\mu_{i2} + \bar{\mu}_{i2})(\mu_{j2} + \bar{\mu}_{j2}) \nu(\mathrm{d}\theta) \right) + \bar{\varepsilon}_{ij},$$

$$\eta_j = \frac{\lambda_j + \phi_j \mathbb{E}[x] + \rho_{ji}\mathbb{E}[u_i]}{c_j},$$

Taking the terms depending on $\mathbb{E}[u_j]$ we have

$$\frac{1}{2}\mathbb{E} \int_0^T c_j \mathbb{E}[u_j]^2 + 2 \left\{ \lambda_j + \phi_j \mathbb{E}[x] + \rho_{ji}\mathbb{E}[u_i] \right\} \mathbb{E}[u_j] \mathrm{d}t$$

$$= \frac{1}{2}\mathbb{E} \int_0^T c_j \left(\mathbb{E}[u_j]^2 + 2\, \eta_j \mathbb{E}[u_j] \right) \mathrm{d}t,$$

and the term $\mathbb{E}[u_j]^2 + 2\, \eta_j \mathbb{E}[u_j]$ is easily expressed as a quadratic form using $(\mathbb{E}[u_j] + \eta_j)^2$ and the reminder η_j^2. Thus, it is concluded that $\mathbb{E}[u_j] = -\eta_j$. The auxiliary variable η_j was given in terms of the expected control input $\mathbb{E}[u_i]$. Replacing the optimal control input, one can solve the values for η as follows:

$$\eta_j + \frac{\rho_{ji}}{c_j}\left(\frac{\lambda_i}{c_i} + \frac{\phi_i}{c_i}\mathbb{E}[x] - \frac{\rho_{ij}}{c_i}\eta_j \right) = \frac{\lambda_j}{c_j} + \frac{\phi_j}{c_j}\mathbb{E}[x],$$

$$\eta_j = \left(1 - \frac{\rho_{ji}}{c_j}\frac{\rho_{ij}}{c_i} \right)^{-1} \left[\frac{\lambda_j}{c_j} - \frac{\rho_{ji}}{c_j}\frac{\lambda_i}{c_i} + \left(\frac{\phi_j}{c_j} - \frac{\rho_{ji}}{c_j}\frac{\phi_i}{c_i} \right) \mathbb{E}[x] \right],$$

$$= \lambda_{1j} + \lambda_{2j}\mathbb{E}[x]$$

and

$$\mathbb{E}[u_j^*] = -\lambda_{1j} - \lambda_{2j}\mathbb{E}[x].$$

Now, let us introduce some new auxiliary variables to expressed the terms depending on the difference $(u_j - \mathbb{E}[u_j])$, *i.e.*,

$$\bar{c}_j = \alpha_i \left(\sigma_{j2}^2 + \int_{\theta \in \Theta} \mu_{j2}^2 \nu(\mathrm{d}\theta) \right),$$

$$\bar{\phi}_j = \left[b_{2j}\alpha_i + \alpha_i \left(\sigma_1 \sigma_{j2} + \int_{\theta \in \Theta} \mu_1 \mu_{j2} \nu(\mathrm{d}\theta) \right) \right],$$

$$\bar{\rho}_{ji} = \left[\alpha_i \left(\sigma_{i2} \sigma_{j2} + \int_{\theta \in \Theta} \mu_{i2} \mu_{j2} \nu(\mathrm{d}\theta) \right) + \varepsilon_{ij} \right],$$

$$\bar{\eta}_j = \frac{\bar{\phi}_j(x - \mathbb{E}[x]) + \bar{\rho}_{ji}(u_i - \mathbb{E}[u_i])}{\bar{c}_j}.$$

Taking the terms depending on $u_j - \mathbb{E}[u_j]$ *we have*

$$\frac{1}{2}\mathbb{E}\int_0^T \bar{c}_j(u_j - \mathbb{E}[u_j])^2 + 2\left\{ \bar{\phi}_j\left(x - \mathbb{E}[x]\right) + \bar{\rho}_{ji}(u_i - \mathbb{E}[u_i]) \right\}(u_j - \mathbb{E}[u_j])\mathrm{d}t$$

$$= \frac{1}{2}\mathbb{E}\int_0^T \bar{c}_j\left((u_j - \mathbb{E}[u_j])^2 + 2 \ (u_j - \mathbb{E}[u_j]) \right)\mathrm{d}t.$$

Then, the term $(u_j - \mathbb{E}[u_j])^2 + 2\ \eta_j(u_j - \mathbb{E}[u_j])$ *can be easily completed with a square* $(u_j - \mathbb{E}[u_j]2 + \eta_j)^2$ *and the reminder* η_j^2. *The auxiliary variable* η_j *was expressed in terms of the subtraction* $(u_i - \mathbb{E}[u_i])$. *Then, replacing the optimal control input yields*

$$\bar{\eta}_j + \frac{\bar{\rho}_{ji}}{\bar{c}_j}\left(\frac{\bar{\phi}_i}{\bar{c}_i}(x - \mathbb{E}[x]) - \frac{\bar{\rho}_{ij}}{\bar{c}_i}\bar{\eta}_j \right) = \frac{\bar{\phi}_j}{\bar{c}_j}(x - \mathbb{E}[x]),$$

$$\bar{\eta}_j = \left(1 - \frac{\bar{\rho}_{ji}}{\bar{c}_j}\frac{\bar{\rho}_{ij}}{\bar{c}_i} \right)^{-1}\left(\frac{\bar{\phi}_j}{\bar{c}_j} - \frac{\bar{\rho}_{ji}}{\bar{c}_j}\frac{\bar{\phi}_i}{\bar{c}_i} \right)(x - \mathbb{E}[x]),$$

$$= \lambda_{3j}(x - \mathbb{E}[x])$$

and

$$u_j^* - \mathbb{E}[u_j^*] = -\lambda_{3j}(x - \mathbb{E}[x]).$$

Once the square completion has been performed for both $\mathbb{E}[u_j]$ *and* $(u_j^* - \mathbb{E}[u_j^*])$, *we replace them into the gap* $\mathbb{E}[L_i(x, s, u_i, u_j)|\mathcal{F}_t^s] - \mathbb{E}[f_i(0, s_0, x_0)]$ *obtaining the following:*

$$\mathbb{E}[L_i(x, s, u_i, u_j)|\mathcal{F}_t^s] - \mathbb{E}[f_i(0, s_0, x_0)] = \frac{1}{2}\mathbb{E}[(q_i - \alpha_i(T))(x(T) - \mathbb{E}[x(T)])^2]$$

$$+ \frac{1}{2}\mathbb{E}[(q_i + \bar{q}_i - \beta_i(T))\mathbb{E}[x(T)]^2] - \mathbb{E}[\gamma_i(T)\mathbb{E}[x(T)]] - \mathbb{E}[\delta_i(T)]$$

$$+ \frac{1}{2}\mathbb{E}\int_0^T \left\{ \dot{\alpha}_i + q_i + 2b_1\alpha_i + \sum_{s' \in \mathcal{S}}(\alpha_i(\cdot, s') - \alpha_i(\cdot, s)) \right.$$

$$+ \alpha_i\left(\sigma_1^2 + \int_{\theta \in \Theta}\mu_1^2\nu(\mathrm{d}\theta)\right) - \frac{\bar{\phi}_j^2}{\bar{c}_j} - \frac{\bar{\rho}_{ji}^2}{\bar{c}_j}\lambda_{3i}^2 - 2\frac{\bar{\phi}_j\bar{\rho}_{ji}}{\bar{c}_j}\lambda_{3i}$$

$$- 2\lambda_{3i}\left[b_{2i}\alpha_i + \alpha_i\left(\sigma_1\sigma_{i2} + \int_{\theta \in \Theta}\mu_1\mu_{i2}\nu(\mathrm{d}\theta)\right)\right]$$

$$\left. + \lambda_{3i}^2\left[\alpha_i\left(\sigma_{i2}^2 + \int_{\theta \in \Theta}\mu_{i2}^2\nu(\mathrm{d}\theta)\right) + r_i\right] \right\}(x - \mathbb{E}[x])^2\,\mathrm{d}t$$

$$+ \frac{1}{2}\mathbb{E}\int_0^T \left\{ \dot{\beta}_i + q_i + \bar{q}_i + 2(b_1 + \bar{b}_1)\beta_i + \sum_{s' \in \mathcal{S}}(\beta_i(\cdot, s') - \beta_i(\cdot, s)) \right.$$

$$+ \alpha_i\left((\sigma_1 + \bar{\sigma}_1)^2 + \int_{\theta \in \Theta}(\mu_1 + \bar{\mu}_1)^2\nu(\mathrm{d}\theta)\right)$$

$$+ \left[\alpha_i\left((\sigma_{i2} + \bar{\sigma}_{i2})^2 + \int_{\theta \in \Theta}(\mu_{i2} + \bar{\mu}_{i2})^2\nu(\mathrm{d}\theta)\right) + r_i + \bar{r}_i\right]\lambda_{2i}^2$$

$$- 2\left[(b_{2i} + \bar{b}_{2i})\beta_i\right.$$

$$\left. + \alpha_i\left((\sigma_1 + \bar{\sigma}_1)(\sigma_{i2} + \bar{\sigma}_{i2}) + \int_{\theta \in \Theta}(\mu_1 + \bar{\mu}_1)(\mu_{i2} + \bar{\mu}_{i2})\nu(\mathrm{d}\theta)\right)\right]\lambda_{2i}$$

$$\left. - \frac{\rho_{ji}^2}{c_j}\lambda_{2i}^2 - \frac{\phi_j^2}{c_j} + 2\frac{\phi_j\rho_{ji}}{c_j}\lambda_{2i} \right\}\mathbb{E}[x]^2\mathrm{d}t$$

$$+ \mathbb{E}\int_0^T \left\{ \dot{\gamma}_i + b_0\beta_i + (b_1 + \bar{b}_1)\gamma_i + \sum_{s' \in \mathcal{S}}(\gamma_i(\cdot, s') - \gamma_i(\cdot, s)) \right.$$

$$+ \alpha_i\left(\sigma_0(\sigma_1 + \bar{\sigma}_1) + \int_{\theta \in \Theta}\mu_0(\mu_1 + \bar{\mu}_1)\nu(\mathrm{d}\theta)\right)$$

$$+ \left[\alpha_i\left((\sigma_{i2} + \bar{\sigma}_{i2})^2 + \int_{\theta \in \Theta}(\mu_{i2} + \bar{\mu}_{i2})^2\nu(\mathrm{d}\theta)\right) + r_i + \bar{r}_i\right]\lambda_{1i}\lambda_{2i}$$

$$- \left[(b_{2i} + \bar{b}_{2i})\beta_i\right.$$

$$\left. + \alpha_i\left((\sigma_1 + \bar{\sigma}_1)(\sigma_{i2} + \bar{\sigma}_{i2}) + \int_{\theta \in \Theta}(\mu_1 + \bar{\mu}_1)(\mu_{i2} + \bar{\mu}_{i2})\nu(\mathrm{d}\theta)\right)\right]\lambda_{1i}$$

$$- \alpha_i\left(\sigma_0(\sigma_{i2} + \bar{\sigma}_{i2}) + \int_{\theta \in \Theta}\mu_0(\mu_{i2} + \bar{\mu}_{i2})\nu(\mathrm{d}\theta)\right)\lambda_{2i}$$

$$- (b_{2i} + \bar{b}_{2i})\gamma_i\lambda_{2i} - \frac{1}{2}\frac{\rho_{ji}^2}{c_j}2\lambda_{1i}\lambda_{2i} - \frac{\lambda_j\phi_j}{c_j} + \frac{\lambda_j\rho_{ji}}{c_j}\lambda_{2i} + \frac{\phi_j\rho_{ji}}{c_j}\lambda_{1i} \Bigg\} \mathbb{E}[x]\mathrm{d}t$$

$$+ \mathbb{E}\int_0^T \Bigg\{ \dot{\delta}_i + b_0\gamma_i + \sum_{s'\in\mathcal{S}}(\delta_i(\cdot, s') - \delta_i(\cdot, s)) + \frac{1}{2}\alpha_i\left(\sigma_0^2 + \int_{\theta\in\Theta}\mu_0^2\nu(\mathrm{d}\theta)\right)$$

$$+ \frac{1}{2}\left[\alpha_i\left((\sigma_{i2} + \bar{\sigma}_{i2})^2 + \int_{\theta\in\Theta}(\mu_{i2} + \bar{\mu}_{i2})^2\nu(\mathrm{d}\theta)\right) + r_i + \bar{r}_i\right]\lambda_{1i}^2$$

$$- (b_{2i} + \bar{b}_{2i})\gamma_i\lambda_{1i} - \alpha_i\left(\sigma_0(\sigma_{i2} + \bar{\sigma}_{i2}) + \int_{\theta\in\Theta}\mu_0(\mu_{i2} + \bar{\mu}_{i2})\nu(\mathrm{d}\theta)\right)\lambda_{1i}$$

$$- \frac{1}{2}\frac{\lambda_j^2}{c_j} - \frac{1}{2}\frac{\rho_{ji}^2}{c_j}\lambda_{1i}^2 + \frac{\lambda_j\rho_{ji}}{c_j}\lambda_{1i} \Bigg\}\mathrm{d}t$$

$$+ \frac{1}{2}\mathbb{E}\int_0^T c_j\left(\mathbb{E}[u_j] + \eta_j\right)^2\mathrm{d}t + \frac{1}{2}\mathbb{E}\int_0^T \bar{c}_j\left(u_j - \mathbb{E}[u_j] + \bar{\eta}_j\right)^2\mathrm{d}t.$$

This long equation is divided into the following groups in order to identify the differential equations for α, β, γ, and δ, together with the boundary conditions and the optimal control inputs:

- *Those terms at the terminal time T*

- *Those terms whose common factors are: $(x - \mathbb{E}[x])^2$, $\mathbb{E}[x]^2$, and $\mathbb{E}[x]$*

- *Those terms independent from the state, mean-field state and control inputs*

- *And the quadratic terms defining the optimal control inputs.*

Thus, the announced terms are obtained by minimizing terms in the latter expression completing the proof.

In Chapter 7, we discussed about a case in which the solution for a two-level hierarchical game problem can coincide with the solution for a non-cooperative game problem in the context of mean-field-type games. Next, we will discuss about the possible relationship between the co-opetitive games studied in 5 and the Berge solution analyzed in this chapter.

8.4 When Berge Solution Corresponds to Co-opetitive Solution for Mean-Field-Type Games

The Berge solution can be interpreted as mutual support. That is why this concept is only applicable to the two-decision-maker or two-team case. Hence, the mutual support can be interpreted as altruism and such behavior can be captured by means of the co-opetition concept as presented in the following Remark:

TABLE 8.1
Simulation parameters

Parameter	Value
T	3
$x(0), \mathbb{E}[x(0)]$	100
r_1, r_2	1
$q_1, q_2, \bar{q}_1, \bar{q}_2$	10
b_0	0.5
b_1	-1
σ_0	2
σ_1	0.1
$\bar{\sigma}_1$	0.1
σ_{12}	2
σ_{22}	10
$\bar{\sigma}_{12}$	2
$\bar{\sigma}_{22}$	10
μ_{12}	1
μ_{22}	2
$\epsilon_{11}, \bar{\epsilon}_{11}, \epsilon_{21}, \bar{\epsilon}_{21}, \epsilon_{12}, \epsilon_{22}, \epsilon_{13}, \epsilon_{23}$	0
$\bar{b}_1, b_{12}, b_{22}, \bar{b}_{12}, \bar{b}_{22}, \mu_0, \mu_1, \bar{\mu}_1, \bar{\mu}_{12}, \bar{\mu}_{22}$	0

Remark 24 *Let us consider a two-decision-maker co-opetitive mean-field-type game as in Chapter 5 with the following co-opetitive parameters:*

$$\lambda = \begin{bmatrix} 0 & 1 \\ 1 & 0 \end{bmatrix}.$$

Then, the co-opetitive mean-field-type equilibrium coincides with the Berge solution.

8.5 Numerical Example

Consider two decision-makers, and a state-dependent drift, state-and-control-dependent diffusion system with control-dependent jumps, and without switching regimes, i.e.,

$$\mathrm{d}x = (b_0 + b_1 x)\mathrm{d}t + \left(\sigma_0 + \sigma_1 x + \bar{\sigma}_1 \bar{x} + \sum_{\ell \in \{i,j\}} \sigma_{\ell 2} u_\ell + \sum_{\ell \in \{i,j\}} \bar{\sigma}_{\ell 2} \bar{u}_\ell \right) \mathrm{d}B$$

$$+ \int_{\Theta} \sum_{\ell \in \{i,j\}} \mu_{\ell 2} u_\ell \tilde{N}(\mathrm{d}t, \mathrm{d}\theta),$$

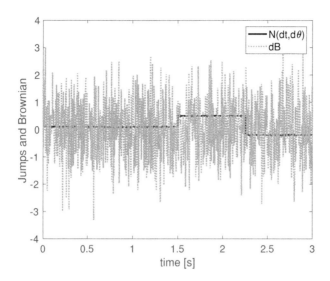

FIGURE 8.2
Considered jumps and Brownian for the numerical example.

$$x(0) = x_0,$$

and the cost function given by

$$L_i(x, u_i, u_j) = \frac{1}{2} q_i x^2(T) + \frac{1}{2} \bar{q}_i \bar{x}^2(T)$$
$$+ \frac{1}{2} \int_0^T \left(q_i x^2 + \bar{q}_i \bar{x}^2 + r_i u_i^2 + \bar{r}_i \bar{u}_i^2 \right) dt,$$

with constant weight parameters q_i, $\bar{q}_i$, r_i, $\bar{r}_i \in \mathbb{R}$. Concretely, consider the parameters that are summarized in Table 8.1. Moreover, Figure 8.2 shows the jumps and Brownian used in the numerical example. On the other hand, Figure 8.3 presents the evolution of the state x and its respective expected value $\mathbb{E}[x]$. Finally, Figures 8.4 and 8.5 show the evolution of the control inputs u_1 and u_2, respectively. It can be observed that due to the fact both term c_ℓ and $\bar{c}_\ell$ depend on the jumps, then the control inputs are also depending on such jumps. Figure 8.6 presents the evolution of α_1 and α_2, which are the differential equations influencing the value for the optimal control inputs since the problem has control-input-independent drift term. It can be seen that the backward trajectories α_1 and α_2 satisfy the terminal boundary conditions.

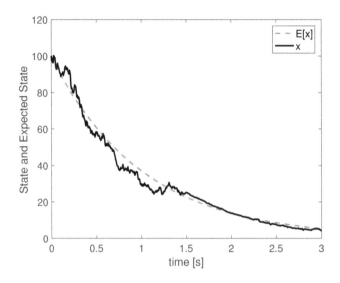

FIGURE 8.3
Evolution of the state $x(t)$ and its expected value $\mathbb{E}[x(t)]$ with initial conditions $x(0) = \mathbb{E}[x(0)] = 100$.

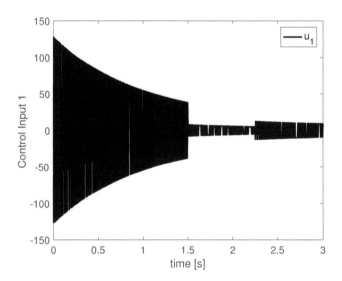

FIGURE 8.4
Evolution of the control input $u_1(t)$.

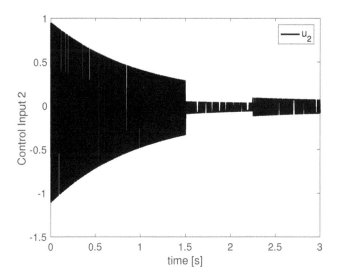

FIGURE 8.5
Evolution of the control input $u_2(t)$.

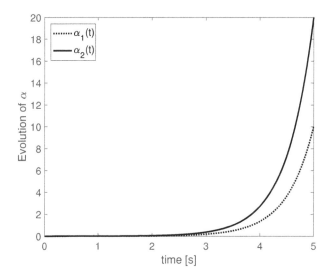

FIGURE 8.6
Evolution of the differential equations $\alpha_1(t)$ and $\alpha_2(t)$.

We have studied the Berge equilibria in mean-field-type games. We have shown that a mean-field-type Berge equilibrium can be semi-explicitly given for linear-quadratic game problems incorporating Jump-Diffusion Process and Regime Switching. Besides, we have considered state, control input, mean state, mean control and switching dependence in all the following terms: drift, diffusion, and jump terms. We have shown that direct method is suitable to solve these types of problems involving the Berge equilibria.

8.6   Exercises

1. Mention an example in nature or any other scenario where we can evidence mutual support.

2. Mention an application from your research field where a Berge configuration can be implemented to improve the performance of the system.

3. What is the difference between the mutual support (Berge solution concept) and the full-cooperation?.

4. After studying Chapters 11 and 12, solve a discrete-time Mutual-Support (Berge) mean-field-type game.

Part IV

Matrix-Valued
Mean-Field-Type Games

9

Matrix-Valued Mean-Field-Type Games

Previous chapters have explored several problems with different game-theoretical solution concepts. We have studied the non-cooperative, coop-erative, co-opetitive, hierarchical, Stackelberg, and Berge type of problems. Besides, we have considered multiple possibilities for the stochastic behavior, i.e., by incorporating Brownian motion, jump process, and switching regimes making the coefficients random. Moreover, we have considered dependency over the diffusion and jump terms on the system state, mean-field state, con-trol inputs, and expected value for the control inputs. Finally, the reader is now familiar with the computation of the semi-explicit solutions using the so-called direct method.

However, even though we have revised a large variety of problem, all of them have been developed for the one-dimensional case, i.e., being x a scalar value. This is clearly a limitation to solve real engineering problems. Nowa-days, the engineering systems are of large-scale nature, implying a large num-ber or large dimension for the system state.

In this chapter, we investigate how the direct method is powerful also to solve matrix-valued problems, i.e., involving a dynamical system whose state and/or control inputs are given by matrices. These results motivate applica-tions, which are associated with graphs. The evolution of graphs can be de-termined by means of matrix-valued dynamics and the graphs can be directly related to networked structures. Here, we introduce new concepts. We present different solution concepts corresponding to non-cooperative, cooperative, and adversarial approaches for both neutral and risk-sensitive scenarios.

The motivation to develop techniques suitable to consider matrix-valued schemes is mainly given by the emergence of network of networks (also known as system of systems) engineering problems. On one hand, the matrix-valued state-space representation becomes a general structure able to capture even scalar and vector states. For instance, the use of diagonal matrices in the matrix-valued state-space representation retrieves the system dynamics corre-sponding to the case where the system states are given by vectors. On the other hand, matrix-valued state-space dynamics is suitable to describe and repre-sent the behavior of coupled networked systems, and/or coupled network of networks. As an example, consider either the multi-exchange currency system under the investment problem or the multi-exchange diversification problem; such problems can be addressed by using matrices instead of vectors in order to mathematically describe the existing relationship among all the currencies.

DOI: 10.1201/9781003098607-9

9.1   Mean-Field-Type Game Set-up

We consider $n \geq 1$ ($n \in \mathbb{N}$) decision-makers over the time horizon $[0, T]$, $T > 0$. Each decision-maker i from the set of decision-makers $\mathcal{N} = \{1, \ldots, n\}$ chooses a matrix-valued strategy U_i over the horizon $[0, T]$. The state satisfies the following matrix-valued linear jump-diffusion-regime switching system of mean-field type:

$$dX = \Big[B_1(X - \mathbb{E}[X]) + (B_1 + \bar{B}_1)\mathbb{E}[X] + \sum_{j \in \mathcal{N}} B_{2j}(U_j - \mathbb{E}[U_j])$$

$$+ \sum_{j \in \mathcal{N}} (B_{2j} + \bar{B}_{2j})\mathbb{E}[U_j] \Big] dt + S_0 dB + \int_{\theta \in \Theta} M_0(\cdot, \theta)\tilde{N}(dt, d\theta), \quad (9.1)$$

where X denotes the system state, $\mathbb{E}[X] = \mathbb{E}[X|\mathcal{F}_t^s]$ the expectation of the system state, and the system parameters are as follows: $B_1(t, s(t))$, $\bar{B}_1(t, s(t))$, $B_{2j}(t, s(t))$, $\bar{B}_{2j}(t, s(t))$, $S_0(t, s(t))$, $M_0(t, s(t)) \in \mathbb{R}^{d \times d}$. Hence, B is a $\mathbb{R}^{d \times d}$ matrix-valued Brownian motion, U_i denotes the control input of the decision-maker $i \in \mathcal{N}$ and $\mathbb{E}[U_i] = \mathbb{E}[U_i|\mathcal{F}_t^s]$ is its expectation. The element s is a regime switching process with transition rates $\tilde{q}_{ss'}$ satisfying $\tilde{q}_{ss'} > 0$,

$$\tilde{q}_{ss} = -\sum_{s' \neq s} \tilde{q}_{ss'}.$$

In addition, $N(t, .)$ is a $\mathbb{R}^{d \times d}$ matrix-valued Poisson random process whose compensated process is

$$\tilde{N}(dt, d\theta) = N(dt, d\theta) - \nu(d\theta)dt,$$

and ν is a matrix of Radon measure over the set of jump sizes Θ. The filtration generated by the regime switching process s is $\mathcal{F}_t^s$.

We can see that the matrix-valued system state dynamics considered here is similar to those considered in Chapters 6, 7, and 8. Apart from the different dimension for the system state, the diffusion and jump coefficients are assumed to be independent from the system state, expectation of the system state, the control inputs, and the expectation of the control inputs.

Besides the state system in (9.1), we associate a matrix-valued cost functional to the decision-maker $i \in \mathcal{N}$ as follows:

$$L_i(X, s, U_1, \ldots, U_n) = \langle Q_i(T, s(T))(X(T) - \mathbb{E}[X(T)]), X(T) - \mathbb{E}[X(T)] \rangle$$

$$+ \langle (Q_i(T, s(T)) + \bar{Q}_i(T, s(T)))\mathbb{E}[X(T)], \mathbb{E}[X(T)] \rangle$$

$$+ \int_0^T \langle Q_i(X - \mathbb{E}[X]), X - \mathbb{E}[X] \rangle + \langle (Q_i + \bar{Q}_i)\mathbb{E}[X], \mathbb{E}[X] \rangle \, dt$$

$$+ \int_0^T \langle R_i(U_i - \mathbb{E}[U_i]), U_i - \mathbb{E}[U_i] \rangle + \langle (R_i + \bar{R}_i)\mathbb{E}[U_i], \mathbb{E}[U_i] \rangle \, dt,$$

$$(9.2)$$

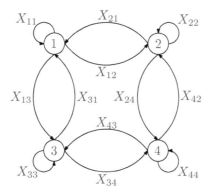

FIGURE 9.1
Example of a matrix-valued application with $d = 4$.

where $\langle A, B \rangle = \text{trace}(A^* B) = \text{trace}(B^* A)$, A^* being the adjoint operator of A (transposition). The weight coefficients of the cost functional $Q_i := Q_i(t, s(t))$, $R_i := R_i(t, s(t))$, $\bar{Q}_i := \bar{Q}_i(t, s(t))$, $\bar{R}_i := \bar{R}_i(t, s(t))$ are possibly time and regime-switching dependent with values in $\mathbb{R}^{d \times d}$.

The reader might wonder about the motivation to study matrix-valued problems and whether it is enough to work with vector system states. It is true that it is possible to map a matrix-value problem into a vector-valued problem. However, depending on the application we are working on, it might be more suitable to pursue a matrix-valued statement. Section 9.1.1 will develop some discussion regarding this question.

9.1.1 Matrix-Valued Applications

The motivation to develop suitable techniques to consider matrix-valued schemes is mainly given by the emergence of network of networks in the engineering problems. The matrix-valued state-space representation becomes a general structure able to represent both scalar and vector states. For instance, the use of diagonal matrices in the matrix-valued state-space representation retrieves the system dynamics corresponding to the case where the states are given by vectors. Matrix-valued state-space dynamics is suitable to describe and represent the behavior of coupled network systems, and/or coupled network of networks. As an example, consider the network in Figure 9.1 where $X \in \mathbb{R}^{4 \times 4}$, and with

$$X_{14} = X_{41} = X_{23} = X_{32} = 0.$$

Thus, $dX(t)$ describes the evolution of the matrix process. One important application of matrix-valued processes is in the multi-currency exchange sector as introduced next.

TABLE 9.1
Exchange rates involving six currencies on October 11th, 2018

	Euro	US Dollar	Australian Dollar	Canadian Dollar	Swiss Franc	Japanese Yen
Euro	1	1.1561	1.6242	1.5055	1.1462	1.2974
US Dollar	0.86497	1	1.4049	1.3022	0.9914	1.1222
Australian Dollar	0.61569	0.7118	1	0.9269	0.70568	0.79878
Canadian Dollar	0.66424	0.76793	1.0789	1	0.76133	0.86177
Swiss Franc	0.87248	1.0087	1.4171	1.3135	1	1.1319
Japanese Yen	0.77079	0.89111	1.2519	1.1604	0.88345	1

A direct application of the matrix-valued system states dynamics is the *Multi-Currency Exchange*, consider $d \geq 1$ assets, blockchain-based tokens or traditional currencies whose prices are modeled by the processes

$$(p_k, k \in \{1, \ldots, d\}).$$

The prices are inter-related because of shared market, shared demand, and market opportunities. At a given time, the unit exchange of currency k to

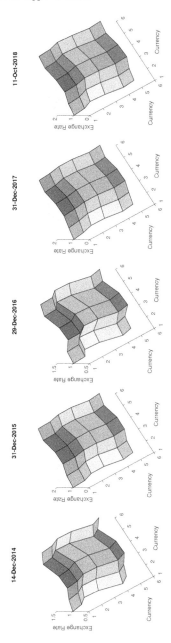

FIGURE 9.2
Exchange matrix for six currencies: Euro, US Dollar, Australian Dollar, Canadian Dollar, Swiss Franc, and Japanese Yen, for five different dates.

currency l with a relative switching cost c_{kl} is

$$X_{kl} = \frac{p_k}{p_l} - c_{kl}.$$

Clearly $X = (X_{kl})_{kl}$ defines a matrix-valued process. Notice that, Figure 9.1 can represent the possible exchange among four currencies [127]. Finally, as a numerical example, we can consider the exchange rates for six currencies as shown in Table 9.1, and Figure 9.2 presents the corresponding evolution of the exchange rates for five different years.

So far in the book, we have been discussing about the risk terms by mainly referring to the variance. Also recall in the introduction we mentioned other type of risk terms such as the skewness or kurtosis. Now, we are going to evolve a bit on the risk concept by introducing two different approaches known as risk-neutral and risk-sensitive mean-field-type games.

9.1.2 Risk-Neutral

We provide basic definitions of the risk-neutral problems and their solution concepts.

Definition 17 (Risk-Neutral Best-Response) *Given the strategies $(U_j, \; j \neq i)$, a risk-neutral best response strategy of the decision-maker $i \in \mathcal{N}$ is a strategy that solves* minimize$_{U_i}$ $\mathbb{E}[L_i(X, s, U_1, \ldots, U_n)]$ *subject to (9.1). The set of risk-neutral best responses of $i \in \mathcal{N}$ is denoted by $rnBR_i(U_{-i})$.*

Definition 18 (Risk-Neutral Nash Equilibrium) *A mean-field-type risk-neutral Nash equilibrium is a strategy profile $(U_j^{\mathrm{rn}}, \; j \in \mathcal{N})$, of all decision-makers such that for every decision-maker $i \in \mathcal{N}$, we have that $U_i^{\mathrm{rn}} \in rnBR_i(U_{-i}^{\mathrm{rn}})$.*

Definition 19 (Risk-Neutral Full-Cooperation) *A mean-field-type risk-neutral fully cooperative solution is a strategy profile $(U_j^{\mathrm{rn,g}}, \; j \in \mathcal{N})$, of all decision-makers such that*

$$\mathbb{E}[L_0(X, s, U_1^{\mathrm{rn,g}}, \ldots, U_n^{\mathrm{rn,g}})] = \underset{(U_1, \ldots, U_n)}{\text{minimize}} \mathbb{E}[L_0(X, s, U_1, \ldots, U_n)],$$

where $L_0(X, s, U_1, \ldots, U_n) := \sum_{j \in \mathcal{N}} L_j(X, s, U_1, \ldots, U_n)$ is the social (global) cost.

The previous definitions are a reminder of the best-response and the solution concepts for the non-cooperative and cooperative game, respectively. Next, we introduce the solution concept in the context of adversarial games. Please refer to the adversarial game problem explained in Chapter 1, Section 1.3.3, Figure 1.10.

Definition 20 (Risk-Neutral Saddle Point Solution) *The set of decision-makers is divided into two teams. A team of defenders and a team of attackers. The defenders set is*

$$I_+ := \{i \in \{1, \ldots, n\} \mid R_i \succ 0, (R_i + \bar{R}_i) \succ 0\}$$

and the attackers set is

$$I_- := \{j \in \mathcal{N} \mid -R_j \succ 0, -(R_j + \bar{R}_j) \succ 0\}.$$

A mean-field-type risk-neutral saddle point is a strategy profile $(U_j^{\mathrm{ad}}, j \in I_+)$, of the team of defenders and $(U_j^{\mathrm{ad}}, j \in I_-)$ of the team of attackers such that

$$L^{\mathrm{ad}}(X, s, (U_i^{\mathrm{ad}})_{i \in I_+}, (U_j)_{j \in I_-}) \leq L^{\mathrm{ad}}(X, s, U^{\mathrm{ad}}),$$
$$\leq L^{\mathrm{ad}}(X, s, (U_i)_{i \in I_+}, (U_j^{\mathrm{ad}})_{j \in I_-})$$

and $L^{\mathrm{ad}}(X, s, U_1^{\mathrm{ad}}, \ldots, U_n^{\mathrm{ad}})$ is the value of the adversarial team (risk-neutral) game problem, where the cost functional is

$$\begin{aligned}
L^{\mathrm{ad}}(X, s, U_1, \ldots, U_n) := & \langle Q(T, s(T))(X(T) - \mathbb{E}[X(T)]), X(T) - \mathbb{E}[X(T)] \rangle \\
& + \langle (Q(T, s(T)) + \bar{Q}(T, s(T)))\mathbb{E}[X(T)], \mathbb{E}[X(T)] \rangle \\
& + \int_0^T \langle Q(X - \mathbb{E}[X]), X - \mathbb{E}[X] \rangle + \langle (Q + \bar{Q})\mathbb{E}[X], \mathbb{E}[X] \rangle \, dt \\
& + \int_0^T \sum_{i \in \mathcal{N}} \langle R_i(U_i - \mathbb{E}[U_i]), U_i - \mathbb{E}[U_i] \rangle \\
& + \sum_{i \in \mathcal{N}} \langle (R_i + \bar{R}_i)\mathbb{E}[U_i], \mathbb{E}[U_i] \rangle \, dt.
\end{aligned}$$

The risk-neutral problems are the same as we have been considered throughout the book. The new concept is the adversarial solution in the latter expression. On the contrary, we introduce next a new class of risk-awareness known as risk-sensitive problems.

9.1.3 Risk-Sensitive

We provide basic definitions of risk-sensitive problems and their solution concepts.

Definition 21 (Risk-Sensitive Best-Response) *Given the strategies $(U_j, j \neq i)$, a risk-sensitive best response strategy of decision-maker $i \in \mathcal{N}$ is a strategy that solves*

$$\underset{U_i}{\text{minimize}} \quad \frac{1}{\lambda_i} \log \left(\mathbb{E}[e^{\lambda_i L_i(X, s, U_1, \ldots, U_n)}] \right)$$

subject to (9.1). The set of risk-sensitive best responses of $i \in \mathcal{N}$ is denoted by $rsBR_i(U_{-i})$.

For $\lambda_i \neq 0$, the risk-sensitive loss functional

$$\frac{1}{\lambda_i} \log \left(\mathbb{E}[e^{\lambda_i L_i(X,s,U_1,\ldots,U_n)}] \right)$$

includes not only the first moment $\mathbb{E}[L_i(X, s, U_1, \ldots, U_n)]$ but also all the higher moments $\mathbb{E}[L_i^k(X, s, U_1, \ldots, U_n)]$, $k \geq 1$.

Definition 22 (Risk-Sensitive Nash Equilibrium)   *A mean-field-type risk-sensitive Nash equilibrium is a strategy profile*

$$(U_j^{rs}, \ j \in \mathcal{N}),$$

of all decision-makers such that for every decision-maker $i \in \mathcal{N}$, and $U_i^{rs} \in rsBR_i(U_{-i}^{rs})$.

Definition 23 (Risk-Sensitive Full-Cooperation) *A mean-field-type risk-sensitive fully cooperative solution is a strategy profile $(U_j^{rs,g}, \ j \in \mathcal{N})$, of all decision-makers such that*

$$\underset{(U_1,\ldots,U_n)}{\text{minimize}} \frac{1}{\lambda} \log \left[\mathbb{E}e^{\lambda L_0(X,s,U_1,\ldots,U_n)} \right] = \frac{1}{\lambda} \log \left[\mathbb{E}e^{\lambda L_0(X,s,U_1^{rs,g},\ldots,U_n^{rs,g})} \right].$$

Below, we introduce the concept of adversarial game for the risk-sensitive class by teams.

Definition 24 (Risk-Sensitive Saddle Point Solution) *The set of decision-makers is divided into two teams. A team of defenders and a team of attackers. The defenders set is*

$$I_+ := \{i \in \mathcal{N} | \ R_i \succ 0, (R_i + \bar{R}_i) \succ 0\}$$

and the attackers set is

$$I_- := \{j \in \mathcal{N} | \ -R_j \succ 0, -(R_j + \bar{R}_j) \succ 0\}.$$

A mean-field-type risk-sensitive saddle point is a strategy profile $(U_j^{ad}, \ j \in I_+)$, of the team of defenders and $(U_j^{ad}, \ j \in I_-)$ of the team of attackers such that

$$\frac{1}{\lambda} \log \left[\mathbb{E}e^{\lambda L^{ad}(X,s,(U_i^{ad})_{i\in I_+},(U_j)_{j\in I_-})} \right] \leq \frac{1}{\lambda} \log \left[\mathbb{E}e^{\lambda L^{ad}(X,s,U_1^{ad},\ldots,U_n^{ad})} \right]$$

$$\leq \frac{1}{\lambda} \log \left[\mathbb{E}e^{\lambda L^{ad}(X,s,(U_i)_{i\in I_+},(U_j^{ad})_{j\in I_-})} \right].$$

In the next sections, we focus on the semi-explicit computation of the risk-neutral and risk-sensitive problems by using the direct method. Here, the reader will see that moving from scalar-valued problems to matrix-valued problems do not affect considerably the procedure. We calculate the Itô's formula for the matrix case and optimize using the square completion using matricial quadratic forms.

9.2 Semi-explicit Solution of the Mean-Field-Type Game Problems: Risk-Neutral Case

We start with the solution for the risk-neutral Nash equilibrium problem in Theorem 4. The structure of this solution together with the differential equations P_i, $\bar{P}_i$, and δ_i, can be compared with respect to the scalar-valued solution that we studied in Chapter 4, and by omitting the jump term M_0.

Theorem 4 *Assume that Q_i, R_i, $(Q_i + \bar{Q}_i)$, $(R_i + \bar{R}_i)$ are symmetric positive definite. Then the matrix-valued mean-field-type (risk-neutral) Nash equilibrium strategy and the (risk-neutral) equilibrium cost are given by:*

$$U_i^{\mathrm{rn}} - \bar{U}_i^{\mathrm{rn}} = -\frac{1}{2} R_i^{-1} B_{2i}^* (P_i^{\mathrm{rn}} + P_i^{\mathrm{rn}})(X - \mathbb{E}[X]),$$

$$\bar{U}_i^{\mathrm{rn}} = -\frac{1}{2}(R_i + \bar{R}_i)^{-1}(B_{2i} + \bar{B}_{2i})^*(\bar{P}_i^{\mathrm{rn}} + \bar{P}_i^{\mathrm{rn}})\mathbb{E}[X],$$

$$L_i^{\mathrm{rn}}(X, s, U_1^{\mathrm{rn}}, \dots, U_n^{\mathrm{rn}}) = \mathbb{E}\langle P_i^{\mathrm{rn}}(0, s(0))(X_0 - \mathbb{E}[X_0]), X_0 - \mathbb{E}[X_0]\rangle$$
$$+ \mathbb{E}\langle \bar{P}_i^{\mathrm{rn}}(0, s(0))\mathbb{E}[X_0], \mathbb{E}[X_0]\rangle$$
$$+ \mathbb{E}[\delta_i^{\mathrm{rn}}(0, s(0))], \ i \in \mathcal{N},$$

where $P_i, \bar{P}_i,$ and δ_i solve the following differential equations:

$$\dot{P}_i = -Q_i - P_i B_1 - B_1^* P_i - \sum_{s' \neq s}(P_i(t, s') - P_i(t, s))\tilde{q}_{ss'}$$

$$+ \frac{1}{4}(P_i^* + P_i)B_{2i}R_i^{-\frac{1}{2}*}R_i^{-\frac{1}{2}}B_{2i}^*(P_i^* + P_i)$$

$$+ \frac{1}{4}\sum_{j \neq i}(P_j^* + P_j)B_{2j}R_j^{-1*}B_{2j}^*(P_i^* + P_i)$$

$$+ \frac{1}{4}\sum_{j \neq i}(P_i^* + P_i)B_{2j}R_j^{-1*}B_{2j}^*(P_j^* + P_j), \tag{9.3a}$$

$$\dot{\bar{P}}_i = -Q_i - \bar{Q}_i - \bar{P}_i(B_1 + \bar{B}_1) - (B_1 + \bar{B}_1)^*\bar{P}_i - \sum_{s' \neq s}(\bar{P}_i(t, s') - \bar{P}_i(t, s))\tilde{q}_{ss'}$$

$$+ \frac{1}{4}(\bar{P}_i^* + \bar{P}_i)(B_{2i} + \bar{B}_{2i})(R_i + \bar{R}_i)^{-\frac{1}{2}*}(R_i + \bar{R}_i)^{-\frac{1}{2}}(B_{2i} + \bar{B}_{2i})^*(\bar{P}_i^* + \bar{P}_i)$$

$$+ \frac{1}{4}\sum_{j \in \mathcal{N}\setminus\{i\}}(\bar{P}_j^* + \bar{P}_j)^*(B_{2j} + \bar{B}_{2j})(R_j + \bar{R}_j)^{-1*}(B_{2j} + \bar{B}_{2j})^*(\bar{P}_i^* + \bar{P}_i)$$

$$+ \frac{1}{4}\sum_{j \in \mathcal{N}\setminus\{i\}}(\bar{P}_i^* + \bar{P}_i)(B_{2j} + \bar{B}_{2j})(R_j + \bar{R}_j)^{-1*}(B_{2j} + \bar{B}_{2j})^*(\bar{P}_j^* + \bar{P}_j),$$
$$\tag{9.3b}$$

$$\dot{\delta}_i = -\frac{1}{2}\langle(P_i^* + P_i)S_0, S_0\rangle - \frac{1}{2}\int_\Theta \langle(P_i^* + P_i)M_0, M_0\nu(d\theta)\rangle$$

$$- \sum_{s' \neq s} (\delta_i(t, s') - \delta_i(t, s)) \tilde{q}_{ss'}, \qquad (9.3c)$$

for all $s \in \mathcal{S}$ and with terminal boundary conditions given by

$$P_i(T, s(T)) = Q_i(T, s(T)),$$
$$\bar{P}_i(T, s(T)) = Q_i(T, s(T)) + \bar{Q}_i(T, s(T)),$$
$$\delta_i(T, s(T)) = 0,$$

whenever these differential equations have a unique solution that does not blow up within $[0, T]$.

In the previous differential equations we have omitted the super-script "rn" to ease the notation. We move on the analysis of matrix-valued problems by finding the semi-explicit solution using the direct method. Here, we require to apply the Itô's formula for an expression involving matrices. Nevertheless, notice that it is the same we have been working with and this should not represent a difficulty. Next, we present the main steps to compute the semi-explicit solution for the risk-neutral mean-field-type game problem.

Proof 20 (Theorem 4) *We have now to propose a new guess functional considering that the system state is given by a matrix (or a vector depending on the case). Nonetheless, we still preserve the same quadratic structure for the cost functional and the same linear form for the system state dynamics. Therefore, we can propose a similar ansatz as the one postulated in Chapter 4 but extended to either the matrix-valued or vector-valued case. Inspired from the form of the cost functional, then let us consider the following ansatz:*

$$F_i(t, X, s) = \langle P_i(X - \mathbb{E}[X]), X - \mathbb{E}[X] \rangle + \langle \bar{P}_i \mathbb{E}[X], \mathbb{E}[X] \rangle + \delta_i.$$

Taking the expectation of Itô's formula for switching regimes and jump terms applied to F_i yields

$$\mathbb{E}[\mathrm{d}F_i(t, X, s)] = \mathbb{E}\Bigg[\langle \dot{P}_i(X - \mathbb{E}[X]), X - \mathbb{E}[X] \rangle + \langle \dot{\bar{P}}_i \mathbb{E}[X], \mathbb{E}[X] \rangle + \dot{\delta}_i$$

$$+ \left\langle (P_i^* + P_i)(X - \mathbb{E}[X]), B_1(X(t) - \mathbb{E}[X(t)]) + \sum_{j \in \mathcal{N}} B_{2j}(U_j - \mathbb{E}[U_j]) \right\rangle$$

$$+ \left\langle (\bar{P}_i^* + \bar{P}_i)\mathbb{E}[X], (B_1 + \bar{B}_1)\mathbb{E}[X(t)] + \sum_{j \in \mathcal{N}} (B_{2j} + \bar{B}_{2j})\mathbb{E}[U_j] \right\rangle$$

$$+ \frac{1}{2} \left\langle (P_i^* + P_i)S_0, S_0 \right\rangle + \frac{1}{2} \int_{\Theta} \left\langle (P_i^* + P_i)M_0, M_0 \nu(\mathrm{d}\theta) \right\rangle \Bigg] \mathrm{d}t$$

$$+ \left\langle \left[\sum_{s' \neq s} (P_i(t, s') - P_i(t, s)) \tilde{q}_{ss'} \right] (X - \mathbb{E}[X]), X - \mathbb{E}[X] \right\rangle \mathrm{d}t$$

$$+ \left\langle \left[\sum_{s' \neq s} (\bar{P}_i(t, s') - \bar{P}_i(t, s)) \tilde{q}_{ss'} \right] \mathbb{E}[X], \mathbb{E}[X] \right\rangle dt$$

$$+ \left[\sum_{s' \neq s} (\delta_i(t, s') - \delta_i(t, s)) \tilde{q}_{ss'} \right] dt.$$

With this computation, we now require to compute the expectation of the difference $\mathbb{E}[L_i(X, s, U_1, \ldots, U_n) - F_i(0, X_0, s_0)]$ *yields*

$$\mathbb{E}[L_i(X, s, U_1, \ldots, U_n) - F_i(0, x_0, s_0)] = \mathbb{E}[0 - \delta_i(T, s(T))]$$

$$+ \mathbb{E} \left\langle (Q_i(T, s(T)) - P_i(T, s(T)))(X(T) - \mathbb{E}[X(T)]), X(T) - \mathbb{E}[X(T)] \right\rangle$$

$$+ \mathbb{E} \left\langle (Q_i(T, s(T)) + \bar{Q}_i(T, s(T)) - \bar{P}_i(T, s(T))) \mathbb{E}[X(T)], \mathbb{E}[X(T)] \right\rangle$$

$$+ \mathbb{E} \int_0^T \left\langle Q_i(X - \mathbb{E}[X]), X - \mathbb{E}[X] \right\rangle dt + \mathbb{E} \int_0^T \left\langle (Q_i + \bar{Q}_i) \mathbb{E}[X], \mathbb{E}[X] \right\rangle dt$$

$$+ \mathbb{E} \int_0^T \left\langle R_i(U_i - \mathbb{E}[U_i]), U_i - \mathbb{E}[U_i] \right\rangle dt + \mathbb{E} \int_0^T \left\langle (R_i + \bar{R}_i) \mathbb{E}[U_i], \mathbb{E}[U_i] \right\rangle dt$$

$$+ \mathbb{E} \int_0^T \left\langle \dot{P}_i(X - \mathbb{E}[X]), X - \mathbb{E}[X] \right\rangle dt + \mathbb{E} \int_0^T \left\langle \dot{\bar{P}}_i \mathbb{E}[X], \mathbb{E}[X] \right\rangle dt$$

$$+ \mathbb{E} \int_0^T \dot{\delta}_i dt$$

$$+ \mathbb{E} \int_0^T \left\langle (P_i^* + P_i)(X - \mathbb{E}[X]), B_1(X - \mathbb{E}[X]) + \sum_{j \in \mathcal{N}} B_{2j}(U_j - \mathbb{E}[U_j]) \right\rangle dt$$

$$+ \mathbb{E} \int_0^T \left\langle (\bar{P}_i^* + \bar{P}_i) \mathbb{E}[X], (B_1 + \bar{B}_1) \mathbb{E}[X] + \sum_{j \in \mathcal{N}} (B_{2j} + \bar{B}_{2j}) \mathbb{E}[U_j] \right\rangle dt$$

$$+ \frac{1}{2} \mathbb{E} \int_0^T \left\langle (P_i^* + P_i) S_0, S_0 \right\rangle dt + \frac{1}{2} \mathbb{E} \int_0^T \int_\Theta \left\langle (P_i^* + P_i) M_0, M_0 \nu(d\theta) \right\rangle dt$$

$$+ \mathbb{E} \int_0^T \left\langle \left[\sum_{s' \neq s} (P_i(t, s') - P_i(t, s)) \tilde{q}_{ss'} \right] (X - \mathbb{E}[X]), X - \mathbb{E}[X] \right\rangle dt$$

$$+ \mathbb{E} \int_0^T \left\langle \left[\sum_{s' \neq s} (\bar{P}_i(t, s') - \bar{P}_i(t, s)) \tilde{q}_{ss'} \right] \mathbb{E}[X], \mathbb{E}[X] \right\rangle dt$$

$$+ \mathbb{E} \int_0^T \left[\sum_{s' \neq s} (\delta_i(t, s') - \delta_i(t, s)) \tilde{q}_{ss'} \right] dt.$$

By grouping terms and separating the control inputs of the decision-maker $i \in$

$\mathcal{N}$ from the control inputs corresponding to other decision-makers $j \in \mathcal{N} \setminus \{i\}$ yields

$$\mathbb{E}[L_i(X, s, U_1, \ldots, U_n) - F_i(0, x_0, s_0)] = \mathbb{E}[0 - \delta_i(T, s(T))]$$

$$+ \mathbb{E}\Big\langle (Q_i(T, s(T)) - P_i(T, s(T)))(X(T) - \mathbb{E}[X(T)]), X(T) - \mathbb{E}[X(T)] \Big\rangle$$

$$+ \mathbb{E}\Big\langle (Q_i(T, s(T)) + \bar{Q}_i(T, s(T)) - \bar{P}_i(T, s(T)))\mathbb{E}[X(T)], \mathbb{E}[X(T)] \Big\rangle$$

$$+ \mathbb{E}\int_0^T \Big\langle \Big[\dot{P}_i + Q_i + B_1^*(P_i^* + P_i)$$

$$+ \sum_{s' \neq s}(P_i(t, s') - P_i(t, s))\tilde{q}_{ss'} \Big](X - \mathbb{E}[X]), X - \mathbb{E}[X] \Big\rangle dt$$

$$+ \mathbb{E}\int_0^T \Big\langle \Big[\dot{\bar{P}}_i + Q_i + \bar{Q}_i + (B_1 + \bar{B}_1)^*(\bar{P}_i^* + \bar{P}_i)$$

$$+ \sum_{s' \neq s}(\bar{P}_i(t, s') - \bar{P}_i(t, s))\tilde{q}_{ss'} \Big]\mathbb{E}[X], \mathbb{E}[X] \Big\rangle dt$$

$$+ \mathbb{E}\int_0^T \dot{\delta}_i dt + \frac{1}{2}\mathbb{E}\int_0^T \langle (P_i^* + P_i)S_0, S_0 \rangle dt$$

$$+ \frac{1}{2}\mathbb{E}\int_0^T \int_{\theta \in \Theta} \langle (P_i^* + P_i)M_0, M_0\nu(d\theta) \rangle dt$$

$$+ \mathbb{E}\int_0^T \Big[\sum_{s' \neq s}(\delta_i(t, s') - \delta_i(t, s))\tilde{q}_{ss'} \Big] dt$$

$$+ \mathbb{E}\int_0^T \langle R_i(U_i - \mathbb{E}[U_i]), U_i - \mathbb{E}[U_i] \rangle dt$$

$$+ \mathbb{E}\int_0^T \langle B_{2i}^*(P_i^* + P_i)(X - \mathbb{E}[X]), U_i - \mathbb{E}[U_i] \rangle dt$$

$$+ \mathbb{E}\int_0^T \langle (R_i + \bar{R}_i)\mathbb{E}[U_i], \mathbb{E}[U_i] \rangle dt$$

$$+ \mathbb{E}\int_0^T \langle (B_{2i} + \bar{B}_{2i})^*(\bar{P}_i^* + \bar{P}_i)\mathbb{E}[X], \mathbb{E}[U_i] \rangle dt$$

$$+ \mathbb{E}\int_0^T \Big\langle (P_i^* + P_i)(X - \mathbb{E}[X]), \sum_{j \in \mathcal{N} \setminus \{i\}} B_{2j}(U_j - \mathbb{E}[U_j]) \Big\rangle dt$$

$$+ \mathbb{E}\int_0^T \Big\langle (\bar{P}_i^* + \bar{P}_i)\mathbb{E}[X], \sum_{j \in \mathcal{N} \setminus \{i\}} (B_{2j} + \bar{B}_{2j})\mathbb{E}[U_j] \Big\rangle dt \tag{9.4}$$

Once the gap between the cost as function of the control inputs $L_i(X, s, U_1, \ldots, U_n)$ and the optimal cost from initial time $t = 0$ uo to terminal time T given by

$F_i(0, x_0, s_0)$, *we optimize over the control inputs. The quadratic terms exhibited here are more involved than those we dealt with in previous chapters. We present the procedure next by using auxiliary variables* K_1, K_2 *and* K_3. *First, we perform the following square completion for the terms involving the control input subtraction* $U_i - \mathbb{E}[U_i]$:

$$\left| K_1[K_2(U_i - \mathbb{E}[U_i]) + K_3(X - \mathbb{E}[X])] \right|^2$$

$$= \Big\langle K_1 K_2(U_i - \mathbb{E}[U_i]) + K_1 K_3(X - \mathbb{E}[X]),$$

$$K_1 K_2(U_i - \mathbb{E}[U_i]) + K_1 K_3(X - \mathbb{E}[X]) \Big\rangle$$

$$= \Big\langle K_1 K_2(U_i - \mathbb{E}[U_i]), K_1 K_2(U_i - \mathbb{E}[U_i]) \Big\rangle$$

$$+ 2\Big\langle K_1 K_3(X - \mathbb{E}[X]), K_1 K_2(U_i - \mathbb{E}[U_i]) \Big\rangle$$

$$+ \Big\langle K_1 K_3(X - \mathbb{E}[X]), K_1 K_3(X - \mathbb{E}[X]) \Big\rangle. \tag{9.5}$$

Then, it is concluded that we should match the following terms to identify the appropriate values for the auxiliary variables K_1, K_2 *and* K_3:

$$\Big\langle K_1 K_2(U_i - \mathbb{E}[U_i]), K_1 K_2(U_i - \mathbb{E}[U_i]) \Big\rangle = \Big\langle R_i(U_i - \mathbb{E}[U_i]), U_i - \mathbb{E}[U_i] \Big\rangle,$$

$$2\Big\langle K_1 K_3(X - \mathbb{E}[X]), K_1 K_2(U_i - \mathbb{E}[U_i]) \Big\rangle = \Big\langle B_{2i}^*(P_i^* + P_i)(X - \mathbb{E}[X]),$$

$$U_i - \mathbb{E}[U_i] \Big\rangle.$$

Thus, we conclude that the appropriate values are given by:

$$K_1 = R_i^{-\frac{1}{2}},$$
$$K_2 = R_i,$$
$$K_3 = \frac{1}{2} B_{2i}^*(P_i^* + P_i),$$

given that we obtain the following according to the quadratic form in (9.5):

$$K_2^* K_1^* K_1 K_2 = R_i,$$
$$K_2^* K_1^* K_1 K_3 = B_{2i}^*(P_i^* + P_i).$$

Using these identified and verified values for K_1, K_2, *and* K_3, *we can write the terms in* (9.4) *that depend on* $(U_i - \mathbb{E}[U_i])$ *as follows:*

$$\mathbb{E} \int_0^T \langle R_i(U_i - \mathbb{E}[U_i]), U_i - \mathbb{E}[U_i] \rangle \, \mathrm{d}t \tag{9.6}$$

$$+ \mathbb{E} \int_0^T \langle B_{2i}^*(P_i^* + P_i)(X - \mathbb{E}[X]), U_i - \mathbb{E}[U_i] \rangle \, dt$$

$$= \mathbb{E} \int_0^T \left| R_i^{\frac{1}{2}} \left[U_i - \mathbb{E}[U_i] + \frac{1}{2} R_i^{-1} B_{2i}^*(P_i^* + P_i)(X - \mathbb{E}[X]) \right] \right|^2 dt$$

$$- \frac{1}{4} \mathbb{E} \int_0^T \langle (P_i^* + P_i) B_{2i} R^{-\frac{1}{2}*} R^{-\frac{1}{2}} B_{2i}^*(P_i^* + P_i)(X - \mathbb{E}[X]), X - \mathbb{E}[X] \rangle \, dt.$$

A similar square completion should be applied to the terms depending on $\mathbb{E}[U_i]$. To this end, we consider the following quadratic expression depending on the auxiliary variables $\tilde{K}_1$, $\tilde{K}_2$, and $\tilde{K}_3$, which should be identified and verified, i.e., from

$$\left| \tilde{K}_1 [\tilde{K}_2 \mathbb{E}[U_i] + \tilde{K}_3 \mathbb{E}[X]] \right|^2$$

yields

$$\left\langle \tilde{K}_1 \tilde{K}_2 (U_i - \mathbb{E}[U_i]), \tilde{K}_1 \tilde{K}_2 (U_i - \mathbb{E}[U_i]) \right\rangle = \left\langle (R_i + \bar{R}_i) \mathbb{E}[U_i], \mathbb{E}[U_i] \right\rangle,$$

$$2 \left\langle \tilde{K}_1 \tilde{K}_3 (X - \mathbb{E}[X]), \tilde{K}_1 \tilde{K}_2 (U_i - \mathbb{E}[U_i]) \right\rangle$$

$$= \left\langle (B_{2i} + \bar{B}_{2i})^* (\bar{P}_i^* + \bar{P}_i) \mathbb{E}[X], \mathbb{E}[U_i] \right\rangle.$$

Thus, we identify that the appropriate values for $\tilde{K}_1$, $\tilde{K}_2$, and $\tilde{K}_3$ are

$$\tilde{K}_1 = (R_i + \bar{R}_i)^{-\frac{1}{2}},$$

$$\tilde{K}_2 = R_i + \bar{R}_i,$$

$$\tilde{K}_3 = \frac{1}{2} (B_{2i} + \bar{B}_{2i})^* (\bar{P}_i^* + \bar{P}_i).$$

From (9.4), one can re-write those terms depending on $\mathbb{E}[U_i]$, i.e.,

$$\mathbb{E} \int_0^T \left\langle (R_i + \bar{R}_i) \mathbb{E}[U_i], \mathbb{E}[U_i] \right\rangle + \langle (B_{2i} + \bar{B}_{2i})^* (\bar{P}_i^* + \bar{P}_i) \mathbb{E}[X], \mathbb{E}[U_i] \rangle \, dt \quad (9.7)$$

$$= \mathbb{E} \int_0^T \left| (R_i + \bar{R}_i)^{\frac{1}{2}} [\mathbb{E}[U_i] + \frac{1}{2} (R_i + \bar{R}_i)^{-1} (B_{2i} + \bar{B}_{2i})^* (\bar{P}_i^* + \bar{P}_i) \mathbb{E}[X]] \right|^2 dt$$

$$- \frac{1}{4} \mathbb{E} \int_0^T \left\langle (\bar{P}_i^* + \bar{P}_i)(B_{2i} + \bar{B}_{2i})(R_i + \bar{R}_i)^{-\frac{1}{2}*} \right.$$

$$\cdot (R_i + \bar{R}_i)^{-\frac{1}{2}} (B_{2i} + \bar{B}_{2i})^* (\bar{P}_i^* + \bar{P}_i) \mathbb{E}[X], \mathbb{E}[X] \right\rangle dt.$$

The square completion procedure has revealed already the optimal control inputs. This result can be directly observed in the quadratic terms $|K_1[K_2(U_i - \mathbb{E}[U_i]) + K_3(X - \mathbb{E}[X])]|^2$ and $\left| \tilde{K}_1 [\tilde{K}_2 \mathbb{E}[U_i] + \tilde{K}_3 \mathbb{E}[X]] \right|^2$ making them null. Thus, one obtains that $(U_i - \mathbb{E}[U_i]) = -K_2^{-1} K_3 (X - \mathbb{E}[X])$ and

$\mathbb{E}[U_i] = -\tilde{K}_2^{-1}\tilde{K}_3\mathbb{E}[X]$. *From* (9.4), *and replacing the square completions in* (9.6)-(9.7), *we obtain that*

$$\mathbb{E}[(L_i(X, s, U_1, \ldots, U_n) - F_i(0))] = \mathbb{E}[0 - \delta_i(T, s(T))]$$

$$+ \mathbb{E}\Big\langle (Q_i(T, s(T)) - P_i(T, s(T)))(X(T) - \mathbb{E}[X(T)]), X(T) - \mathbb{E}[X(T)]\Big\rangle$$

$$+ \mathbb{E}\langle ((Q_i(T, s(T)) + \bar{Q}_i(T, s(T)) - \bar{P}_i(T, s(T)))\mathbb{E}[X(T)], \mathbb{E}[X(T)]\rangle$$

$$+ \mathbb{E}\int_0^T \Big\langle \Big[\dot{P}_i + Q_i + B_1^*(P_i^* + P_i)$$

$$+ \sum_{s' \neq s}(P_i(t, s') - P_i(t, s))\tilde{q}_{ss'} \Big](X - \mathbb{E}[X]), X - \mathbb{E}[X] \Big\rangle dt$$

$$+ \mathbb{E}\int_0^T \Big\langle \Big[\dot{\bar{P}}_i + Q_i + \bar{Q}_i + (B_1 + \bar{B}_1)^*(\bar{P}_i^* + \bar{P}_i)$$

$$+ \sum_{s' \neq s}(\bar{P}_i(t, s') - \bar{P}_i(t, s))\tilde{q}_{ss'} \Big]\mathbb{E}[X], \mathbb{E}[X] \Big\rangle dt$$

$$+ \mathbb{E}\int_0^T \dot{\delta}_i dt + \frac{1}{2}\mathbb{E}\int_0^T \langle (P_i^* + P_i)S_0, S_0\rangle dt$$

$$+ \frac{1}{2}\mathbb{E}\int_0^T \int_\Theta \langle (P_i^* + P_i)M_0, M_0\nu(d\theta)\rangle dt$$

$$+ \mathbb{E}\int_0^T \Big[\sum_{s' \neq s}(\delta_i(t, s') - \delta_i(t, s))\tilde{q}_{ss'} \Big] dt$$

$$+ \mathbb{E}\int_0^T \Big| R_i^{\frac{1}{2}}\Big[U_i - \mathbb{E}[U_i] + \frac{1}{2}R_i^{-1}B_{2i}^*(P_i^* + P_i)(X - \mathbb{E}[X]) \Big] \Big|^2 dt$$

$$- \frac{1}{4}\mathbb{E}\int_0^T \Big\langle (P_i^* + P_i)B_{2i}R^{-\frac{1}{2}*}R^{-\frac{1}{2}}B_{2i}^*(P_i^* + P_i) \cdot$$

$$\cdot (X - \mathbb{E}[X]), X - \mathbb{E}[X] \Big\rangle dt$$

$$+ \mathbb{E}\int_0^T \Big| (R_i + \bar{R}_i)^{\frac{1}{2}} \cdot$$

$$\cdot \Big[\mathbb{E}[U_i] + \frac{1}{2}(R_i + \bar{R}_i)^{-1}(B_{2i} + \bar{B}_{2i})^*(\bar{P}_i^* + \bar{P}_i)\mathbb{E}[X] \Big] \Big|^2 dt$$

$$- \frac{1}{4}\mathbb{E}\int_0^T \Big\langle (\bar{P}_i^* + \bar{P}_i)(B_{2i} + \bar{B}_{2i})(R_i + \bar{R}_i)^{-\frac{1}{2}*} \cdot$$

$$\cdot (R_i + \bar{R}_i)^{-\frac{1}{2}}(B_{2i} + \bar{B}_{2i})^*(\bar{P}_i^* + \bar{P}_i)\mathbb{E}[X], \mathbb{E}[X] \Big\rangle dt$$

$$+ \mathbb{E} \int_0^T \frac{1}{2} \Big\langle (P_i^* + P_i)(X - \mathbb{E}[X]),$$

$$\sum_{j \in \mathcal{N} \backslash \{i\}} B_{2j} R_j^{-1} B_{2j}^* (P_j^* + P_j)(X - \mathbb{E}[X]) \Big\rangle dt$$

$$+ \mathbb{E} \int_0^T \frac{1}{2} \Big\langle (\bar{P}_i^* + \bar{P}_i)\mathbb{E}[X],$$

$$\sum_{j \in \mathcal{N} \backslash \{i\}} (B_{2j} + \bar{B}_{2j})(R_j + \bar{R}_j)^{-1}(B_{2j} + \bar{B}_{2j})^* (\bar{P}_j^* + \bar{P}_j)\mathbb{E}[X] \Big\rangle dt.$$

By grouping terms one arrives at the following:

$$\mathbb{E}[L_i(X, s, U_1, \ldots, U_n) - F_i(0, x_0, s_0)] = \mathbb{E}[0 - \delta_i(T, s(T))]$$

$$+ \mathbb{E}\Big\langle (Q_i(T, s(T)) - P_i(T, s(T)))(X(T) - \mathbb{E}[X(T)]), X(T) - \mathbb{E}[X(T)] \Big\rangle$$

$$+ \mathbb{E}\Big\langle (Q_i(T, s(T)) + \bar{Q}_i(T, s(T)) - \bar{P}_i(T, s(T)))\mathbb{E}[X(T)], \mathbb{E}[X(T)] \Big\rangle$$

$$+ \mathbb{E} \int_0^T \Big\langle \Big[\dot{P}_i + Q_i + B_1^*(P_i^* + P_i)$$

$$+ \sum_{s' \neq s} (P_i(t, s') - P_i(t, s))\tilde{q}_{ss'}$$

$$- \frac{1}{4}(P_i^* + P_i)B_{2i}R_i^{-\frac{1}{2}*}R_i^{-\frac{1}{2}}B_{2i}^*(P_i^* + P_i)$$

$$+ \frac{1}{2} \sum_{j \in \mathcal{N} \backslash \{i\}} (P_j^* + P_j)B_{2j}R_j^{-1*}B_{2j}^*(P_i^* + P_i) \Big](X - \mathbb{E}[X]), X - \mathbb{E}[X] \Big\rangle dt$$

$$+ \mathbb{E} \int_0^T \Big\langle \Big[\dot{\bar{P}}_i + Q_i + \bar{Q}_i + (B_1 + \bar{B}_1)^*(\bar{P}_i^* + \bar{P}_i)$$

$$+ \sum_{s' \neq s} (\bar{P}_i(t, s') - \bar{P}_i(t, s))\tilde{q}_{ss'}$$

$$- \frac{1}{4}(\bar{P}_i^* + \bar{P}_i)(B_{2i} + \bar{B}_{2i})(R_i + \bar{R}_i)^{-\frac{1}{2}*}(R_i + \bar{R}_i)^{-\frac{1}{2}}(B_{2i} + \bar{B}_{2i})^*(\bar{P}_i^* + \bar{P}_i)$$

$$+ \frac{1}{2} \sum_{j \in \mathcal{N} \backslash \{i\}} (\bar{P}_j^* + \bar{P}_j)^*(B_{2j} + \bar{B}_{2j})(R_j + \bar{R}_j)^{-1*}(B_{2j} + \bar{B}_{2j})^*(\bar{P}_i^* + \bar{P}_i) \Big].$$

$$\cdot \mathbb{E}[X], \mathbb{E}[X] \Big\rangle dt$$

$$+ \mathbb{E} \int_0^T \Big[\dot{\delta}_i + \frac{1}{2}\langle (P_i^* + P_i)S_0, S_0 \rangle + \frac{1}{2} \int_\Theta \langle (P_i^* + P_i)M_0, M_0 \nu(d\theta) \rangle$$

$$+ \sum_{s' \neq s} (\delta_i(t, s') - \delta_i(t, s)) \tilde{q}_{ss'} \Bigg] dt$$

$$+ \mathbb{E} \int_0^T \left| R_i^{\frac{1}{2}} \left[U_i - \mathbb{E}[U_i] + \frac{1}{2} R_i^{-1} B_{2i}^* (P_i^* + P_i)(X - \mathbb{E}[X]) \right] \right|^2 dt$$

$$+ \mathbb{E} \int_0^T \Bigg| (R_i + \bar{R}_i)^{\frac{1}{2}} \cdot$$

$$\cdot \left[\mathbb{E}[U_i] + \frac{1}{2} (R_i + \bar{R}_i)^{-1} (B_{2i} + \bar{B}_{2i})^* (\bar{P}_i^* + \bar{P}_i) \mathbb{E}[X] \right] \Bigg|^2 dt$$

The latter matrix-valued long expression for the gap

$$\mathbb{E}[L_i(X, s, U_1, \ldots, U_n) - F_i(0, x_0, s_0)]$$

is divided into the following groups:

- *Those terms evaluated at the terminal time T, which define the boundary conditions for the matrix-valued differential equations P_i, $\bar{P}_i$, and δ_i*

- *Those terms involved in a quadratic form of $(X - \mathbb{E}[X])$ and $\mathbb{E}[X]$, defining the differential equations for P_i and $\bar{P}_i$, respectively*

- *Those terms whose common factor is given by $\mathbb{E}[X]$ defining the differential equation δ_i*

- *And those terms depending on the control inputs $(U_i - \mathbb{E}[U_i])$ and $\mathbb{E}[U_i]$ defining the optimal control inputs.*

Finally, we make the process identification and minimize terms obtaining the announced result.

Under the symmetric matrix assumption above, it is easy to check that if P is a solution then P^* is also a solution. Therefore $P_i^*(t, s) = P_i(t, s)$, $(t, s) \in [0, T] \times \mathcal{S}$. From the state system (9.1), the conditional expected matrix $\mathbb{E}[X(t)] := \mathbb{E}[X(t) | \mathcal{F}_t^s]$, where $\mathcal{F}^s$ is the natural filtration of the regime switching process s up to t, solves the following system:

$$d\mathbb{E}[X] = \left[(B_1 + \bar{B}_1)\mathbb{E}[X] + \sum_{j \in \mathcal{N}} (B_{2j} + \bar{B}_{2j})\mathbb{E}[U_j] \right] dt,$$

$$\mathbb{E}[X](0) = \mathbb{E}[X_0].$$

which means that

$$\mathbb{E}[\dot{X}] = \left[(B_1 + \bar{B}_1) \right.$$

$$-\frac{1}{2}\sum_{j\in\mathcal{N}}(B_{2j}+\bar{B}_{2j})(R_j+\bar{R}_j)^{-1}(B_{2j}+\bar{B}_{2j})^*(\bar{P}_j^*+\bar{P}_j)\Bigg]\mathbb{E}[X],$$

$$\mathbb{E}[X(0)] = \mathbb{E}[X_0],$$

which will be used for feedback in the optimal strategy. Observe that $\mathbb{E}[X]$ appears in the optimal control law. Next, we provide a semi-explicit solution to the full-cooperation case.

Corollary 1 *Assume that*

$$Q_0 := \sum_{j\in\mathcal{N}} Q_j,$$

$$Q_0 + \bar{Q}_0 := \sum_{j\in\mathcal{N}}[Q_j + \bar{Q}_j],$$

R_i, and $R_i+\bar{R}_i$, are symmetric positive definite. The fully-cooperative solution of the problem

$$\begin{cases} \underset{U_1,\dots,U_n}{\text{minimize}}\ \mathbb{E}\left[\sum_{j\in\mathcal{N}} L_j(X,s,U_1,\dots,U_n)\right], \\[2ex] \text{subject to} \\[1ex] dX = \Big[B_1(X - \mathbb{E}[X]) + (B_1 + \bar{B}_1)\mathbb{E}[X] + \sum_{j\in\mathcal{N}} B_{2j}(U_j - \mathbb{E}[U_j]) \\[1ex] \qquad + \sum_{j\in\mathcal{N}}(B_{2j} + \bar{B}_{2j})\mathbb{E}[U_j]\Big]dt + S_0 dB + \int_{\theta\in\Theta} M_0(\cdot,\theta)\tilde{N}(dt, d\theta), \\[1ex] X(0) = X_0. \end{cases}$$

is given by

$$U_i^{\mathrm{rn,g}} - \mathbb{E}[U_i^{\mathrm{rn,g}}] = -R_i^{-1}B_{2i}^* P_0^{\mathrm{rn,g}}(X - \mathbb{E}[X]),$$

$$\mathbb{E}[U_i^{\mathrm{rn,g}}] = -(R_i + \bar{R}_i)^{-1}(B_{2i} + \bar{B}_{2i})^* \bar{P}_0^{\mathrm{rn,g}}\mathbb{E}[X],$$

$$L_0^{\mathrm{rn,g}}(X, s, U_1^{\mathrm{rn,g}},\dots,U_n^{\mathrm{rn,g}}) = \mathbb{E}\langle P_0^{\mathrm{rn,g}}(0, s(0))(X_0 - \mathbb{E}[X_0]), X_0 - \mathbb{E}[X_0]\rangle$$
$$+ \mathbb{E}\langle \bar{P}_0^{\mathrm{rn,g}}(0, s(0))\mathbb{E}[X_0], \mathbb{E}[X_0]\rangle + \mathbb{E}[\delta_0^{\mathrm{rn,g}}(0, s(0))],$$

where $P_0, \bar{P}_0$, and δ_0 solve the following differential equations:

$$\dot{P}_0 = -Q_0 - P_0 B_1 - B_1^* P_0 - \sum_{s'\neq s}(P_0(t, s') - P_0(t, s))\tilde{q}_{ss'}$$

$$\qquad + P_0\Big[\sum_{i\in\mathcal{N}} B_{2i} R_i^{-1} B_{2i}^*\Big]P_0, \tag{9.8a}$$

$$\dot{\bar{P}}_0 = -Q_0 - \bar{Q}_0 - \bar{P}_0(B_1 + \bar{B}_1) - (B_1 + \bar{B}_1)^* \bar{P}_0$$

$$- \sum_{s' \neq s} (\bar{P}_0(t, s') - \bar{P}_0(t, s)) \tilde{q}_{ss'}$$

$$+ \bar{P}_0 \Big[\sum_{i \in \mathcal{N}} (B_{2i} + \bar{B}_{2i})(R_i + \bar{R}_i)^{-1}(B_{2i} + \bar{B}_{2i})^* \Big] \bar{P}_0, \qquad (9.8b)$$

$$\dot{\delta}_0 = -\langle P_0 S_0, S_0 \rangle - \int_\Theta \langle P_0 M_0, M_0 \nu(\mathrm{d}\theta) \rangle$$

$$- \sum_{s' \neq s} (\delta_0(t, s') - \delta_0(t, s)) \tilde{q}_{ss'}, \qquad (9.8c)$$

for all $s \in \mathcal{S}$ with terminal boundary conditions

$$P_0(T, s) = Q_0(T, s),$$
$$\bar{P}_0(T, s) = Q_0(T, s) + \bar{Q}_0(T, s),$$
$$\delta_0(T, s) = 0.$$

Notice that these Riccati equations have positive solution $P_0, \bar{P}_0$, and δ_0, and there is no blow up in the time interval $[0, T]$.

Notice that, we have omitted the super-script "rn, g" in the previous differential equations to ease the notation.

Proof 21 (Corollary 1) *The proof is immediate from Theorem 4 by one single team and with a choice vector of matrices $U = (U_i)_{i \in \mathcal{N}}$.*

We present now the solution corresponding to the adversarial mean-field-type game problem in Corollary 2. It is relevant to note that the adversarial game can be stated using the same structure as a fully-cooperative game and by setting the weight parameters in the cost functional appropriately.

Corollary 2 *Assume that $Q, Q + \bar{Q}, R_i$, and $R_i + \bar{R}_i$ are symmetric positive definite for $i \in \mathcal{N}_+$ and $-R_j, -(R_j + \bar{R}_j)$ are symmetric positive definite for $j \in \mathcal{N}_-$. We assume that $\mathcal{N}_+ \cup \mathcal{N}_- = \mathcal{N}$, and $\mathcal{N}_+ \cap \mathcal{N}_- = \emptyset$.*

The adversarial game problem of the team attackers $\mathcal{N}_-$ and the team of defenders $\mathcal{N}_+$ has a saddle and it is given by

$$U_i^{\mathrm{ad}} - \mathbb{E}[U_i^{\mathrm{ad}}] = -R_i^{-1} B_{2i}^* P^{\mathrm{ad}}(X - \mathbb{E}[X]),$$
$$\mathbb{E}[U_i^{\mathrm{ad}}] = -(R_i + \bar{R}_i)^{-1}(B_{2i} + \bar{B}_{2i})^* \bar{P}^{\mathrm{ad}} \mathbb{E}[X], \ i \in \mathcal{N}_+,$$
$$V_j^{\mathrm{ad}} - V_j^{\mathrm{ad}} = -R_j^{-1} B_{2j}^* P^{\mathrm{ad}}(X - \mathbb{E}[X]), \ j \in \mathcal{N}_-,$$
$$\bar{V}_j^{\mathrm{ad}} = -(R_j + \bar{R}_j)^{-1}(B_{2j} + \bar{B}_{2j})^* \bar{P}^{\mathrm{ad}} \mathbb{E}[X], \ j \in \mathcal{N}_-,$$
$$L^{\mathrm{ad}}(X, s, U_1^{\mathrm{ad}}, \dots, U_n^{\mathrm{ad}}) = \mathbb{E}\langle P^{\mathrm{ad}}(0, s(0))(X_0 - \mathbb{E}[X_0]), X_0 - \mathbb{E}[X_0] \rangle$$
$$+ \mathbb{E}\langle \bar{P}^{\mathrm{ad}}(0, s(0)) \mathbb{E}[X - 0], \mathbb{E}[X_0] \rangle + \mathbb{E}[\delta^{\mathrm{ad}}(0, s(0))],$$

where $P^{\mathrm{ad}}, \bar{P}^{\mathrm{ad}}$, and δ^{ad} solve the following differential equations:

$$\dot{P} = -Q - PB_1 - B_1^* P - \sum_{s' \neq s} (P(t, s') - P(t, s)) \tilde{q}_{ss'}$$

$$+ P \Big[\sum_{i \in \mathcal{N}_+} B_{2i} R_i^{-1} B_{2i}^* + \sum_{j \in \mathcal{N}_-} B_{2j} R_j^{-1} B_{2j}^* \Big] P, \qquad (9.9\mathrm{a})$$

$$\dot{\bar{P}} = -Q - \bar{Q} - \bar{P}(B_1 + \bar{B}_1) - (B_1 + \bar{B}_1)^* \bar{P} - \sum_{s' \neq s} (\bar{P}(t, s') - \bar{P}(t, s)) \tilde{q}_{ss'}$$

$$+ \bar{P} \Big[\sum_{i \in \mathcal{N}_+} (B_{2i} + \bar{B}_{2i})(R_i + \bar{R}_i)^{-1}(B_{2i} + \bar{B}_{2i})^*$$

$$+ \sum_{j \in \mathcal{N}_-} (B_{2j} + \bar{B}_{2j})(R_j + \bar{R}_j)^{-1}(B_{2j} + \bar{B}_{2j})^* \Big] \bar{P}, \qquad (9.9\mathrm{b})$$

$$\dot{\delta} = -\langle P S_0, S_0 \rangle - \int_\Theta \langle P M_0, M_0 \nu(\mathrm{d}\theta) \rangle - \sum_{s' \neq s} (\delta(t, s') - \delta(t, s)) \tilde{q}_{ss'}, \qquad (9.9\mathrm{c})$$

for all $s \in \mathcal{S}$ with terminal boundary conditions

$$P(T, s) = Q(T, s),$$
$$\bar{P}(T, s) = Q(T, s) + \bar{Q}(T, s),$$
$$\delta(T, s) = 0.$$

Notice that, we have omitted the super-script "ad" in the previous differential equations to ease the notation.

Proof 22 (Corollary 2) *The proof is immediate from Theorem 4 by considering two adversarial teams and with choice vector of matrices $U_+ = (U_i)_{i \in \mathcal{N}_+}$ and $U_- = (U_i)_{i \in \mathcal{N}_-}$, respectively.*

Notice that, the Riccati equations in (9.9) have positive definite solution P^{ad} if in addition

$$\Big[\sum_{i \in \mathcal{N}_+} B_{2i} R_i^{-1} B_{2i}^* + \sum_{j \in \mathcal{N}_-} B_{2j} R_j^{-1} B_{2j}^* \Big] \succ 0,$$

which does not blow up within $[0, T]$, and positive solution $\bar{P}^{\mathrm{ad}}$ if in addition

$$\Big[\sum_{i \in \mathcal{N}_+} (B_{2i} + \bar{B}_{2i})(R_i + \bar{R}_i)^{-1}(B_{2i} + \bar{B}_{2i})^*$$

$$+ \sum_{j \in \mathcal{N}_-} (B_{2j} + \bar{B}_{2j})(R_j + \bar{R}_j)^{-1}(B_{2j} + \bar{B}_{2j})^* \Big] \succ 0,$$

within $[0, T]$. Next, we study the risk-sensitive case and point out some facts regarding the comparison of its solution with respect to the risk-neutral case as the risk-sensitivity index vanishes.

9.3 Semi-explicit Solution of the Mean-Field-Type Game Problems: Risk-Sensitive Case

A risk averse decision-maker (with cost functional) is a decision-maker who prefers higher cost with known risks rather than lower cost with unknown risks. In other words, among various control strategies giving the same cost with different levels of risks, this decision-maker always prefers the alternative with the lowest risk. When $M_0 \neq 0$, the exponential martingale of compensated Poisson random process times a linear process yields to an exponential non-quadratic terms, and

$$\mathbb{E}\left[\exp\left[\int_0^T \int_\Theta \langle (P_i^* + P_i)(X - \mathbb{E}[X]), M_0 \tilde{N}(\mathrm{d}t, \mathrm{d}\theta) \rangle \right]\right]$$

has an exponential non-quadratic term. Therefore we consider the risk-sensitive case when M_0 vanishes (no Poisson jump).

When $\lambda_i > 0$ the decision-maker i is risk-averse and when $\lambda_i < 0$ is risk-seeking. As λ_i goes to zero, decision-maker i becomes a risk-neutral decision-maker as in the previous section. The best-response problem of the decision-maker $i \in \mathcal{N}$ is well-posed only for $\lambda_i \leq \bar{\lambda}_i$ where $\bar{\lambda}_i$ will be determined from the solution region of the differential system derived below.

Theorem 5 *Let us assume that the weight parameters in the cost functional Q_i, R_i, $Q_i + \bar{Q}_i$, $R_i + \bar{R}_i$ are symmetric positive definite and we do not consider Poisson jump, i.e., $M_0 = 0$. Then the matrix-valued mean-field-type (risk-sensitive) Nash equilibrium strategy and the (risk-sensitive) equilibrium cost are given by:*

$$U_i^{\mathrm{rs}} - \bar{U}_i^{\mathrm{rs}} = -R_i^{-1}B_{2i}^* P_i^{\mathrm{rs}}(X - \mathbb{E}[X]),$$
$$\bar{U}_i^{\mathrm{rs}} = -(R_i + \bar{R}_i)^{-1}(B_{2i} + \bar{B}_{2i})^* \bar{P}_i^{\mathrm{rs}} \mathbb{E}[X],$$
$$L_i^{\mathrm{rs}}(X, s, U_1^{\mathrm{rs}}, \dots, U_n^{\mathrm{rs}}) = \frac{1}{\lambda_i} \log \mathbb{E} \exp\{\lambda_i[\langle P_i^{\mathrm{rs}}(0, s(0))(X_0 - \mathbb{E}[X_0]), X_0 - \mathbb{E}[X_0] \rangle$$
$$+ \langle \bar{P}_i^{\mathrm{rs}}(0, s(0)) \mathbb{E}[X_0], \mathbb{E}[X_0] \rangle + \delta_i^{\mathrm{rs}}(0, s(0))]\}, i \in \{1, \dots, n\},$$

where P_i^{rs}, $\bar{P}_i^{\mathrm{rs}}$, and δ_i^{rs} solve the following differential equations:

$$\dot{P}_i = -Q_i - P_i B_1^* - B_1 P_i - \sum_{s' \neq s}(P_i(t, s') - P_i(t, s))\tilde{q}_{ss'}$$

$$+ P_i[B_{2i}R_i^{-1}B_{2i}^* - 2\lambda_i S_0 S_0^*]P_i + \sum_{j \in \mathcal{N}\setminus\{i\}} P_j B_{2j} R_j^{-1*} B_{2j}^* P_i$$

$$+ \sum_{j \in \mathcal{N}\setminus\{i\}} P_i B_{2j} R_j^{-1*} B_{2j}^* P_j, \tag{9.10a}$$

$$\dot{P_i} = -Q_i - \bar{Q}_i - \bar{P}_i(B_1 + \bar{B}_1)^* - (B_1 + \bar{B}_1)\bar{P}_i - \sum_{s' \neq s}(\bar{P}_i(t, s') - \bar{P}_i(t, s))\tilde{q}_{ss'}$$

$$+ \bar{P}_i[(B_{2i} + \bar{B}_{2i})(R_i + \bar{R}_i)^{-1}(B_{2i} + \bar{B}_{2i})^*]\bar{P}_i$$

$$+ \sum_{j \in \mathcal{N}\backslash\{i\}} \bar{P}_j(B_{2j} + \bar{B}_{2j})(R_j + \bar{R}_j)^{-1*}(B_{2j} + \bar{B}_{2j})^*\bar{P}_i$$

$$+ \sum_{j \in \mathcal{N}\backslash\{i\}} \bar{P}_i(B_{2j} + \bar{B}_{2j})(R_j + \bar{R}_j)^{-1*}(B_{2j} + \bar{B}_{2j})^*\bar{P}_j, \qquad (9.10b)$$

$$\dot{\delta_i} = -\langle P_i S_0, S_0\rangle - \sum_{s' \neq s}(\delta_i(t, s') - \delta_i(t, s))\tilde{q}_{ss'}, \qquad (9.10c)$$

for all $s \in \mathcal{S}$ with terminal boundary conditions

$$P_i(T, s) = Q_i(T, s),$$
$$\bar{P}_i(T, s) = Q_i(T, s) + \bar{Q}_i(T, s),$$
$$\delta_i(T, s) = 0,$$

whenever these differential system of equations have a unique solution that does not blow up in $[0, T]$.

In the previous Theorem, we have omitted the super-script "rs" in the differential equations to ease notation. Next, we explain the additional consideration that should be taken into account for the risk-sensitive game problem.

Proof 23 (Theorem 5) *The martingale term in the Itô's formula is*

$$2\lambda_i \int_0^T \langle S_0^* P_i(X - \mathbb{E}[X]), \mathrm{d}B\rangle.$$

By adding and removing the term

$$2\lambda_i^2 \int_0^T \langle P_i S_0 S_0^* P_i(X - \mathbb{E}[X]), (X - \mathbb{E}[X])\rangle \mathrm{d}t$$

to the Itô's formula, we re-organize the exponential quadratic terms. Using that relation we have

$$\mathbb{E}\left[\exp\left\{2\lambda_i \int_0^T \langle S_0^* P_i(X - \mathbb{E}[X]), \mathrm{d}B\rangle\right.\right.$$
$$\left.\left. - 2\lambda_i^2 \int_0^T \langle P_i S_0 S_0^* P_i(X - \mathbb{E}[X]), (X - \mathbb{E}[X])\rangle \mathrm{d}t\right\}\right] = 1,$$

and matching that term in P_i, we arrive at the announced result. This completes the proof.

As all λ_i vanish, the matrix-valued differential system (9.10) becomes the risk-neutral system in (9.13) above, and the risk-sensitive optimal strategy coincides with the risk-neutral one. A bound for $\bar{\lambda}_i$ can be obtained from the positivity condition of the matrices as follows:

$$[B_{2i}R_i^{-1}B_{2i}^* - 2\lambda_i S_0 S_0^*] \succ 0. \tag{9.11}$$

Note that $\bar{P}_i^{\mathrm{rs}} = \bar{P}_i^{\mathrm{rn}}$ because this coefficient is associated with the term $\langle \mathbb{E}[X], \mathbb{E}[X] \rangle$ which is independent of the Brownian motion.

Consider a finite population of decision-makers $\mathcal{N} := \{1, \dots n\}$ classified as follows:

- Risk-neutral decision-makers: $\mathcal{N}_0 = \{i \in \mathcal{N} | \ \lambda_i \to 0\}$

- Risk-averse decision-makers: $\mathcal{N}_+ = \{i \in \mathcal{N} | \ \lambda_i > 0\}$

- Risk-seeking decision-makers: $\mathcal{N}_- = \{i \in \mathcal{N} | \ \lambda_i < 0\}$.

In Chapter 5, we analyzed situation in which the decision-makers were heterogeneous regarding the cooperative or competitive behavior they take by means of co-opetitive parameters. Here, we also allow decision-makers be heterogeneous but in a different sense. We allow the decision-makers behave in a different manner regarding how they interpret and treat the risk, i.e., risk-neutral, risk-averse or risk-seeking. This analysis is shown below in Corollary 3.

Corollary 3 *A mixture of risk-neutral, risk-seeking and risk-averse are obtained solving the following system:*

$$\begin{cases} i \in \mathcal{N}_0 : \\ \dot{P}_i + Q_i + P_i B_1^* + B_1 P_i + \displaystyle\sum_{s' \neq s}(P_i(t, s') - P_i(t, s))\tilde{q}_{ss'} \\ -P_i B_{2i} R_i^{-1} B_{2i}^* P_i - \displaystyle\sum_{j \in \mathcal{N} \backslash \{i\}} P_j B_{2j} R_j^{-1*} B_{2j}^* P_i \\ - \displaystyle\sum_{j \in \mathcal{N} \backslash \{i\}} P_i B_{2j} R_j^{-1*} B_{2j}^* P_j = 0, \end{cases} \tag{9.12a}$$

$$\begin{cases} i \in \mathcal{N}_+ : \\ \dot{P}_i + Q_i + P_i B_1^* + B_1 P_i + \displaystyle\sum_{s' \neq s}(P_i(t, s') - P_i(t, s))\tilde{q}_{ss'} \\ -P_i \left[B_{2i} R_i^{-1} B_{2i}^* - 2\lambda_i S_0 S_0^* \right] P_i - \displaystyle\sum_{j \in \mathcal{N} \backslash \{i\}} P_j B_{2j} R_j^{-1*} B_{2j}^* P_i \\ - \displaystyle\sum_{j \in \mathcal{N} \backslash \{i\}} P_i B_{2j} R_j^{-1*} B_{2j}^* P_j = 0, \end{cases} \tag{9.12b}$$

$$\begin{cases} i \in \mathcal{N}_- : \\ \dot{P}_i + Q_i + P_i B_1^* + B_1 P_i + \sum_{s' \neq s} (P_i(t,s') - P_i(t,s)) \tilde{q}_{ss'} \\ \quad - P_i [B_{2i} R_i^{-1} B_{2i}^* + 2(-\lambda_i) S_0 S_0^*] P_i - \sum_{j \in \mathcal{N} \setminus \{i\}} P_j B_{2j} R_j^{-1*} B_{2j}^* P_i \qquad (9.12c) \\ \quad - \sum_{j \in \mathcal{N} \setminus \{i\}} P_i B_{2j} R_j^{-1*} B_{2j}^* P_j = 0, \end{cases}$$

with the terminal boundary condition

$$P_i(T,s) = Q_i(T,s), \quad \forall\, i \in \mathcal{N}, \; s \in \mathcal{S}.$$

The fully-cooperative risk-neutral problem whose solution has been presented in Corollary 1, is extended to the risk-sensitive scenario.

Corollary 4 *We have the following weight parameters obtained as the sum of all the weights for all the decision-makers as shown next. Here, assume that*

$$Q_0 := \sum_{i \in \mathcal{N}} Q_i,$$

$$Q_0 + \bar{Q}_0 := \sum_{i \in \mathcal{N}} [Q_i, + \bar{Q}_i],$$

and $R_i, R_i + \bar{R}_i$ are symmetric positive definite. The risk-sensitive fully-cooperative solution of the problem

$$\begin{cases} \underset{U_1,\dots,U_n}{\text{minimize}} \; \dfrac{1}{\lambda} \log \mathbb{E} \left[e^{\lambda \sum_{j \in \mathcal{N}} L_j(X,s,U_1,\dots,U_n)} \right], \\[2mm] \text{subject to} \\[2mm] dX = \Big[B_1(X - \mathbb{E}[X]) + (B_1 + \bar{B}_1)\mathbb{E}[X] + \sum_{j \in \mathcal{N}} B_{2j}(U_j - \mathbb{E}[U_j]) \\ \qquad + \sum_{j \in \mathcal{N}} (B_{2j} + \bar{B}_{2j})\mathbb{E}[U_j] \Big] dt + S_0 dB, \\ X(0) = X_0. \end{cases}$$

is given by

$$U_i^{\mathrm{rs,g}} - \mathbb{E}[U_i^{\mathrm{rs,g}}] = -R_i^{-1} B_{2i}^* P_0^{\mathrm{rs,g}}(X - \mathbb{E}[X]),$$
$$\mathbb{E}[U_i^{\mathrm{rs,g}}] = -(R_i + \bar{R}_i)^{-1}(B_{2i} + \bar{B}_{2i})^* \bar{P}_0^{\mathrm{rs,g}} \mathbb{E}[X],$$
$$L_0^{\mathrm{rs,g}}(X,s,U_1^{\mathrm{rs,g}},\dots,U_n^{\mathrm{rs,g}}) = \frac{1}{\lambda} \log \mathbb{E} \exp\{\lambda[\langle P_0^{\mathrm{rs,g}}(0,s(0))(X_0 - \mathbb{E}[X_0]), X_0 - \mathbb{E}[X_0]\rangle$$
$$+ \langle \bar{P}_0^{\mathrm{rs,g}}(0,s(0))\mathbb{E}[X_0], \mathbb{E}[X_0]\rangle + \delta_0^{\mathrm{rs,g}}(0,s(0))]\}, i \in \mathcal{N},$$

where $P_0^{\mathrm{rs,g}}, \bar{P}_0^{\mathrm{rs,g}}$, and $\delta_0^{\mathrm{rs,g}}$ solve the following differential equations:

$$\dot{P}_0 = -Q_0 - P_0 B_1 - B_1^* P_0 - \sum_{s' \neq s} (P_0(t,s') - P_0(t,s)) \tilde{q}_{ss'}$$

$$+ P_0 \Big[\sum_{i \in \mathcal{N}} B_{2i} R_i^{-1} B_{2i}^* - 2\lambda S_0 S_0^* \Big] P_0, \tag{9.13a}$$

$$\dot{\bar{P}}_0 = -Q_0 - \bar{Q}_0 - \bar{P}_0 (B_1 + \bar{B}_1) - (B_1 + \bar{B}_1)^* \bar{P}_0$$
$$- \sum_{s' \neq s} (\bar{P}_0(t, s') - \bar{P}_0(t, s)) \tilde{q}_{ss'}$$
$$+ \bar{P}_0 \Big[\sum_{i \in \mathcal{N}} (B_{2i} + \bar{B}_{2i})(R_i + \bar{R}_i)^{-1}(B_{2i} + \bar{B}_{2i})^* \Big] \bar{P}_0, \tag{9.13b}$$

$$\dot{\delta}_0 = -\langle P_0 S_0, S_0 \rangle - \sum_{s' \neq s} (\delta_0(t, s') - \delta_0(t, s)) \tilde{q}_{ss'}, \tag{9.13c}$$

for all $s \in \mathcal{S}$ with terminal boundary condition

$$P_0(T, s) = Q_0(T, s),$$
$$\bar{P}_0(T, s) = Q_0(T, s) + \bar{Q}_0(T, s),$$
$$\delta_0(T, s) = 0.$$

Please note that we have removed the super-script "rs, g" in the differential equations in the previous Corollary. We follow the analysis by computing a condition for the term λ in the previous presented Corollary.

Remark 25 (Shared-Risk Situation) *Notice that, the bound $\bar{\lambda}$ is obtained from the positivity condition*

$$\sum_{i \in \mathcal{N}} B_{2i} R_i^{-1} B_{2i}^* - 2\lambda S_0 S_0^* \succ 0.$$

Hence, the risk condition is relaxed thanks to the full-cooperation in comparison with the non-cooperative risk consideration, i.e.,

$$\bar{\lambda} = \sup \left\{ \lambda \,\Big|\, \begin{array}{c} \text{full-cooperation risk-sensitive} \\ \text{problem is well-posed} \end{array} \right\},$$
$$\bar{\lambda}_i = \sup \left\{ \lambda_i \,\Big|\, \begin{array}{c} \text{non-cooperative risk-sensitive} \\ \text{problem is well-posed} \end{array} \right\},$$

where

$$B_{2i} R_i^{-1} B_{2i}^* \succ 2\lambda_i S_0 S_0, \quad (\text{from } (9.11)),$$
$$\sum_{i \in \mathcal{N}} B_{2i} R_i^{-1} B_{2i}^* \succ 2 \sum_{i \in \mathcal{N}} \lambda_i S_0 S_0,$$

therefore, it is concluded that

$$\sum_{i \in \mathcal{N}} \bar{\lambda}_i \leq \bar{\lambda}.$$

Full-cooperation increases the well-posedness domain by means of shared-risk.

Finally, we analyze the adversarial settings for the risk-sensitive mean-field-type game problem. This is easily computed by using the fully-cooperative solution and by assigning appropriate signs over the weight parameters as shown in Corollary 5.

Corollary 5 *Assume that Q, $Q + \bar{Q}$, R_i, $R_i + \bar{R}_i$ are symmetric positive definite for $i \in \mathcal{N}_+$ and $-R_j$, $-(R_j + \bar{R}_j)$ are symmetric positive definite for $j \in \mathcal{N}_-$. We assume that $\mathcal{N}_+ \cup \mathcal{N}_- = \mathcal{N}$ and $\mathcal{N}_+ \cap \mathcal{N}_- = \emptyset$.*

The adversarial risk-sensitive game problem of the team attackers $\mathcal{N}_-$ and the team of defenders $\mathcal{N}_+$ has a saddle and it is given by

$$U_i^{\text{rs,ad}} - \mathbb{E}[U_i^{\text{rs,ad}}] = -R_i^{-1}B_{2i}^* P^{\text{rs,ad}}(X - \mathbb{E}[X]),$$

$$\mathbb{E}[U_i^{\text{rs,ad}}] = -(R_i + \bar{R}_i)^{-1}(B_{2i} + \bar{B}_{2i})^* \bar{P}^{\text{rs,ad}}\mathbb{E}[X], \quad i \in \mathcal{N}_+,$$

$$V_j^{\text{rs,ad}} - \bar{V}_j^{\text{rs,ad}} = -R_j^{-1}B_{2j}^* P^{\text{rs,ad}}(X - \mathbb{E}[X]),$$

$$\bar{V}_j^{\text{rs,ad}} = -(R_j + \bar{R}_j)^{-1}(B_{2j} + \bar{B}_{2j})^* \bar{P}^{\text{rs,ad}}\mathbb{E}[X], \quad j \in \mathcal{N}_-,$$

$$L^{\text{rs,ad}}(X, s, U_1^{\text{rs,ad}}, \dots, U_n^{\text{rs,ad}}) = \mathbb{E}\langle P^{\text{rs,ad}}(0, s(0))(X_0 - \mathbb{E}[X_0]), X_0 - \mathbb{E}[X_0]\rangle$$
$$+ \mathbb{E}\langle \bar{P}^{\text{rs,ad}}(0, s(0))\mathbb{E}[X_0], \mathbb{E}[X_0]\rangle + \mathbb{E}[\delta^{\text{rs,ad}}(0, s(0))],$$

where $P^{\text{rs,ad}}$, $\bar{P}^{\text{rs,ad}}$, and $\delta^{\text{rs,ad}}$ solve the following differential equations:

$$\dot{P} = -Q - PB_1 - B_1^* P - \sum_{s' \neq s}(P(t, s') - P(t, s))\tilde{q}_{ss'}$$

$$+ P\left[\sum_{i \in \mathcal{N}_+} B_{2i}R_i^{-1}B_{2i}^* + \sum_{j \in \mathcal{N}_-} B_{2j}R_j^{-1}B_{2j}^* - 2\lambda S_0 S_0^*\right]P, \qquad (9.14a)$$

$$\dot{\bar{P}} = -Q - \bar{Q} - \bar{P}(B_1 + \bar{B}_1) - (B_1 + \bar{B}_1)^* \bar{P} - \sum_{s' \neq s}(\bar{P}(t, s') - \bar{P}(t, s))\tilde{q}_{ss'}$$

$$+ \bar{P}\left[\sum_{i \in \mathcal{N}_+}(B_{2i} + \bar{B}_{2i})(R_i + \bar{R}_i)^{-1}(B_{2i} + \bar{B}_{2i})^*\right.$$

$$\left. + \sum_{j \in \mathcal{N}_-}(B_{2j} + \bar{B}_{2j})(R_j + \bar{R}_j)^{-1}(B_{2j} + \bar{B}_{2j})^* - 2\lambda S_0 S_0^*\right]\bar{P}, \qquad (9.14b)$$

$$\dot{\delta} = -\langle PS_0, S_0\rangle - \sum_{s' \neq s}(\delta(t, s') - \delta(t, s))\tilde{q}_{ss'}, \qquad (9.14c)$$

for all $s \in \mathcal{S}$ and with the terminal boundary conditions:

$$P(T, s) = Q(T, s),$$
$$\bar{P}(T, s) = Q(T, s) + \bar{Q}(T, s),$$
$$\delta(T, s) = 0.$$

We have removed the super-script "rs, ad" to ease the notation.

Notice that, these risk-sensitive adversarial Riccati equations have positive definite solution $P^{\text{rs,ad}}$ if in addition

$$\left[\sum_{i \in \mathcal{N}_+} B_{2i} R_i^{-1} B_{2i}^* + \sum_{j \in \mathcal{N}_-} B_{2j} R_j^{-1} B_{2j}^* - 2\lambda S_0 S_0^* \right] \succ 0,$$

which does not blow up within $[0, T]$.

9.4 Numerical Examples

Example 1: Matrix-Valued Continuous-Time Non-cooperative Scenario

Consider a two-player matrix-valued non-cooperative mean-field-type game interacting throughout the following system dynamics:

$$
\begin{aligned}
dX = \Bigg[&\begin{pmatrix} -1 & 0.1 \\ 0.2 & -1.5 \end{pmatrix} (X - \mathbb{E}[X]) + \begin{pmatrix} -1.5 & 0.1 \\ 0.2 & -2.5 \end{pmatrix} \mathbb{E}[X] \\
&+ \begin{pmatrix} 1 & 0.5 \\ 0.5 & 1 \end{pmatrix} (U_1 - \mathbb{E}[U_1]) - \begin{pmatrix} 2 & 1 \\ 1 & 2 \end{pmatrix} (U_2 - \mathbb{E}[U_2]) \\
&+ \begin{pmatrix} 2 & 0.5 \\ 0.5 & 2 \end{pmatrix} \mathbb{E}[U_1] - \begin{pmatrix} 1.5 & 1 \\ 1 & 1.5 \end{pmatrix} \mathbb{E}[U_2] \Bigg] dt + 2 dB.
\end{aligned}
\tag{9.15}
$$

We consider a terminal time is $T = 2\,\text{s}$. Then, the cost functional of the first player is given by:

$$
\begin{aligned}
L_1(X, U_1, U_2) = &\left\langle \begin{pmatrix} 1 & 0 \\ 0 & 1 \end{pmatrix} (X(2) - \mathbb{E}[X(2)]), X(2) - \mathbb{E}[X(2)] \right\rangle \\
&+ \left\langle \begin{pmatrix} 2 & 0 \\ 0 & 2 \end{pmatrix} \mathbb{E}[X(2)], \mathbb{E}[X(2)] \right\rangle \\
&+ \int_0^2 \left\langle \begin{pmatrix} 1 & 0 \\ 0 & 1 \end{pmatrix} (X - \mathbb{E}[X]), X - \mathbb{E}[X] \right\rangle dt \\
&+ \int_0^2 \left\langle \begin{pmatrix} 2 & 0 \\ 0 & 2 \end{pmatrix} \mathbb{E}[X], \mathbb{E}[X] \right\rangle dt \\
&+ \int_0^2 \left\langle \begin{pmatrix} 1 & 0 \\ 0 & 1 \end{pmatrix} (U_1 - \mathbb{E}[U_1]), U_1 - \mathbb{E}[U_1] \right\rangle dt \\
&+ \int_0^2 \left\langle \begin{pmatrix} 2 & 0 \\ 0 & 2 \end{pmatrix} \mathbb{E}[U_1], \mathbb{E}[U_1] \right\rangle dt,
\end{aligned}
$$

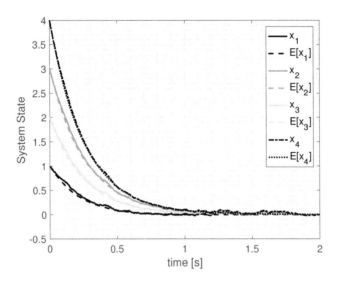

FIGURE 9.3
Evolution of the system state and its expectation for the matrix-value continuous-time non-cooperative scenario.

and the cost function for the second player is:

$$
\begin{aligned}
L_2(X, U_1, U_2) = & \left\langle \begin{pmatrix} 2 & 0 \\ 0 & 2 \end{pmatrix} (X(2) - \mathbb{E}[X(2)]), X(2) - \mathbb{E}[X(2)] \right\rangle \\
& + \left\langle \begin{pmatrix} 4 & 0 \\ 0 & 4 \end{pmatrix} \mathbb{E}[X(2)], \mathbb{E}[X(2)] \right\rangle \\
& + \int_0^2 \left\langle \begin{pmatrix} 2 & 0 \\ 0 & 2 \end{pmatrix} (X - \mathbb{E}[X]), X - \mathbb{E}[X] \right\rangle dt \\
& + \int_0^2 \left\langle \begin{pmatrix} 4 & 0 \\ 0 & 4 \end{pmatrix} \mathbb{E}[X], \mathbb{E}[X] \right\rangle dt \\
& + \int_0^2 \left\langle \begin{pmatrix} 2 & 0 \\ 0 & 2 \end{pmatrix} (U_2 - \mathbb{E}[U_2]), U_2 - \mathbb{E}[U_2] \right\rangle dt \\
& + \int_0^2 \left\langle \begin{pmatrix} 4 & 0 \\ 0 & 4 \end{pmatrix} \mathbb{E}[U_2], \mathbb{E}[U_2] \right\rangle dt.
\end{aligned}
$$

The considered initial conditions are given by:

$$
X_0 = \mathbb{E}[X_0] = \begin{pmatrix} 1 & 2 \\ 3 & 4 \end{pmatrix}.
$$

Figure 9.3 shows the evolution of the system states X and their expectation $\mathbb{E}[X]$. On the other hand, Figures 9.4 and 9.5 present the evolution of the

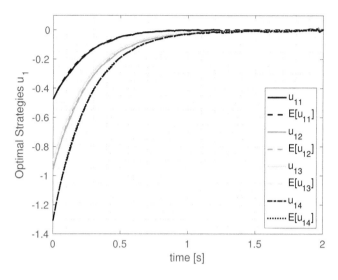

FIGURE 9.4
Evolution of the first player strategies and their expectation for the matrix-value continuous-time non-cooperative scenario.

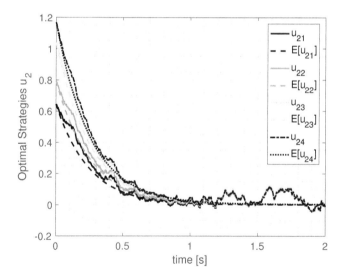

FIGURE 9.5
Evolution of the second player strategies and their expectation for the matrix-value continuous-time non-cooperative scenario.

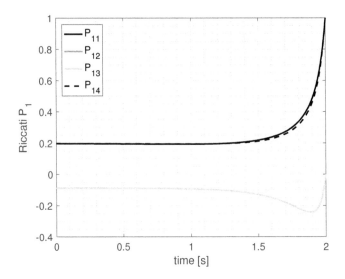

FIGURE 9.6

Evolution of the equation P_1 for the matrix-value continuous-time non-cooperative scenario.

optimal strategies and their expectation U_1, $\mathbb{E}[U_1]$, U_2, and $\mathbb{E}[U_2]$, respectively. The optimal strategies are given by

$$U_1^* = -\begin{pmatrix} \frac{1}{2} & \frac{1}{8} \\ \frac{1}{8} & \frac{1}{2} \end{pmatrix}(\bar{P}_1^* + \bar{P}_1)\mathbb{E}[X] - \begin{pmatrix} \frac{1}{2} & \frac{1}{4} \\ \frac{1}{4} & \frac{1}{2} \end{pmatrix}(P_1^* + P_1)(X - \mathbb{E}[X]),$$

$$U_2^* = -\begin{pmatrix} \frac{3}{16} & \frac{1}{8} \\ \frac{1}{8} & \frac{3}{16} \end{pmatrix}(\bar{P}_2^* + \bar{P}_2)\mathbb{E}[X] - \begin{pmatrix} \frac{1}{2} & \frac{1}{4} \\ \frac{1}{4} & \frac{1}{2} \end{pmatrix}(P_2^* + P_2)(X - \mathbb{E}[X]),$$

Figures 9.6 and 9.7 show the evolution of the Riccati equations P_1 and P_2, i.e.,

$$\dot{P}_1 = -\begin{pmatrix} 1 & 0 \\ 0 & 1 \end{pmatrix} - P_1\begin{pmatrix} -1 & \frac{1}{10} \\ \frac{1}{5} & -\frac{3}{2} \end{pmatrix} - \begin{pmatrix} -1 & \frac{1}{5} \\ \frac{1}{10} & -\frac{3}{2} \end{pmatrix}P_1$$
$$+ \frac{1}{4}(P_1^* + P_1)\begin{pmatrix} \frac{5}{4} & 1 \\ 1 & \frac{5}{4} \end{pmatrix}(P_1^* + P_1)$$
$$+ \frac{1}{4}(P_2^* + P_2)\begin{pmatrix} \frac{5}{2} & 2 \\ 2 & \frac{5}{2} \end{pmatrix}(P_1^* + P_1) + \frac{1}{4}(P_1^* + P_1)\begin{pmatrix} \frac{5}{2} & 2 \\ 2 & \frac{5}{2} \end{pmatrix}(P_2^* + P_2),$$

$$\dot{P}_2 = -\begin{pmatrix} 2 & 0 \\ 0 & 2 \end{pmatrix} - P_2\begin{pmatrix} -1 & \frac{1}{10} \\ \frac{1}{5} & -\frac{3}{2} \end{pmatrix} - \begin{pmatrix} -1 & \frac{1}{5} \\ \frac{1}{10} & -\frac{3}{2} \end{pmatrix}P_2$$
$$+ \frac{1}{4}(P_2^* + P_2)\begin{pmatrix} \frac{5}{2} & 2 \\ 2 & \frac{5}{2} \end{pmatrix}(P_2^* + P_2)$$

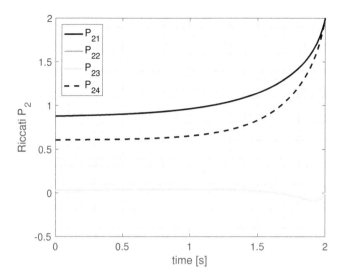

FIGURE 9.7
Evolution of the equation P_2 for the matrix-value continuous-time non-cooperative scenario.

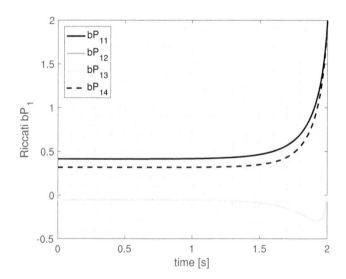

FIGURE 9.8
Evolution of the equation $\bar{P}_1$ for the matrix-value continuous-time non-cooperative scenario.

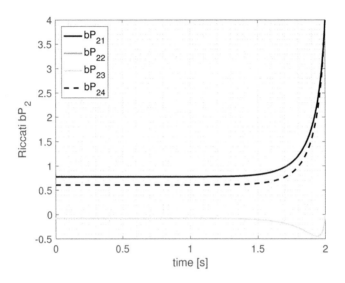

FIGURE 9.9
Evolution of the equation $\bar{P}_2$ for the matrix-value continuous-time non-cooperative scenario.

$$+ \frac{1}{4}(P_1^* + P_1) \begin{pmatrix} \frac{5}{4} & 1 \\ 1 & \frac{5}{4} \end{pmatrix} (P_2^* + P_2) + \frac{1}{4}(P_2^* + P_2) \begin{pmatrix} \frac{5}{4} & 1 \\ 1 & \frac{5}{4} \end{pmatrix} (P_1^* + P_1).$$

Figures 9.8 and 9.9 show the evolution of the Riccati equations $\bar{P}_1$ and $\bar{P}_2$, i.e.,

$$\dot{P}_1 = - \begin{pmatrix} 2 & 0 \\ 0 & 2 \end{pmatrix} - P_1 \begin{pmatrix} -\frac{3}{2} & \frac{1}{10} \\ \frac{1}{5} & -\frac{5}{2} \end{pmatrix} - \begin{pmatrix} -\frac{3}{2} & \frac{1}{5} \\ \frac{1}{10} & -\frac{5}{2} \end{pmatrix} P_1$$

$$+ \frac{1}{4}(\bar{P}_1^* + \bar{P}_1) \begin{pmatrix} \frac{5}{8} & \frac{1}{2} \\ \frac{1}{2} & \frac{5}{8} \end{pmatrix} (\bar{P}_1^* + \bar{P}_1) + \frac{1}{4}(\bar{P}_2^* + \bar{P}_2)^* \begin{pmatrix} \frac{5}{16} & \frac{1}{4} \\ \frac{1}{4} & \frac{5}{16} \end{pmatrix} (\bar{P}_1^* + \bar{P}_1)$$

$$+ \frac{1}{4}(\bar{P}_1^* + \bar{P}_1) \begin{pmatrix} \frac{5}{16} & \frac{1}{4} \\ \frac{1}{4} & \frac{5}{16} \end{pmatrix} (\bar{P}_2^* + \bar{P}_2),$$

$$\dot{P}_2 = - \begin{pmatrix} 4 & 0 \\ 0 & 4 \end{pmatrix} - P_2 \begin{pmatrix} -\frac{3}{2} & \frac{1}{10} \\ \frac{1}{5} & -\frac{5}{2} \end{pmatrix} - \begin{pmatrix} -\frac{3}{2} & \frac{1}{5} \\ \frac{1}{10} & -\frac{5}{2} \end{pmatrix} P_2$$

$$+ \frac{1}{4}(\bar{P}_2^* + \bar{P}_2) \begin{pmatrix} \frac{5}{16} & \frac{1}{4} \\ \frac{1}{4} & \frac{5}{16} \end{pmatrix} (\bar{P}_2^* + \bar{P}_2) + \frac{1}{4}(\bar{P}_1^* + \bar{P}_1)^* \begin{pmatrix} \frac{5}{8} & \frac{1}{2} \\ \frac{1}{2} & \frac{5}{8} \end{pmatrix} (\bar{P}_2^* + \bar{P}_2)$$

$$+ \frac{1}{4}(\bar{P}_2^* + \bar{P}_2) \begin{pmatrix} \frac{5}{8} & \frac{1}{2} \\ \frac{1}{2} & \frac{5}{8} \end{pmatrix} (\bar{P}_1^* + \bar{P}_1).$$

The evolution of δ_1, and δ_2 are presented in Figure 9.10. These differential equations are given by $\dot{\delta}_1 = -2\,\text{trace}(P_1^* + P_1)$, and $\dot{\delta}_2 = -2\,\text{trace}(P_2^* + P_2)$. Finally, Figure 9.11 shows the optimal costs for players 1 and 2, which are

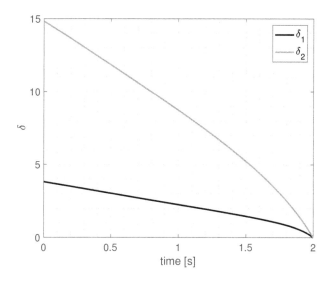

FIGURE 9.10
Evolution of the Riccati equations δ_1, and δ_2 for the matrix-value continuous-time non-cooperative scenario.

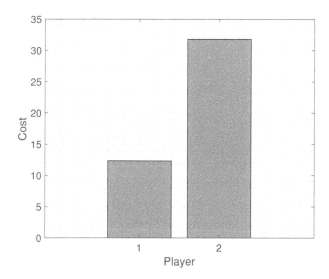

FIGURE 9.11
Optimal cost function for each player for the matrix-value continuous-time non-cooperative scenario.

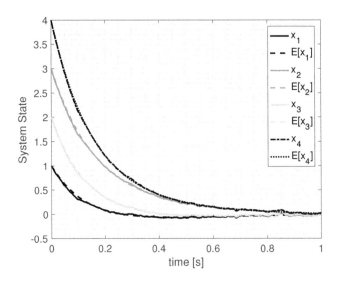

FIGURE 9.12
Evolution of the system state and its expectation for the matrix-value
continuous-time fully-cooperative scenario.

given by

$$
\begin{aligned}
L_1(X, U_1^*, U_2^*) &= \mathbb{E}\langle P_1(0)(X_0 - \mathbb{E}[X_0]), X_0 - \mathbb{E}[X_0]\rangle \\
&\quad + \mathbb{E}\langle \bar{P}_1(0)\mathbb{E}[X_0], \mathbb{E}[X_0]\rangle + \mathbb{E}[\delta_1(0)], \\
L_2(X, U_1^*, U_2^*) &= \mathbb{E}\langle P_2(0)(X_0 - \mathbb{E}[X_0]), X_0 - \mathbb{E}[X_0]\rangle \\
&\quad + \mathbb{E}\langle \bar{P}_2(0)\mathbb{E}[X_0], \mathbb{E}[X_0]\rangle + \mathbb{E}[\delta_2(0)].
\end{aligned}
$$

Example 2: Matrix-Valued Continuous-Time Cooperative Scenario

Consider a two-player matrix-valued fully-cooperative mean-field-type game
interacting throughout the same system dynamics of the previous example in
(9.15). We consider a terminal time is $T = 1$ s. Then, the cost functional of
the first player is given by:

$$
\begin{aligned}
L_1(X, U_1, U_2) &= \left\langle \begin{pmatrix} 1 & \frac{1}{10} \\ \frac{1}{5} & \frac{9}{10} \end{pmatrix} (X(1) - \mathbb{E}[X(1)]), X(1) - \mathbb{E}[X(1)] \right\rangle \\
&\quad + \left\langle \begin{pmatrix} 3 & \frac{3}{10} \\ \frac{3}{5} & \frac{12}{5} \end{pmatrix} \mathbb{E}[X(1)], \mathbb{E}[X(1)] \right\rangle \\
&\quad + \int_0^1 \left\langle \begin{pmatrix} 1 & \frac{1}{10} \\ \frac{1}{5} & \frac{9}{10} \end{pmatrix} (X - \mathbb{E}[X]), X - \mathbb{E}[X] \right\rangle \mathrm{d}t
\end{aligned}
$$

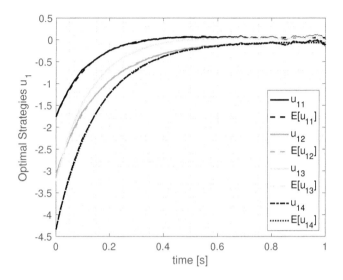

FIGURE 9.13
Evolution of the first player strategies and their expectation for the matrix-value continuous-time fully-cooperative scenario.

$$+ \int_0^1 \left\langle \begin{pmatrix} 3 & \frac{3}{10} \\ \frac{3}{5} & \frac{12}{5} \end{pmatrix} \mathbb{E}[X], \mathbb{E}[X] \right\rangle dt$$

$$+ \int_0^1 \left\langle \begin{pmatrix} 1 & 0 \\ 0 & 1 \end{pmatrix} (U_1 - \mathbb{E}[U_1]), U_1 - \mathbb{E}[U_1] \right\rangle dt$$

$$+ \int_0^1 \left\langle \begin{pmatrix} 2 & 0 \\ 0 & 2 \end{pmatrix} \mathbb{E}[U_1], \mathbb{E}[U_1] \right\rangle dt,$$

and the cost function for the second player is:

$$L_2(X, U_1, U_2) = \left\langle \begin{pmatrix} 2 & \frac{1}{3} \\ \frac{2}{5} & \frac{3}{5} \end{pmatrix} (X(1) - \mathbb{E}[X(1)]), X(1) - \mathbb{E}[X(1)] \right\rangle$$

$$+ \left\langle \begin{pmatrix} 6 & \frac{3}{5} \\ \frac{6}{5} & \frac{24}{5} \end{pmatrix} \mathbb{E}[X(1)], \mathbb{E}[X(1)] \right\rangle$$

$$+ \int_0^1 \left\langle \begin{pmatrix} 2 & \frac{1}{3} \\ \frac{2}{5} & \frac{3}{5} \end{pmatrix} (X - \mathbb{E}[X]), X - \mathbb{E}[X] \right\rangle dt$$

$$+ \int_0^1 \left\langle \begin{pmatrix} 6 & \frac{3}{5} \\ \frac{6}{5} & \frac{24}{5} \end{pmatrix} \mathbb{E}[X], \mathbb{E}[X] \right\rangle dt$$

$$+ \int_0^1 \left\langle \begin{pmatrix} 2 & 0 \\ 0 & 2 \end{pmatrix} (U_2 - \mathbb{E}[U_2]), U_2 - \mathbb{E}[U_2] \right\rangle dt$$

$$+ \int_0^1 \left\langle \begin{pmatrix} 4 & 0 \\ 0 & 4 \end{pmatrix} \mathbb{E}[U_2], \mathbb{E}[U_2] \right\rangle dt.$$

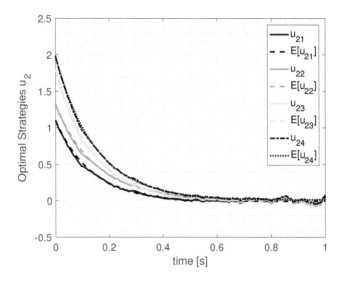

FIGURE 9.14
Evolution of the second player strategies and their expectation for the matrix-value continuous-time fully-cooperative scenario.

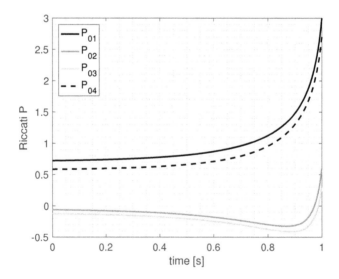

FIGURE 9.15
Evolution of the equation P_0 for the matrix-value continuous-time fully-cooperative scenario.

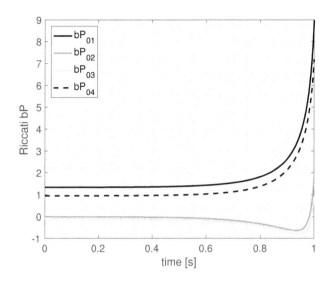

FIGURE 9.16
Evolution of the equation $\bar{P}_0$ for the matrix-value continuous-time fully-cooperative scenario.

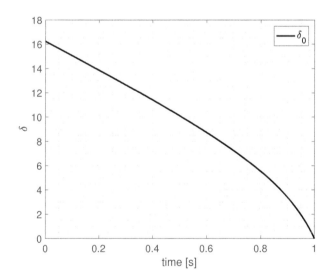

FIGURE 9.17
Evolution of the Riccati equation δ_0 for the matrix-value continuous-time fully-cooperative scenario.

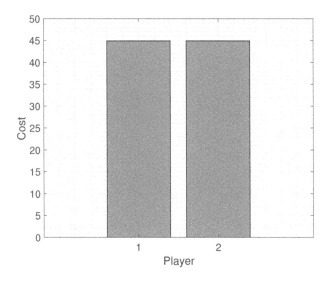

FIGURE 9.18
Optimal cost function for each player for the matrix-value continuous-time
fully-cooperative scenario.

Figure 9.12 shows the evolution of the system states X and their respective
expectation $\mathbb{E}[X]$. It can be seen that the system states approach to zero
according to the minimization of the cost functions. The optimal strategies
are given by

$$U_1^* = -\begin{pmatrix} 1 & \frac{1}{4} \\ \frac{1}{4} & 1 \end{pmatrix} \bar{P}_0 \mathbb{E}[X] - \begin{pmatrix} 1 & \frac{1}{2} \\ \frac{1}{2} & 1 \end{pmatrix} P_0 (X - \mathbb{E}[X]),$$

$$U_2^* = \begin{pmatrix} \frac{3}{8} & \frac{1}{4} \\ \frac{1}{4} & \frac{3}{8} \end{pmatrix} \bar{P}_0 \mathbb{E}[X] + \begin{pmatrix} 1 & \frac{1}{2} \\ \frac{1}{2} & 1 \end{pmatrix} P_0 (X - \mathbb{E}[X]),$$

whose evolutions are shown in Figures 9.13 and 9.14. Now, the evolution of
the corresponding Riccati equations P_0 and $\bar{P}_0$ are shown in Figures 9.15 and
9.16, which are given by

$$\dot{P}_0 = -\begin{pmatrix} 3 & \frac{3}{10} \\ \frac{3}{5} & \frac{27}{10} \end{pmatrix} - P_0 \begin{pmatrix} -1 & \frac{1}{10} \\ \frac{1}{5} & -\frac{3}{2} \end{pmatrix} - \begin{pmatrix} -1 & \frac{1}{5} \\ \frac{1}{10} & -\frac{3}{2} \end{pmatrix} P_0$$

$$+ P_0 \left[\begin{pmatrix} \frac{5}{4} & 1 \\ 1 & \frac{5}{4} \end{pmatrix} + \begin{pmatrix} \frac{5}{2} & 2 \\ 2 & \frac{5}{2} \end{pmatrix} \right] P_0,$$

$$\dot{\bar{P}}_0 = -\begin{pmatrix} 9 & \frac{9}{10} \\ \frac{9}{5} & \frac{36}{5} \end{pmatrix} - \bar{P}_0 \begin{pmatrix} -\frac{3}{2} & \frac{1}{10} \\ \frac{1}{5} & -\frac{5}{2} \end{pmatrix} - \begin{pmatrix} -\frac{3}{2} & \frac{1}{5} \\ \frac{1}{10} & -\frac{5}{2} \end{pmatrix} \bar{P}_0$$

$$+ \bar{P}_0 \left[\begin{pmatrix} 2.125 & 1 \\ 1 & 2.125 \end{pmatrix} + \begin{pmatrix} 0.8125 & 0.75 \\ 0.75 & 0.8125 \end{pmatrix} \right] \bar{P}_0,$$

finally, the evolution of the differential equation $\dot{\delta}_0$ is presented in Figure 9.17, and it is $\dot{\delta}_0 = -9 \, \text{trace}[P_0]$. The optimal cost functions

$$L_0(X, U_1^*, U_2^*) = \mathbb{E}\langle P_0(0)(X_0 - \mathbb{E}[X_0]), X_0 - \mathbb{E}[X_0]\rangle$$
$$+ \mathbb{E}\langle \bar{P}_0(0)\mathbb{E}[X_0], \mathbb{E}[X_0]\rangle + \mathbb{E}[\delta_0(0)],$$

are presented in Figure 9.18.

This chapter has addressed the matrix-valued problem setup. This approach is more general than the scalar-valued and the vector-valued cases. Moreover, consider a matrix-valued problem involving matrices in $\mathbb{R}^{d \times d}$. This problem can be mapped into a vector problem in $\mathbb{R}^{[d^2]}$. Hence, even though the problem is expressed in terms of vectors, the Riccati equations, e.g., $\dot{P}$, are matrices of dimension $d^2 \times d^2$. In contrast, when the problem is addressed as a matrix-valued problem, the obtained Riccati equations, e.g., $\dot{P}$, are matrices of dimension $d \times d$. Notice that, there is a big implication from the computation point of view. The matrix-valued scenario in discrete-time is presented in chapter 12.

9.5 Exercises

1. Consider a mean-field-type control problem with scalar-valued system state and vector-valued control input. The system dynamics are

$$dx = \left[b_1 x + \bar{b}_1 \mathbb{E}[x] + B_2 U + \bar{B}_2 \mathbb{E}[U] \right] dt + \sigma_0 dB, \qquad (9.16)$$

and the cost functional is as follows:

$$L(U) = \frac{1}{2} q(T) x(T)^2 + \frac{1}{2} \bar{q}(T) \mathbb{E}[x(T)]^2 \qquad (9.17)$$

$$+ \frac{1}{2} \int_0^T \left(q x^2 + \bar{q} \mathbb{E}[x]^2 + \langle RU, U \rangle + \langle \bar{R}\mathbb{E}[U], \mathbb{E}[U] \rangle \right) dt,$$

with diagonal weight matrices R and $\bar{R}$.

2. Explain the relationship between the previous mean-field-type control problem and a scalar-valued cooperative mean-field-type game problem (see Chapter 4).

10

A Class of Constrained Matrix-Valued Mean-Field-Type Games

At this point in the book, the reader has studied a large spectrum of game problems including scalar and matrix-valued states. Additionally, we have found semi-explicit solutions for all the problems by following the easy and friendly direct method. Nevertheless, we have not taken into consideration constraints.

In this chapter, we study an alternative in order to deal with the inputs constraints by means of the use of auxiliary variables, which guarantee the evolution within the feasible set. More specifically, we have used the well known projection dynamics. Hence, we show how a constrained problem can be transformed into a non-constrained problem that can be solved as in the previous chapters.

The method we address here to tackle the constraints is applicable to particular cases. Note that the general consideration of constraints for either the system states or control input for this kind of mean-field-type games requires further analytically study and/or algorithmic approaches. It is needed to optimize the Hamiltonian we refer to in Chapter 1 while satisfying the stated constraints. The incorporation of constraints could also be handled by the use of barrier functions and/or other well studied methods.

10.1 Constrained Mean-Field-Type Game Set-up

Let us consider a set of $n \geq 2$ ($n \in \mathbb{N}$) decision-makers denoted by $\mathcal{N} = \{1, \ldots, n\}$, which are strategically interacting through a system given by

$$\mathrm{d}x = \left(Ax + \bar{A}\mathbb{E}[x] + \sum_{i \in \mathcal{N}} B_i u_i + \sum_{i \in \mathcal{N}} \bar{B}_i \mathbb{E}[u_i]\right) \mathrm{d}t + S \mathrm{d}B_r^x, \qquad (10.1)$$

where the vector $x \in \mathbb{R}^d$ denotes the system states, the vector $\mathbb{E}[x] \in \mathbb{R}^d$ is the expected value of the system states, $u_j \in \mathbb{R}^{q_j}$ corresponds to the control inputs for all the decision-maker $j \in \mathcal{N}$, $B_r^x \in \mathbb{R}^s$ represents a standard Brownian motion, and $A, \bar{A}, B_1, \ldots, B_n$, and S are the state-space matrices with suitable dimensions. Now, let us consider a set of m disjoint coupled control input

DOI: 10.1201/9781003098607-10

constraints $\mathcal{C} = \{1, \ldots, m\}$. Each constraint $\ell \in \mathcal{C}$ involves components from the decision-makers.

For the sake of simplicity, and without loss of generality, let the control inputs be organized according to the m existing constraints, i.e.,

$$\begin{pmatrix} u_1 \\ u_2 \\ \vdots \\ u_n \end{pmatrix} = \begin{pmatrix} u^1 \\ u^2 \\ \vdots \\ u^m \end{pmatrix}. \tag{10.2}$$

Then, the constraints are

$$\langle u^\ell, \mathbb{1}_{|u^\ell|} \rangle = \pi_\ell, \ \forall \ell \in \mathcal{C}, \tag{10.3}$$

where $\mathbb{1}_{|u^\ell|}$ denotes the vector of $|u^\ell|$ unitary inputs. On the other hand, each decision-maker has a cost functional

$$\begin{aligned} L_i(x, u_1, \ldots, u_n) &= \langle Q_i(T)(x(T) - \mathbb{E}[x(T)]), x(T) - \mathbb{E}[x(T)] \rangle \\ &+ \langle (Q_i(T) + \bar{Q}_i(T))\mathbb{E}[x(T)], \mathbb{E}[x(T)] \rangle \\ &+ \int_0^T \langle Q_i(x - \mathbb{E}[x]), x - \mathbb{E}[x] \rangle + \langle (Q_i + \bar{Q}_i)\mathbb{E}[x], \mathbb{E}[x] \rangle \, dt \\ &+ \int_0^T \langle R_i(u_i - \mathbb{E}[u_i]), u_i - \mathbb{E}[u_i] \rangle + \langle (R_i + \bar{R}_i)\mathbb{E}[u_i], \mathbb{E}[u_i] \rangle \, dt. \end{aligned} \tag{10.4}$$

The constrained mean-field-type game problem is introduced next.

Problem 19 *The Constrained Mean-Field-Type Game Problem is given by*

$$\begin{cases} \underset{u_i}{\text{minimize}} \ L_i(x, u_1, \ldots, u_n), \\[2mm] \text{subject to} \\[2mm] dx = \left(Ax + \bar{A}\mathbb{E}[x] + \sum_{i \in \mathcal{N}} B_i u_i + \sum_{i \in \mathcal{N}} \bar{B}_i \mathbb{E}[u_i] \right) dt + S d B_r^x, \\ \langle u^\ell, \mathbb{1}_{|u^\ell|} \rangle = \pi_\ell, \ \forall \ell \in \mathcal{C}, \\ x(0) \triangleq x_0 \end{cases}$$

where the weight parameters in the cost functional are Q_i, $(Q_i + \bar{Q}_i) \succeq 0$, and R_i, $(R_i + \bar{R}_i) \succ 0$, for all $i \in \mathcal{N}$.

Once the statement of the constrained mean-field-type game problem has been presented, then we define the constrained best-response strategy and the generalized-Nash equilibrium next in Definitions 25 and 26, respectively.

Definition 25 (Constrained Best-Response Strategy) *A minimizer* u_i^* *is a Best-Response strategy of the decision-maker* $i \in \mathcal{N}$ *given the strategies of the other players*

$$u_{-i} = (u_1, \ldots, u_{i-1}, u_{i+1}, \ldots, u_n)$$

if it solves the constrained Problem 19. Besides, the set of constrained best-response strategies of the decision-maker $i \in \mathcal{N}$ *is denoted by* $\mathrm{BR}_i(u_{-i})$.

In the same way we were defining Nash equilibria by using the best-response concept in previous chapters, we now define the generalized-Nash equilibria next.

Definition 26 (Generalized-Nash Equilibrium) *The strategy profile* $u_1^*, \ldots, u_n^*$ *is a generalized-Nash equilibrium if* $u_i^* \in \mathrm{BR}_i(u_{-i})$, *for all* $i \in \mathcal{N}$ *satisfying the constraints in* (10.3).

In order to deal with the constraints in (10.3), we introduce some auxiliary dynamics under which the feasible set is invariant. The used dynamics belongs to the family of evolutionary dynamics as reported, e.g., in [41], [128], [129].

10.1.1 Auxiliary Dynamics

We introduce auxiliary dynamics for each constraint $\ell \in \mathcal{C}$ involving the decision-makers, i.e., there are as many auxiliary dynamics as constraints in the mean-field-type game Problem 19. The evolution of the auxiliary variables is given by the projection dynamics [130] plus a Brownian motion, i.e.,

$$\mathrm{d}u^\ell = L^\ell v^\ell \mathrm{d}t + \beta^\ell L^\ell \mathrm{d}B_r^\ell, \ \forall \ell \in \mathcal{C}, \tag{10.5}$$

where $\beta^\ell > 0$, for all $\ell \in \mathcal{C}$, and L^ℓ corresponds to the Laplacian matrix of a complete graph involving $|u^\ell|$ nodes, i.e.,

$$\begin{cases} A^\ell = \mathbb{1}_{|u^\ell|}^\top \mathbb{1}_{|u^\ell|} - \mathbf{I}_{|u^\ell|}, \\ D^\ell = \mathrm{diag}[\mathbb{1}_{|u^\ell|}^\top A^\ell], \\ L^\ell = D^\ell - A^\ell, \end{cases}$$

where D^ℓ and A^ℓ correspond to the degree and adjacency matrices of a complete graph, respectively. Hence, the motivation for using the dynamics in (10.5) is the fact that, the trajectories of the auxiliary variables evolve within the feasible sets considered in the mean-field-type game problem as presented in Lemma 1.

Lemma 1 (Invariance of the feasible set) *The feasible set* $\{u^\ell \in \mathbb{R}^{|u^\ell|} : \langle u^\ell, \mathbb{1}_{|u^\ell|} \rangle = \pi_\ell\}$ *according to* (10.3) *is almost surely invariant under the auxiliary dynamics in* (10.5).

Proof 24 (Lemma 1) *The claim immediately follows from the fact that the Laplacian matrix* L^ℓ *satisfies that* $\langle L^\ell, \mathbb{1}_{|u^\ell|} \rangle = \mathbf{0}$.

10.1.2   Augmented Formulation of the Constrained MFTG

Taking advantage of the invariance of the feasible region under the auxiliary dynamics presented in Lemma 1, we reformulate the constrained mean-field-type game problem in order to deal with the coupled constraints in (10.3). To this end, let

$$
d\begin{pmatrix} x \\ u^1 \\ \vdots \\ u^m \end{pmatrix} = \begin{pmatrix} A & \cdots & B_n \\ 0 & \cdots & 0 \end{pmatrix} \begin{pmatrix} x \\ u^1 \\ \vdots \\ u^m \end{pmatrix} dt + \begin{pmatrix} \bar{A} & \cdots & \bar{B}_n \\ 0 & \cdots & 0 \end{pmatrix} \mathbb{E}\begin{pmatrix} x \\ u^1 \\ \vdots \\ u^m \end{pmatrix} dt
$$

$$
+ \begin{pmatrix} 0 & \cdots & 0 \\ L^1 & \cdots & 0 \\ \vdots & \ddots & \vdots \\ 0 & \cdots & L^m \end{pmatrix} \begin{pmatrix} v^1 \\ v^2 \\ \vdots \\ v^m \end{pmatrix} dt + \begin{pmatrix} S & \cdots & 0 \\ 0 & \ddots & 0 \\ 0 & \cdots & \beta^m L^m \end{pmatrix} dB_r^z.
$$

Considering the same grouping of the control inputs for all the n players into m constraints, the new variables $v^1, \ldots, v^m$ are grouped equivalently as $v_1, \ldots, v_n$, i.e.,

$$
\begin{pmatrix} v_1 \\ v_2 \\ \vdots \\ v_n \end{pmatrix} = \begin{pmatrix} v^1 \\ v^2 \\ \vdots \\ v^m \end{pmatrix},
$$

the transformed dynamics are equivalent to

$$
dz = \left(\tilde{A}z + \bar{\tilde{A}}\mathbb{E}[z] + \sum_{i \in \mathcal{N}} \tilde{B}_i v_i \right) dt + \check{S}dB_r^z. \tag{10.6}
$$

Notice that the dynamics in (10.6) contains the same system model as in (10.1) and they also guarantee the satisfaction of the constraints in (10.3) thanks to the auxiliary dynamics in (10.5), together with Lemma 1. Hence, the cost function for each decision-maker $i \in \mathcal{N}$ is expressed as

$$
\begin{aligned}
\tilde{L}_i(x, v_1, \ldots, v_n) =& \left\langle \tilde{Q}_i(T)(z(T) - \mathbb{E}[z(T)]), z(T) - \mathbb{E}[z(T)] \right\rangle \\
&+ \left\langle (\tilde{Q}_i(T) + \bar{\tilde{Q}}_i(T))\mathbb{E}[z(T)], \mathbb{E}[z(T)] \right\rangle \\
&+ \int_0^T \left\langle \tilde{Q}_i(z - \mathbb{E}[z]), z - \mathbb{E}[z] \right\rangle + \left\langle (\tilde{Q}_i + \bar{\tilde{Q}}_i)\mathbb{E}[z], \mathbb{E}[z] \right\rangle dt \\
&+ \int_0^T \left\langle \tilde{R}_i(v_i - \mathbb{E}[v_i]), v_i - \mathbb{E}[v_i] \right\rangle + \left\langle \bar{\tilde{R}}_i \mathbb{E}[v_i], \mathbb{E}[v_i] \right\rangle dt,
\end{aligned} \tag{10.7}
$$

The new mean-field-type game with control inputs v is stated next in Problem 20.

Problem 20 *The Transformed Unconstrained Mean-Field-Type Game Problem using auxiliary dynamics is*

$$
\begin{cases}
\text{minimize}_{v_i} \ \tilde{L}_i(x, v_1, \ldots, v_n), \\[2mm]
\text{subject to} \\[2mm]
dz = \left(\tilde{A}z + \bar{\tilde{A}}\mathbb{E}[z] + \sum_{i \in \mathcal{N}} \tilde{B}_i v_i \right) dt + \tilde{S}dB_r^z, \\
z(0) = (x(0)^\top \quad u(0)^\top)^\top \triangleq z_0, \\
u_i(0), \ \forall i \in \mathcal{N}, \ \textit{such that} \\
u^\ell \in \left\{ y \in \mathbb{R}^{|u^\ell|} : \langle y, \mathbb{1}_{|u^\ell|} \rangle = \pi_\ell \right\}, \ \forall \ell \in \mathcal{C},
\end{cases}
$$

where the weight parameters for the cost functional are as follows:

$$
\tilde{Q}_i = \begin{pmatrix} Q_i & \mathbf{0} \\ \mathbf{0} & \check{R}_i \end{pmatrix}, \quad
\bar{\tilde{Q}}_i = \begin{pmatrix} \bar{Q}_i & \mathbf{0} \\ \mathbf{0} & \bar{\check{R}}_i \end{pmatrix}, \quad \tilde{R}_i \succ 0,
$$

being $\check{R}_i$ $(\bar{\check{R}}_i)$ a diagonal block matrix with R_i $(\bar{R}_i)$ in the ith diagonal entry, and $\mathbf{0}$ elsewhere.

The Problem 20 can solved in a semi-explicit manner without considering constraints over the new control inputs v. Such solution is computed next.

10.2 Semi-explicit Solution of the Constrained Mean-Field-Type Game Problem

The semi-explicit solutions for the mean-field-type games Problem 20 is introduced in Proposition 17.

Proposition 17 *Assume that $\tilde{Q}_i$, $\tilde{R}_i$, $\tilde{Q}_i + \bar{\tilde{Q}}_i$ are positive definite. Then, the non-cooperative mean-field-type Nash equilibrium strategy and the equilibrium cost corresponding are given by:*

$$
v_i^* - \mathbb{E}[v_i^*] = -\frac{1}{2}\tilde{R}_i^{-1}\tilde{B}_i^*(P_i^* + P_i)(z - \mathbb{E}[z]),
$$

$$
\mathbb{E}[v_i^*] = -\frac{1}{2}\tilde{R}_i^{-1}\tilde{B}_i^*(\bar{P}_i^* + \bar{P}_i)\mathbb{E}[z],
$$

$$
\tilde{L}_i(x, v_1^*, \ldots, v_n^*) = \langle P_i(0)(z_0 - \mathbb{E}[z_0]), z_0 - \mathbb{E}[z_0] \rangle \\
+ \langle \bar{P}_i(0)\mathbb{E}[z_0], \mathbb{E}[z_0] \rangle + \delta_i(0),
$$

where P_i, $\bar{P}_i$, and δ_i solve the following differential equations:

$$
\dot{P}_i = -\tilde{Q}_i - \tilde{A}^*(P_i^* + P_i) + \frac{1}{4}(P_i^* + P_i)\tilde{B}_i\tilde{R}_i^{-\frac{1}{2}*}\tilde{R}_i^{-\frac{1}{2}}\tilde{B}_i^*(P_i^* + P_i)
$$

$$+ \frac{1}{2} \sum_{j \neq i} (P_j^* + P_j) \tilde{B}_j \tilde{R}_j^{-1*} \tilde{B}_j^* (P_i^* + P_i),$$

$$\dot{\bar{P}}_i = -\tilde{Q}_i - \bar{\tilde{Q}}_i - (\tilde{A} + \bar{\tilde{A}})^* (\bar{P}_i^* + \bar{P}_i) + \frac{1}{4} (\bar{P}_i^* + \bar{P}_i) \tilde{B}_i \tilde{R}_i^{-\frac{1}{2}*} \tilde{R}_i^{-\frac{1}{2}} \tilde{B}_i^* (\bar{P}_i^* + \bar{P}_i)$$

$$+ \frac{1}{2} \sum_{j \neq i} (\bar{P}_j^* + \bar{P}_j)^* \tilde{B}_j \tilde{R}_j^{-1*} \tilde{B}_j^* (\bar{P}_i^* + \bar{P}_i) = 0,$$

$$\dot{\delta}_i = -\frac{1}{2} \left\langle (P_i^* + P_i) \tilde{S}, \tilde{S} \right\rangle,$$

with terminal boundary conditions:

$$P_i(T) = \tilde{Q}_i(T),$$
$$\bar{P}_i(T) = \tilde{Q}_i(T) + \bar{\tilde{Q}}_i(T),$$
$$\delta_i(T) = 0;$$

whenever these differential equations have a unique solution that does not blow up within $[0, T]$.

Proof 25 (Proposition 17) *This proof is straight-forward by following the same procedure as in the proof of Theorem 4 in Chapter 9. Therefore, we do not develop the proof in this book given the big similarities with Theorem 4.*

In this chapter, we have presented an alternative way to deal with coupled constraints by means of evolutionary dynamics. Other type of constraints can be studied by performing a projection over the feasible region or by introducing barrier functions in the cost functional. The latter approach would require the modification of the guess functional in an appropriate manner.

10.3   Exercise

1. Propose a discrete-time mean-field-type game problem involving input coupled constraints and compute the corresponding semi-explicit solution by using the invariance set property of some evolutionary dynamics.

    Hint: Use the projection dynamics in discrete time in order to transformed the constrained discrete-time problem into a standard non-constrained discrete-time mean-field-type game (see e.g., [129]).

Part V

Discrete-Time
Mean-Field-Type Games

11

One-Dimensional Discrete-Time Mean-Field-Type Games

The previous part of this book, Part III, has been devoted to continuous-time mean-field-type games by using the direct method in its continuous version as presented in Figure 1.17 in the introductory Chapter 1.

In fact, the last discrete-time problem we addressed was in Part II where we solved the discrete-time optimal control problem, the difference game problem, and then the discrete-time mean-field game problem. In the next two chapters, we will focus back on the discrete-time game problems.

For the discrete-time version, we solve mainly the non-cooperative, fully-cooperative, and adversarial game problems. We do not discuss other type of games such as the co-opetitive, Stackelberg, hierarchical, or Berge games. Based on the discrete solutions for the non-cooperative and fully-cooperative cases, and the concepts we have presented for all the game-theoretical solution concepts in continuous-time, the reader can easily propose and solve the discrete-time counterpart cases.

In this new Part V, we treat the discrete-time mean-field-type games, which are solved in a semi-explicit way by means of the discrete version of the direct method (see the summary scheme in Figure 1.18). We start by presenting the scalar or equivalently known as one-dimensional case. Finally, some illustrative examples are shown to illustrate the results.

11.1   Discrete-Time Mean-Field-Type Game Set-up

Let us consider a set of $n \geq 2$ $(n \in \mathbb{N})$ decision-makers given by $\mathcal{N} = \{1, \ldots, n\}$. The decision-makers interact during a time window denoted by $[k..k + N - 1]$. The decision-maker $i \in \mathcal{N}$ selects a control action denoted by $u_{i,\ell}$ at time $\ell \in T_k$. The system state dynamics is

$$x_{\ell+1} = b_0 + b_1 x_\ell + \bar{b}_1 \mathbb{E}[x_\ell]$$
$$+ \sum_{j \in \mathcal{N}} b_{2j} u_{j,\ell} + \sum_{j \in \mathcal{N}} \bar{b}_{2j} \mathbb{E}[u_{j,\ell}] + \sigma_0 w_\ell, \qquad (11.1)$$

DOI: 10.1201/9781003098607-11

with $\ell \in [k..k + N - 1]$, and where $b_0, b_1, \bar{b}_1, b_{2j}, \bar{b}_{2j}, \sigma_0 \in \mathbb{R}$, for all $j \in \mathcal{N}$ are system parameters. On the other hand, $w_\ell \in \mathbb{R}$ denotes a stochastic disturbance with zero mean and unitary variance. Moreover, each decision-maker has an associated cost functional denoted by $L_i(x, u_1, \ldots, u_n)$ involving the system state from (11.1). The sequence of optimal control inputs for the ith decision-maker is

$$(u_{i,k}, \ldots, u_{i,k+N-1}).$$

When the optimal control sequences, for all $j \in \mathcal{N}$, are applied to the system state in (11.1), then we obtain a sequence for the system state

$$(x_k, \ldots, x_{k+N}).$$

The cost functional is composed by a running cost $g_i(x, u_1, \ldots, u_n)$ and a terminal cost given by $h_i(x_{k+N})$, where

$$h_i(x_{k+N}) = q_{i,k+N} x_{k+N}^2 + \bar{q}_{i,k+N} \mathbb{E}[x_{k+N}]^2, \tag{11.2a}$$

$$g_i(x, u_1, \ldots, u_n) = \sum_{\ell=k}^{k+N-1} \left[q_{i,\ell} x_\ell^2 + \bar{q}_{i,\ell} \mathbb{E}[x_\ell]^2 \right.$$
$$\left. + r_{i,\ell} u_{i,\ell}^2 + \bar{r}_{i,\ell} \mathbb{E}[u_{i,\ell}]^2 \right], \tag{11.2b}$$

and the weight parameters in the cost functional are $q_{i,\ell}, q_{i,\ell} + \bar{q}_{i,\ell} \geq 0$ and $r_{i,\ell}, r_{i,\ell} + \bar{r}_{i,\ell} > 0$, for all $i \in \mathcal{N}$, with time step ℓ and game stage k. Thus, the cost functional for the ith decision-maker is given by:

$$L_i(x, u_1, \ldots, u_n) = h_i(x_{k+N}) + g_i(x, u_1, \ldots, u_n). \tag{11.3}$$

We formulate below the two main game problems corresponding to the non-cooperative and fully-cooperative scenarios.

Problem 21 (Non-Cooperative Approach) *The non-cooperative mean-field-type game problem for each $i \in \mathcal{N}$ is given by*

$$\begin{cases} \underset{u_1, \ldots, u_n}{\text{minimize}} \, \mathbb{E}\left[L_i(x, u_1, \ldots, u_n) \right], \\ \\ \text{subject to} \\ \\ x_{\ell+1} = b_0 + b_1 x_\ell + \bar{b}_1 \mathbb{E}[x_\ell] + \sum_{j \in \mathcal{N}} b_{2j} u_{j,\ell} + \sum_{j \in \mathcal{N}} \bar{b}_{2j} \mathbb{E}[u_{j,\ell}] + \sigma_0 w_\ell, \\ \forall \ell \in [k..k + N - 1], \end{cases} \tag{11.4}$$

where the initial condition x_k is given, and with weight parameters in the cost functional given by $q_{i,\ell} \geq 0$, $q_{i,\ell} + \bar{q}_{i,\ell} \geq 0$, $r_{i,\ell} > 0$, $r_{i,\ell} + \bar{r}_{i,\ell} > 0$, for all $\ell \in [k..k + N - 1]$.

In contrast, the cooperative game approach is as follows:

Problem 22 (Fully-Cooperative Approach) *The fully-cooperative mean-field-type game problem is given by*

$$
\begin{cases}
L_0(x, u_1, \ldots, u_n) = \displaystyle\sum_{i \in \mathcal{N}} L_i(x, u_1, \ldots, u_n), \\[2mm]
\underset{u_1, \ldots, u_n}{\text{minimize}} \ \mathbb{E}[L_0(x, u_1, \ldots, u_n)], \\[4mm]
\text{subject to} \\[4mm]
x_{\ell+1} = b_0 + b_1 x_\ell + \bar{b}_1 \mathbb{E}[x_\ell] + \displaystyle\sum_{j \in \mathcal{N}} b_{2j} u_{j,\ell} + \sum_{j \in \mathcal{N}} \bar{b}_{2j} \mathbb{E}[u_{j,\ell}] + \sigma_0 w_\ell, \\[4mm]
\forall \ell \in [k..k+N-1],
\end{cases}
\tag{11.5}
$$

where the initial condition x_k is given, and with the weight parameters in the cost functional are as follows:

$$
q_{0,\ell} = \sum_{i \in \mathcal{N}} q_{i,\ell} \geq 0,
$$

$$
q_{0,\ell} + \bar{q}_{0,\ell} = \sum_{i \in \mathcal{N}} q_{i,\ell} + \bar{q}_{i,\ell} \geq 0,
$$

$r_{i,\ell} > 0$, $r_{i,\ell} + \bar{r}_{i,\ell} > 0$, *for all time instants $\ell \in [k..k+N-1]$.*

The discrete-time mean-field-type games also incorporate risk terms. This incorporation of risk terms is the same as we studied in Chapter 4 in (4.4). Remark 26 highlights the fact we have variance terms in the discrete-time counterpart as well.

Remark 26 *The expectation of the cost function in (11.3) includes mean-variance terms for both the system states and the control inputs, i.e., $\mathbb{E}[x_\ell]$, $\mathbb{E}[u_{i,\ell}]$,*

$$
\text{var}[x_\ell] = \mathbb{E}[(x_\ell - \mathbb{E}[x_\ell])^2],
$$

$$
\text{var}[u_{i,\ell}] = \mathbb{E}[(u_{i,\ell} - \mathbb{E}[u_{i,\ell}])^2], \ i \in \mathcal{N},
$$

for all $\ell \in [k..k+N-1]$.

Likewise, we conveniently express the evolution of the system state depending on the subtraction $(x_\ell - \mathbb{E}[x_\ell])$ and the expected term $\mathbb{E}[x_\ell]$ as shown in Remark 27.

Remark 27 *The system in (11.1) can be written as*

$$
x_{\ell+1} = b_0 + b_1(x_\ell - \mathbb{E}[x_\ell]) + (b_1 + \bar{b}_1)\mathbb{E}[x_\ell] + \sum_{j \in \mathcal{N}} b_{2j}(u_{j,\ell} - \mathbb{E}[u_{j,\ell}])
$$

$$+ \sum_{j \in \mathcal{N}} (b_{2j} + \bar{b}_{2j}) \mathbb{E}[u_{j,\ell}] + \sigma_0 w_\ell, \tag{11.6a}$$

and

$$\mathbb{E}[x_{\ell+1}] = b_0 + (b_1 + \bar{b}_1) \mathbb{E}[x_\ell] + \sum_{j \in \mathcal{N}} (b_{2j} + \bar{b}_{2j}) \mathbb{E}[u_{j,\ell}], \tag{11.6b}$$

corresponds to the evolution of the expected state $\mathbb{E}[x_{\ell+1}]$.

Each decision-maker plays a best response strategy, and the objective consists of reaching a Nash equilibrium. Both concepts are defined next in the context of discrete-time games. We still use the same standard definitions presented in the previous chapters adjusted to the problem treated in this Chapter.

Definition 27 *(Mean-Field-Type Best-Response Sequence) For the non-cooperative approach, any control input sequence* $(u^*_{i,k}, \ldots, u^*_{i,k+N-1})$, *such that*

$$(u^*_{i,k}, \ldots, u^*_{i,k+N-1}) \in \arg \min_{u_1, \ldots, u_n} \mathbb{E}\left[L_i(x, u_1, \ldots, u_n)\right],$$

is a mean-field-type best-response strategic sequence of the decision-maker $i \in \mathcal{N}$ *against the strategies*

$$(u_1, \ldots, u_{i-1}, u_{i+1}, \ldots, u_n)$$

selected from the other decision-makers $\mathcal{N} \setminus \{i\}$.

Thus, the set of mean-field-type Nash equilibria is defined by using the mean-field-type best-response concept, i.e.,

Definition 28 *(Mean-Field-Type Nash equilibrium) Any control input strategic sequence profile* $(u^*_1, \ldots, u^*_n)$ *such that the sequence* $(u^*_{i,k}, \ldots, u^*_{i,k+N})$ *is a best response against* $(u_1, \ldots, u_{i-1}, u_{i+1}, \ldots, u_n)$, *for all* $i \in \mathcal{N}$.

We compute then the semi-explicit solution by applying the discrete-time direct method. In this case, it is not required to compute the integration formula we have been using in the continuous-time game problems, but we simply apply the telescopic sum.

11.2 Semi-explicit Solution of the Discrete-Time Non-Cooperative Mean-Field-Type Game Problem

The main result of this chapter is presented in Proposition 18 referring to the semi-explicit computation of the Nash equilibrium. This result is more suitable to be implemented in real applications than the continuous-time solution. This is supported by the fact that the existing current technology easily allow the implementation of difference equations as we obtain in this result. In this regard, this solution can even be considered as explicit solution given that it is not require to use solvers to compute ODEs.

Proposition 18 *The explicit solution of Problem 21 is given by the following optimal control inputs:*

$$u_{i,\ell}^* = \mathbb{E}[u_{i,\ell}^*] - \eta_{i,\ell}(x_\ell - \mathbb{E}[x_\ell]), \tag{11.7a}$$

$$\mathbb{E}[u_{i,\ell}^*] = -\bar{\eta}_{i,\ell}\mathbb{E}[x_\ell] - \hat{\eta}_{i,\ell}, \tag{11.7b}$$

with auxiliary variables $\eta_{i,\ell}$, $\bar{\eta}_{i,\ell}$, $\hat{\eta}_{i,\ell}$, $c_{i,\ell}$, and $\bar{c}_{i,\ell}$ as follows:

$$\eta_{i,\ell} = \frac{\alpha_{i,\ell+1}}{c_{i,\ell}}\left(-b_{2i}\sum_{j\in\mathcal{N}\setminus\{i\}}b_{2j}\eta_{j,\ell} + b_1 b_{2i}\right),$$

$$\bar{\eta}_{i,\ell} = \frac{\beta_{i,\ell+1}}{\bar{c}_{i,\ell}}\left((b_1+\bar{b}_1)(b_{2i}+\bar{b}_{2i}) - \sum_{j\in\mathcal{N}\setminus\{i\}}(b_{2i}+\bar{b}_{2i})(b_{2j}+\bar{b}_{2j})\bar{\eta}_{j,\ell}\right),$$

$$\hat{\eta}_{i,\ell} = \frac{(b_{2i}+\bar{b}_{2i})}{\bar{c}_{i,\ell}}\left(\beta_{i,\ell+1}b_0 - \beta_{i,\ell+1}\sum_{j\in\mathcal{N}\setminus\{i\}}(b_{2j}+\bar{b}_{2j})\hat{\eta}_{j,\ell} + \frac{\gamma_{i,\ell+1}}{2}\right),$$

$$c_{i,\ell} = r_{i,\ell} + \alpha_{i,\ell+1}b_{2i}^2,$$

$$\bar{c}_{i,\ell} = r_{i,\ell} + \bar{r}_{i,\ell} + \beta_{i,\ell+1}(b_{2i}+\bar{b}_{2i})^2,$$

for all time instants $\ell \in [k..k+N-1]$, $i \in \mathcal{N}$, and the best-response cost for each decision-maker $i \in \mathcal{N}$ at from time instant k up to N is

$$\mathbb{E}[L_i(x, u_1^*, \ldots, u_n^*)] = \alpha_{i,k}\mathbb{E}[(x_k - \mathbb{E}[x_k])^2]$$
$$+ \beta_{i,k}\mathbb{E}[x_k]^2 + \gamma_{i,k}\mathbb{E}[x_k] + \delta_{i,k}, \tag{11.8}$$

where $\alpha_{i,\ell}$, $\beta_{i,\ell}$, $\gamma_{i,\ell}$, and $\delta_{i,\ell}$ solve the following difference backward equations:

$$\alpha_{i,\ell} = \alpha_{i,\ell+1}\left(b_1 - \sum_{j\in\mathcal{N}\setminus\{i\}}b_{2j}\eta_{j,\ell}\right)^2 + q_{i,\ell} - c_{i,\ell}\eta_{i,\ell}^2,$$

$$\beta_{i,\ell} = \beta_{i,\ell+1}\left(b_1 + \bar{b}_1 - \sum_{j\in\mathcal{N}\setminus\{i\}}(b_{2j}+\bar{b}_{2j})\bar{\eta}_{j,\ell}\right)^2 + q_{i,\ell} + \bar{q}_{i,\ell} - \bar{c}_{i,\ell}\bar{\eta}_{i,\ell}^2,$$

$$\gamma_{i,\ell} = \left(\gamma_{i,\ell+1} + 2\beta_{i,\ell+1}\left[b_0 - \sum_{j\in\mathcal{N}\setminus\{i\}}(b_{2j}+\bar{b}_{2j})\hat{\eta}_{j,\ell}\right]\right)(b_1+\bar{b}_1)$$

$$+ \left(2\beta_{i,\ell+1}\left[\sum_{j\in\mathcal{N}\setminus\{i\}}(b_{2j}+\bar{b}_{2j})\hat{\eta}_{j,\ell}\right] - \gamma_{i,\ell+1} - 2\beta_{i,\ell+1}b_0\right).$$

$$\cdot \left[\sum_{j \in \mathcal{N} \setminus \{i\}} (b_{2j} + \bar{b}_{2j}) \bar{\eta}_{j,\ell} \right] - 2\bar{c}_{i,\ell} \bar{\eta}_{i,\ell} \hat{\eta}_{i,\ell},$$

$$\delta_{i,\ell} = \delta_{i,\ell+1} + \alpha_{i,\ell+1} \sigma_0^2 \mathbb{E} w_\ell^2 + \bar{c}_{i,\ell} \hat{\eta}_{i,\ell}^2 + \beta_{i,\ell+1} \left[\sum_{j \in \mathcal{N} \setminus \{i\}} (b_{2j} + \bar{b}_{2j}) \hat{\eta}_{j,\ell} - b_0 \right]^2$$

$$- \gamma_{i,\ell+1} \sum_{j \in \mathcal{N} \setminus \{i\}} (b_{2j} + \bar{b}_{2j}) \hat{\eta}_{j,\ell},$$

for all $\ell \in [k..k + N - 1], i \in \mathcal{N}$ with the terminal boundary conditions:

$$\alpha_{i,k+N} = q_{i,k+N},$$
$$\beta_{i,k+N} = q_{i,k+N} + \bar{q}_{i,k+N},$$
$$\gamma_{i,k+N} = 0,$$
$$\delta_{i,k+N} = 0.$$

Next, we present the step-by-step computation of the explicit solution for Problem 21. The reader will notice that the procedure is even easier than the one for the continuous-time version. However, the price to pay for this simplicity is the length of the required algebraic operations.

Proof 26 (Proposition 18) *According to Remark 26, we re-write the cost functional in terms of variance terms, i.e.,*

$$\mathbb{E}[L_i(x, u_1, \ldots, u_n)] = q_{i,k+N} \mathbb{E}[(x_{k+N} - \mathbb{E}[x_{k+N}])^2]$$
$$+ (q_{i,k+N} + \bar{q}_{i,k+N}) \mathbb{E}[x_{k+N}]^2$$
$$+ \mathbb{E} \sum_{\ell=k}^{k+N-1} q_{i,\ell} (x_\ell - \mathbb{E}[x_\ell])^2 + \mathbb{E} \sum_{\ell=k}^{k+N-1} (q_{i,\ell} + \bar{q}_{i,\ell}) \mathbb{E}[x_\ell]^2$$
$$+ \mathbb{E} \sum_{\ell=k}^{k+N-1} r_{i,\ell} (u_{i,\ell} - \mathbb{E}[u_{i,\ell}])^2 + \mathbb{E} \sum_{\ell=k}^{k+N-1} (r_{i,\ell} + \bar{r}_{i,\ell}) \mathbb{E}[u_{i,\ell}]^2,$$

The first step consists of propose an appropriate guess functional for the optimal cost from time k up to terminal time N. Inspired by the structure of the cost function we propose a guess functional (ansatz), i.e.,

$$f_{i,\ell} = \alpha_{i,\ell} (x_\ell - \mathbb{E}[x_\ell])^2 + \beta_{i,\ell} \mathbb{E}[x_\ell]^2 + \gamma_{i,\ell} \mathbb{E}[x_\ell] + \delta_{i,\ell}, \qquad (11.9)$$

for all $\ell \in [k..k + N - 1]$, where $\alpha_{i,\ell}$, $\beta_{i,\ell}$, $\gamma_{i,\ell}$, and $\delta_{i,\ell}$ are deterministic difference equations function of time.

Now, we apply the telescopic sum decomposition given by

$$f_{i,k+N} = f_{i,k} + \sum_{\ell=k}^{k+N-1} (f_{i,\ell+1} - f_{i,\ell}).$$

in order to derive the difference between the guess functional and the cost functionals. Hence, we replace the ansatz for time instant $k + 1$ in terms of $\alpha_{i,k+1}$, $\beta_{i,k+1}$, $\gamma_{i,k+1}$, $\delta_{i,k+1}$, and the system state.

$$
f_{i,k+N} - f_{i,k} = \sum_{\ell=k}^{k+N-1} \alpha_{i,\ell+1} (x_{\ell+1} - \mathbb{E}[x_{\ell+1}])^2
$$
$$
- \sum_{\ell=k}^{k+N-1} \alpha_{i,\ell} (x_\ell - \mathbb{E}[x_\ell])^2
$$
$$
+ \sum_{\ell=k}^{k+N-1} [\beta_{i,\ell+1} \mathbb{E}[x_{\ell+1}]^2 - \beta_{i,\ell} \mathbb{E}[x_\ell]^2]
$$
$$
+ \sum_{\ell=k}^{k+N-1} [\gamma_{i,\ell+1} \mathbb{E}[x_{\ell+1}] - \gamma_{i,\ell} \mathbb{E}[x_\ell]]
$$
$$
+ \sum_{\ell=k}^{k+N-1} [\delta_{i,\ell+1} - \delta_{i,\ell}].
$$

Therefore, we now compute the gap $\mathbb{E}[L_i(x, u_1, \ldots, u_n) - f_{i,k}]$. We require to find the optimal control inputs such that the cost $\mathbb{E}[L_i(x, u_1, \ldots, u_n)]$ matches with the optimal cost from time instant k up to N given by the ansatz $\mathbb{E}[f_{i,k}]$, i.e.,

$$
\mathbb{E}\Big[L_i(x, u_1, \ldots, u_n) - \alpha_{i,k}(x_k - \mathbb{E}[x_k])^2 - \beta_{i,k}\mathbb{E}[x_k]^2 - \gamma_{i,k}\mathbb{E}[x_k] - \delta_{i,k} \Big]
$$
$$
= -\alpha_{i,k+N}\mathbb{E}[(x_{k+N} - \mathbb{E}[x_{k+N}])^2] - \beta_{i,k+N}\mathbb{E}[x_{k+N}]^2
$$
$$
- \gamma_{i,k+N}\mathbb{E}[x_{k+N}] - \delta_{i,k+N}
$$
$$
+ q_{i,k+N}\mathbb{E}[(x_{k+N} - \mathbb{E}[x_{k+N}])^2] + (q_{i,k+N} + \bar{q}_{i,k+N})\mathbb{E}[x_{k+N}]^2
$$
$$
+ \mathbb{E}\sum_{\ell=k}^{k+N-1} q_{i,\ell}(x_\ell - \mathbb{E}[x_\ell])^2 + \mathbb{E}\sum_{\ell=k}^{k+N-1}(q_{i,\ell} + \bar{q}_{i,\ell})\mathbb{E}[x_\ell]^2
$$
$$
+ \mathbb{E}\sum_{\ell=k}^{k+N-1} r_{i,\ell}(u_{i,\ell} - \mathbb{E}[u_{i,\ell}])^2 + \mathbb{E}\sum_{\ell=k}^{k+N-1}(r_{i,\ell} + \bar{r}_{i,\ell})\mathbb{E}[u_{i,\ell}]^2
$$
$$
+ \mathbb{E}\sum_{\ell=k}^{k+N-1} \alpha_{i,\ell+1}(x_{\ell+1} - \mathbb{E}[x_{\ell+1}])^2 - \mathbb{E}\sum_{\ell=k}^{k+N-1} \alpha_{i,\ell}(x_\ell - \mathbb{E}[x_\ell])^2
$$
$$
+ \mathbb{E}\sum_{\ell=k}^{k+N-1} [\beta_{i,\ell+1}\mathbb{E}[x_{\ell+1}]^2 - \beta_{i,\ell}\mathbb{E}[x_\ell]^2]
$$
$$
+ \mathbb{E}\sum_{\ell=k}^{k+N-1} [\gamma_{i,\ell+1}\mathbb{E}[x_{\ell+1}] - \gamma_{i,\ell}\mathbb{E}[x_\ell]]
$$

$$+ \mathbb{E} \sum_{\ell=k}^{k+N-1} [\delta_{i,\ell+1} - \delta_{i,\ell}].$$

From the previous expression, we need to calculate some terms. For the sake of organization and good explanation, we compute these terms separately before replacing in the gap $\mathbb{E}[L_i(x, u_1, \ldots, u_n) - f_{i,k}]$. Let us start by expressing the terms $(x_{\ell+1} - \mathbb{E}[x_{\ell+1}])^2$ and $\mathbb{E}[x_{\ell+1}]^2$ as function of $(x_\ell - \mathbb{E}[x_\ell])$, $(u_{i,\ell} - \mathbb{E}[u_{i,\ell}])$, $\mathbb{E}[x_\ell]$, and $\mathbb{E}[u_{i,\ell}]$. Then, according to Remark 27,

$$\mathbb{E}[x_{\ell+1}] = (b_1 + \bar{b}_1)\mathbb{E}[x_\ell] + (b_{2i} + \bar{b}_{2i})\mathbb{E}[u_{i,\ell}] + b_0 + \sum_{j \in \mathcal{N} \backslash \{i\}} (b_{2j} + \bar{b}_{2j})\mathbb{E}[u_{j,\ell}],$$

and then the square $\mathbb{E}[x_{\ell+1}]^2$ is

$$\mathbb{E}[x_{\ell+1}]^2 = (b_1 + \bar{b}_1)^2 \mathbb{E}[x_\ell]^2 + [b_0]^2 + (b_{2i} + \bar{b}_{2i})^2 \mathbb{E}[u_{i,\ell}]^2$$
$$+ \left[\sum_{j \in \mathcal{N} \backslash \{i\}} (b_{2j} + \bar{b}_{2j})\mathbb{E}[u_{j,\ell}] \right]^2$$
$$+ 2 \sum_{j \in \mathcal{N} \backslash \{i\}} (b_{2i} + \bar{b}_{2i})(b_{2j} + \bar{b}_{2j})\mathbb{E}[u_{j,\ell}]\mathbb{E}[u_{i,\ell}]$$
$$+ 2 \sum_{j \in \mathcal{N}} (b_1 + \bar{b}_1)(b_{2j} + \bar{b}_{2j})\mathbb{E}[u_{j,\ell}]\mathbb{E}[x_\ell],$$
$$+ 2(b_1 + \bar{b}_1)\mathbb{E}[x_\ell]b_0 + 2\sum_{j \in \mathcal{N}} (b_{2j} + \bar{b}_{2j})\mathbb{E}[u_{j,\ell}]b_0.$$

On the hand, by computing the difference between $x_{\ell+1}$ and its expected value $\mathbb{E}[x_{\ell+1}]$ yields

$$x_{\ell+1} - \mathbb{E}[x_{\ell+1}] = b_1(x_\ell - \mathbb{E}[x_\ell]) + \sum_{j \in \mathcal{N}} b_{2j}(u_{j,\ell} - \mathbb{E}[u_{j,\ell}]) + \sigma_0 w_\ell,$$

and the square of this difference is as follows:

$$(x_{\ell+1} - \mathbb{E}[x_{\ell+1}])^2 = b_1^2(x_\ell - \mathbb{E}[x_\ell])^2 + b_{2i}^2(u_{i,\ell} - \mathbb{E}[u_{i,\ell}])^2$$
$$+ \left[\sum_{j \in \mathcal{N} \backslash \{i\}} b_{2j}(u_{j,\ell} - \mathbb{E}[u_{j,\ell}]) \right]^2$$
$$+ 2b_{2i}(u_{i,\ell} - \mathbb{E}[u_{i,\ell}]) \sum_{j \in \mathcal{N} \backslash \{i\}} b_{2j}(u_{j,\ell} - \mathbb{E}[u_{j,\ell}]) + \sigma^2 w_\ell^2$$
$$+ 2 \sum_{j \in \mathcal{N}} b_1(x_\ell - \mathbb{E}[x_\ell])b_{2j}(u_{j,\ell} - \mathbb{E}[u_{j,\ell}])$$
$$+ 2b_1(x_\ell - \mathbb{E}[x_\ell])\sigma_0 w_\ell + 2\sum_{j \in \mathcal{N}} b_{2j}(u_{j,\ell} - \mathbb{E}[u_{j,\ell}])\sigma_0 w_\ell.$$

Now, replacing the previously computed values $(x_{\ell+1} - \mathbb{E}[x_{\ell+1}])^2$ and $\mathbb{E}[x_{\ell+1}]^2$ in the gap $\mathbb{E}[L_i(x, u_1, \ldots, u_n)] - f_{i,k}]$ one arrives at the next expression:

$$\mathbb{E}[L_i(x, u_1, \ldots, u_n)] - \alpha_{i,k}(x_k - \mathbb{E}[x_k])^2 - \beta_{i,k}\mathbb{E}[x_k]^2 - \gamma_{i,k}\mathbb{E}[x_k] - \delta_{i,k}]$$

$$= [q_{i,k+N} - \alpha_{i,k+N}]\mathbb{E}[(x_{k+N} - \mathbb{E}[x_{k+N}])^2]$$

$$+ [q_{i,k+N} + \bar{q}_{i,k+N} - \beta_{i,k+N}]\mathbb{E}[x_{k+N}]^2$$

$$- \gamma_{i,k+N}\mathbb{E}[x_{k+N}] - \delta_{i,k+N}$$

$$+ \mathbb{E} \sum_{\ell=k}^{k+N-1} \left(q_{i,\ell} + \alpha_{i,\ell+1}b_1^2 - \alpha_{i,\ell}\right)(x_\ell - \mathbb{E}[x_\ell])^2$$

$$+ \mathbb{E} \sum_{\ell=k}^{k+N-1} \left(q_{i,\ell} + \bar{q}_{i,\ell} + \beta_{i,\ell+1}(b_1 + \bar{b}_1)^2 - \beta_{i,\ell}\right)\mathbb{E}[x_\ell]^2$$

$$+ \mathbb{E} \sum_{\ell=k}^{k+N-1} \left(\gamma_{i,\ell+1}(b_1 + \bar{b}_1) - \gamma_{i,\ell} + 2\beta_{i,\ell+1}(b_1 + \bar{b}_1)b_0\right)\mathbb{E}[x_\ell]$$

$$+ \mathbb{E} \sum_{\ell=k}^{k+N-1} \left(r_{i,\ell} + \alpha_{i,\ell+1}b_{2i}^2\right)(u_{i,\ell} - \mathbb{E}[u_{i,\ell}])^2$$

$$+ \mathbb{E} \sum_{\ell=k}^{k+N-1} 2\alpha_{i,\ell+1}\left(b_{2i} \sum_{j\in\mathcal{N}\backslash\{i\}} b_{2j}(u_{j,\ell} - \mathbb{E}[u_{j,\ell}])\right.$$

$$\left. + b_1(x_\ell - \mathbb{E}[x_\ell])b_{2i}\right)(u_{i,\ell} - \mathbb{E}[u_{i,\ell}])$$

$$+ \mathbb{E} \sum_{\ell=k}^{k+N-1} \left(r_{i,\ell} + \bar{r}_{i,\ell} + \beta_{i,\ell+1}(b_{2i} + \bar{b}_{2i})^2\right)\mathbb{E}[u_{i,\ell}]^2$$

$$+ \mathbb{E} \sum_{\ell=k}^{k+N-1} 2\Big(\beta_{i,\ell+1}\Big((b_{2i} + \bar{b}_{2i})(b_1 + \bar{b}_1)\mathbb{E}[x_\ell] + (b_{2i} + \bar{b}_{2i})b_0$$

$$+ \sum_{j\in\mathcal{N}\backslash\{i\}} (b_{2i} + \bar{b}_{2i})(b_{2j} + \bar{b}_{2j})\mathbb{E}[u_{j,\ell}]\Big) + \frac{1}{2}\gamma_{i,\ell+1}(b_{2i} + \bar{b}_{2i})\Big)\mathbb{E}[u_{i,\ell}]$$

$$+ \mathbb{E} \sum_{\ell=k}^{k+N-1} \alpha_{i,\ell+1}\left[\sum_{j\in\mathcal{N}\backslash\{i\}} b_{2j}(u_{j,\ell} - \mathbb{E}[u_{j,\ell}])\right]^2$$

$$+ \mathbb{E} \sum_{\ell=k}^{k+N-1} 2\alpha_{i,\ell+1} \sum_{j\in\mathcal{N}\backslash\{i\}} b_1(x_\ell - \mathbb{E}[x_\ell])b_{2j}(u_{j,\ell} - \mathbb{E}[u_{j,\ell}])$$

$$+ \mathbb{E} \sum_{\ell=k}^{k+N-1} \beta_{i,\ell+1}\left(\sum_{j\in\mathcal{N}\backslash\{i\}} (b_{2j} + \bar{b}_{2j})\mathbb{E}[u_{j,\ell}]\right)^2$$

$$+ \mathbb{E} \sum_{\ell=k}^{k+N-1} \left(2\beta_{i,\ell+1}[(b_1 + \bar{b}_1)\mathbb{E}[x_\ell] + b_0] + \gamma_{i,\ell+1} \right) \cdot$$

$$\cdot \sum_{j \in \mathcal{N}\backslash\{i\}} (b_{2j} + \bar{b}_{2j})\mathbb{E}[u_{j,\ell}]$$

$$+ \mathbb{E} \sum_{\ell=k}^{k+N-1} \alpha_{i,\ell+1}\sigma_0^2 w_\ell^2 + \mathbb{E} \sum_{\ell=k}^{k+N-1} \delta_{i,\ell+1}$$

$$- \mathbb{E} \sum_{\ell=k}^{k+N-1} \delta_{i,\ell} + \mathbb{E} \sum_{\ell=k}^{k+N-1} \beta_{i,\ell+1}[b_0]^2.$$

Before optimizing over the control inputs, we consider some changes of variables introduced in the statement of Proposition 18. Then, we perform square completion for the quadratic terms depending on the control inputs of the ith decision-maker. First, for the terms involving $(u_{i,\ell} - \mathbb{E}[u_{i,\ell}])^2$ yields

$$c_{i,\ell} = r_{i,\ell} + \alpha_{i,\ell+1}b_{2i}^2, \tag{11.10a}$$

$$\eta_{i,\ell} = \frac{\alpha_{i,\ell+1}}{c_{i,\ell}} \left(-b_{2i} \sum_{j \in \mathcal{N}\backslash\{i\}} b_{2j}\eta_{j,\ell} + b_1 b_{2i} \right). \tag{11.10b}$$

Equivalently, using

$$\phi_{ij} = \frac{\alpha_{i,\ell+1}}{c_{i,\ell}} b_{2i} b_{2j},$$

$$\varrho_i = \frac{\alpha_{i,\ell+1}}{c_{i,\ell}} b_1 b_{2i},$$

the parameters $\eta_{1,\ell}, \ldots, \eta_{n,\ell}$ solve the following equality:

$$\begin{bmatrix} 1 & \phi_{12} & \cdots & \phi_{1n} \\ \phi_{21} & 1 & \cdots & \phi_{2n} \\ \vdots & \ddots & \ddots & \vdots \\ \phi_{n1} & \cdots & \phi_{n(n-1)} & 1 \end{bmatrix} \begin{bmatrix} \eta_{1,\ell} \\ \eta_{2,\ell} \\ \vdots \\ \eta_{n,\ell} \end{bmatrix} = \begin{bmatrix} \varrho_1 \\ \varrho_2 \\ \vdots \\ \varrho_n \end{bmatrix}.$$

Then, it follows that:

$$c_{i,\ell}[(u_{i,\ell} - \mathbb{E}[u_{i,\ell}])^2 + 2(u_{i,\ell} - \mathbb{E}[u_{i,\ell}])\eta_{i,\ell}(x_\ell - \mathbb{E}[x_\ell])]$$
$$= c_{i,\ell}(u_{i,\ell} - \mathbb{E}[u_{i,\ell}] + \eta_{i,\ell}(x_\ell - \mathbb{E}[x_\ell]))^2 - c_{i,\ell}\eta_{i,\ell}^2(x_\ell - \mathbb{E}[x_\ell])^2,$$

from which it is concluded the optimal control input for the subtraction $u_{i,\ell} - \mathbb{E}[u_{i,\ell}]$ is in the following state-feedback form:

$$u_{i,\ell} - \mathbb{E}[u_{i,\ell}] = -\eta_{i,\ell} \left(x_\ell - \mathbb{E}[x_\ell] \right), \quad \forall i \in \mathcal{N}.$$

On the other hand, for the terms involving the expected value $\mathbb{E}[u_{i,\ell}]$, consider the following auxiliary variables

$$\bar{c}_{i,\ell} = r_{i,\ell} + \bar{r}_{i,\ell} + \beta_{i,\ell+1}(b_{2i} + \bar{b}_{2i})^2, \tag{11.11a}$$

$$\bar{\eta}_{i,\ell} = \frac{1}{\bar{c}_{i,\ell}}\left(\beta_{i,\ell+1}(b_1 + \bar{b}_1)(b_{2i} + \bar{b}_{2i})\right. \tag{11.11b}$$

$$\left. - \beta_{i,\ell+1}\sum_{j\in\mathcal{N}\backslash\{i\}}(b_{2i} + \bar{b}_{2i})(b_{2j} + \bar{b}_{2j})\bar{\eta}_{j,\ell}\right), \tag{11.11c}$$

$$\hat{\eta}_{i,\ell} = -\frac{1}{\bar{c}_{i,\ell}}\beta_{i,\ell+1}\sum_{j\in\mathcal{N}\backslash\{i\}}(b_{2i} + \bar{b}_{2i})(b_{2j} + \bar{b}_{2j})\hat{\eta}_{j,\ell}$$

$$+ \frac{\gamma_{i,\ell+1}}{2\bar{c}_{i,\ell}}(b_{2i} + \bar{b}_{2i}) + \frac{\beta_{i,\ell+1}}{\bar{c}_{i,\ell}}(b_{2i} + \bar{b}_{2i})b_0. \tag{11.11d}$$

Using the terms

$$\bar{\phi}_{ij} = \hat{\phi}_{ij} = \frac{1}{\bar{c}_{i,\ell}}\beta_{i,\ell+1}(b_{2i} + \bar{b}_{2i})(b_{2j} + \bar{b}_{2j}),$$

$$\bar{\varrho}_i = \frac{1}{\bar{c}_{i,\ell}}\beta_{i,\ell+1}(b_1 + \bar{b}_1)(b_{2i} + \bar{b}_{2i}),$$

$$\hat{\varrho}_i = \frac{1}{2\bar{c}_{i,\ell}}\gamma_{i,\ell+1}(b_{2i} + \bar{b}_{2i}) + \frac{\beta_{i,\ell+1}}{\bar{c}_{i,\ell}}(b_{2i} + \bar{b}_{2i})b_0,$$

then $\bar{\eta}_{1,\ell}, \ldots, \bar{\eta}_{n,\ell}$, and $\hat{\eta}_{1,\ell}, \ldots, \hat{\eta}_{n,\ell}$ solve the following equalities:

$$\begin{bmatrix} 1 & \bar{\phi}_{12} & \cdots & \bar{\phi}_{1n} \\ \bar{\phi}_{21} & 1 & \cdots & \bar{\phi}_{2n} \\ \vdots & \ddots & \ddots & \vdots \\ \bar{\phi}_{n1} & \cdots & \bar{\phi}_{n(n-1)} & 1 \end{bmatrix}\begin{bmatrix} \bar{\eta}_{1,\ell} \\ \bar{\eta}_{2,\ell} \\ \vdots \\ \bar{\eta}_{n,\ell} \end{bmatrix} = \begin{bmatrix} \bar{\varrho}_1 \\ \bar{\varrho}_2 \\ \vdots \\ \bar{\varrho}_n \end{bmatrix},$$

$$\begin{bmatrix} 1 & \hat{\phi}_{12} & \cdots & \hat{\phi}_{1n} \\ \hat{\phi}_{21} & 1 & \cdots & \hat{\phi}_{2n} \\ \vdots & \ddots & \ddots & \vdots \\ \hat{\phi}_{n1} & \cdots & \hat{\phi}_{n(n-1)} & 1 \end{bmatrix}\begin{bmatrix} \hat{\eta}_{1,\ell} \\ \hat{\eta}_{2,\ell} \\ \vdots \\ \hat{\eta}_{n,\ell} \end{bmatrix} = \begin{bmatrix} \hat{\varrho}_1 \\ \hat{\varrho}_2 \\ \vdots \\ \hat{\varrho}_n \end{bmatrix}.$$

Next, it follows that the square completion for terms depending on $\mathbb{E}[u_{i,\ell}]$ is

$$\bar{c}_{i,\ell}\left[\mathbb{E}[u_{i,\ell}]^2 + 2\mathbb{E}[u_{i,\ell}]\left(\bar{\eta}_{i,\ell}\mathbb{E}[x_\ell] + \hat{\eta}_{i,\ell}\right)\right] = \bar{c}_{i,\ell}\left(\mathbb{E}[u_{i,\ell}] + \bar{\eta}_{i,\ell}\mathbb{E}[x_\ell] + \hat{\eta}_{i,\ell}\right)^2$$
$$- \bar{c}_{i,\ell}\hat{\eta}_{i,\ell}^2 - \bar{c}_{i,\ell}\bar{\eta}_{i,\ell}^2\mathbb{E}[x_\ell]^2$$
$$- 2\bar{c}_{i,\ell}\bar{\eta}_{i,\ell}\mathbb{E}[x_\ell]\hat{\eta}_{i,\ell},$$

from which it is concluded that the optimal expected control input is of the following state-feedback form:

$$\mathbb{E}[u_{i,\ell}] = -\bar{\eta}_{i,\ell}\mathbb{E}[x_\ell] - \hat{\eta}_{i,\ell}, \quad \forall i \in \mathcal{N}.$$

As the last step, we replace back all the terms obtained from the optimization process by means of the two square completion we just performed above in the gap $\mathbb{E}[L_i(x, u_1, \ldots, u_n) - f_{i,k}]$. *One arrives at:*

$$\mathbb{E}[L_i(x, u_1, \ldots, u_n) - \alpha_{i,k}(x_k - \mathbb{E}[x_k])^2 - \beta_{i,k}\mathbb{E}[x_k]^2 - \gamma_{i,k}\mathbb{E}[x_k] - \delta_{i,k}]$$

$$= [q_{i,k+N} - \alpha_{i,k+N}]\mathbb{E}[(x_\ell - \mathbb{E}[x_{k+N}])^2]$$

$$+ [q_{i,k+N} + \bar{q}_{i,k+N} - \beta_{i,k+N}]\mathbb{E}[x_{k+N}]^2 - \gamma_{i,k+N}\mathbb{E}[x_{k+N}] - \delta_{i,k+N}$$

$$+ \mathbb{E}\sum_{\ell=k}^{k+N-1} \left(q_{i,\ell} + \alpha_{i,\ell+1}b_1^2 - \alpha_{i,\ell} - c_{i,\ell}\eta_{i,\ell}^2 + \alpha_{i,\ell+1}\left[\sum_{j\in\mathcal{N}\setminus\{i\}} b_{2j}\eta_{j,\ell}\right]^2 \right.$$

$$\left. - 2\alpha_{i,\ell+1}\sum_{j\in\mathcal{N}\setminus\{i\}} b_1 b_{2j}\eta_{j,\ell} \right)(x_\ell - \mathbb{E}[x_\ell])^2$$

$$+ \mathbb{E}\sum_{\ell=k}^{k+N-1} \left(q_{i,\ell} + \bar{q}_{i,\ell} + \beta_{i,\ell+1}(b_1 + \bar{b}_1)^2 - \beta_{i,\ell} - \bar{c}_{i,\ell}\bar{\eta}_{i,\ell}^2 \right.$$

$$+ \beta_{i,\ell+1}\left[\sum_{j\in\mathcal{N}\setminus\{i\}} (b_{2j} + \bar{b}_{2j})\bar{\eta}_{j,\ell}\right]^2$$

$$\left. - 2\beta_{i,\ell+1}(b_1 + \bar{b}_1)\sum_{j\in\mathcal{N}\setminus\{i\}} (b_{2j} + \bar{b}_{2j})\bar{\eta}_{j,\ell} \right)\mathbb{E}[x_\ell]^2$$

$$+ \mathbb{E}\sum_{\ell=k}^{k+N-1} \left(\gamma_{i,\ell+1}(b_1 + \bar{b}_1) - \gamma_{i,\ell} + 2\beta_{i,\ell+1}(b_1 + \bar{b}_1)b_0 - 2\bar{c}_{i,\ell}\bar{\eta}_{i,\ell}\hat{\eta}_{i,\ell} \right.$$

$$+ 2\beta_{i,\ell+1}\left[\sum_{j\in\mathcal{N}\setminus\{i\}} (b_{2j} + \bar{b}_{2j})\bar{\eta}_{j,\ell}\right]\left[\sum_{j\in\mathcal{N}\setminus\{i\}} (b_{2j} + \bar{b}_{2j})\hat{\eta}_{j,\ell}\right]$$

$$- 2\beta_{i,\ell+1}b_0\sum_{j\in\mathcal{N}\setminus\{i\}} (b_{2j} + \bar{b}_{2j})\bar{\eta}_{j,\ell} - \gamma_{i,\ell+1}\sum_{j\in\mathcal{N}\setminus\{i\}} (b_{2j} + \bar{b}_{2j})\bar{\eta}_{j,\ell}$$

$$\left. - 2\beta_{i,\ell+1}(b_1 + \bar{b}_1)\sum_{j\in\mathcal{N}\setminus\{i\}} (b_{2j} + \bar{b}_{2j})\hat{\eta}_{j,\ell} \right)\mathbb{E}[x_\ell]$$

$$+ \mathbb{E}\sum_{\ell=k}^{k+N-1} \left(\delta_{i,\ell+1} - \delta_{i,\ell} + \alpha_{i,\ell+1}\sigma_0^2 w_\ell^2 \right.$$

$$- 2\beta_{i,\ell+1}b_0\sum_{j\in\mathcal{N}\setminus\{i\}} (b_{2j} + \bar{b}_{2j})\hat{\eta}_{j,\ell}$$

$$+ \beta_{i,\ell+1}\left[\sum_{j\in\mathcal{N}\setminus\{i\}} (b_{2j} + \bar{b}_{2j})\hat{\eta}_{j,\ell}\right]^2 + \beta_{i,\ell+1}[b_0]^2$$

$$- \gamma_{i,\ell+1} \sum_{j \in \mathcal{N} \setminus \{i\}} (b_{2j} + \bar{b}_{2j}) \hat{\eta}_{j,\ell} + \bar{c}_{i,\ell} \hat{\eta}_{i,\ell}^2 \Big)$$

$$+ \mathbb{E} \sum_{\ell=k}^{k+N-1} c_{i,\ell} \left(u_{i,\ell} - \mathbb{E}[u_{i,\ell}] + \eta_{i,\ell}(x_\ell - \mathbb{E}[x_\ell]) \right)^2$$

$$+ \mathbb{E} \sum_{\ell=k}^{k+N-1} \bar{c}_{i,\ell} \left(\mathbb{E}[u_{i,\ell}] + \bar{\eta}_{i,\ell} \mathbb{E}[x_\ell] + \hat{\eta}_{i,\ell} \right)^2 .$$

The right-hand side in the latter expression is divided into different groups as explained next:

- *Those terms at the terminal time $k + N$ defining the boundary conditions for α_i, β_i, γ_i and δ_i*

- *Those terms whose common factor is given by $(x_\ell - \mathbb{E}[x_\ell])^2$, defining the difference equation α_i*

- *Those terms whose common factor is given by $\mathbb{E}[x_\ell]^2$, defining the difference equation β_i*

- *Those terms whose common factor is given by $\mathbb{E}[x_\ell]$, defining the difference equation γ_i*

- *Those terms independent from the system state, control inputs, and their expectation, defining the difference equation δ_i*

- *And those terms that define the optimal control inputs.*

Finally, we perform the process identification, obtaining the announced results and completing the proof.

Following the same reasoning we applied in Chapter 4, we continue by solving the fully-cooperative mean-field-type game in discrete-time. Recall that the solution of the cooperative case corresponds to the mean-field-type control problem.

11.3 Semi-explicit Solution of the Discrete-Time Cooperative Mean-Field-Type Game Problem

Proposition 19 presents the explicit solution for the cooperative game problem, or equivalently, the solution for a mean-field-type control. We observe that the same α_0, β_0, γ_0, and δ_0 are used for all the decision-makers.

Proposition 19 *The explicit solution of Problem in (11.5) is given by the following optimal control inputs and optimal cost:*

$$u_\ell - \mathbb{E}[u_\ell] = -K_2^{-1}K_3(x_\ell - \mathbb{E}[x_\ell]),$$

$$\mathbb{E}[u_\ell] = -\tilde{K}_2^{-1}\tilde{K}_3\left(\beta_{0,\ell+1}\left[(b_1 + \bar{b}_1)\mathbb{E}[x_\ell] + b_0\right] + \frac{1}{2}\gamma_{0,\ell+1}\right),$$

$$\mathbb{E}[L_0(x, u_1^*, \ldots, u_n^*)] = \alpha_{0,k}(x_k - \mathbb{E}[x_k])^2 + \beta_{0,k}\mathbb{E}[x_k]^2 + \gamma_{0,k}\mathbb{E}[x_k] + \delta_{0,k},$$

with the variables:

$$K_1 = \left(R_\ell + \alpha_{0,\ell+1}b_2 b_2^\top\right)^{-\frac{1}{2}},$$
$$K_2 = R_\ell + \alpha_{0,\ell+1}b_2 b_2^\top,$$
$$K_3 = \alpha_{0,\ell+1}b_1 b_2,$$
$$\tilde{K}_1 = \left(R_\ell + \bar{R}_\ell + \beta_{0,\ell+1}(b_2 + \bar{b}_2)(b_2 + \bar{b}_2)^\top\right)^{-\frac{1}{2}},$$
$$\tilde{K}_2 = R_\ell + \bar{R}_\ell + \beta_{0,\ell+1}(b_2 + \bar{b}_2)(b_2 + \bar{b}_2)^\top,$$
$$\tilde{K}_3 = (b_2 + \bar{b}_2),$$

and where α, β, γ, and δ solve the following difference equations:

$$\alpha_{0,\ell} = \alpha_{0,\ell+1}b_1^2 + q_{0,\ell} - K_3^\top K_1^\top K_1 K_3,$$
$$\beta_{0,\ell} = \beta_{0,\ell+1}(b_1 + \bar{b}_1)^2 + (q_{0,\ell} + \bar{q}_{0,\ell}) - K_3^\top \tilde{K}_1^\top K_1 \tilde{K}_3 \beta_{0,\ell+1}^2 (b_1 + \bar{b}_1)^2,$$
$$\gamma_{0,\ell} = \gamma_{0,\ell+1}(b_1 + \bar{b}_1) + 2\beta_{0,\ell+1}(b_1 + \bar{b}_1)b_0$$
$$\quad - 2\beta_{0,\ell+1}(b_1 + \bar{b}_1)\tilde{K}_3^\top \tilde{K}_1^\top \tilde{K}_1 \tilde{K}_3\left(\beta_{0,\ell+1}b_0 + \frac{1}{2}\gamma_{0,\ell+1}\right),$$
$$\delta_{0,\ell} = \delta_{0,\ell+1} + \alpha_{0,\ell+1}\sigma_0^2 + \beta_{0,\ell+1}[b_0]^2$$
$$\quad - \tilde{K}_3^\top \tilde{K}_1^\top \tilde{K}_1 \tilde{K}_3\left(\beta_{0,\ell+1}^2[b_0]^2 + \frac{1}{4}\gamma_{0,\ell+1}^2 + \beta_{0,\ell+1}\gamma_{0,\ell+1}b_0\right) + \gamma_{0,\ell+1}b_0,$$

for all $\ell \in [k..k + N - 1]$ and with the terminal boundary conditions

$$\alpha_{0,k+N} = q_{0,k+N},$$
$$\beta_{0,k+N} = q_{0,k+N} + \bar{q}_{0,k+N},$$
$$\gamma_{0,k+N} = 0,$$
$$\delta_{0,k+N} = 0.$$

The computation of the explicit solution for the fully-cooperative mean-field-type game problem is developed next by following the discrete-time direct method.

Proof 27 (Proposition 19) *We develop this proof following the same reasoning applied in Proposition 18, so we postulate the same structure for the guess functional. Then, let us consider the following ansatz:*

$$f_{0,\ell} = \alpha_{0,\ell}(x_\ell - \mathbb{E}[x_\ell])^2 + \beta_{0,\ell}\mathbb{E}[x_\ell]^2 + \gamma_{0,\ell}\mathbb{E}[x_\ell] + \delta_{0,\ell}, \tag{11.12}$$

*for all $\ell \in [k..k + N - 1]$, being $\alpha_{0,\ell}$, $\beta_{0,\ell}$, $\gamma_{0,\ell}$, and $\delta_{0,\ell}$ deterministic differ-
ence equations function of time. Let us introduce the following matrices to
simplify notation in the cost functional as shown below:*

$$R_\ell = \mathrm{diag}([r_{1,\ell} \quad \ldots \quad r_{n,\ell}]),$$
$$\bar{R}_\ell = \mathrm{diag}([\bar{r}_{1,\ell} \quad \ldots \quad \bar{r}_{n,\ell}]),$$
$$b_2 = [b_{21} \quad \ldots \quad b_{2n}]^\top,$$
$$\bar{b}_2 = [\bar{b}_{21} \quad \ldots \quad \bar{b}_{2n}]^\top,$$
$$u_\ell = [u_{1,\ell} \quad \ldots \quad u_{n,\ell}]^\top,$$
$$\mathbb{E}[u_\ell] = [\mathbb{E}[u_{1,\ell}] \quad \ldots \quad \mathbb{E}[u_{n,\ell}]]^\top.$$

*Thus, the terms in the cost function in $\mathbb{E}[L_0(x, u_1, \ldots, u_n)]$ can be re-written
as follows:*

$$\sum_{j \in \mathcal{N}} r_{j,\ell}(u_{j,\ell} - \mathbb{E}[u_{j,\ell}])^2 = \langle R_\ell(u_\ell - \mathbb{E}[u_\ell]), u_\ell - \mathbb{E}[u_\ell] \rangle,$$

$$\sum_{j \in \mathcal{N}} (r_{j,\ell} + \bar{r}_{j,\ell})\mathbb{E}[u_{j,\ell}]^2 = \langle (R_\ell + \bar{R}_\ell)\mathbb{E}[u_\ell], \mathbb{E}[u_\ell] \rangle,$$

$$\sum_{j \in \mathcal{N}} b_{2j}(u_{j,k} - \mathbb{E}[u_{j,k}]) = \langle b_2, (u_\ell - \mathbb{E}[u_\ell]) \rangle,$$

$$\sum_{j \in \mathcal{N}} (b_{2j} + \bar{b}_{2j})\mathbb{E}[u_{j,k}] = \langle (b_2 + \bar{b}_2), \mathbb{E}[u_\ell] \rangle.$$

*The gap between the cost function $\mathbb{E}[L_0(x, u_1, \ldots, u_n)]$ depending on the con-
trol inputs and the optimal cost $f_{0,k}$ from initial time k up to terminal time
N is given by:*

$$\mathbb{E}[L_0(x, u_1, \ldots, u_n)] - \alpha_{0,k}(x_k - \mathbb{E}[x_k])^2 - \beta_{0,k}\mathbb{E}[x_k]^2 - \gamma_{0,k}\mathbb{E}[x_k] - \delta_{0,k}]$$
$$= -\alpha_{0,k+N}\mathbb{E}[(x_\ell - \mathbb{E}[x_{k+N}])^2] - \beta_{0,k+N}\mathbb{E}[x_{k+N}]^2$$
$$- \gamma_{0,k+N}\mathbb{E}[x_{k+N}] - \delta_{0,N}$$
$$+ q_{0,k+N}\mathbb{E}[(x_{k+N} - \mathbb{E}[x_{k+N}])^2] + (q_{0,k+N} + \bar{q}_{0,k+N})\mathbb{E}[x_{k+N}]^2$$
$$+ \mathbb{E}\sum_{\ell=k}^{k+N-1} q_{0,\ell}(x_\ell - \mathbb{E}[x_\ell])^2 + \mathbb{E}\sum_{\ell=k}^{k+N-1}(q_{0,\ell} + \bar{q}_{0,\ell})\mathbb{E}[x_\ell]^2$$
$$+ \mathbb{E}\sum_{\ell=k}^{k+N-1} \langle R_\ell(u_\ell - \mathbb{E}[u_\ell]), u_\ell - \mathbb{E}[u_\ell] \rangle$$
$$+ \mathbb{E}\sum_{\ell=k}^{k+N-1} \langle (R_\ell + \bar{R}_\ell)\mathbb{E}[u_\ell], \mathbb{E}[u_\ell] \rangle$$
$$+ \mathbb{E}\sum_{\ell=k}^{k+N-1} \alpha_{0,\ell+1}(x_{\ell+1} - \mathbb{E}[x_{\ell+1}])^2 - \mathbb{E}\sum_{\ell=k}^{k+N-1} \alpha_{0,\ell}(x_\ell - \mathbb{E}[x_\ell])^2$$

$$+ \mathbb{E} \sum_{\ell=k}^{k+N-1} \beta_{0,\ell+1}(\mathbb{E}[x_{\ell+1}])^2 - \mathbb{E} \sum_{\ell=k}^{k+N-1} \beta_{0,\ell}(\mathbb{E}[x_\ell])^2$$

$$+ \mathbb{E} \sum_{\ell=k}^{k+N-1} \gamma_{0,\ell+1}\mathbb{E}[x_{\ell+1}] - \mathbb{E} \sum_{\ell=k}^{k+N-1} \gamma_{0,\ell}\mathbb{E}[x_\ell]$$

$$+ \mathbb{E} \sum_{\ell=k}^{k+N-1} \delta_{0,\ell+1} - \mathbb{E} \sum_{\ell=k}^{k+N-1} \delta_{0,\ell}.$$

We express separately the terms $(x_{\ell+1} - \mathbb{E}[x_{\ell+1}])^2$ *and* $\mathbb{E}[x_{\ell+1}]^2$ *as function of* $(x_\ell - \mathbb{E}[x_\ell])$, $(u_\ell - \mathbb{E}[u_\ell])$, $\mathbb{E}[x_\ell]$, *and* $\mathbb{E}[u_\ell]$. *For the square term* $\mathbb{E}[x_{\ell+1}]^2$ *we have*

$$\mathbb{E}[x_{\ell+1}]^2 = (b_1 + \bar{b}_1)^2 \mathbb{E}[x_\ell]^2 + \left\langle (b_2 + \bar{b}_2)(b_2 + \bar{b}_2)^\top \mathbb{E}[u_\ell], \mathbb{E}[u_\ell] \right\rangle$$
$$+ [b_0]^2 + 2(b_1 + \bar{b}_1)\mathbb{E}[x_\ell] \left\langle (b_2 + \bar{b}_2), \mathbb{E}[u_\ell] \right\rangle$$
$$+ 2(b_1 + \bar{b}_1)\mathbb{E}[x_\ell]b_0 + 2b_0 \left\langle (b_2 + \bar{b}_2), \mathbb{E}[u_\ell] \right\rangle.$$

For the square difference $(x_{\ell+1} - \mathbb{E}[x_{\ell+1}])^2$, *we have*

$$(x_{\ell+1} - \mathbb{E}[x_{\ell+1}])^2 = b_1^2(x_\ell - \mathbb{E}[x_\ell])^2 + \left\langle b_2 b_2^\top (u_\ell - \mathbb{E}[u_\ell]), u_\ell - \mathbb{E}[u_\ell] \right\rangle + \sigma_0^2 w_\ell^2$$
$$+ 2b_1(x_\ell - \mathbb{E}[x_\ell]) \left\langle b_2, (u_\ell - \mathbb{E}[u_\ell]) \right\rangle + 2b_1(x_\ell - \mathbb{E}[x_\ell])\sigma_0 w_\ell$$
$$+ 2 \left\langle b_2, (u_\ell - \mathbb{E}[u_\ell]) \right\rangle \sigma_0 w_\ell,$$

and finally, for the term $\mathbb{E}[x_{k+1}]$ *we have*

$$\mathbb{E}[x_{k+1}] = (b_1 + \bar{b}_1)\mathbb{E}[x_k] + \left\langle (b_2 + \bar{b}_2), \mathbb{E}[u_\ell] \right\rangle + b_0.$$

With the previous terms, it follows that

$$\mathbb{E}[L_0(x, u_1, \ldots, u_n)] - \alpha_{0,k}(x_k - \mathbb{E}[x_k])^2 - \beta_{0,k}\mathbb{E}[x_k]^2 - \gamma_{0,k}\mathbb{E}[x_k] - \delta_{0,k}]$$
$$= -\alpha_{0,k+N}\mathbb{E}[(x_\ell - \mathbb{E}[x_{k+N}])^2] - \beta_{0,k+N}\mathbb{E}[x_{k+N}]^2$$
$$- \gamma_{0,k+N}\mathbb{E}[x_{k+N}] - \delta_{0,N}$$
$$+ q_{0,k+N}\mathbb{E}[(x_{k+N} - \mathbb{E}[x_{k+N}])^2] + (q_{0,k+N} + \bar{q}_{0,k+N})\mathbb{E}[x_{k+N}]^2$$
$$+ \mathbb{E} \sum_{\ell=k}^{k+N-1} q_{0,\ell}(x_\ell - \mathbb{E}[x_\ell])^2 + \mathbb{E} \sum_{\ell=k}^{k+N-1} (q_{0,\ell} + \bar{q}_{0,\ell})\mathbb{E}[x_\ell]^2$$
$$+ \mathbb{E} \sum_{\ell=k}^{k+N-1} \alpha_{0,\ell+1}b_1^2(x_\ell - \mathbb{E}[x_\ell])^2$$
$$+ \mathbb{E} \sum_{\ell=k}^{k+N-1} \left\langle R_\ell(u_\ell - \mathbb{E}[u_\ell]), u_\ell - \mathbb{E}[u_\ell] \right\rangle$$
$$+ \mathbb{E} \sum_{\ell=k}^{k+N-1} \left\langle \alpha_{0,\ell+1}b_2 b_2^\top (u_\ell - \mathbb{E}[u_\ell]), u_\ell - \mathbb{E}[u_\ell] \right\rangle$$

$$+ \mathbb{E} \sum_{\ell=k}^{k+N-1} \langle 2\alpha_{0,\ell+1}b_1(x_\ell - \mathbb{E}[x_\ell])b_2, u_\ell - \mathbb{E}[u_\ell] \rangle + \mathbb{E} \sum_{\ell=k}^{k+N-1} \alpha_{0,\ell+1}\sigma_0^2 w_\ell^2$$

$$- \mathbb{E} \sum_{\ell=k}^{k+N-1} \alpha_{0,\ell}(x_\ell - \mathbb{E}[x_\ell])^2 + \mathbb{E} \sum_{\ell=k}^{k+N-1} \beta_{0,\ell+1}(b_1 + \bar{b}_1)^2 \mathbb{E}[x_\ell]^2$$

$$+ \mathbb{E} \sum_{\ell=k}^{k+N-1} \langle (R_\ell + \bar{R}_\ell)\mathbb{E}[u_\ell], \mathbb{E}[u_\ell] \rangle$$

$$+ \mathbb{E} \sum_{\ell=k}^{k+N-1} \langle \beta_{0,\ell+1}(b_2 + \bar{b}_2)(b_2 + \bar{b}_2)^\top \mathbb{E}[u_\ell], \mathbb{E}[u_\ell] \rangle$$

$$+ \mathbb{E} \sum_{\ell=k}^{k+N-1} \langle 2\beta_{0,\ell+1}(b_1 + \bar{b}_1)\mathbb{E}[x_\ell](b_2 + \bar{b}_2), \mathbb{E}[u_\ell] \rangle$$

$$+ \mathbb{E} \sum_{\ell=k}^{k+N-1} \left\langle 2\left(\beta_{0,\ell+1}b_0 + \frac{1}{2}\gamma_{0,\ell+1}\right)(b_2 + \bar{b}_2), \mathbb{E}[u_\ell] \right\rangle$$

$$+ \mathbb{E} \sum_{\ell=k}^{k+N-1} \beta_{0,\ell+1}\left([b_0]^2 + 2(b_1 + \bar{b}_1)\mathbb{E}[x_\ell]b_0\right) - \mathbb{E} \sum_{\ell=k}^{k+N-1} \beta_{0,\ell}(\mathbb{E}[x_\ell])^2$$

$$+ \mathbb{E} \sum_{\ell=k}^{k+N-1} \gamma_{0,\ell+1}(b_1 + \bar{b}_1)\mathbb{E}[x_k] + \mathbb{E} \sum_{\ell=k}^{k+N-1} \gamma_{0,\ell+1}b_0$$

$$- \mathbb{E} \sum_{\ell=k}^{k+N-1} \gamma_{0,\ell}\mathbb{E}[x_\ell] + \mathbb{E} \sum_{\ell=k}^{k+N-1} \delta_{0,\ell+1} - \mathbb{E} \sum_{\ell=k}^{k+N-1} \delta_{0,\ell}.$$

We optimize over the terms depending on the difference $(u_\ell - \mathbb{E}[u_\ell])$ and the control input $\mathbb{E}[u_\ell]$. We perform first the following square completion for the terms involving the subtraction $(u_\ell - \mathbb{E}[u_\ell])$:

$$\left| K_1[K_2(u_\ell - \mathbb{E}[u_\ell]) + K_3(x_\ell - \mathbb{E}[x_\ell])] \right|^2$$
$$- \left\langle K_1 K_3(x_\ell - \mathbb{E}[x_\ell]), K_1 K_3(x_\ell - \mathbb{E}[x_\ell]) \right\rangle$$
$$= \left\langle K_1 K_2(u_\ell - \mathbb{E}[u_\ell]), K_1 K_2(u_\ell - \mathbb{E}[u_\ell]) \right\rangle$$
$$+ 2\left\langle K_1 K_3(x_\ell - \mathbb{E}[x_\ell]), K_1 K_2(u_\ell - \mathbb{E}[u_\ell]) \right\rangle.$$

Then, it is concluded that

$$\left\langle \left(R_\ell + \alpha_{0,\ell+1}b_2 b_2^\top\right)(u_\ell - \mathbb{E}[u_\ell]), u_\ell - \mathbb{E}[u_\ell] \right\rangle$$
$$= \langle K_2^\top K_1^\top K_1 K_2(u_\ell - \mathbb{E}[u_\ell]), u_\ell - \mathbb{E}[u_\ell] \rangle,$$
$$2\langle \alpha_{0,\ell+1}b_1 b_2(x_\ell - \mathbb{E}[x_\ell]), u_\ell - \mathbb{E}[u_\ell] \rangle$$
$$= 2\langle K_2^\top K_1^\top K_1 K_3(x_\ell - \mathbb{E}[x_\ell]), u_\ell - \mathbb{E}[u_\ell] \rangle,$$

and the control-input-dependent terms (depending on $(u_\ell - \mathbb{E}[u_\ell]))$ from the gap $\mathbb{E}[L_0(x, u_1, \ldots, u_n) - f_{0,\ell}]$ can be expressed as follows using the square completion:

$$\langle R_\ell(u_\ell - \mathbb{E}[u_\ell]), u_\ell - \mathbb{E}[u_\ell]\rangle$$
$$+ \langle \alpha_{0,\ell+1} b_2 b_2^\top (u_\ell - \mathbb{E}[u_\ell]), u_\ell - \mathbb{E}[u_\ell]\rangle$$
$$+ 2\langle \alpha_{0,\ell+1} b_1(x_\ell - \mathbb{E}[x_\ell]) b_2, u_\ell - \mathbb{E}[u_\ell]\rangle$$
$$= |K_1[K_2(u_\ell - \mathbb{E}[u_\ell]) + K_3(x_\ell - \mathbb{E}[x_\ell])]|^2$$
$$- \langle K_1 K_3(x_\ell - \mathbb{E}[x_\ell]), K_1 K_3(x_\ell - \mathbb{E}[x_\ell])\rangle,$$

being the variables K_1, K_2, and K_3 as follows:

$$K_1 = \left(R_\ell + \alpha_{0,\ell+1} b_2 b_2^\top\right)^{-\frac{1}{2}},$$
$$K_2 = R_\ell + \alpha_{0,\ell+1} b_2 b_2^\top,$$
$$K_3 = \alpha_{0,\ell+1} b_1 b_2,$$

where

$$\left\langle K_1 K_3(x_\ell - \mathbb{E}[x_\ell]), K_1 K_3(x_\ell - \mathbb{E}[x_\ell])\right\rangle = K_3^\top K_1^\top K_1 K_3(x_\ell - \mathbb{E}[x_\ell])^2.$$

Similarly, optimizing for the terms involving $\mathbb{E}[u_\ell]$ we have

$$\left|\tilde{K}_1[\tilde{K}_2\mathbb{E}[u_\ell] + \tilde{K}_3\Xi]\right|^2 = \left\langle \tilde{K}_1\tilde{K}_2\mathbb{E}[u_\ell], \tilde{K}_1\tilde{K}_2\mathbb{E}[u_\ell]\right\rangle + 2\left\langle \tilde{K}_1\tilde{K}_3\Xi, \tilde{K}_1\tilde{K}_2\mathbb{E}[u_\ell]\right\rangle$$
$$+ \left\langle \tilde{K}_1\tilde{K}_3\Xi, \tilde{K}_1\tilde{K}_3\Xi\right\rangle,$$

$$\Xi = \beta_{0,\ell+1}\left[(b_1 + \bar{b}_1)\mathbb{E}[x_\ell] + b_0\right] + \frac{1}{2}\gamma_{0,\ell+1},$$

where it is concluded that

$$\left\langle \left(R_\ell + \bar{R}_\ell + \beta_{0,\ell+1}(b_2 + \bar{b}_2)(b_2 + \bar{b}_2)^\top\right)\mathbb{E}[u_\ell], \mathbb{E}[u_\ell]\right\rangle$$
$$= \langle \tilde{K}_2^\top \tilde{K}_1^\top \tilde{K}_1 \tilde{K}_2\mathbb{E}[u_\ell], \mathbb{E}[u_\ell]\rangle, 2\left\langle (b_2 + \bar{b}_2)\Xi, \mathbb{E}[u_\ell]\right\rangle$$
$$= 2\left\langle \tilde{K}_2^\top \tilde{K}_1^\top \tilde{K}_1 \tilde{K}_3\Xi, \mathbb{E}[u_\ell]\right\rangle,$$

and the control-input-dependent terms (depending on $\mathbb{E}[u_\ell]$) from the gap $\mathbb{E}[L_0(x, u_1, \ldots, u_n) - f_{0,\ell}]$ can be expressed as follows using the square completion:

$$\left\langle \left(R_\ell + \bar{R}_\ell + \beta_{0,\ell+1}(b_2 + \bar{b}_2)(b_2 + \bar{b}_2)^\top\right)\mathbb{E}[u_\ell], \mathbb{E}[u_\ell]\right\rangle + 2\left\langle (b_2 + \bar{b}_2)\Xi, \mathbb{E}[u_\ell]\right\rangle$$
$$= \left|\tilde{K}_1\left[\tilde{K}_2\mathbb{E}[u_\ell] + \tilde{K}_3\Xi\right]\right|^2 - \left\langle \tilde{K}_1\tilde{K}_3\Xi, \tilde{K}_1\tilde{K}_3\Xi\right\rangle,$$

being the variables $\tilde{K}_1$, $\tilde{K}_2$, and $\tilde{K}_3$ are as follows:

$$\tilde{K}_1 = \left(R_\ell + \bar{R}_\ell + \beta_{0,\ell+1}(b_2 + \bar{b}_2)(b_2 + \bar{b}_2)^\top\right)^{-\frac{1}{2}},$$

$$\tilde{K}_2 = R_\ell + \bar{R}_\ell + \beta_{0,\ell+1}(b_2 + \bar{b}_2)(b_2 + \bar{b}_2)^\top,$$
$$\tilde{K}_3 = (b_2 + \bar{b}_2).$$

Hence,

$$\left\langle \tilde{K}_1 \tilde{K}_3 \left(\beta_{0,\ell+1} \left[(b_1 + \bar{b}_1)\mathbb{E}[x_\ell] + b_0 \right] + \frac{1}{2}\gamma_{0,\ell+1} \right), \right.$$

$$\left. \tilde{K}_1 \tilde{K}_3 \left(\beta_{0,\ell+1} \left[(b_1 + \bar{b}_1)\mathbb{E}[x_\ell] + b_0 \right] + \frac{1}{2}\gamma_{0,\ell+1} \right) \right\rangle$$

$$= \tilde{K}_3^\top \tilde{K}_1^\top \tilde{K}_1 \tilde{K}_3 \beta_{0,\ell+1}^2 (b_1 + \bar{b}_1)^2 \mathbb{E}[x_\ell]^2 + \tilde{K}_3^\top \tilde{K}_1^\top \tilde{K}_1 \tilde{K}_3 \beta_{0,\ell+1}^2 [b_0]^2$$

$$+ \frac{1}{4}\tilde{K}_3^\top \tilde{K}_1^\top \tilde{K}_1 \tilde{K}_3 \gamma_{0,\ell+1}^2 + 2\tilde{K}_3^\top \tilde{K}_1^\top \tilde{K}_1 \tilde{K}_3 \beta_{0,\ell+1}^2 (b_1 + \bar{b}_1)b_0 \mathbb{E}[x_\ell]$$

$$+ \tilde{K}_3^\top \tilde{K}_1^\top \tilde{K}_1 \tilde{K}_3 \beta_{0,\ell+1}\gamma_{0,\ell+1}(b_1 + \bar{b}_1)\mathbb{E}[x_\ell]$$

$$+ \tilde{K}_3^\top \tilde{K}_1^\top \tilde{K}_1 \tilde{K}_3 \beta_{0,\ell+1}\gamma_{0,\ell+1}b_0.$$

Now, we replace the terms obtained from the optimization over the control inputs in the gap $\mathbb{E}[L_0(x, u_1, \ldots, u_n) - f_{0,k}]$. *After that, one arrives at*

$$\mathbb{E}\Big[L_0(x, u_1, \ldots, u_n) - \alpha_{0,k}(x_k - \mathbb{E}[x_k])^2 - \beta_{0,k}\mathbb{E}[x_k]^2 - \gamma_{0,k}\mathbb{E}[x_k] - \delta_{0,k} \Big]$$

$$= -\alpha_{0,k+N}\mathbb{E}[(x_\ell - \mathbb{E}[x_{k+N}])^2] - \beta_{0,k+N}\mathbb{E}[x_{k+N}]^2$$

$$- \gamma_{0,k+N}\mathbb{E}[x_{k+N}] - \delta_{0,N}$$

$$+ q_{0,k+N}\mathbb{E}[(x_{k+N} - \mathbb{E}[x_{k+N}])^2] + (q_{0,k+N} + \bar{q}_{0,k+N})\mathbb{E}[x_{k+N}]^2$$

$$+ \mathbb{E}\sum_{\ell=k}^{k+N-1} q_{0,\ell}(x_\ell - \mathbb{E}[x_\ell])^2 + \mathbb{E}\sum_{\ell=k}^{k+N-1}(q_{0,\ell} + \bar{q}_{0,\ell})\mathbb{E}[x_\ell]^2$$

$$+ \mathbb{E}\sum_{\ell=k}^{k+N-1} \alpha_{0,\ell+1}b_1^2(x_\ell - \mathbb{E}[x_\ell])^2 + \mathbb{E}\sum_{\ell=k}^{k+N-1}\alpha_{0,\ell+1}\sigma_0^2 w_\ell^2$$

$$- \mathbb{E}\sum_{\ell=k}^{k+N-1} \alpha_{0,\ell}(x_\ell - \mathbb{E}[x_\ell])^2 + \mathbb{E}\sum_{\ell=k}^{k+N-1}\beta_{0,\ell+1}(b_1 + \bar{b}_1)^2\mathbb{E}[x_\ell]^2$$

$$+ \mathbb{E}\sum_{\ell=k}^{k+N-1} \beta_{0,\ell+1}\Big([b_0]^2 + 2(b_1 + \bar{b}_1)\mathbb{E}[x_\ell|b_0] \Big)$$

$$- \mathbb{E}\sum_{\ell=k}^{k+N-1} \beta_{0,\ell}(\mathbb{E}[x_\ell])^2 + \mathbb{E}\sum_{\ell=k}^{k+N-1}\gamma_{0,\ell+1}(b_1 + \bar{b}_1)\mathbb{E}[x_k]$$

$$+ \mathbb{E}\sum_{\ell=k}^{k+N-1} \gamma_{0,\ell+1}b_0 - \mathbb{E}\sum_{\ell=k}^{k+N-1}\gamma_{0,\ell}\mathbb{E}[x_\ell]$$

$$+ \mathbb{E}\sum_{\ell=k}^{k+N-1} \delta_{0,\ell+1} - \mathbb{E}\sum_{\ell=k}^{k+N-1}\delta_{0,\ell}$$

$$- \mathbb{E} \sum_{\ell=k}^{k+N-1} K_3^\top K_1^\top K_1 K_3 (x_\ell - \mathbb{E}[x_\ell])^2$$

$$- \mathbb{E} \sum_{\ell=k}^{k+N-1} \tilde{K}_3^\top \tilde{K}_1^\top \tilde{K}_1 \tilde{K}_3 \beta_{0,\ell+1}^2 (b_1 + \bar{b}_1)^2 \mathbb{E}[x_\ell]^2$$

$$- \mathbb{E} \sum_{\ell=k}^{k+N-1} \tilde{K}_3^\top \tilde{K}_1^\top \tilde{K}_1 \tilde{K}_3 \beta_{0,\ell+1}^2 [b_0]^2 - \frac{1}{4} \mathbb{E} \sum_{\ell=k}^{k+N-1} \tilde{K}_3^\top \tilde{K}_1^\top \tilde{K}_1 \tilde{K}_3 \gamma_{0,\ell+1}^2$$

$$- 2\mathbb{E} \sum_{\ell=k}^{k+N-1} \tilde{K}_3^\top \tilde{K}_1^\top \tilde{K}_1 \tilde{K}_3 \beta_{0,\ell+1}^2 (b_1 + \bar{b}_1) b_0 \mathbb{E}[x_\ell]$$

$$- \mathbb{E} \sum_{\ell=k}^{k+N-1} \tilde{K}_3^\top \tilde{K}_1^\top \tilde{K}_1 \tilde{K}_3 \beta_{0,\ell+1} \gamma_{0,\ell+1} (b_1 + \bar{b}_1) \mathbb{E}[x_\ell]$$

$$- \mathbb{E} \sum_{\ell=k}^{k+N-1} \tilde{K}_3^\top \tilde{K}_1^\top \tilde{K}_1 \tilde{K}_3 \beta_{0,\ell+1} \gamma_{0,\ell+1} b_0$$

$$+ \mathbb{E} \sum_{\ell=k}^{k+N-1} \left| K_1 \Big[K_2 (u_\ell - \mathbb{E}[u_\ell]) + K_3 (x_\ell - \mathbb{E}[x_\ell]) \Big] \right|^2$$

$$+ \mathbb{E} \sum_{\ell=k}^{k+N-1} \left| \tilde{K}_1 \Big[\tilde{K}_2 \mathbb{E}[u_\ell] \right.$$

$$\left. + \tilde{K}_3 \Big(\beta_{0,\ell+1} \big[(b_1 + \bar{b}_1) \mathbb{E}[x_\ell] + b_0 \big] + \frac{1}{2} \gamma_{0,\ell+1} \Big) \Big] \right|^2 .$$

The latter long expression is then divided into several groups:

- *Those terms evaluated at the terminal time $k+N$, which define the boundary conditions for the backward difference equations α_0, β_0, γ_0, and δ_0*

- *Those terms with common factor of the quadratic form $((x_\ell - \mathbb{E}[x_\ell])^2)$. From this procedure one obtains the difference equation α_0*

- *Those terms with common factor of the quadratic form $\mathbb{E}[x_\ell]^2$, obtaining the difference equation β_0*

- *Those terms whose common factor is $\mathbb{E}[x_\ell]$. From this procedure one obtains the difference equation γ_0*

- *Those terms independent from the system state, control inputs, expected state, and expected control inputs. We obtain from here the difference equation δ_0*

- *And those quadratic terms depending on the control inputs defining the optimal solution.*

We make this process identification to obtain the announced result completing the proof.

11.4 Exercises

1. Solve the following discrete-time mean-field-type control problem. Consider the system dynamics given by

$$x_{\ell+1} = b_0 + b_1 x_\ell + \bar{b}_1 \mathbb{E}[x_\ell] + b_2 u_\ell + \bar{b}_2 \mathbb{E}[u_\ell] + \sigma_0 w_\ell,$$

and with a cost functional given by

$$L(x, u) = q_{k+N} x_{k+N}^2 + \bar{q}_{k+N} \mathbb{E}[x_{k+N}]^2$$
$$+ \sum_{\ell=k}^{k+N-1} \left[q_\ell x_\ell^2 + \bar{q}_\ell \mathbb{E}[x_\ell]^2 + r_\ell u_\ell^2 + \bar{r}_\ell \mathbb{E}[u_\ell]^2 \right], \quad (11.13)$$

with $q_\ell, q_\ell + \bar{q}_\ell \geq 0$ and $r_\ell, r_\ell + \bar{r}_\ell > 0$, time step ℓ from time k.

Hint: An appropriate guess functional is given by

$$f_\ell = \alpha_\ell (x_\ell - \mathbb{E}[x_\ell])^2 + \beta_\ell \mathbb{E}[x_\ell]^2 + \gamma_\ell \mathbb{E}[x_\ell] + \delta_\ell.$$

2. Solve the following discrete-time mean-field-type control problem. Consider the system dynamics given by

$$x_{\ell+1} = \left(b_0 + b_1 x_\ell + \bar{b}_1 \mathbb{E}[x_\ell] + b_2 u_\ell + \bar{b}_2 \mathbb{E}[u_\ell] \right)$$
$$+ \left(\sigma_0 + \sigma_1 x_\ell + \bar{\sigma}_1 \mathbb{E}[x_\ell] + \sigma_2 u_\ell + \bar{\sigma}_2 \mathbb{E}[u_\ell] \right) w_\ell,$$

and with the same cost functional in (11.13).

Hint: An appropriate guess functional is the same as in the previous exercise.

12

Matrix-Valued Discrete-Time Mean-Field-Type Games

Now we address the computation of semi-explicit solutions for the matrix-valued discrete-time mean-field-type game problems. The treatment of matrix-valued system states is suitable to tackle problems over graphs as we justify in Chapter 9. Besides, the matrix-valued approach becomes more general since vector-valued problems can be also considered by means of diagonal matrices.

We use the direct method in discrete time as presented in the previous chapter in order to solve the game problem. We study the non-cooperative, fully-cooperative and adversarial game cases. The square completion technique for the quadratic forms appearing in this chapter are treated in the same manner as we did in Chapter 9. Over the end of the chapter, some numerical examples illustrate the performance of the closed-loop solutions.

With the analysis presented in this chapter, we finish the Part V. Indeed, we actually end our exposition of the the mean-field-type game problems and their respective semi-explicit solutions using the direct method either in continuous or discrete time. In the next and last part of this book, Part VI, we will study some learning approaches using all the theoretical lectures we have presented up to now. Then, we will focus on several engineering applications. By completing the study of this chapter material, we expect that the reader has now enough tools and knowledge about different game-theoretical settings to do research in the field and also to apply this theory to its particular engineering interests.

12.1 Discrete-Time Mean-Field-Type Game Set-up

Consider a mean-field-type game involving $n \geq 2$ $(n \in \mathbb{N})$ decision-makers, $\mathcal{N} = \{1, \ldots, n\}$ denotes the set of decision-makers. Besides, consider a system whose dynamics are given by the following difference equation:

$$X_{k+1} = B_1 X_k + \bar{B}_1 \mathbb{E}[X_k] + \sum_{j \in \mathcal{N}} B_{2j} U_{j,k} + \sum_{j \in \mathcal{N}} \bar{B}_{2j} \mathbb{E}[U_{j,k}] + S W_k, \quad (12.1)$$

DOI: 10.1201/9781003098607-12

where $X \in \mathbb{R}^{d \times d}$ represents the matrix-valued system states, $\mathbb{E}[X] \in \mathbb{R}^{d \times d}$ corresponds to the expected value of the system states, and $B_1, \bar{B}_1 \in \mathbb{R}^{d \times d}$ are system matrices associated with the state. On the other hand, $U_j \in \mathbb{R}^{r \times d}$ denotes the control input, $\mathbb{E}[U_j] \in \mathbb{R}^{r \times d}$ corresponds to its expected value, and $B_{2j}, \bar{B}_{2j} \in \mathbb{R}^{d \times r}$ are system matrices, for all decision-makers $j \in \mathcal{N}$. Finally, $W \in \mathbb{R}^{s \times d}$ denotes a matrix of disturbances with $S \in \mathbb{R}^{d \times s}$.

There is a cost functional $L_i(X, U_1, \ldots, U_n)$ corresponding to each decision-maker $i \in \mathcal{N}$, i.e.,

$$L_i(X, U_1, \ldots, U_n) = \langle Q_{i,k+N} X_{k+N}, X_{k+N} \rangle + \langle \bar{Q}_{i,k+N} \mathbb{E}[X_{k+N}], \mathbb{E}[X_{k+N}] \rangle$$
$$+ \sum_{\ell=k}^{k+N-1} \langle Q_{i,\ell} X_\ell, X_\ell \rangle + \sum_{\ell=k}^{k+N-1} \langle \bar{Q}_{i,\ell} \mathbb{E}[X_\ell], \mathbb{E}[X_\ell] \rangle$$
$$+ \sum_{\ell=k}^{k+N-1} \langle R_{i,\ell} U_{i,\ell}, U_{i,\ell} \rangle + \sum_{\ell=k}^{k+N-1} \langle \bar{R}_{i,\ell} \mathbb{E}[U_{i,\ell}], \mathbb{E}[U_{i,\ell}] \rangle,$$

where $\langle A, B \rangle = \mathrm{trace}(A^* B) = \mathrm{trace}(B^* A)$, and A^* is the adjoint operator of A (transposition).

Next, we present the statement of the non-cooperative game problem for matrix-valued system state.

Problem 23 (Non-cooperative Scenario) *The non-cooperative mean-field-type game problem is given by*

$$\begin{cases} \underset{U_i}{\mathrm{minimize}} \ \mathbb{E}[L_i(X, U_1, \ldots, U_n)], \\[2ex] \text{subject to} \\[2ex] X_{k+1} = B_1 X_k + \bar{B}_1 \mathbb{E}[X_k] + \sum_{j \in \mathcal{N}} B_{2j} U_{j,k} + \sum_{j \in \mathcal{N}} \bar{B}_{2j} \mathbb{E}[U_{j,k}] + S W_k, \\[2ex] X_k \triangleq X_0, \end{cases}$$

where the weight parameters in the cost functional satisfy Q_i, $Q_i + \bar{Q}_i \succeq 0$, and R_i, $R_i + \bar{R}_i \succ 0$, for all $i \in \mathcal{N}$.

In Lemma 2 we re-formulate the non-cooperative game problem in terms of the differences $(X - \mathbb{E}[X])$, and $(U_j - \mathbb{E}[U_j])$; and the expected values $\mathbb{E}[X]$, and $\mathbb{E}[U_j]$, for all $j \in \mathcal{N}$. Thus, one can observe the risk-terms appearing in the cost functional expectation $\mathbb{E}[L_i(X, U_1, \ldots, U_n)]$.

Lemma 2 *Problem 23 is equivalent to the following problem:*

$$\underset{U_i}{\mathrm{minimize}} \ \mathbb{E}[L_i(X, U_1, \ldots, U_n)]$$
$$= \left\langle Q_{i,k+N}(X_{k+N} - \mathbb{E}[X_{k+N}]), X_{k+N} - \mathbb{E}[X_{k+N}] \right\rangle$$

$$+ \Big\langle (Q_{i,k+N} + \bar{Q}_{i,k+N}) \mathbb{E}[X_{k+N}], \mathbb{E}[X_{k+N}] \Big\rangle$$

$$+ \sum_{\ell=k}^{k+N-1} \Big\langle Q_{i,\ell}(X_\ell - \mathbb{E}[X_\ell]), X_\ell - \mathbb{E}[X_\ell] \Big\rangle$$

$$+ \sum_{\ell=k}^{k+N-1} \Big\langle (Q_{i,\ell} + \bar{Q}_{i,\ell}) \mathbb{E}[X_\ell], \mathbb{E}[X_\ell] \Big\rangle$$

$$+ \sum_{\ell=k}^{k+N-1} \Big\langle R_{i,\ell}(U_{i,\ell} - \mathbb{E}[U_{i,\ell}]), U_{i,\ell} - \mathbb{E}[U_{i,\ell}] \Big\rangle$$

$$+ \sum_{\ell=k}^{k+N-1} \Big\langle (R_{i,\ell} + \bar{R}_{i,\ell}) \mathbb{E}[U_{i,\ell}], \mathbb{E}[U_{i,\ell}] \Big\rangle,$$

subject to

$$X_{k+1} = B_1(X_k - \mathbb{E}[X_k]) + (B_1 + \bar{B}_1)\mathbb{E}[X_k] + \sum_{j \in \mathcal{N}} B_{2j}(U_{j,k} - \mathbb{E}[U_{j,k}])$$

$$+ \sum_{j \in \mathcal{N}} (B_{2j} + \bar{B}_{2j})\mathbb{E}[U_{j,k}] + SW_k,$$

where matrices Q_i, $(Q_i + \bar{Q}_i) \succeq 0$, *and* R_i, $(R_i + \bar{R}_i) \succ 0$.

The equivalence between Problem 23 and the one shown in Lemma 2 is easily proven by means of the decomposition presented below. Notice that, this is the same decomposition with which we prove the risk-awareness in the mean-field-type games in Chapter 4 obtaining the variance-dependent terms in (4.4)

Proof 28 (Lemma 2) *The equivalence is directly obtained by re-writing the cost function in terms of* $(X - \mathbb{E}[X])$ *and* $\mathbb{E}[X]$. *This is performed by using the decomposition*

$$X_k = X_k - \mathbb{E}[X_k] + \mathbb{E}[X_k],$$
$$U_{j,k} = U_{j,k} - \mathbb{E}[U_{j,k}] + \mathbb{E}[U_{j,k}], \ \forall j \in \mathcal{N},$$

and noticing that both

$$\mathbb{E}[\mathbb{E}[X_k](X_k - \mathbb{E}[X_k])] = 0,$$
$$\mathbb{E}[\mathbb{E}[U_{j,k}](U_{j,k} - \mathbb{E}[U_{j,k}])] = 0, \ \forall j \in \mathcal{N},$$

since $X_k - \mathbb{E}[X_k] \perp \mathbb{E}[X_k]$ *and* $U_{j,k} - \mathbb{E}[U_{j,k}] \perp \mathbb{E}[U_{j,k}]$. *Hence, the evolution of the expectation of the system state is given by*

$$\mathbb{E}[X_{k+1}] = (B_1 + \bar{B}_1)\mathbb{E}[X_k] + \sum_{j \in \mathcal{N}} (B_{2j} + \bar{B}_{2j})\mathbb{E}[U_{j,k}].$$

Once the Non-Cooperative Mean-Field-Type Game has been defined in Problem 23, then we define both the best-response strategy and the Nash equilibrium next in Definitions 29 and 30 for the matrix-valued case.

Definition 29 (Best-Response Strategy) *A minimizer U_i^{op} is named a best-response strategy of the decision-maker $i \in \mathcal{N}$ given the strategies of the other decision-makers*

$$U_{-i} = (U_1, \ldots, U_{i-1}, U_{i+1}, \ldots, U_n)$$

if it solves Problem 23. Besides, the set of best-response strategies, corresponding to the decision-maker $i \in \mathcal{N}$, is denoted by $\mathrm{BR}_i(U_{-i})$.

Using the best-response strategies, we define the set of mean-field-type Nash equilibria next.

Definition 30 (Mean-Field-Type Nash Equilibrium) *The strategic profile $U_1^{\mathrm{op}}, \ldots, U_n^{\mathrm{op}}$ is a Nash equilibrium is for all $i \in \mathcal{N}$, $U_i^{\mathrm{op}} \in \mathrm{BR}_i(U_{-i}^{\mathrm{op}})$.*

Regarding the fully-cooperative approach, we have the following Problem 24 for the matrix-valued discrete-time case.

Problem 24 (Fully-Cooperative Scenario) *The Full-Cooperative Mean-Field-Type Game Problem is given by*

$$\begin{cases} L_0(X, U_1, \ldots, U_n) = \sum_{i \in \mathcal{N}} L_i(X, U_1, \ldots, U_n), \\ \underset{U_1, \ldots, U_n}{\mathrm{minimize}} \ L_0(X, U_1, \ldots, U_n), \\ \text{subject to} \\ X_{k+1} = B_1 X_k + \bar{B}_1 \mathbb{E}[X_k] + \sum_{j \in \mathcal{N}} B_{2j} U_{j,k} + \sum_{j \in \mathcal{N}} \bar{B}_{2j} \mathbb{E}[U_{j,k}] + S W_k, \\ X_k \triangleq X_0, \end{cases}$$

where the weight parameters for the fully-cooperative game problem are as follows:

$$Q_0 = \sum_{i \in \mathcal{N}} Q_i \succeq 0,$$

$$Q_0 + \bar{Q}_0 = \sum_{i \in \mathcal{N}} (Q_i + \bar{Q}_i) \succeq 0,$$

and R_i, $R_i + \bar{R}_i \succ 0$, for all $i \in \mathcal{N}$.

Finally, we study the Adversarial/Robust Mean-Field-Type Game as stated next. Let us remember that this adversarial situation can be represented in the same form as of the fully-cooperative game problem and by imposing the appropriate signs to the weight parameters in the cost functional. Thus, the min-max problem can be unified into a minimization problem.

Problem 25 (Adversarial Scenario) *Let us consider that all players* $\mathcal{N}$ *are divided into two teams, i.e.,* $\mathcal{I}_- \cup \mathcal{I}_+ = \mathcal{N}$. *These teams* $\mathcal{I}_-$ *and* $\mathcal{I}_+$ *are known as the attackers and defenders, respectively. The Adversarial Mean-Field-Type Game Problem is given by*

$$
\begin{cases}
L^{\mathrm{ad}}(X, U_1, \ldots, U_n) = \displaystyle\sum_{i \in \mathcal{N}} L_i(X, U_1, \ldots, U_n), \\[2mm]
\underset{\{U_i\}_{i \in \mathcal{I}_+}}{\text{minimize}} \ \underset{\{U_j\}_{j \in \mathcal{I}_-}}{\text{maximize}} \ L^{\mathrm{ad}}(X, U_1, \ldots, U_n), \\[4mm]
\text{subject to} \\[2mm]
X_{k+1} = B_1 X_k + \bar{B}_1 \mathbb{E}[X_k] + \displaystyle\sum_{j \in \mathcal{N}} B_{2j} U_{j,k} + \sum_{j \in \mathcal{N}} \bar{B}_{2j} \mathbb{E}[U_{j,k}] + SW_k, \\[4mm]
X_k \triangleq X_0,
\end{cases}
$$

where the matrices are Q_i, $Q_i + \bar{Q}_i \succeq 0$, *and* R_i, $R_i + \bar{R}_i \succ 0$, *for all defender* $i \in \mathcal{I}_+$, *and* $-Q_i$, $-(Q_i + \bar{Q}_i) \succeq 0$, *and* $-R_i$, $-(R_i + \bar{R}_i) \succ 0$, *for all attacker* $i \in \mathcal{I}_-$.

Semi-explicit solutions for the aforementioned problems together with the respective proof are introduced in the coming section. As we have been doing throughout the book, we compute the solutions by means of the simple direct method.

12.2 Semi-explicit Solution of the Discrete-Time Mean-Field-Type Game Problem

Theorem presents the semi-explicit solution for the non-cooperative game problem with matrix-valued state and in discrete time. For the sake of organization in the exposition of the solution, we introduced some auxiliary variables.

Theorem 6 (Non-cooperative Scenario) *The explicit solution of Problem 23 is given by the following optimal control inputs and cost functional:*

$$
U_{i,\ell}^{\mathrm{op}} - \mathbb{E}[U_{i,\ell}]^{\mathrm{op}} = -[B_{2i}^* P_{i,\ell+1} B_{2i} + R_{i,\ell}]^{-1} \Phi_{i3,\ell}(X_\ell - \mathbb{E}[X_\ell]),
$$

$$
\mathbb{E}[U_{i,\ell}]^{\mathrm{op}} = -\Big[(B_{2i} + \bar{B}_{2i})^* \bar{P}_{i,\ell+1}(B_{2i} + \bar{B}_{2i}) + (R_{i,\ell} + \bar{R}_{i,\ell})\Big]^{-1} \cdot
$$

$$
\cdot \, \Xi_{i3,\ell} \mathbb{E}[X_\ell], \quad \forall \ell \in [k..k + N - 1],
$$

$$
L_i(X, U_1^{\mathrm{op}}, \ldots, U_n^{\mathrm{op}}) = \langle P_{i,k}(X_k - \bar{X}_k), X_k - \bar{X}_k \rangle + \langle \bar{P}_{i,k} \bar{X}_k, \bar{X}_k \rangle + \delta_{i,k}.
$$

Let us consider the variables:

$$\tilde{\Phi}_{ij} = B_{2i}^* P_{i,k+1} B_{2j} [B_{2j}^* P_{j,k+1} B_{2j} + R_{j,k}]^{-1},$$

$$\tilde{\Xi}_{ij,k} = (B_{2i} + \bar{B}_{2i})^* \bar{P}_{i,k+1} (B_{2j} + \bar{B}_{2j}) \cdot$$
$$\cdot \left[(B_{2j} + \bar{B}_{2j})^* \bar{P}_{j,k+1} (B_{2j} + \bar{B}_{2j}) + (R_{j,k} + \bar{R}_{j,k}) \right]^{-1},$$

and with

$$\begin{pmatrix} I & \tilde{\Phi}_{12,k} & \cdots & \tilde{\Phi}_{1n,k} \\ \tilde{\Phi}_{21,k} & I & \cdots & \tilde{\Phi}_{2n,k} \\ \vdots & \ddots & \ddots & \vdots \\ \tilde{\Phi}_{n1,k} & \cdots & \tilde{\Phi}_{n(n-1),k} & I \end{pmatrix} \begin{pmatrix} \Phi_{13,k} \\ \Phi_{23,k} \\ \vdots \\ \Phi_{n3,k} \end{pmatrix} = \begin{pmatrix} B_1^* P_{1,k+1} A \\ B_2^* P_{2,k+1} A \\ \vdots \\ B_n^* P_{n,k+1} A \end{pmatrix},$$

$$(12.2\text{a})$$

$$\begin{pmatrix} I & \tilde{\Xi}_{12,k} & \cdots & \tilde{\Xi}_{1n,k} \\ \tilde{\Xi}_{21,k} & I & \cdots & \tilde{\Xi}_{2n,k} \\ \vdots & \ddots & \ddots & \vdots \\ \tilde{\Xi}_{n1,k} & \cdots & \tilde{\Xi}_{n(n-1),k} & I \end{pmatrix} \begin{pmatrix} \Xi_{13,k} \\ \Xi_{23,k} \\ \vdots \\ \Xi_{n3,k} \end{pmatrix}$$

$$= \begin{pmatrix} (B_1 + \bar{B}_1)^* \bar{P}_{1,k+1} (B_1 + \bar{B}_1) \\ (B_2 + \bar{B}_2)^* \bar{P}_{2,k+1} (B_1 + \bar{B}_1) \\ \vdots \\ (B_n + \bar{B}_n)^* \bar{P}_{n,k+1} (B_1 + \bar{B}_1) \end{pmatrix}.$$

$$(12.2\text{b})$$

In addition, $P_{i,k}$, $\bar{P}_{i,k}$, and $\delta_{i,k}$ solve the following difference equations:

$$P_{i,k} = Q_{i,k} + B_1^* P_{i,k+1} B_1 - B_1^* P_{i,k+1} \sum_{j \neq i} B_{2j} \Phi_{j2,k}^{-1} \Phi_{j3,k}$$

$$- \sum_{j \neq i} \Phi_{j3,k}^* [\Phi_{j2,k}^{-1}]^* B_{2j}^* P_{i,k+1} A$$

$$+ \sum_{j \neq i} \Phi_{j3,k}^* [\Phi_{j2,k}^{-1}]^* B_{2j}^* P_{i,k+1} \sum_{j \neq i} B_{2j} \Phi_{j2,k}^{-1} \Phi_{j3,k} - \Phi_{i3,k}^* \Phi_{i1,k}^* \Phi_{i1,k} \Phi_{i3,k},$$

$$\bar{P}_{i,k} = (Q_{i,k} + \bar{Q}_{i,k}) + (B_1 + \bar{B}_1)^* \bar{P}_{i,k+1} (B_1 + \bar{B}_1)$$

$$- (B_1 + \bar{B}_1)^* \bar{P}_{i,k+1} \sum_{j \neq i} (B_{2j} + \bar{B}_{2j}) \Xi_{j2,k}^{-1} \Xi_{j3,k}$$

$$- \sum_{j \neq i} \Xi_{j3,k}^* [\Xi_{j2,k}^{-1}]^* (B_{2j} + \bar{B}_{2j})^* \bar{P}_{i,k+1} (B_1 + \bar{B}_1)$$

$$+ \sum_{j \neq i} \Xi_{j3,k}^* [\Xi_{j2,k}^{-1}]^* (B_{2j} + \bar{B}_{2j})^* \bar{P}_{i,k+1} \sum_{j \neq i} (B_{2j} + \bar{B}_{2j}) \Xi_{j2,k}^{-1} \Xi_{j3,k}$$

$$- \Xi_{i3,k}^* \Xi_{i1,k}^* \Xi_{i1,k} \Xi_{i3,k},$$

$$\delta_{i,k} = \delta_{i,k+1} + \mathbb{E} \langle P_{i,k+1} SW_k, SW_k \rangle,$$

and with the terminal conditions

$$P_{i,k+N} = Q_{i,k+N},$$
$$\bar{P}_{i,k+N} = Q_{i,k+N} + \bar{Q}_{i,k+N},$$
$$\delta_{i,k+N} = 0.$$

Hence,

$$\Phi_{i1,k} = \Phi_{i2,k}^{-\frac{1}{2}},$$
$$\Phi_{i2,k} = [B_{2i}^* P_{i,k+1} B_{2i} + R_{i,k}],$$
$$\Xi_{i1,k} = \Xi_{i2,k}^{-\frac{1}{2}},$$
$$\Xi_{i2,k} = (B_{2i} + \bar{B}_{2i})^* \bar{P}_{i,k+1}(B_{2i} + \bar{B}_{2i}) + (R_{i,k} + \bar{R}_{i,k}),$$

are the variables appearing in the difference equations for both $P_{i,k}$ and $\bar{P}_{i,k}$.

Proof 29 (Theorem 6) *Let us postulate an ansatz for the optimal cost from time k up to terminal time N. Inspired from the structure of the cost functional, we propose the following guess functional:*

$$F_{i,\ell} = \langle P_{i,\ell}(X_\ell - \mathbb{E}[X_\ell]), X_\ell - \mathbb{E}[X_\ell] \rangle + \langle \bar{P}_{i,\ell}\mathbb{E}[X_\ell], \mathbb{E}[X_\ell] \rangle + \delta_{i,\ell}.$$

The telescopic sum applied to the guess functional $F_{i,\ell}$ is

$$F_{i,k+N} = F_{i,k} + \sum_{\ell=k}^{k+N-1} (F_{i,\ell+1} - F_{i,\ell}).$$

We express the guess functional $F_{i,\ell+1}$ in terms of $X_{\ell+1}$ and $\mathbb{E}[X_{\ell+1}]$, i.e.,

$$F_{i,k+N} - F_{i,k} = \sum_{\ell=k}^{k+N-1} \langle P_{i,\ell+1}(X_{\ell+1} - \mathbb{E}[X_{\ell+1}]), X_{\ell+1} - \mathbb{E}[X_{\ell+1}] \rangle$$
$$+ \sum_{\ell=k}^{k+N-1} \langle \bar{P}_{i,\ell+1}\mathbb{E}[X_{\ell+1}], \mathbb{E}[X_{\ell+1}] \rangle + \sum_{\ell=k}^{k+N-1} \delta_{i,\ell+1}$$
$$- \sum_{\ell=k}^{k+N-1} \langle P_{i,\ell}(X_\ell - \mathbb{E}[X_\ell]), X_\ell - \mathbb{E}[X_\ell] \rangle$$
$$- \sum_{\ell=k}^{k+N-1} \langle \bar{P}_{i,\ell}\mathbb{E}[X_\ell], \mathbb{E}[X_\ell] \rangle - \sum_{\ell=k}^{k+N-1} \delta_{i,\ell}. \qquad (12.3)$$

We compute or conveniently re-write some terms from the latter expression in a separated way, e.g., $X_{\ell+1}$, $\mathbb{E}[X_{\ell+1}]$, $\mathbb{E}[X_{\ell+1}]$, and $X_{\ell+1} - \mathbb{E}[X_{\ell+1}]$.

$$X_{\ell+1} = B_1(X_\ell - \mathbb{E}[X_\ell]) + (B_1 + \bar{B}_1)\mathbb{E}[X_\ell]$$

$$+ \sum_{j \in \mathcal{N}} B_{2j}(U_{j,\ell} - \mathbb{E}[U_{j,\ell}])$$

$$+ \sum_{j \in \mathcal{N}} (B_{2j} + \bar{B}_{2j})\mathbb{E}[U_{j,\ell}] + SW_\ell,$$

$$\mathbb{E}[X_{\ell+1}] = (B_1 + \bar{B}_1)\mathbb{E}[X_\ell] + \sum_{j \in \mathcal{N}} (B_{2j} + \bar{B}_{2j})\mathbb{E}[U_{j,\ell}],$$

$$X_{\ell+1} - \mathbb{E}[X_{\ell+1}] = B_1(X_\ell - \mathbb{E}[X_\ell]) + \sum_{j \in \mathcal{N}} B_{2j}(U_{j,\ell} - \mathbb{E}[U_{j,\ell}]) + SW_\ell.$$

We can then compute the quadratic term

$$\langle P_{i,\ell+1}(X_{\ell+1} - \mathbb{E}[X_{\ell+1}]), X_{\ell+1} - \mathbb{E}[X_{\ell+1}] \rangle$$

in (12.3) is as follows:

$$\left\langle P_{i,\ell+1}(X_{\ell+1} - \mathbb{E}[X_{\ell+1}]), X_{\ell+1} - \mathbb{E}[X_{\ell+1}] \right\rangle$$

$$= \left\langle B_1^* P_{i,\ell+1} B_1(X_\ell - \mathbb{E}[X_\ell]), X_\ell - \mathbb{E}[X_\ell] \right\rangle$$

$$+ \left\langle B_{2i}^* P_{i,\ell+1} B_1(X_\ell - \mathbb{E}[X_\ell]), U_{i,\ell} - \mathbb{E}[U_{i,\ell}] \right\rangle$$

$$+ \left\langle P_{i,\ell+1} B_1(X_\ell - \mathbb{E}[X_\ell]), \sum_{j \in \mathcal{N}\setminus\{i\}} B_{2j}(U_{j,\ell} - \mathbb{E}[U_{j,\ell}]) \right\rangle$$

$$+ \left\langle P_{i,\ell+1} B_1(X_\ell - \mathbb{E}[X_\ell]), SW_\ell \right\rangle$$

$$+ \langle B_1^* P_{i,\ell+1} B_{2i}(U_{i,\ell} - \mathbb{E}[U_{i,\ell}]), X_\ell - \mathbb{E}[X_\ell] \rangle$$

$$+ \left\langle B_1^* P_{i,\ell+1} \sum_{j \in \mathcal{N}\setminus\{i\}} B_{2j}(U_{j,\ell} - \mathbb{E}[U_{j,\ell}]), X_\ell - \mathbb{E}[X_\ell] \right\rangle$$

$$+ \langle B_{2i}^* P_{i,\ell+1} B_{2i}(U_{i,\ell} - \mathbb{E}[U_{i,\ell}]), U_{i,\ell} - \mathbb{E}[U_{i,\ell}] \rangle$$

$$+ \left\langle B_{2i}^* P_{i,\ell+1} \sum_{j \in \mathcal{N}\setminus\{i\}} B_{2j}(U_{j,\ell} - \mathbb{E}[U_{j,\ell}]), U_{i,\ell} - \mathbb{E}[U_{i,\ell}] \right\rangle$$

$$+ \left\langle P_{i,\ell+1} B_{2i}(U_{i,\ell} - \mathbb{E}[U_{i,\ell}]), \sum_{j \in \mathcal{N}\setminus\{i\}} B_{2j}(U_{j,\ell} - \mathbb{E}[U_{j,\ell}]) \right\rangle$$

$$+ \left\langle P_{i,\ell+1} \sum_{j \in \mathcal{N}\setminus\{i\}} B_{2j}(U_{j,\ell} - \mathbb{E}[U_{j,\ell}]), \sum_{j \in \mathcal{N}\setminus\{i\}} B_{2j}(U_{j,\ell} - \mathbb{E}[U_{j,\ell}]) \right\rangle$$

$$+ \langle P_{i,\ell+1} B_{2i}(U_{i,\ell} - \mathbb{E}[U_{i,\ell}]), SW_\ell \rangle$$

$$+ \left\langle P_{i,\ell+1} \sum_{j \in \mathcal{N}\setminus\{i\}} B_{2j}(U_{j,\ell} - \mathbb{E}[U_{j,\ell}]), SW_\ell \right\rangle$$

$$+ \langle A^* P_{i,\ell+1} SW_\ell, X_\ell - \mathbb{E}[X_\ell] \rangle + \langle B_{2i}^* P_{i,\ell+1} SW_\ell, U_{i,\ell} - \mathbb{E}[U_{i,\ell}] \rangle$$

$$+ \left\langle P_{i,\ell+1} SW_\ell, \sum_{j \in \mathcal{N} \backslash \{i\}} B_{2j}(U_{j,\ell} - \mathbb{E}[U_{j,\ell}]) \right\rangle + \langle P_{i,\ell+1} SW_\ell, SW_\ell \rangle.$$

Taken into account that

$$\mathbb{E} \langle P_{i,\ell+1} B_1 (X_\ell - \mathbb{E}[X_\ell]), SW_\ell \rangle = 0,$$
$$\mathbb{E} \langle P_{i,\ell+1} B_{2i}(U_{i,\ell} - \mathbb{E}[U_{i,\ell}]), SW_\ell \rangle = 0,$$
$$\mathbb{E} \left\langle P_{i,\ell+1} \sum_{j \in \mathcal{N} \backslash \{i\}} B_{2j}(U_{j,\ell} - \mathbb{E}[U_{j,\ell}]), SW_\ell \right\rangle = 0,$$
$$\mathbb{E} \langle B_1^* P_{i,\ell+1} SW_\ell, X_\ell - \mathbb{E}[X_\ell] \rangle = 0,$$
$$\mathbb{E} \langle B_{2i}^* P_{i,\ell+1} SW_\ell, U_{i,\ell} - \mathbb{E}[U_{i,\ell}] \rangle = 0,$$
$$\mathbb{E} \left\langle P_{i,\ell+1} SW_\ell, \sum_{j \in \mathcal{N} \backslash \{i\}} B_{2j}(U_{j,\ell} - \mathbb{E}[U_{j,\ell}]) \right\rangle = 0,$$

with the expectation yields

$$\mathbb{E} \left\langle P_{i,\ell+1}(X_{\ell+1} - \mathbb{E}[X_{\ell+1}]), X_{\ell+1} - \mathbb{E}[X_{\ell+1}] \right\rangle$$
$$= \mathbb{E} \langle B_1^* P_{i,\ell+1} B_1 (X_\ell - \mathbb{E}[X_\ell]), X_\ell - \mathbb{E}[X_\ell] \rangle$$
$$+ \mathbb{E} \langle B_{2i}^* P_{i,\ell+1} B_{2i}(U_{i,\ell} - \mathbb{E}[U_{i,\ell}]), U_{i,\ell} - \mathbb{E}[U_{i,\ell}] \rangle$$
$$+ \mathbb{E} \langle B_1^* P_{i,\ell+1} B_{2i}(U_{i,\ell} - \mathbb{E}[U_{i,\ell}]), X_\ell - \mathbb{E}[X_\ell] \rangle$$
$$+ \mathbb{E} \langle B_{2i}^* P_{i,\ell+1} B_1 (X_\ell - \mathbb{E}[X_\ell]), U_{i,\ell} - \mathbb{E}[U_{i,\ell}] \rangle$$
$$+ \mathbb{E} \left\langle P_{i,\ell+1} B_{2i}(U_{i,\ell} - \mathbb{E}[U_{i,\ell}]), \sum_{j \in \mathcal{N} \backslash \{i\}} B_{2j}(U_{j,\ell} - \mathbb{E}[U_{j,\ell}]) \right\rangle$$
$$+ \mathbb{E} \left\langle B_{2i}^* P_{i,\ell+1} \sum_{j \in \mathcal{N} \backslash \{i\}} B_{2j}(U_{j,\ell} - \mathbb{E}[U_{j,\ell}]), U_{i,\ell} - \mathbb{E}[U_{i,\ell}] \right\rangle$$
$$+ \mathbb{E} \left\langle B_1^* P_{i,\ell+1} \sum_{j \in \mathcal{N} \backslash \{i\}} B_{2j}(U_{j,\ell} - \mathbb{E}[U_{j,\ell}]), X_\ell - \mathbb{E}[X_\ell] \right\rangle$$
$$+ \mathbb{E} \left\langle P_{i,\ell+1} B_1 (X_\ell - \mathbb{E}[X_\ell]), \sum_{j \in \mathcal{N} \backslash \{i\}} B_{2j}(U_{j,\ell} - \mathbb{E}[U_{j,\ell}]) \right\rangle$$
$$+ \mathbb{E} \left\langle P_{i,\ell+1} \sum_{j \in \mathcal{N} \backslash \{i\}} B_{2j}(U_{j,\ell} - \mathbb{E}[U_{j,\ell}]), \sum_{j \in \mathcal{N} \backslash \{i\}} B_{2j}(U_{j,\ell} - \mathbb{E}[U_{j,\ell}]) \right\rangle$$
$$+ \mathbb{E} \langle P_{i,\ell+1} SW_\ell, SW_\ell \rangle. \tag{12.4}$$

On the other hand and similarly, we calculate the quadratic form $\langle \bar{P}_{i,\ell+1} \mathbb{E}[X_{\ell+1}], \mathbb{E}[X_{\ell+1}] \rangle$ *also appearing in* (12.3).

$$\left\langle \bar{P}_{i,\ell+1} \mathbb{E}[X_{\ell+1}], \mathbb{E}[X_{\ell+1}] \right\rangle = \left\langle (B_1 + \bar{B}_1)^* \bar{P}_{i,\ell+1}(B_1 + \bar{B}_1) \mathbb{E}[X_\ell], \mathbb{E}[X_\ell] \right\rangle$$

$$+ \left\langle (B_{2i} + \bar{B}_{2i})^* \bar{P}_{i,\ell+1}(B_{2i} + \bar{B}_{2i})\mathbb{E}[U_{i,\ell}], \mathbb{E}[U_{i,\ell}] \right\rangle$$
$$+ \left\langle \bar{P}_{i,\ell+1}(B_1 + \bar{B}_1)\mathbb{E}[X_\ell], (B_{2i} + \bar{B}_{2i})\mathbb{E}[U_{i,\ell}] \right\rangle$$
$$+ \left\langle \bar{P}_{i,\ell+1}(B_{2i} + \bar{B}_{2i})\mathbb{E}[U_{i,\ell}], (B_1 + \bar{B}_1)\mathbb{E}[X_\ell] \right\rangle$$
$$+ \left\langle \bar{P}_{i,\ell+1}(B_{2i} + \bar{B}_{2i})\mathbb{E}[U_{i,\ell}], \sum_{j \in \mathcal{N} \setminus \{i\}} (B_{2j} + \bar{B}_{2j})\mathbb{E}[U_{j,\ell}] \right\rangle$$
$$+ \left\langle \bar{P}_{i,\ell+1} \sum_{j \in \mathcal{N} \setminus \{i\}} (B_{2j} + \bar{B}_{2j})\mathbb{E}[U_{j,\ell}], (B_{2i} + \bar{B}_{2i})\mathbb{E}[U_{i,\ell}] \right\rangle$$
$$+ \left\langle \bar{P}_{i,\ell+1} \sum_{j \in \mathcal{N} \setminus \{i\}} (B_{2j} + \bar{B}_{2j})\mathbb{E}[U_{j,\ell}], (B_1 + \bar{B}_1)\mathbb{E}[X_\ell] \right\rangle$$
$$+ \left\langle \bar{P}_{i,\ell+1}(B_1 + \bar{B}_1)\mathbb{E}[X_\ell], \sum_{j \in \mathcal{N} \setminus \{i\}} (B_{2j} + \bar{B}_{2j})\mathbb{E}[U_{j,\ell}] \right\rangle$$
$$+ \left\langle \bar{P}_{i,\ell+1} \sum_{j \in \mathcal{N} \setminus \{i\}} (B_{2j} + \bar{B}_{2j})\mathbb{E}[U_{j,\ell}], \sum_{j \in \mathcal{N} \setminus \{i\}} (B_{2j} + \bar{B}_{2j})\mathbb{E}[U_{j,\ell}] \right\rangle.$$

$$(12.5)$$

Then, using the telescopic sum in (12.3) and the terms in (12.4) and (12.5), the difference $\mathbb{E}[L_i(X, U_1, \ldots, U_n) - F_{i,k}]$ is expressed as follows:

$$\mathbb{E}[L_i(X, U_1, \ldots, U_n) - F_{i,k}] = \mathbb{E}\left\langle Q_{i,k+N}(X_{k+N} - \mathbb{E}[X_{k+N}]), X_{k+N} - \mathbb{E}[X_{k+N}] \right\rangle$$
$$+ \mathbb{E}\left\langle (Q_{i,k+N} + \bar{Q}_{i,k+N})\mathbb{E}[X_{k+N}], \mathbb{E}[X_{k+N}] \right\rangle$$
$$- \mathbb{E}\left\langle P_{i,k+N}(X_{k+N} - \mathbb{E}[X_{k+N}]), X_{k+N} - \mathbb{E}[X_{k+N}] \right\rangle$$
$$- \mathbb{E}\left\langle \bar{P}_{i,k+N}\mathbb{E}[X_{k+N}], \mathbb{E}[X_{k+N}] \right\rangle - \mathbb{E}\delta_{i,k+N}$$
$$+ \mathbb{E} \sum_{\ell=k}^{k+N-1} \left\langle B_1^* P_{i,\ell+1} B_1(X_\ell - \mathbb{E}[X_\ell]), X_\ell - \mathbb{E}[X_\ell] \right\rangle$$
$$+ \mathbb{E} \sum_{\ell=k}^{k+N-1} \left\langle B_{2i}^* P_{i,\ell+1} B_{2i}(U_{i,\ell} - \mathbb{E}[U_{i,\ell}]), U_{i,\ell} - \mathbb{E}[U_{i,\ell}] \right\rangle$$
$$+ \mathbb{E} \sum_{\ell=k}^{k+N-1} \left\langle B_1^* P_{i,\ell+1} B_{2i}(U_{i,\ell} - \mathbb{E}[U_{i,\ell}]), X_\ell - \mathbb{E}[X_\ell] \right\rangle$$
$$+ \mathbb{E} \sum_{\ell=k}^{k+N-1} \left\langle B_{2i}^* P_{i,\ell+1} B_1(X_\ell - \mathbb{E}[X_\ell]), U_{i,\ell} - \mathbb{E}[U_{i,\ell}] \right\rangle$$
$$+ \mathbb{E} \sum_{\ell=k}^{k+N-1} \left\langle P_{i,\ell+1} B_{2i}(U_{i,\ell} - \mathbb{E}[U_{i,\ell}]), \sum_{j \in \mathcal{N} \setminus \{i\}} B_{2j}(U_{j,\ell} - \mathbb{E}[U_{j,\ell}]) \right\rangle$$

$$+ \mathbb{E} \sum_{\ell=k}^{k+N-1} \left\langle B_{2i}^* P_{i,\ell+1} \sum_{j \in \mathcal{N} \setminus \{i\}} B_{2j}(U_{j,\ell} - \mathbb{E}[U_{j,\ell}]), U_{i,\ell} - \mathbb{E}[U_{i,\ell}] \right\rangle$$

$$+ \mathbb{E} \sum_{\ell=k}^{k+N-1} \left\langle B_1^* P_{i,\ell+1} \sum_{j \in \mathcal{N} \setminus \{i\}} B_{2j}(U_{j,\ell} - \mathbb{E}[U_{j,\ell}]), X_\ell - \mathbb{E}[X_\ell] \right\rangle$$

$$+ \mathbb{E} \sum_{\ell=k}^{k+N-1} \left\langle P_{i,\ell+1} B_1(X_\ell - \mathbb{E}[X_\ell]), \sum_{j \in \mathcal{N} \setminus \{i\}} B_{2j}(U_{j,\ell} - \mathbb{E}[U_{j,\ell}]) \right\rangle$$

$$+ \mathbb{E} \sum_{\ell=k}^{k+N-1} \left\langle P_{i,\ell+1} \sum_{j \in \mathcal{N} \setminus \{i\}} B_{2j}(U_{j,\ell} - \mathbb{E}[U_{j,\ell}]), \sum_{j \in \mathcal{N} \setminus \{i\}} B_{2j}(U_{j,\ell} - \mathbb{E}[U_{j,\ell}]) \right\rangle$$

$$+ \mathbb{E} \sum_{\ell=k}^{k+N-1} \left\langle P_{i,\ell+1} SW_\ell, SW_\ell \right\rangle$$

$$+ \mathbb{E} \sum_{\ell=k}^{k+N-1} \left\langle \bar{P}_{i,\ell+1}(B_1 + \bar{B}_1)^*(B_1 + \bar{B}_1)\mathbb{E}[X_\ell], \mathbb{E}[X_\ell] \right\rangle$$

$$+ \mathbb{E} \sum_{\ell=k}^{k+N-1} \left\langle (B_{2i} + \bar{B}_{2i})^* \bar{P}_{i,\ell+1}(B_{2i} + \bar{B}_{2i})\mathbb{E}[U_{i,\ell}], \mathbb{E}[U_{i,\ell}] \right\rangle$$

$$+ \mathbb{E} \sum_{\ell=k}^{k+N-1} \left\langle \bar{P}_{i,\ell+1}(B_1 + \bar{B}_1)\mathbb{E}[X_\ell], (B_{2i} + \bar{B}_{2i})\mathbb{E}[U_{i,\ell}] \right\rangle$$

$$+ \mathbb{E} \sum_{\ell=k}^{k+N-1} \left\langle \bar{P}_{i,\ell+1}(B_{2i} + \bar{B}_{2i})\mathbb{E}[U_{i,\ell}], (B_1 + \bar{B}_1)\mathbb{E}[X_\ell] \right\rangle$$

$$+ \mathbb{E} \sum_{\ell=k}^{k+N-1} \left\langle \bar{P}_{i,\ell+1}(B_{2i} + \bar{B}_{2i})\mathbb{E}[U_{i,\ell}], \sum_{j \in \mathcal{N} \setminus \{i\}} (B_{2j} + \bar{B}_{2j})\mathbb{E}[U_{j,\ell}] \right\rangle$$

$$+ \mathbb{E} \sum_{\ell=k}^{k+N-1} \left\langle \bar{P}_{i,\ell+1} \sum_{j \in \mathcal{N} \setminus \{i\}} (B_{2j} + \bar{B}_{2j})\mathbb{E}[U_{j,\ell}], (B_{2i} + \bar{B}_{2i})\mathbb{E}[U_{i,\ell}] \right\rangle$$

$$+ \mathbb{E} \sum_{\ell=k}^{k+N-1} \left\langle \bar{P}_{i,\ell+1} \sum_{j \in \mathcal{N} \setminus \{i\}} (B_{2j} + \bar{B}_{2j})\mathbb{E}[U_{j,\ell}], (B_1 + \bar{B}_1)\mathbb{E}[X_\ell] \right\rangle$$

$$+ \mathbb{E} \sum_{\ell=k}^{k+N-1} \left\langle \bar{P}_{i,\ell+1}(B_1 + \bar{B}_1)\mathbb{E}[X_\ell], \sum_{j \in \mathcal{N} \setminus \{i\}} (B_{2j} + \bar{B}_{2j})\mathbb{E}[U_{j,\ell}] \right\rangle$$

$$+ \mathbb{E} \sum_{\ell=k}^{k+N-1} \left\langle \bar{P}_{i,\ell+1} \sum_{j \in \mathcal{N} \setminus \{i\}} (B_{2j} + \bar{B}_{2j})\mathbb{E}[U_{j,\ell}], \sum_{j \in \mathcal{N} \setminus \{i\}} (B_{2j} + \bar{B}_{2j})\mathbb{E}[U_{j,\ell}] \right\rangle$$

$$- \mathbb{E} \sum_{\ell=k}^{k+N-1} \left\langle P_{i,\ell}(X_\ell - \mathbb{E}[X_\ell]), X_\ell - \mathbb{E}[X_\ell] \right\rangle$$

$$- \mathbb{E} \sum_{\ell=k}^{k+N-1} \langle \bar{P}_{i,\ell} \mathbb{E}[X_\ell], \mathbb{E}[X_\ell] \rangle$$

$$+ \mathbb{E} \sum_{\ell=k}^{k+N-1} \langle Q_{i,\ell}(X_\ell - \mathbb{E}[X_\ell]), X_\ell - \mathbb{E}[X_\ell] \rangle$$

$$+ \mathbb{E} \sum_{\ell=k}^{k+N-1} \langle (Q_{i,\ell} + \bar{Q}_{i,\ell}) \mathbb{E}[X_\ell], \mathbb{E}[X_\ell] \rangle$$

$$+ \mathbb{E} \sum_{\ell=k}^{k+N-1} \delta_{i,\ell+1} - \mathbb{E} \sum_{\ell=k}^{k+N-1} \delta_{i,\ell}$$

$$+ \mathbb{E} \sum_{\ell=k}^{k+N-1} \langle R_{i,\ell}(U_{i,\ell} - \mathbb{E}[U_{i,\ell}]), U_{i,\ell} - \mathbb{E}[U_{i,\ell}] \rangle$$

$$+ \mathbb{E} \sum_{\ell=k}^{k+N-1} \langle (R_{i,\ell} + \bar{R}_{i,\ell}) \mathbb{E}[U_{i,\ell}], \mathbb{E}[U_{i,\ell}] \rangle, \tag{12.6}$$

Performing square completion for the difference $(U_{i,\ell} - \mathbb{E}[U_{i,\ell}])$ *we have*

$$\left| \Phi_{i1,\ell} \Phi_{i2,\ell}[(U_{i,\ell} - \mathbb{E}[U_{i,\ell}]) + \Phi_{i2,\ell}^{-1} \Phi_{i3,\ell}(X_\ell - \mathbb{E}[X_\ell])] \right|^2$$

$$= \langle \Phi_{i1,\ell} \Phi_{i2,\ell}(U_{i,\ell} - \mathbb{E}[U_{i,\ell}]), \Phi_{i1,\ell} \Phi_{i2,\ell}(U_{i,\ell} - \mathbb{E}[U_{i,\ell}]) \rangle$$

$$+ \langle \Phi_{i1,\ell} \Phi_{i2,\ell}(U_{i,\ell} - \mathbb{E}[U_{i,\ell}]), \Phi_{i1,\ell} \Phi_{i3,\ell}(X_\ell - \mathbb{E}[X_\ell]) \rangle$$

$$+ \langle \Phi_{i1,\ell} \Phi_{i3,\ell}(X_\ell - \mathbb{E}[X_\ell]), \Phi_{i1,\ell} \Phi_{i2,\ell}(U_{i,\ell} - \mathbb{E}[U_{i,\ell}]) \rangle$$

$$+ \langle \Phi_{i1,\ell} \Phi_{i3,\ell}(X_\ell - \mathbb{E}[X_\ell]), \Phi_{i1,\ell} \Phi_{i3,\ell}(X_\ell - \mathbb{E}[X_\ell]) \rangle$$

Thus, at this point we conclude that

$$(U_{j,\ell} - \mathbb{E}[U_{j,\ell}]) = -\Phi_{j2,\ell}^{-1} \Phi_{j3,\ell}(X_\ell - \mathbb{E}[X_\ell]), \ \forall j \in \mathcal{N},$$

and the control inputs for all the other decision-makers $j \in \mathcal{N} \setminus \{i\}$ *can be expressed in the state-feedback form, i.e.,*

$$B_1(X_\ell - \mathbb{E}[X_\ell]) + \sum_{j \in \mathcal{N} \setminus \{i\}} B_{2j}(U_{j,\ell} - \mathbb{E}[U_{j,\ell}])$$

$$= \left(B_1 - \sum_{j \in \mathcal{N} \setminus \{i\}} B_{2j} \Phi_{j2,\ell}^{-1} \Phi_{j3,\ell} \right)(X_\ell - \mathbb{E}[X_\ell]),$$

$$\left\langle P_{i,\ell+1}[B_1(X_\ell - \mathbb{E}[X_\ell]) + \sum_{j \in \mathcal{N} \setminus \{i\}} B_{2j}(U_{j,\ell} - \mathbb{E}[U_{j,\ell}])], B_{2i}(U_{i,\ell} - \mathbb{E}[U_{i,\ell}]) \right\rangle$$

$$= \left\langle B_{2i}^* P_{i,\ell+1}[B_1 - \sum_{j \in \mathcal{N} \setminus \{i\}} B_{2j} \Phi_{j2,\ell}^{-1} \Phi_{j3,\ell}](X_\ell - \mathbb{E}[X_\ell]), (U_{i,\ell} - \mathbb{E}[U_{i,\ell}]) \right\rangle.$$

Hence we identify the variables $\Phi_{i1,\ell}$, $\Phi_{i2,\ell}$, *and* $\Phi_{i3,\ell}$ *from*

$$\Phi_{i2,\ell}^* \Phi_{i1,\ell}^* \Phi_{i1,\ell} \Phi_{i2,\ell} = B_{2i}^* P_{i,\ell+1} B_{2i} + R_{i,\ell},$$

$$\Phi_{i2,\ell}^* \Phi_{i1,\ell}^* \Phi_{i1,\ell} \Phi_{i3,\ell} = B_{2i}^* P_{i,\ell+1} \left(B_1 - \sum_{j \in \mathcal{N} \setminus \{i\}} B_{2j} \Phi_{j2,\ell}^{-1} \Phi_{j3,\ell} \right).$$

We consider the following parameters:

$$\Phi_{i1,\ell} = [B_{2i}^* P_{i,\ell+1} B_{2i} + R_{i,\ell}]^{-\frac{1}{2}},$$
$$\Phi_{i2,\ell} = B_{2i}^* P_{i,\ell+1} B_{2i} + R_{i,\ell},$$

$$\Phi_{i3,\ell} = B_{2i}^* P_{i,\ell+1} \left(B_1 - \sum_{j \in \mathcal{N} \setminus \{i\}} B_{2j} [B_{2j}^* P_{j,\ell+1} B_{2j} + R_{j,\ell}]^{-1} \Phi_{j3,\ell} \right).$$

Thus, the terms depending on the subtractions $(U_{j,\ell} - \mathbb{E}[U_{j,\ell}])$, *for all* $j \in \mathcal{N}$, *from the gap* $\mathbb{E}[L_i(X, U_1, \dots, U_n)] - F_{i,k}$ *are re-written as follows:*

$$\langle [B_{2i}^* P_{i,\ell+1} B_{2i} + R_{i,\ell}](U_{i,\ell} - \mathbb{E}[U_{i,\ell}]), U_{i,\ell} - \mathbb{E}[U_{i,\ell}] \rangle$$

$$+ \left\langle P_{i,\ell+1} B_{2i}(U_{i,\ell} - \mathbb{E}[U_{i,\ell}]), B_1(X_\ell - \mathbb{E}[X_\ell]) + \sum_{j \in \mathcal{N} \setminus \{i\}} B_{2j}(U_{j,\ell} - \mathbb{E}[U_{j,\ell}]) \right\rangle$$

$$+ \left\langle P_{i,\ell+1}[B_1(X_\ell - \mathbb{E}[X_\ell]) + \sum_{j \in \mathcal{N} \setminus \{i\}} B_{2j}(U_{j,\ell} - \mathbb{E}[U_{j,\ell}])], B_{2i}(U_{i,\ell} - \mathbb{E}[U_{i,\ell}]) \right\rangle$$

$$= |\Phi_{i1,\ell}[\Phi_{i2,\ell}(U_{i,\ell} - \mathbb{E}[U_{i,\ell}]) + \Phi_{i3,\ell}(X_\ell - \mathbb{E}[X_\ell])]|^2$$
$$- \langle \Phi_{i1,\ell} \Phi_{i3,\ell}(X_\ell - \mathbb{E}[X_\ell]), \Phi_{i1,\ell} \Phi_{i3,\ell}(X_\ell - \mathbb{E}[X_\ell]) \rangle. \tag{12.7}$$

Similarly, we perform optimization for those terms depending on $\mathbb{E}[U_{i,\ell}]$. *We have*

$$\left| \Xi_{i1,\ell} \Xi_{i2,\ell}[\mathbb{E}[U_{i,\ell}] + \Xi_{i2,\ell}^{-1} \Xi_{i3,\ell} \mathbb{E}[X_\ell]] \right|^2 = \langle \Xi_{i1,\ell} \Xi_{i2,\ell} \mathbb{E}[U_{i,\ell}], \Xi_{i1,\ell} \Xi_{i2,\ell} \mathbb{E}[U_{i,\ell}] \rangle$$

$$+ \langle \Xi_{i1,\ell} \Xi_{i2,\ell} \mathbb{E}[U_{i,\ell}], \Xi_{i1,\ell} \Xi_{i3,\ell} \mathbb{E}[X_\ell] \rangle$$

$$+ \langle \Xi_{i1,\ell} \Xi_{i3,\ell} \mathbb{E}[X_\ell], \Xi_{i1,\ell} \Xi_{i2,\ell} \mathbb{E}[U_{i,\ell}] \rangle$$

$$+ \langle \Xi_{i1,\ell} \Xi_{i3,\ell} \mathbb{E}[X_\ell], \Xi_{i1,\ell} \Xi_{i3,\ell} \mathbb{E}[X_\ell] \rangle$$

Thus, it is concluded that

$$\mathbb{E}[U_{j,\ell}] = -\Xi_{j2,\ell}^{-1} \Xi_{j3,\ell} \mathbb{E}[X_\ell], \ \forall j \in \mathcal{N},$$

and the terms depending on the control input $\mathbb{E}[U_{j,\ell}]$, *for all* $j \in \mathcal{N} \setminus \{i\}$, *can be written in the state-feedback form, i.e.,*

$$(B_1 + \bar{B}_1)\mathbb{E}[X_\ell] + \sum_{j \in \mathcal{N} \setminus \{i\}} (B_{2j} + \bar{B}_{2j})\mathbb{E}[U_{j,\ell}]$$

$$= \left((B_1 + \bar{B}_1) - \sum_{j \in \mathcal{N} \setminus \{i\}} (B_{2j} + \bar{B}_{2j}) \Xi_{j2,\ell}^{-1} \Xi_{j3,\ell} \right) \mathbb{E}[X_\ell].$$

And the variables $\Xi_{i1,\ell}$, $\Xi_{i2,\ell}$, *and* $\Xi_{i3,\ell}$ *are as follows:*

$$\Xi_{i1,\ell} = [(B_{2i} + \bar{B}_{2i})^* \bar{P}_{i,\ell+1}(B_{2i} + \bar{B}_{2i}) + (R_{i,\ell} + \bar{R}_{i,\ell})]^{-\frac{1}{2}},$$

$$\Xi_{i2,\ell} = (B_{2i} + \bar{B}_{2i})^* \bar{P}_{i,\ell+1}(B_{2i} + \bar{B}_{2i}) + (R_{i,\ell} + \bar{R}_{i,\ell}),$$

$$\Xi_{i3,\ell} = (B_{2i} + \bar{B}_{2i})^* \bar{P}_{i,\ell+1}\left((B_1 + \bar{B}_1) - \sum_{j \in \mathcal{N} \setminus \{i\}} (B_{2j} + \bar{B}_{2j}) \cdot \right.$$

$$\left. \cdot \left[(B_{2j} + \bar{B}_{2j})^* \bar{P}_{j,\ell+1}(B_{2j} + \bar{B}_{2j}) + (R_{j,\ell} + \bar{R}_{j,\ell}) \right]^{-1} \Xi_{j3,\ell} \right).$$

The terms depending on the expectation $\mathbb{E}[U_{j,\ell}]$, *for all* $j \in \mathcal{N}$, *from the gap* $\mathbb{E}[L_i(X, U_1, \ldots, U_n) - F_{i,k}]$ *are re-written as follows:*

$$\left\langle [(B_{2i} + \bar{B}_{2i})^* \bar{P}_{i,\ell+1}(B_{2i} + \bar{B}_{2i}) + (R_{i,\ell} + \bar{R}_{i,\ell})]\mathbb{E}[U_{i,\ell}], \mathbb{E}[U_{i,\ell}] \right\rangle$$

$$+ \left\langle \bar{P}_{i,\ell+1}(B_{2i} + \bar{B}_{2i})\mathbb{E}[U_{i,\ell}], (B_1 + \bar{B}_1)\mathbb{E}[X_\ell] + \sum_{j \in \mathcal{N} \setminus \{i\}} (B_{2j} + \bar{B}_{2j})\mathbb{E}[U_{j,\ell}] \right\rangle$$

$$+ \left\langle \bar{P}_{i,\ell+1}[(B_1 + \bar{B}_1)\mathbb{E}[X_\ell] + \sum_{j \in \mathcal{N} \setminus \{i\}} (B_{2j} + \bar{B}_{2j})\mathbb{E}[U_{j,\ell}]], (B_{2i} + \bar{B}_{2i})\mathbb{E}[U_{i,\ell}] \right\rangle$$

$$= |\Xi_{i1,\ell}[\Xi_{i2,\ell}\mathbb{E}[U_{i,\ell}] + \Xi_{i3,\ell}\mathbb{E}[X_\ell]]|^2 - \left\langle \Xi_{i1,\ell}\Xi_{i3,\ell}\mathbb{E}[X_\ell], \Xi_{i1,\ell}\Xi_{i3,\ell}\mathbb{E}[X_\ell] \right\rangle.$$
$$(12.8)$$

Finally, we obtain that from (12.6), *and replacing the terms in* (12.7) *and* (12.8) *that*

$$\mathbb{E}[L_i(X, U_1, \ldots, U_n) - F_{i,k}] = \mathbb{E}[0 - \delta_{i,k+N}]$$

$$+ \mathbb{E}\left\langle (Q_{i,k+N} - P_{i,k+N})(X_{k+N} - \mathbb{E}[X_{k+N}]), X_{k+N} - \mathbb{E}[X_{k+N}] \right\rangle$$

$$+ \mathbb{E}\left\langle (Q_{i,k+N} + \bar{Q}_{i,k+N} - \bar{P}_{i,k+N})\mathbb{E}[X_{k+N}], \mathbb{E}[X_{k+N}] \right\rangle$$

$$+ \mathbb{E}\sum_{\ell=k}^{k+N-1} \left\langle Q_{i,\ell}(X_\ell - \mathbb{E}[X_\ell]), X_\ell - \mathbb{E}[X_\ell] \right\rangle$$

$$+ \mathbb{E}\sum_{\ell=k}^{k+N-1} \left\langle B_1^* P_{i,\ell+1} B_1(X_\ell - \mathbb{E}[X_\ell]), X_\ell - \mathbb{E}[X_\ell] \right\rangle$$

$$- \mathbb{E}\sum_{\ell=k}^{k+N-1} \left\langle B_1^* P_{i,\ell+1} \sum_{j \in \mathcal{N} \setminus \{i\}} B_{2j} \Phi_{j2,\ell}^{-1} \Phi_{j3,\ell}(X_\ell - \mathbb{E}[X_\ell]), X_\ell - \mathbb{E}[X_\ell] \right\rangle$$

$$- \mathbb{E}\sum_{\ell=k}^{k+N-1} \left\langle P_{i,\ell+1} B_1(X_\ell - \mathbb{E}[X_\ell]), \sum_{j \in \mathcal{N} \setminus \{i\}} B_{2j} \Phi_{j2,\ell}^{-1} \Phi_{j3,\ell}(X_\ell - \mathbb{E}[X_\ell]) \right\rangle$$

$$+ \mathbb{E} \sum_{\ell=k}^{k+N-1} \left\langle P_{i,\ell+1} \sum_{j \in \mathcal{N} \setminus \{i\}} B_{2j} \Phi_{j2,\ell}^{-1} \Phi_{j3,\ell} (X_\ell - \mathbb{E}[X_\ell]), \right.$$

$$\left. \sum_{j \in \mathcal{N} \setminus \{i\}} B_{2j} \Phi_{j2,\ell}^{-1} \Phi_{j3,\ell} (X_\ell - \mathbb{E}[X_\ell]) \right\rangle$$

$$- \mathbb{E} \sum_{\ell=k}^{k+N-1} \left\langle P_{i,\ell} (X_\ell - \mathbb{E}[X_\ell]), X_\ell - \mathbb{E}[X_\ell] \right\rangle$$

$$- \mathbb{E} \sum_{\ell=k}^{k+N-1} \left\langle \Phi_{i1,\ell} \Phi_{i3,\ell} (X_\ell - \mathbb{E}[X_\ell]), \Phi_{i1,\ell} \Phi_{i3,\ell} (X_\ell - \mathbb{E}[X_\ell]) \right\rangle$$

$$+ \mathbb{E} \sum_{\ell=k}^{k+N-1} \left\langle (Q_{i,\ell} + \bar{Q}_{i,\ell}) \mathbb{E}[X_\ell], \mathbb{E}[X_\ell] \right\rangle$$

$$+ \mathbb{E} \sum_{\ell=k}^{k+N-1} \left\langle \bar{P}_{i,\ell+1} (B_1 + \bar{B}_1)^* (B_1 + \bar{B}_1) \mathbb{E}[X_\ell], \mathbb{E}[X_\ell] \right\rangle$$

$$- \mathbb{E} \sum_{\ell=k}^{k+N-1} \left\langle \bar{P}_{i,\ell+1} \sum_{j \in \mathcal{N} \setminus \{i\}} (B_{2j} + \bar{B}_{2j}) \Xi_{j2,\ell}^{-1} \Xi_{j3,\ell} \mathbb{E}[X_\ell], (B_1 + \bar{B}_1) \mathbb{E}[X_\ell] \right\rangle$$

$$+ \mathbb{E} \sum_{\ell=k}^{k+N-1} \left\langle \bar{P}_{i,\ell+1} \sum_{j \in \mathcal{N} \setminus \{i\}} (B_{2j} + \bar{B}_{2j}) \Xi_{j2,\ell}^{-1} \Xi_{j3,\ell} \mathbb{E}[X_\ell], \right.$$

$$\left. \sum_{j \in \mathcal{N} \setminus \{i\}} (B_{2j} + \bar{B}_{2j}) \Xi_{j2,\ell}^{-1} \Xi_{j3,\ell} \mathbb{E}[X_\ell] \right\rangle$$

$$- \mathbb{E} \sum_{\ell=k}^{k+N-1} \left\langle \bar{P}_{i,\ell} \mathbb{E}[X_\ell], \mathbb{E}[X_\ell] \right\rangle$$

$$- \mathbb{E} \sum_{\ell=k}^{k+N-1} \left\langle \bar{P}_{i,\ell+1} (B_1 + \bar{B}_1) \mathbb{E}[X_\ell], \sum_{j \in \mathcal{N} \setminus \{i\}} (B_{2j} + \bar{B}_{2j}) \Xi_{j2,\ell}^{-1} \Xi_{j3,\ell} \mathbb{E}[X_\ell] \right\rangle$$

$$- \mathbb{E} \sum_{\ell=k}^{k+N-1} \left\langle \Xi_{i1,\ell} \Xi_{i3,\ell} \mathbb{E}[X_\ell], \Xi_{i1,\ell} \Xi_{i3,\ell} \mathbb{E}[X_\ell] \right\rangle$$

$$+ \mathbb{E} \sum_{\ell=k}^{k+N-1} (\delta_{i,\ell+1} - \delta_{i,\ell}) + \mathbb{E} \sum_{\ell=k}^{k+N-1} \left\langle P_{i,\ell+1} SW_\ell, SW_\ell \right\rangle$$

$$+ \mathbb{E} \sum_{\ell=k}^{k+N-1} |\Phi_{i1,\ell} [\Phi_{i2,\ell} (U_{i,\ell} - \mathbb{E}[U_{i,\ell}]) + \Phi_{i3,\ell} (X_\ell - \mathbb{E}[X_\ell])]|^2$$

$$+ \mathbb{E} \sum_{\ell=k}^{k+N-1} |\Xi_{i1,\ell} [\Xi_{i2,\ell} \mathbb{E}[U_{i,\ell}] + \Xi_{i3,\ell} \mathbb{E}[X_\ell]]|^2 .$$

As a last step, we group the previous expression into the following terms:

- *Terms evaluated at the terminal time $k+N$ from which we infer the terminal values for the difference equations P_i, $\bar{P}_i$ and δ_i*

- *Those terms depending on the quadratic form of the subtraction $(X - \mathbb{E}[X])$. Thus, we find the difference equation P_i*

- *Those terms whose common factor is the quadratic form of the expected state $\mathbb{E}[X]$. These terms define the difference equation $\bar{P}_i$*

- *State-and-control-input independent terms defining the difference equation δ_i*

- *And the quadratic terms depending on the control inputs from which we infer the optimal strategies.*

Then, minimizing the terms in $\mathbb{E}[L_i(X, U_1, \ldots, U_n) - F_{i,k}]$ we obtain the announced result.

On the other hand, for the full-cooperative scenario we have the following result:

Theorem 7 (Full-Cooperative Scenario) *Let us consider the following change of variables:*

$$V_\ell = (U_{1,\ell} \quad \cdots \quad U_{n,\ell}),$$
$$\mathbb{E}[V_\ell] = (\mathbb{E}[U_{1,\ell}] \quad \cdots \quad \bar{U}_{n,\ell}),$$

the concatenated matrices

$$B_2 = (B_{21} \quad \cdots \quad B_{2n}),$$
$$\bar{B}_2 = (\bar{B}_{21} \quad \cdots \quad \bar{B}_{2n}),$$

and the block diagonal matrices

$$R_\ell = \mathrm{diag}(R_{1,\ell} \quad \cdots \quad R_{n,\ell}),$$
$$\bar{R}_\ell = \mathrm{diag}(\bar{R}_{1,\ell} \quad \cdots \quad \bar{R}_{n,\ell}).$$

The explicit solution of Problem 24 is given by

$$V_\ell^{\mathrm{op}} - \mathbb{E}[V_\ell]^{\mathrm{op}} = -[R_\ell + B_2^* P_{0,\ell+1} B_2]^{-1} B_2^* P_{0,\ell+1} A(X_\ell - \mathbb{E}[X_\ell]),$$
$$\mathbb{E}[V_\ell]^{\mathrm{op}} = -\left[(R_\ell + \bar{R}_\ell) + (B_2 + \bar{B}_2)^* \bar{P}_{0,\ell+1}(B_2 + \bar{B}_2)\right]^{-1} \cdot$$
$$\cdot (B_2 + \bar{B}_2)^* \bar{P}_{0,\ell+1}(B_1 + \bar{B}_1)\mathbb{E}[X_\ell],$$
$$\forall \ell \in [k..k+N-1],$$
$$L_0(X, U_1^{\mathrm{op}}, \ldots, U_n^{\mathrm{op}}) = \langle P_{0,k}(X_k - \bar{X}_k), X_k - \bar{X}_k \rangle + \langle \bar{P}_{0,k}\bar{X}_k, \bar{X}_k \rangle + \delta_{0,k},$$

where P_0, $\bar{P}_0$, and δ_0 solve the following difference equations:

$$P_{0,k} = Q_{0,k} + B_1^* P_{0,k+1} B_1 - \Phi_{03,k}^* \Phi_{01,k}^* \Phi_{01,k} \Phi_{03,k}$$

$$\bar{P}_{0,k} = (Q_{0,k} + \bar{Q}_{0,k}) + (B_1 + \bar{B}_1)^* \bar{P}_{0,k+1}(B_1 + \bar{B}_1) - \Xi_{03,k}^* \Xi_{01,k}^* \Xi_{01,k} \Xi_{03,k}$$

$$\delta_{0,k} = \delta_{0,k+1} + \mathbb{E} \langle P_{0,k+1} SW_k, SW_k \rangle,$$

with terminal conditions

$$P_{0,k+N} = Q_{0,k+N},$$
$$\bar{P}_{0,k+N} = Q_{0,k+N} + \bar{Q}_{0,k+N},$$
$$\delta_{0,k+N} = 0.$$

Moreover, the weight matrices in the cost functional are

$$Q_0 = \sum_{j \in \mathcal{N}} Q_j,$$

$$\bar{Q}_0 = \sum_{j \in \mathcal{N}} \bar{Q}_j,$$

Finally, the auxiliary variables are

$$\Phi_{01,\ell} = [R_\ell + B_2^* P_{0,\ell+1} B_2]^{-\frac{1}{2}},$$
$$\Phi_{02,\ell} = R_\ell + B_2^* P_{0,\ell+1} B_2,$$
$$\Phi_{03,\ell} = B_2^* P_{0,\ell+1} B_1,$$
$$\Xi_{01,\ell} = [(R_\ell + \bar{R}_\ell) + (B_2 + \bar{B}_2)^* \bar{P}_{0,\ell+1}(B_2 + \bar{B}_2)]^{-\frac{1}{2}},$$
$$\Xi_{02,\ell} = (R_\ell + \bar{R}_\ell) + (B_2 + \bar{B}_2)^* \bar{P}_{0,\ell+1}(B_2 + \bar{B}_2),$$
$$\Xi_{03,\ell} = (B_2 + \bar{B}_2)^* \bar{P}_{0,\ell+1}(B_1 + \bar{B}_1),$$

which appear in the difference equations $P_{0,k}$ and $\bar{P}_{0,k}$.

Proof 30 (Theorem 7) *We use the same guess functional as in Theorem 6 because of the same structure in the cost functional and in the system dynamics, i.e.,*

$$F_{0,\ell} = \langle P_{0,\ell}(X_\ell - \mathbb{E}[X_\ell]), X_\ell - \mathbb{E}[X_\ell] \rangle + \langle \bar{P}_{0,\ell} \mathbb{E}[X_\ell], \mathbb{E}[X_\ell] \rangle + \delta_{0,\ell}.$$

The telescopic sum applied to the ansatz returns the following expression:

$$F_{0,k+N} - F_{0,k} = \sum_{\ell=k}^{k+N-1} \langle P_{0,\ell+1}(X_{\ell+1} - \mathbb{E}[X_{\ell+1}]), X_{\ell+1} - \mathbb{E}[X_{\ell+1}] \rangle$$

$$+ \sum_{\ell=k}^{k+N-1} \langle \bar{P}_{0,\ell+1} \mathbb{E}[X_{\ell+1}], \mathbb{E}[X_{\ell+1}] \rangle + \sum_{\ell=k}^{k+N-1} \delta_{0,\ell+1}$$

$$- \sum_{\ell=k}^{k+N-1} \langle P_{0,\ell}(X_\ell - \mathbb{E}[X_\ell]), X_\ell - \mathbb{E}[X_\ell] \rangle$$

$$-\sum_{\ell=k}^{k+N-1} \langle \bar{P}_{0,\ell}\mathbb{E}[X_\ell], \mathbb{E}[X_\ell] \rangle - \sum_{\ell=k}^{k+N-1} \delta_{0,\ell}. \tag{12.9}$$

Let us consider the cluster of control inputs in order to treat the cooperative game as a control problem

$$V_\ell = (U_{1,\ell} \quad \dots \quad U_{n,\ell}),$$
$$\bar{V}_\ell = (\mathbb{E}[U_{1,\ell}] \quad \dots \quad \mathbb{E}[U_{n,\ell}]),$$

and the concatenated matrices

$$B_2 = (B_{21} \quad \dots \quad B_{2n}),$$
$$\bar{B}_2 = (\bar{B}_{21} \quad \dots \quad \bar{B}_{2n}).$$

Now, we compute separately some required terms appearing in the telescopic sum.

$$X_{\ell+1} = B_1(X_\ell - \mathbb{E}[X_\ell]) + (B_1 + \bar{B}_1)\mathbb{E}[X_\ell]$$
$$+ B_2(V_\ell - \mathbb{E}[V_\ell]) + (B_2 + \bar{B}_2)\mathbb{E}[V_\ell] + SW_\ell,$$
$$\mathbb{E}[X_{\ell+1}] = (B_1 + \bar{B}_1)\mathbb{E}[X_\ell] + (B_2 + \bar{B}_2)\mathbb{E}[V_\ell],$$
$$X_{\ell+1} - \mathbb{E}[X_{\ell+1}] = B_1(X_\ell - \mathbb{E}[X_\ell]) + B_2(V_\ell - \mathbb{E}[V_\ell]) + SW_\ell.$$

Then, the quadratic term $\langle P_{0,\ell+1}(X_{\ell+1} - \mathbb{E}[X_{\ell+1}]), X_{\ell+1} - \mathbb{E}[X_{\ell+1}] \rangle$ *appearing in the telescopic sum in (12.9) is*

$$\left\langle P_{0,\ell+1}(X_{\ell+1} - \mathbb{E}[X_{\ell+1}]), X_{\ell+1} - \mathbb{E}[X_{\ell+1}] \right\rangle$$
$$= \langle B_1^* P_{0,\ell+1} B_1(X_\ell - \mathbb{E}[X_\ell]), (X_\ell - \mathbb{E}[X_\ell]) \rangle$$
$$+ \langle B_1^* P_{0,\ell+1} B_2(V_\ell - \mathbb{E}[V_\ell]), (X_\ell - \mathbb{E}[X_\ell]) \rangle$$
$$+ \langle B_1^* P_{0,\ell+1} SW_\ell, (X_\ell - \mathbb{E}[X_\ell]) \rangle$$
$$+ \langle B_2^* P_{0,\ell+1} B_1(X_\ell - \mathbb{E}[X_\ell]), (V_\ell - \mathbb{E}[V_\ell]) \rangle$$
$$+ \langle B_2^* P_{0,\ell+1} B_2(V_\ell - \mathbb{E}[V_\ell]), (V_\ell - \mathbb{E}[V_\ell]) \rangle$$
$$+ \langle B_2^* P_{0,\ell+1} SW_\ell, (V_\ell - \mathbb{E}[V_\ell]) \rangle + \langle P_{0,\ell+1} B_1(X_\ell - \mathbb{E}[X_\ell]), SW_\ell \rangle$$
$$+ \langle P_{0,\ell+1} B_2(V_\ell - \mathbb{E}[V_\ell]), SW_\ell \rangle + \langle P_{0,\ell+1} SW_\ell, SW_\ell \rangle.$$

Taking into consideration the nullity values for the expected terms

$$\mathbb{E}\langle P_{0,\ell+1} B_1(X_\ell - \mathbb{E}[X_\ell]), SW_\ell \rangle = 0,$$
$$\mathbb{E}\langle P_{0,\ell+1} B_2(V_\ell - \mathbb{E}[V_\ell]), SW_\ell \rangle = 0,$$
$$\mathbb{E}\langle B_1^* P_{0,\ell+1} SW_\ell, X_\ell - \mathbb{E}[X_\ell] \rangle = 0,$$
$$\mathbb{E}\langle B_2^* P_{0,\ell+1} SW_\ell, V_\ell - \mathbb{E}[V_\ell] \rangle = 0,$$

then, it follows that the expectation of the quadratic term is

$$\mathbb{E}\left\langle P_{0,\ell+1}(X_{\ell+1} - \mathbb{E}[X_{\ell+1}]), X_{\ell+1} - \mathbb{E}[X_{\ell+1}] \right\rangle$$

$$
\begin{aligned}
&= \mathbb{E}\left\langle B_1^* P_{0,\ell+1} B_1 (X_\ell - \mathbb{E}[X_\ell]), (X_\ell - \mathbb{E}[X_\ell]) \right\rangle \\
&+ \mathbb{E}\left\langle B_1^* P_{0,\ell+1} B_2 (V_\ell - \mathbb{E}[V_\ell]), (X_\ell - \mathbb{E}[X_\ell]) \right\rangle \\
&+ \mathbb{E}\left\langle B_2^* P_{0,\ell+1} B_1 (X_\ell - \mathbb{E}[X_\ell]), (V_\ell - \mathbb{E}[V_\ell]) \right\rangle \\
&+ \mathbb{E}\left\langle B_2^* P_{0,\ell+1} B_2 (V_\ell - \mathbb{E}[V_\ell]), (V_\ell - \mathbb{E}[V_\ell]) \right\rangle \\
&+ \mathbb{E}\left\langle P_{0,\ell+1} SW_\ell, SW_\ell \right\rangle .
\end{aligned}
\tag{12.10}
$$

On the other hand, from the other quadratic term $\left\langle \bar{P}_{0,\ell+1} \mathbb{E}[X_{\ell+1}], \mathbb{E}[X_{\ell+1}] \right\rangle$ *appearing also in the telescopic sum in* (12.9) *yields*

$$
\begin{aligned}
\left\langle \bar{P}_{0,\ell+1} \mathbb{E}[X_{\ell+1}], \mathbb{E}[X_{\ell+1}] \right\rangle &= \left\langle (B_1 + \bar{B}_1)^* \bar{P}_{0,\ell+1} (B_1 + \bar{B}_1) \mathbb{E}[X_\ell], \mathbb{E}[X_\ell] \right\rangle \\
&+ \left\langle (B_1 + \bar{B}_1)^* \bar{P}_{0,\ell+1} (B_2 + \bar{B}_2) \mathbb{E}[V_\ell], \mathbb{E}[X_\ell] \right\rangle \\
&+ \left\langle (B_2 + \bar{B}_2)^* \bar{P}_{0,\ell+1} (B_1 + \bar{B}_1) \mathbb{E}[X_\ell], \mathbb{E}[V_\ell] \right\rangle \\
&+ \left\langle (B_2 + \bar{B}_2)^* \bar{P}_{0,\ell+1} (B_2 + \bar{B}_2) \mathbb{E}[V_\ell], \mathbb{E}[V_\ell] \right\rangle .
\end{aligned}
\tag{12.11}
$$

From a combination of the equations in (12.9), *and the terms* (12.10) *and* (12.11), *one can compute the gap between the cost that is function of the control inputs and the optimal cost from time instant k up to terminal time N given by* $\mathbb{E}[L_0(X,V) - F_{0,k}]$, *where* $L_0(X,V) = L_0(X, U_1, \ldots, U_n)$. *To this end, we also use the block diagonal matrices* $R_\ell = \mathrm{diag}(R_{1,\ell} \quad \cdots \quad R_{n,\ell})$ *and* $\bar{R}_\ell = \mathrm{diag}(\bar{R}_{1,\ell} \quad \cdots \quad \bar{R}_{n,\ell})$, *as follows:*

$$
\begin{aligned}
\mathbb{E}[L_0(X,V)] - F_{0,k} &= \mathbb{E}\langle Q_{0,k+N}(X_{k+N} - \mathbb{E}[X_{k+N}]), X_{k+N} - \mathbb{E}[X_{k+N}] \rangle \\
&+ \mathbb{E}\langle (Q_{0,k+N} + \bar{Q}_{0,k+N}) \mathbb{E}[X_{k+N}], \mathbb{E}[X_{k+N}] \rangle \\
&- \mathbb{E}\langle P_{0,k+N}(X_{k+N} - \mathbb{E}[X_{k+N}]), X_{k+N} - \mathbb{E}[X_{k+N}] \rangle \\
&- \mathbb{E}\langle \bar{P}_{0,k+N} \mathbb{E}[X_{k+N}], \mathbb{E}[X_{k+N}] \rangle - \mathbb{E}[\delta_{0,k+N}] \\
&+ \mathbb{E} \sum_{\ell=k}^{k+N-1} \langle Q_{0,\ell}(X_\ell - \mathbb{E}[X_\ell]), X_\ell - \mathbb{E}[X_\ell] \rangle \\
&+ \mathbb{E} \sum_{\ell=k}^{k+N-1} \langle (Q_{0,\ell} + \bar{Q}_{0,\ell}) \mathbb{E}[X_\ell], \mathbb{E}[X_\ell] \rangle \\
&+ \mathbb{E} \sum_{\ell=k}^{k+N-1} \langle B_1^* P_{0,\ell+1} B_1 (X_\ell - \mathbb{E}[X_\ell]), X_\ell - \mathbb{E}[X_\ell] \rangle \\
&+ \mathbb{E} \sum_{\ell=k}^{k+N-1} \langle P_{0,\ell+1} SW_\ell, SW_\ell \rangle \\
&+ \mathbb{E} \sum_{\ell=k}^{k+N-1} \langle (B_1 + \bar{B}_1)^* \bar{P}_{0,\ell+1} (B_1 + \bar{B}_1) \mathbb{E}[X_\ell], \mathbb{E}[X_\ell] \rangle \\
&+ \mathbb{E} \sum_{\ell=k}^{k+N-1} \delta_{0,\ell+1} - \mathbb{E} \sum_{\ell=k}^{k+N-1} \langle P_{0,\ell}(X_\ell - \mathbb{E}[X_\ell]), X_\ell - \mathbb{E}[X_\ell] \rangle
\end{aligned}
$$

$$-\mathbb{E}\sum_{\ell=k}^{k+N-1}\langle\bar{P}_{0,\ell}\mathbb{E}[X_\ell],\mathbb{E}[X_\ell]\rangle-\mathbb{E}\sum_{\ell=k}^{k+N-1}\delta_{0,\ell}$$

$$+\mathbb{E}\sum_{\ell=k}^{k+N-1}\langle R_\ell(V_\ell-\mathbb{E}[V_\ell]),V_\ell-\mathbb{E}[V_\ell]\rangle$$

$$+\mathbb{E}\sum_{\ell=k}^{k+N-1}\langle B_2^*P_{0,\ell+1}B_2(V_\ell-\mathbb{E}[V_\ell]),V_\ell-\mathbb{E}[V_\ell]\rangle$$

$$+\mathbb{E}\sum_{\ell=k}^{k+N-1}\langle B_1^*P_{0,\ell+1}B_2(V_\ell-\mathbb{E}[V_\ell]),X_\ell-\mathbb{E}[X_\ell]\rangle$$

$$+\mathbb{E}\sum_{\ell=k}^{k+N-1}\langle B_2^*P_{0,\ell+1}B_1(X_\ell-\mathbb{E}[X_\ell]),V_\ell-\mathbb{E}[V_\ell]\rangle$$

$$+\mathbb{E}\sum_{\ell=k}^{k+N-1}\langle (R_\ell+\bar{R}_\ell)\mathbb{E}[V_\ell],\mathbb{E}[V_\ell]\rangle$$

$$+\mathbb{E}\sum_{\ell=k}^{k+N-1}\langle (B+\bar{B})^*\bar{P}_{0,\ell+1}(B_2+\bar{B}_2)\mathbb{E}[V_\ell],\mathbb{E}[V_\ell]\rangle$$

$$+\mathbb{E}\sum_{\ell=k}^{k+N-1}\langle (B_1+\bar{B}_1)^*\bar{P}_{0,\ell+1}(B_2+\bar{B}_2)\mathbb{E}[V_\ell],\mathbb{E}[X_\ell]\rangle$$

$$+\mathbb{E}\sum_{\ell=k}^{k+N-1}\langle (B_2+\bar{B}_2)^*\bar{P}_{0,\ell+1}(B_1+\bar{B}_1)\mathbb{E}[X_\ell],\mathbb{E}[V_\ell]\rangle. \quad (12.12)$$

We continue the computational task by optimizing over the control input subtraction $V_\ell-\mathbb{E}[V_\ell]$. Performing square completion we have

$$\left|\Phi_{01,\ell}\Phi_{02,\ell}[(V_\ell-\mathbb{E}[V_\ell])+\Phi_{02,\ell}^{-1}\Phi_{03,\ell}(X_\ell-\mathbb{E}[X_\ell])]\right|^2$$

$$=\langle\Phi_{01,\ell}\Phi_{02,\ell}(V_\ell-\mathbb{E}[V_\ell]),\Phi_{01,\ell}\Phi_{02,\ell}(V_\ell-\mathbb{E}[V_\ell])\rangle$$
$$+\langle\Phi_{01,\ell}\Phi_{02,\ell}(V_\ell-\mathbb{E}[V_\ell]),\Phi_{01,\ell}\Phi_{03,\ell}(X_\ell-\mathbb{E}[X_\ell])\rangle$$
$$+\langle\Phi_{01,\ell}\Phi_{03,\ell}(X_\ell-\mathbb{E}[X_\ell]),\Phi_{01,\ell}\Phi_{02,\ell}(V_\ell-\mathbb{E}[V_\ell])\rangle$$
$$+\langle\Phi_{01,\ell}\Phi_{03,\ell}(X_\ell-\mathbb{E}[X_\ell]),\Phi_{01,\ell}\Phi_{03,\ell}(X_\ell-\mathbb{E}[X_\ell])\rangle,$$

From which we infer that the following equalities should be satisfied:

$$\Phi_{02,\ell}^*\Phi_{01,\ell}^*\Phi_{01,\ell}\Phi_{02,\ell}=R_\ell+B_2^*P_{0,\ell+1}B_2,$$
$$\Phi_{02,\ell}^*\Phi_{01,\ell}^*\Phi_{01,\ell}\Phi_{03,\ell}=B_2^*P_{0,\ell+1}B_1,$$

and from which is deduced that

$$\Phi_{01,\ell}=[R_\ell+B_2^*P_{0,\ell+1}B_2]^{-\frac{1}{2}},$$

$$\Phi_{02,\ell} = R_\ell + B_2^* P_{0,\ell+1} B_2,$$
$$\Phi_{03,\ell} = B_2^* P_{0,\ell+1} B_1.$$

Similarly, for the square completion involving the expected term $\mathbb{E}[V_\ell]$, *we have*

$$\left| \Xi_{01,\ell} \Xi_{02,\ell} [\mathbb{E}[V_\ell] + \Xi_{02,\ell}^{-1} \Xi_{03,\ell} \mathbb{E}[X_\ell]] \right|^2 = \langle \Xi_{01,\ell} \Xi_{02,\ell} \mathbb{E}[V_\ell], \Xi_{01,\ell} \Xi_{02,\ell} \mathbb{E}[V_\ell] \rangle$$
$$+ \langle \Xi_{01,\ell} \Xi_{02,\ell} \mathbb{E}[V_\ell], \Xi_{01,\ell} \Xi_{03,\ell} \mathbb{E}[X_\ell] \rangle$$
$$+ \langle \Xi_{01,\ell} \Xi_{03,\ell} \mathbb{E}[X_\ell], \Xi_{01,\ell} \Xi_{02,\ell} \mathbb{E}[V_\ell] \rangle$$
$$+ \langle \Xi_{01,\ell} \Xi_{03,\ell} \mathbb{E}[X_\ell], \Xi_{01,\ell} \Xi_{03,\ell} \mathbb{E}[X_\ell] \rangle$$

from which is deduced that

$$\Xi_{01,\ell} = [(R_\ell + \bar{R}_\ell) + (B_2 + \bar{B}_2)^* \bar{P}_{0,\ell+1} (B_2 + \bar{B}_2)]^{-\frac{1}{2}},$$
$$\Xi_{02,\ell} = (R_\ell + \bar{R}_\ell) + (B_2 + \bar{B}_2)^* \bar{P}_{0,\ell+1} (B_2 + \bar{B}_2),$$
$$\Xi_{03,\ell} = (B_2 + \bar{B}_2)^* \bar{P}_{0,\ell+1} (B_1 + \bar{B}_1).$$

Replacing the square completion terms in the gap $\mathbb{E}[L_0(X, V) - F_{0,k}]$ *shown in* (12.12) *yields*

$$\mathbb{E}[L_0(X, V) - F_{0,k}] = \mathbb{E}\langle Q_{0,k+N}(X_{k+N} - \mathbb{E}[X_{k+N}]), X_{k+N} - \mathbb{E}[X_{k+N}] \rangle$$
$$+ \mathbb{E}\langle (Q_{0,k+N} + \bar{Q}_{0,k+N}) \mathbb{E}[X_{k+N}], \mathbb{E}[X_{k+N}] \rangle$$
$$- \mathbb{E}\langle P_{0,k+N}(X_{k+N} - \mathbb{E}[X_{k+N}]), X_{k+N} - \mathbb{E}[X_{k+N}] \rangle$$
$$- \mathbb{E}\langle \bar{P}_{0,k+N} \mathbb{E}[X_{k+N}], \mathbb{E}[X_{k+N}] \rangle - \mathbb{E}[\delta_{0,k+N}]$$
$$+ \mathbb{E} \sum_{\ell=k}^{k+N-1} \langle Q_{0,\ell}(X_\ell - \mathbb{E}[X_\ell]), X_\ell - \mathbb{E}[X_\ell] \rangle$$
$$+ \mathbb{E} \sum_{\ell=k}^{k+N-1} \langle B_1^* P_{0,\ell+1} B_1 (X_\ell - \mathbb{E}[X_\ell]), X_\ell - \mathbb{E}[X_\ell] \rangle$$
$$- \mathbb{E} \sum_{\ell=k}^{k+N-1} \langle P_{0,\ell}(X_\ell - \mathbb{E}[X_\ell]), X_\ell - \mathbb{E}[X_\ell] \rangle$$
$$- \mathbb{E} \sum_{\ell=k}^{k+N-1} \langle \Phi_{03,\ell}^* \Phi_{01,\ell}^* \Phi_{01,\ell} \Phi_{03,\ell}(X_\ell - \mathbb{E}[X_\ell]), (X_\ell - \mathbb{E}[X_\ell]) \rangle$$
$$+ \mathbb{E} \sum_{\ell=k}^{k+N-1} \langle (Q_{0,\ell} + \bar{Q}_{0,\ell}) \mathbb{E}[X_\ell], \mathbb{E}[X_\ell] \rangle$$
$$+ \mathbb{E} \sum_{\ell=k}^{k+N-1} \langle (B_1 + \bar{B}_1)^* \bar{P}_{0,\ell+1} (B_1 + \bar{B}_1) \mathbb{E}[X_\ell], \mathbb{E}[X_\ell] \rangle$$
$$- \mathbb{E} \sum_{\ell=k}^{k+N-1} \langle \bar{P}_{0,\ell} \mathbb{E}[X_\ell], \mathbb{E}[X_\ell] \rangle$$

$$- \mathbb{E} \sum_{\ell=k}^{k+N-1} \langle \Xi_{01,\ell} \Xi_{03,\ell} \mathbb{E}[X_\ell], \Xi_{01,\ell} \Xi_{03,\ell} \mathbb{E}[X_\ell] \rangle$$

$$+ \mathbb{E} \sum_{\ell=k}^{k+N-1} \langle P_{0,\ell+1} SW_\ell, SW_\ell \rangle$$

$$+ \mathbb{E} \sum_{\ell=k}^{k+N-1} \delta_{0,\ell+1} - \mathbb{E} \sum_{\ell=k}^{k+N-1} \delta_{0,\ell}$$

$$+ \mathbb{E} \sum_{\ell=k}^{k+N-1} \left| \Phi_{01,\ell} \Phi_{02,\ell} [(V_\ell - \mathbb{E}[V_\ell]) + \Phi_{02,\ell}^{-1} \Phi_{03,\ell} (X_\ell - \mathbb{E}[X_\ell])] \right|^2$$

$$+ \mathbb{E} \sum_{\ell=k}^{k+N-1} \left| \Xi_{01,\ell} \Xi_{02,\ell} [\mathbb{E}[V_\ell] + \Xi_{02,\ell}^{-1} \Xi_{03,\ell} \mathbb{E}[X_\ell]] \right|^2. \qquad (12.13)$$

We group terms by making a process identification, i.e.,

- *Those terms evaluated at the terminal time $k+N$, which impose the terminal values for the difference equations P_0, $\bar{P}_0$, and δ_0*

- *Terms whose common factors are given by the quadratic form of $(X - \mathbb{E}[X])$ and $\mathbb{E}[X]$. We obtain then the difference equations P_0 and $\bar{P}_0$, respectively*

- *State-and-control-input independent terms defining the difference equation δ_0*

- *The quadratic terms that determine the optimal control inputs.*

Finally, minimizing the terms to match $\mathbb{E}[L_0(X,V)]$ with $\mathbb{E}[F_{0,k}]$, we obtain the announced result.

We conclude this chapter by showing the result for the adversarial mean-field-type game problem, which can be conveniently stated as a cooperative game by modifying the sign of the attackers in the cost functional.

Theorem 8 (Adversarial Scenario) *Let $R, \bar{R}$ be symmetric and consider the following change of variables corresponding to the sets $\mathcal{I}_+$ and $\mathcal{I}_-$:*

$$V_\ell^+ = (U_{j,\ell})_{j\in\mathcal{I}_+}, \quad V_\ell^- = (U_{j,\ell})_{j\in\mathcal{I}_-},$$
$$\bar{V}_\ell^+ = (\mathbb{E}[U_{j,\ell}])_{j\in\mathcal{I}_+}, \quad \bar{V}_\ell^- = (\mathbb{E}[U_{j,\ell}])_{j\in\mathcal{I}_-},$$

the concatenated matrices

$$B_2^+ = (B_{2j,\ell})_{j\in\mathcal{I}_+}, \quad \bar{B}_2^+ = (\bar{B}_{2j,\ell})_{j\in\mathcal{I}_+},$$
$$B_2^- = (B_{2j,\ell})_{j\in\mathcal{I}_-}, \quad \bar{B}_2^- = (\bar{B}_{2j,\ell})_{j\in\mathcal{I}_-},$$

and the block diagonal matrices

$$R_\ell^+ = \mathrm{diag}([R_{j,\ell}]_{j\in\mathcal{I}_+}), \quad \bar{R}_\ell^+ = \mathrm{diag}([\bar{R}_{j,\ell}]_{j\in\mathcal{I}_+}),$$

$$R_\ell^- = \mathrm{diag}([R_{j,\ell}]_{j \in \mathcal{I}_-}), \quad \bar{R}_\ell^- = \mathrm{diag}([\bar{R}_{j,\ell}]_{j \in \mathcal{I}_-}).$$

Now, let

$$R_\ell = \mathrm{diag}(R_\ell^+, R_\ell^-),$$
$$\bar{R}_\ell = \mathrm{diag}(\bar{R}_\ell^+, \bar{R}_\ell^-),$$
$$V_\ell = (V_\ell^{+\top}, V_\ell^{-\top})^\top,$$
$$\bar{V}_\ell = (\bar{V}_\ell^+, \bar{V}_\ell^-),$$

and the concatenated matrices

$$B_2 = (B_2^+, B_2^-),$$
$$\bar{B}_2 = (\bar{B}_2^+, \bar{B}_2^-).$$

Then explicit solution of Problem 25 is given by

$$\begin{pmatrix} V_\ell^{+\mathrm{op}} - \mathbb{E}[V_\ell]^{+\mathrm{op}} \\ V_\ell^{-\mathrm{op}} - \mathbb{E}[V_\ell]^{-\mathrm{op}} \end{pmatrix} = -\left[\begin{pmatrix} R_\ell^+ & 0 \\ 0 & R_\ell^- \end{pmatrix} + \begin{pmatrix} B_2^+ \\ B_2^- \end{pmatrix} P_{ad,\ell+1} \begin{pmatrix} B_2^+ & B_2^- \end{pmatrix} \right]^{-1} \cdot$$
$$\cdot \begin{pmatrix} B_2^+ \\ B_2^- \end{pmatrix} P_{ad,\ell+1} B_1 (X_\ell - \mathbb{E}[X_\ell]),$$

$$\begin{pmatrix} \mathbb{E}[V_\ell]^{+\mathrm{op}} \\ \mathbb{E}[V_\ell]^{-\mathrm{op}} \end{pmatrix} = -\left(\begin{pmatrix} R_\ell^+ + \bar{R}_\ell^+ & 0 \\ 0 & R_\ell^- + \bar{R}_\ell^- \end{pmatrix} \right.$$
$$\left. + \begin{pmatrix} B_2^+ + \bar{B}_2^+ \\ B_2^- + \bar{B}_2^- \end{pmatrix} \bar{P}_{ad,\ell+1} \begin{pmatrix} B_2^+ + \bar{B}_2^+ & B_2^- + \bar{B}_2^- \end{pmatrix} \right)^{-1} \cdot$$
$$\cdot \begin{pmatrix} B_2^+ + \bar{B}_2^+ \\ B_2^- + \bar{B}_2^- \end{pmatrix} \bar{P}_{ad,\ell+1} (B_1 + \bar{B}_1) \mathbb{E}[X_\ell],$$
$$\forall \ell \in [k..k+N-1],$$
$$L_{ad}(X, V^{+\mathrm{op}}, V^{-\mathrm{op}}) = \langle P_{ad,k}(X_k - \bar{X}_k), X_k - \bar{X}_k \rangle + \langle \bar{P}_{ad,k} \bar{X}_k, \bar{X}_k \rangle$$
$$+ \delta_{ad},$$

where P_{ad}, $\bar{P}_{ad}$, *and* δ_{ad} *are given by*

$$P_{ad,k} = Q_{ad,k} + B_1^* P_{ad,k+1} B_1$$
$$- B_1^* P_{0,k+1} \begin{pmatrix} B_2^+ & B_2^- \end{pmatrix} \left(\begin{pmatrix} R_\ell^+ & 0 \\ 0 & R_\ell^- \end{pmatrix} + \begin{pmatrix} B_2^+ \\ B_2^- \end{pmatrix} P_{ad,\ell+1} \begin{pmatrix} B_2^+ & B_2^- \end{pmatrix} \right)^{-1} \cdot$$
$$\cdot \begin{pmatrix} B_2^+ \\ B_2^- \end{pmatrix} P_{0,k+1} B_1$$

$$\bar{P}_{ad,k} = (Q_{ad,k} + \bar{Q}_{ad,k}) + (B_1 + \bar{B}_1)^* \bar{P}_{ad,k+1} (B_1 + \bar{B}_1)$$
$$- (B_1 + \bar{B}_1)^* \bar{P}_{0,\ell+1} \begin{pmatrix} B_2^+ + \bar{B}_2^+ & B_2^- + \bar{B}_2^- \end{pmatrix} \left(\begin{pmatrix} R_\ell^+ + \bar{R}_\ell^+ & 0 \\ 0 & R_\ell^- + \bar{R}_\ell^- \end{pmatrix} \right.$$

$$+ \begin{pmatrix} B_2^+ + \bar{B}_2^+ \\ B_2^- + \bar{B}_2^- \end{pmatrix} \bar{P}_{ad,\ell+1} \begin{pmatrix} B_2^+ + \bar{B}_2^+ & B_2^- + \bar{B}_2^- \end{pmatrix} \Bigg)^{-1}.$$

$$\cdot \begin{pmatrix} B_2^+ + \bar{B}_2^+ \\ B_2^- + \bar{B}_2^- \end{pmatrix} \bar{P}_{0,\ell+1}(B_1 + \bar{B}_1)$$

$$\delta_{ad,k} = \delta_{ad,k+1} + \mathbb{E} \langle P_{ad,k+1} SW_k, SW_k \rangle,$$

with terminal conditions

$$P_{ad,k+N} = Q_{ad,k+N},$$
$$\bar{P}_{ad,k+N} = Q_{ad,k+N} + \bar{Q}_{ad,k+N},$$
$$\delta_{ad,k+N} = 0,$$

where matrices

$$Q_{ad} = \sum_{i \in \mathcal{N}} Q_i,$$

$$Q_{ad} + \bar{Q}_{ad,k} = \sum_{i \in \mathcal{N}} (Q_i + \bar{Q}_i).$$

Moreover, recall Q_i, $Q_i + \bar{Q}_i \succeq 0$, and R_i, $R_i + \bar{R}_i \succ 0$, for all $i \in \mathcal{I}_+$, and $-Q_i$, $-(Q_i + \bar{Q}_i) \succeq 0$, and $-R_i$, $-(R_i + \bar{R}_i) \succ 0$, for all $i \in \mathcal{I}_-$.

Proof 31 (Theorem 8) *The announced result is straight-forwardly obtained by following the same procedure as in the proof of Theorem 7.*

12.3   Numerical Examples

Similar to what was presented in Chapter 4, we perform numerical results for the two main game scenarios: non-cooperative and cooperative game problem.

Example 1: Matrix-Valued Discrete-Time Non-cooperative Scenario

Consider a non-cooperative two-player matrix-valued mean-field-type game problem under the following difference equation describing the evolution of the system states:

$$X_{k+1} = \begin{pmatrix} -2 & \frac{1}{10} & 0 \\ \frac{1}{10} & -2 & 0 \\ 0 & \frac{1}{10} & -2 \end{pmatrix} X_k + \begin{pmatrix} \frac{1}{2} & 0 & 0 \\ 0 & \frac{1}{2} & 0 \\ 0 & 0 & \frac{1}{2} \end{pmatrix} \bar{B}_1 \mathbb{E}[X_k]$$

$$+ \begin{pmatrix} \frac{1}{2} & 0 & 0 \\ 0 & \frac{1}{2} & 0 \\ 0 & 0 & \frac{1}{2} \end{pmatrix} U_{1,k} + \begin{pmatrix} \frac{1}{5} & 0 & 0 \\ 0 & \frac{1}{5} & 0 \\ 0 & 0 & \frac{1}{5} \end{pmatrix} \mathbb{E}[U_{1,k}] + U_{2,k}$$

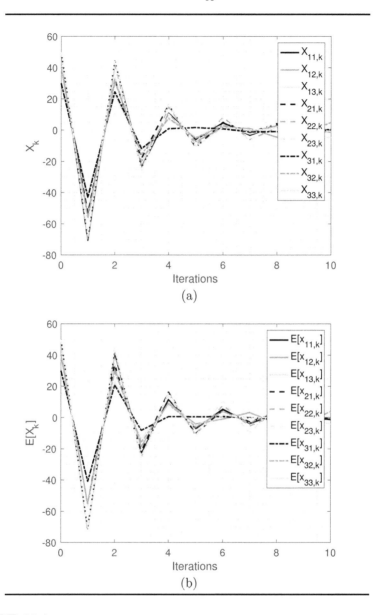

FIGURE 12.1
Evolution of the (a) system state and (b) its expectation for the matrix-value discrete-time non-cooperative scenario.

$$+ \begin{pmatrix} \frac{1}{3} & 0 & 0 \\ 0 & \frac{1}{3} & 0 \\ 0 & 0 & \frac{1}{3} \end{pmatrix} \mathbb{E}[U_{2,k}] + W_k.$$

Let us consider the following parameters for the cost functional:

$$
Q_1 = \begin{pmatrix} 1 & 0 & 0 \\ 0 & 1 & 0 \\ 0 & 0 & 1 \end{pmatrix}, \qquad
\bar{Q}_1 = \begin{pmatrix} \frac{1}{2} & 0 & 0 \\ 0 & \frac{1}{2} & 0 \\ 0 & 0 & \frac{1}{2} \end{pmatrix},
$$

$$
Q_2 = \begin{pmatrix} 2 & 0 & 0 \\ 0 & 2 & 0 \\ 0 & 0 & 2 \end{pmatrix}, \qquad
\bar{Q}_2 = \begin{pmatrix} 1 & 0 & 0 \\ 0 & 1 & 0 \\ 0 & 0 & 1 \end{pmatrix},
$$

$$
R_1 = \bar{R}_1 = \begin{pmatrix} 2 & 0 & 0 \\ 0 & 2 & 0 \\ 0 & 0 & 2 \end{pmatrix}, \qquad
R_2 = \bar{R}_2 = \begin{pmatrix} 5 & 0 & 0 \\ 0 & 5 & 0 \\ 0 & 0 & 5 \end{pmatrix},
$$

with terminal time instant $N = 10$, and considering the following initial conditions for the system states:

$$
X_0 = \mathbb{E}[X_0] = \begin{pmatrix} 50 & 40 & 30 \\ 40 & 50 & 40 \\ 30 & 40 & 50 \end{pmatrix},
$$

Figures 12.1(a) and 12.1(b) show the optimal trajectory of the system states X and their expectation $\mathbb{E}[X]$, respectively. It can be seen that the system states are driven to zero according to the minimization of the cost functional. On the other hand, Figures 12.2(a) and 12.2(b), and 12.3(a) and 12.3(b) show the evolution of the optimal control inputs for both decision-makers and their expectation, respectively.

Figures 12.4 and 12.5 present the evolution of the Riccati equations for the first decision-maker P_1 and $\bar{P}_1$. Likewise, Figures 12.6 and 12.7 show the evolution of the Riccati equations corresponding to the second decision-maker. Notice that, it can be verified that the boundary conditions are satisfied.

The evolution of $\delta_{1,k}$ and $\delta_{2,k}$ are exhibited in Figure 12.8. Finally, the optimal costs,

$$
L_1(X, U^{\mathrm{op}}) = \left\langle \bar{P}_{1,k} \begin{pmatrix} 50 & 40 & 30 \\ 40 & 50 & 40 \\ 30 & 40 & 50 \end{pmatrix}, \begin{pmatrix} 50 & 40 & 30 \\ 40 & 50 & 40 \\ 30 & 40 & 50 \end{pmatrix} \right\rangle + \delta_{1,k},
$$

$$
L_2(X, U^{\mathrm{op}}) = \left\langle \bar{P}_{2,k} \begin{pmatrix} 50 & 40 & 30 \\ 40 & 50 & 40 \\ 30 & 40 & 50 \end{pmatrix}, \begin{pmatrix} 50 & 40 & 30 \\ 40 & 50 & 40 \\ 30 & 40 & 50 \end{pmatrix} \right\rangle + \delta_{2,k},
$$

are shown in Figure 12.9.

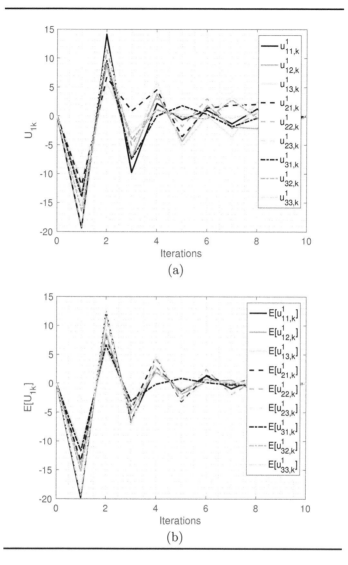

FIGURE 12.2
Evolution of the (a) first player strategies and (b) their expectation for the matrix-value discrete-time non-cooperative scenario.

Example 2: Matrix-Valued Discrete-Time Fully-cooperative Scenario

Consider a cooperative two-player matrix-valued mean-field-type game under the following difference equation describing the evolution of the system states:

$$X_{k+1} = \begin{pmatrix} -2 & \frac{1}{10} & 0 \\ \frac{1}{10} & -2 & 0 \\ 0 & \frac{1}{10} & -2 \end{pmatrix} X_k + \begin{pmatrix} \frac{1}{2} & 0 & 0 \\ 0 & \frac{1}{2} & 0 \\ 0 & 0 & \frac{1}{2} \end{pmatrix} \bar{B}_1 \mathbb{E}[X_k]$$

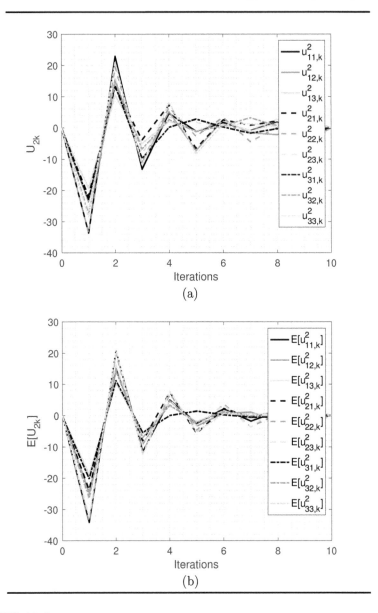

FIGURE 12.3
Evolution of the (a) second player strategies and (b) their expectation for the matrix-value discrete-time non-cooperative scenario.

$$+ \begin{pmatrix} \frac{1}{2} & 0 & 0 \\ 0 & \frac{1}{2} & 0 \\ 0 & 0 & \frac{1}{2} \end{pmatrix} U_{1,k} + \begin{pmatrix} \frac{1}{2} & 0 & 0 \\ 0 & \frac{1}{2} & 0 \\ 0 & 0 & \frac{1}{2} \end{pmatrix} \mathbb{E}[U_{1,k}] + U_{2,k}$$

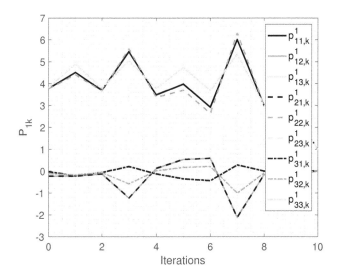

FIGURE 12.4
Evolution of the equation P_1 for the matrix-value discrete-time non-cooperative scenario.

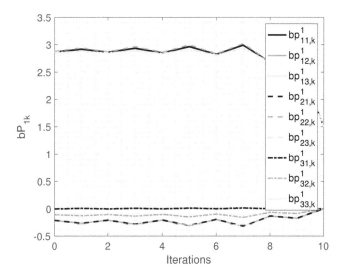

FIGURE 12.5
Evolution of the equation P_2 for the matrix-value discrete-time non-cooperative scenario.

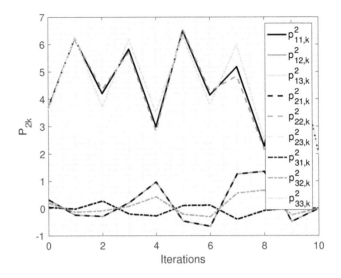

FIGURE 12.6
Evolution of the equation $\bar{P}_1$ for the matrix-value discrete-time non-cooperative scenario.

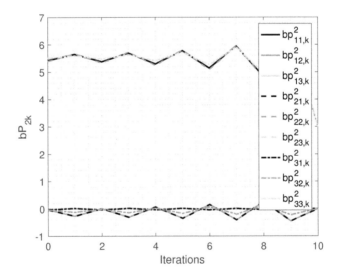

FIGURE 12.7
Evolution of the equation $\bar{P}_2$ for the matrix-value discrete-time non-cooperative scenario.

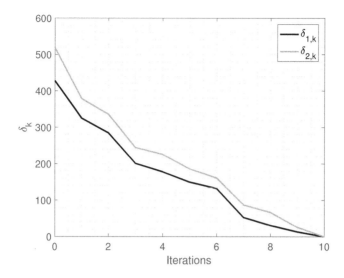

FIGURE 12.8
Evolution of the Riccati equations δ_1, and δ_2 for the matrix-value discrete-time non-cooperative scenario.

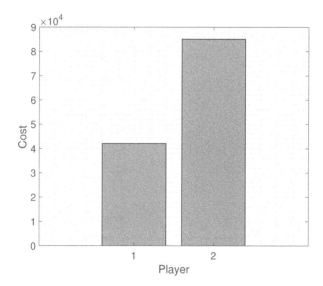

FIGURE 12.9
Optimal cost function for each player for the matrix-value discrete-time non-cooperative scenario.

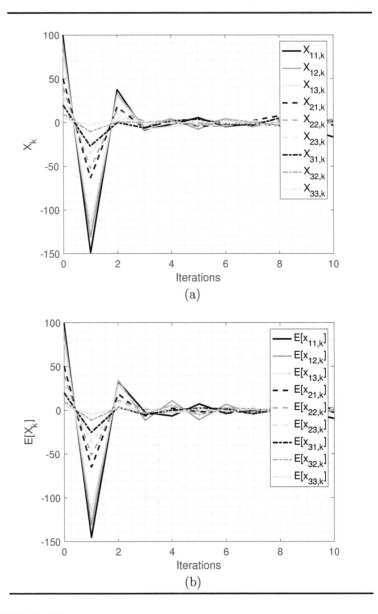

FIGURE 12.10
Evolution of the (a) system state and (b) its expectation for the matrix-value
discrete-time fully-cooperative scenario.

$$+ \begin{pmatrix} \frac{1}{2} & 0 & 0 \\ 0 & \frac{1}{2} & 0 \\ 0 & 0 & \frac{1}{2} \end{pmatrix} \mathbb{E}[U_{2,k}] + \begin{pmatrix} 2 & 0 & 0 \\ 0 & 2 & 0 \\ 0 & 0 & 2 \end{pmatrix} W_k,$$

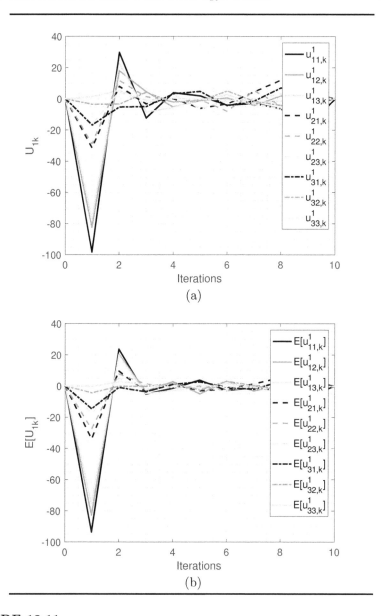

FIGURE 12.11
Evolution of the (a) first player strategies and (b) their expectation for the matrix-value discrete-time fully-cooperative scenario.

Let us consider the following parameters for the cost functional:

$$Q_1 = \bar{Q}_1 = \begin{pmatrix} 2 & 0 & 0 \\ 0 & 2 & 0 \\ 0 & 0 & 2 \end{pmatrix}, \qquad Q_2 = \bar{Q}_2 = \begin{pmatrix} 1 & 0 & 0 \\ 0 & 1 & 0 \\ 0 & 0 & 1 \end{pmatrix},$$

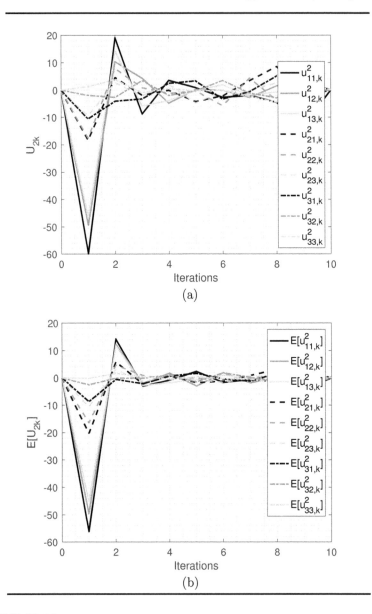

FIGURE 12.12
Evolution of the (a) second player strategies and (b) their expectation for the
matrix-value discrete-time fully-cooperative scenario.

$$R_1 = \bar{R}_1 = \begin{pmatrix} 2 & 0 & 0 \\ 0 & 2 & 0 \\ 0 & 0 & 2 \end{pmatrix}, \qquad R_2 = \bar{R}_2 = \begin{pmatrix} 5 & 0 & 0 \\ 0 & 5 & 0 \\ 0 & 0 & 5 \end{pmatrix},$$

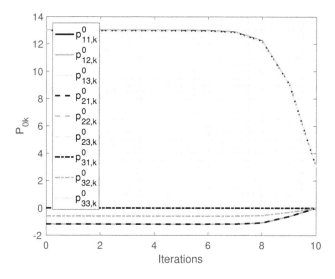

FIGURE 12.13
Evolution of the equation P_0 for the matrix-value discrete-time fully-cooperative scenario.

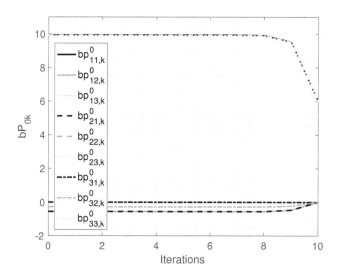

FIGURE 12.14
Evolution of the equation $\bar{P}_0$ for the matrix-value discrete-time fully-cooperative scenario.

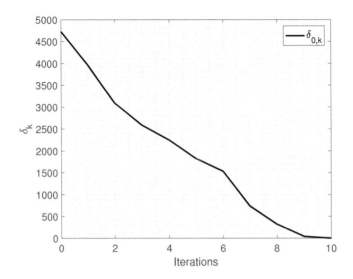

FIGURE 12.15
Evolution of the Riccati equation δ_0 for the matrix-value discrete-time fully-cooperative scenario.

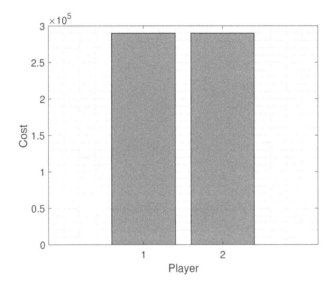

FIGURE 12.16
Optimal cost function for each player for the matrix-value discrete-time fully-cooperative scenario.

with terminal time instant $N = 10$, and considering the following initial conditions for the system states:

$$X_0 = \mathbb{E}[X_0] = \begin{pmatrix} 100 & 90 & 80 \\ 50 & 40 & 30 \\ 20 & 10 & 5 \end{pmatrix}.$$

Figures 12.10(a) and 12.10(b) present the evolution of the system states X and their expectation $\mathbb{E}[X]$, respectively. The evolution of the optimal control inputs U_1, U_2, and their expectation $\mathbb{E}[U_1]$ and $\mathbb{E}[U_2]$ are presented in Figures 12.11(a), 12.11(b), 12.12(a), and 12.12(b), respectively. The evolution of the Riccati equations P_0 and $\bar{P}_0$ is shown in Figures 12.13 and 12.14, respectively. It can be observed that the boundary conditions for both P_0 and $\bar{P}_0$ are satisfied. Moreover, Figure 12.15 exhibits the evolution of the difference equation δ_0 with null boundary condition. Finally, the optimal cost for the cooperative matrix-valued mean-field-type game is shown in Figure 12.16.

12.4   Exercises

1. Solve the following matrix-valued discrete-time mean-field-type control problem. Consider the system dynamics given by

$$X_{k+1} = B_1 X_k + \bar{B}_1 \mathbb{E}[X_k] + B_2 U_k + \bar{B}_2 \mathbb{E}[U_k] + S_0 W_k,$$

and with a cost functional given by

$$
\begin{aligned}
L(X, U) = {} & \langle Q_{k+N} X_{k+N}, X_{k+N} \rangle + \langle \bar{Q}_{k+N} \mathbb{E}[X_{k+N}], \mathbb{E}[X_{k+N}] \rangle \\
& + \sum_{\ell=k}^{k+N-1} \langle Q_\ell X_\ell, X_\ell \rangle + \sum_{\ell=k}^{k+N-1} \langle \bar{Q}_\ell \mathbb{E}[X_\ell], \mathbb{E}[X_\ell] \rangle \\
& + \sum_{\ell=k}^{k+N-1} \langle R_\ell U_\ell, U_\ell \rangle + \sum_{\ell=k}^{k+N-1} \langle \bar{R}_\ell \mathbb{E}[U_\ell], \mathbb{E}[U_\ell] \rangle, \quad (12.14)
\end{aligned}
$$

where $\langle A, B \rangle = \operatorname{trace}(A^* B) = \operatorname{trace}(B^* A)$, A^* being the adjoint operator of A (transposition). Matrices Q, $(Q + \bar{Q}) \succeq 0$, and R, $(R + \bar{R}) \succ 0$.

Hint: An appropriate guess functional is given by

$$F_\ell = \langle P_\ell(X_\ell - \mathbb{E}[X_\ell]), X_\ell - \mathbb{E}[X_\ell] \rangle + \langle \bar{P}_\ell \mathbb{E}[X_\ell], \mathbb{E}[X_\ell] \rangle + \delta_\ell.$$

2. Solve the following matrix-valued discrete-time mean-field-type control problem with state-and-control-input-dependent noise. Consider the system dynamics given by

$$X_{k+1} = \left(B_1 X_k + \bar{B}_1 \mathbb{E}[X_k] + B_2 U_k + \bar{B}_2 \mathbb{E}[U_k]\right)$$
$$+ \left(S_1 X_k + \bar{S}_1 \mathbb{E}[X_k] + S_2 U_k + \bar{S}_2 \mathbb{E}[U_k]\right) W_k,$$

and with the same cost functional as in (12.14).

Hint: An appropriate guess functional is the same as the once considered in the previous exercise.

Part VI

Learning Approaches and Applications

13

Constrained Mean-Field-Type Games: Stationary Case

In [131], the author considered a $n-$decision-maker *deterministic* game, in which the constraints for each decision-maker, as well as his payoff function, may depend on the strategy of every decision-maker. The existence of a generalized Nash equilibrium point for such a game was shown. By requiring appropriate component-wise convexity in the cost functions, the author proved that there is a unique equilibrium point for every strictly convex game in a convex compact domain. He also proposed a learning algorithm to select generalized Nash and normalized equilibrium. However, the stochastic setting is not examined in [131].

The constrained mean-field-type games are discussed here for finite-decision–maker static games. The discussion developed in this chapter also contributes to the clarification that, the existence of mean-field terms do not necessarily imply a large number of decision-makers in the strategic interaction. We introduce, formulate and analyze the constrained mean-field-type static games. The difference between the deterministic and the proposed games in this chapter are highlighted by showing that the mean-field-type games consider the risk within their analysis. To this end, mean-field terms are added to the cost functionals of the players and constraints over these terms are established. Thus, we extend the game-theoretical concepts of best-response strategies, generalized Nash equilibrium, variational equilibrium, and the potential games, to the static mean-field-type games with constraints.

13.1 Constrained Games

To highlight the differences between the standard constrained deterministic static games, and the proposed novel constrained and static mean-field-type games, we present an illustrative example showing the mean-variance terms appearing under the mean-field-type framework. We start with a constrained deterministic game.

DOI: 10.1201/9781003098607-13

13.1.1   A Constrained Deterministic Game

Let $\mathcal{N} = \{1, \ldots, n\}$ denotes the set of $n \geq 2$ decision-makers. Decision-maker $i \in \mathcal{N}$ is seeking to minimize the cost function $L_i(x)$ subject to some constraints as follows:

$$\underset{x_i}{\text{minimize }} L_i(x) = (x_i + 1)^2 + \sum_{j \in \mathcal{N} \setminus \{i\}} x_j^2, \qquad (13.1a)$$

subject to

$$x_i \geq 0, \qquad (13.1b)$$

$$\sum_{j \in \mathcal{N}} x_j \geq 1. \qquad (13.1c)$$

Here, under this static framework, x_i is selected by the decision-maker $i \in \mathcal{N}$. We would like to investigate the best reaction of a decision-maker given the strategy of the other decision-makers. First observe that the feasible set comprising the constraints (13.1b) and (13.1c) is not empty, and it is convex. However, it is not a compact set. The non-compactness of the constraint set does not add any difficulty because the cost function $L_i(x)$ is coercive. We compute the best-response strategy for decision-maker $i \in \mathcal{N}$ under the constraints in (13.1b) and (13.1c). To this end, let

$$b = \max \left\{ 0, 1 - \sum_{j \in \mathcal{N} \setminus \{i\}} x_j \right\}.$$

Then, for some fixed given strategies x_j, $j \in \mathcal{N} \setminus \{i\}$ we compute the Lagrangian function, i.e.,

$$\mathcal{L}_i(x_i, \mu_i) = (x_i + 1)^2 + \mu_i (b - x_i),$$

and the corresponding Karush-Kuhn-Tucker conditions, i.e.,

$$\frac{\partial \mathcal{L}_i(x_i, \mu_i)}{\partial x_i} = 0, \qquad (13.2a)$$

$$-x_i \leq -b, \qquad (13.2b)$$

$$\mu_i \geq 0, \qquad (13.2c)$$

$$\mu_i (b - x_i) = 0. \qquad (13.2d)$$

Thus, from (13.2a) we have that $x_i = (\mu_i - 2)/2$. It follows that from (13.2b) it is concluded that $\mu_i \geq 2(b + 1)$. Therefore, according to (13.2c), the constraint in (13.2b) should be always satisfied holding equality, i.e., the constrained best-response set for the decision-maker $i \in \mathcal{N}$ is

$$\text{BR}_i \left(\{x_j\}_{j \in \mathcal{N} \setminus \{i\}} \right) = \left\{ \max \left(0, 1 - \sum_{j \in \mathcal{N} \setminus \{i\}} x_j \right) \right\}. \qquad (13.3)$$

Using the constrained best-response strategy, one can define the constrained-equilibrium concept known as Generalized Nash Equilibrium (GNE), which is defined next.

Definition 31 *A Generalized Nash Equilibrium (GNE) is a strategy profile* $(x_i)_{i \in \mathcal{N}}$ *such that the constraints (13.1b) and (13.1c) are satisfied, and* $x_i \in$ BR$_i\left(\{x_j\}_{j \in \mathcal{N} \setminus \{i\}}\right)$, *for any* $i \in \mathcal{N}$.

According to Definition 31 and the solution obtained in (13.3), the set of GNE is the $(n-1)$–dimensional simplex is given by

$$\Delta_{n-1} = \left\{ (x_1, \ldots, x_n) | x_i \geq 0, i \in \mathcal{N}, \sum_{j \in \mathcal{N}} x_j = 1 \right\}.$$

In particular, there is a continuum of GNE (there are infinite equilibria) and the selection of the efficient equilibrium will be discussed below.

13.1.2 A Constrained Mean-Field-Type Game

In order to point out the fundamental differences between the standard deterministic static games in front of the mean-field-type games in the stationary regime, we modify the example presented above by incorporating a random variable $w = (w_i)_{i \in \mathcal{N}}$, with the following mean and variance:

$$\mathbb{E}[w_i] = 1,$$
$$\mathrm{var}(w_i) = \sigma^2.$$

Then, the problem in (13.1) becomes a stochastic game, i.e.,

$$L_i(x, w_i, D_{w_i}) = (x_i + w_i)^2 + \sum_{j \in \mathcal{N} \setminus \{i\}} x_j^2 + \sum_{j \in \mathcal{N}} \varepsilon_{ij}(x_j - \mathbb{E}[x_j])^2,$$

and

$$\underset{x_i}{\mathrm{minimize}} \ \mathbb{E}\left[L_i(x, w_i, D_{w_i})\right], \tag{13.4a}$$

subject to

$$\mathbb{E}[x_i] \geq 0, \tag{13.4b}$$

$$\sum_{j \in \mathcal{N}} \mathbb{E}[x_j] \geq 1, \tag{13.4c}$$

where $w_i \in \mathbb{R}$ is a random variable with certain distribution denoted by D_{w_i}, for all $i \in \mathcal{N}$.

Remark 28 *Note that if,* $\varepsilon_{ij} = 0$, *for all* $i, j \in \mathcal{N}$, *and* $\sigma = 0$, *then we retrieve the deterministic problem in (13.1).*

Lemma 3 *The stochastic problem presented in (13.4) is equivalent to problem (13.5) stated next*

$$
\begin{aligned}
\operatorname*{minimize}_{x_i} \mathbb{E}\left[L_i(x, w_i, D_{w_i})\right] = {} & (1 + \varepsilon_{ii})\mathrm{var}(x_i) + \mathrm{var}(w_i) \\
& + 2\mathrm{cov}(x_i, w_i) + \mathbb{E}[x_i]^2 + \mathbb{E}[w_i]^2 + 2\mathbb{E}[x_i]\mathbb{E}[w_i] \\
& + \sum_{j \in \mathcal{N} \setminus \{i\}} (1 + \varepsilon_{ij})\mathrm{var}(x_j) + \mathbb{E}[x_j]^2, \qquad (13.5)
\end{aligned}
$$

subject to the same constraints (13.4b) and (13.4c).

The proof of Lemma 3 is developed in the same manner as we did in Chapter 4 to obtain the risk-dependent terms in (4.5), or as in Proposition 16 presented in Chapter 6.

Proof 32 (Lemma 3) *We use the decompositions $x_i = x_i - \mathbb{E}[x_i] + \mathbb{E}[x_i]$ and $w_i = w_i - \mathbb{E}[w_i] + \mathbb{E}[w_i]$ in the cost function (13.1a). Then,*

$$
\begin{aligned}
\mathbb{E}[(x_i - w_i)^2] = \mathbb{E}\Big[& (x_i - \mathbb{E}[x_i])^2 + \mathbb{E}[x_i]^2 + (w_i - \mathbb{E}[w_i])^2 \\
& + \mathbb{E}[w_i]^2 + 2x_i w_i + 2(x_i - \mathbb{E}[x_i])(w_i - \mathbb{E}[w_i])\Big],
\end{aligned}
$$

noticing that both terms

$$
\begin{aligned}
\mathbb{E}[\mathbb{E}[x_i](x_i - \mathbb{E}[x_i])] &= 0, \\
\mathbb{E}[\mathbb{E}[w_i](w_i - \mathbb{E}[w_i])] &= 0,
\end{aligned}
$$

since $(x_i - \mathbb{E}[x_i]) \perp \mathbb{E}[x_i]$ as well as $(w_i - \mathbb{E}[w_i]) \perp \mathbb{E}[w_i]$; and

$$
\begin{aligned}
\mathbb{E}[\mathbb{E}[x_i](w_i - \mathbb{E}[w_i])] &= 0, \\
\mathbb{E}[\mathbb{E}[w_i](x_i - \mathbb{E}[x_i])] &= 0,
\end{aligned}
$$

obtaining the problem in (13.1).

Once the new structure for the stochastic problem is obtained, it can be decomposed into two sub-problems as follows. Let us consider the following cost functional:

$$
\begin{aligned}
L_i^1(y, w_i) = {} & (1 + \varepsilon_{ii})\mathbb{E}[y_i^2] + 2\mathbb{E}[y_i(w_i - \mathbb{E}[w_i])] \\
& + \sum_{j \in \mathcal{N} \setminus \{i\}} (1 + \varepsilon_{ij})\mathbb{E}[y_j^2],
\end{aligned}
$$

then the first problem is stated as

$$
\operatorname*{minimize}_{y_i} L_i^1(y, w), \qquad (13.6a)
$$

subject to

$$\mathbb{E}[y_i] = 0. \tag{13.6b}$$

And the second cost functional is as follows:

$$L_i^2(\mathbb{E}[x]) = (\mathbb{E}[x_i] + 1)^2 + \sum_{j \in \mathcal{N} \setminus \{i\}} \mathbb{E}[x_j]^2, \tag{13.7}$$

and the second problem is

$$\underset{\mathbb{E}[x_i]}{\text{minimize}} \ L_i^2(\mathbb{E}[x]), \tag{13.8a}$$

subject to

$$\mathbb{E}[x_i] \geq 0, \tag{13.8b}$$

$$\sum_{j \in \mathcal{N}} \mathbb{E}[x_j] \geq 1. \tag{13.8c}$$

The optimization in the first problem (13.4) can be performed by making square-completion terms. For the sake of simplification in the notation, let us introduce the variable $\xi_{ij} = (1 + \varepsilon_{ij})$, arriving at

$$\xi_{ii}\mathbb{E}[y_i^2] + 2\mathbb{E}[y_i(w_i - \mathbb{E}[w_i])] + \sum_{j \in \mathcal{N} \setminus \{i\}} \xi_{ij}\mathbb{E}[y_j^2]$$
$$= \xi_{ii}\left(\mathbb{E}[y_i^2] + 2\mathbb{E}\left[y_i\frac{(w_i - \mathbb{E}[w_i])}{\xi_{ii}}\right]\right)$$
$$+ \sum_{j \in \mathcal{N} \setminus \{i\}} \xi_{ij}\mathbb{E}[y_j^2],$$

where

$$\mathbb{E}\left[y_i + \frac{w_i - \mathbb{E}[w_i]}{1 + \varepsilon_{ii}}\right]^2 = \mathbb{E}[y_i^2] + 2\mathbb{E}\left[y_i\frac{(w_i - \mathbb{E}[w_i])}{\xi_{ii}}\right] + \left(\frac{w_i - \mathbb{E}[w_i]}{1 + \varepsilon_{ii}}\right)^2.$$

Then, we conclude that

$$\xi_{ii}\mathbb{E}[y_i^2] + 2\mathbb{E}[y_i(w_i - \mathbb{E}[w_i])] + \sum_{j \in \mathcal{N} \setminus \{i\}} \xi_{ij}\mathbb{E}[y_j^2] = \xi_{ii}\mathbb{E}\left[y_i + \frac{w_i - \mathbb{E}[w_i]}{\xi_{ii}}\right]^2$$
$$+ \mathbb{E}\sum_{j \in \mathcal{N} \setminus \{i\}} \xi_{ij}y_j^2 - \frac{\sigma^2}{\xi_{ii}}.$$

And by minimizing terms the optimal decision variable is obtained

$$x_i^* = \mathbb{E}[x_i^*] - \frac{w_i - \mathbb{E}[w_i]}{1 + \varepsilon_{ii}}.$$

Since $\mathbb{E}[x^*] \in \Delta_{n-1}$ according to the solution of the deterministic game approach in (13.3); note that, the set of GNE of the mean-field-type game is a simplex perturbed by a random noise, i.e.,

$$x^* \in \Delta_{n-1} - \{\vec{N}_{\mathrm{GNE}}\},$$

with the ith component of $\vec{N}_{\mathrm{GNE}}$ as follows:

$$N_i = -\frac{w_i - \mathbb{E}[w_i]}{1 + \varepsilon_{ii}}, \ \forall i \in \mathcal{N}.$$

Remark 29 *Notice that the cost is reduced by*

$$\frac{\sigma^2}{1 + \varepsilon_{ii}}$$

when the mean-field-type game approach is adopted.

13.1.3   Constrained Variational Equilibrium

In the previous two problems, i.e., the deterministic and mean-field-type games, the set of GNE is infinite. It is desirable to refine the set of equilibria by introducing the concept of variational equilibrium, which is defined below.

Definition 32 *the strategic profile*

$$x^* = \mathbb{E}[x^*] + y^*$$

is a variational equilibrium of the mean-field-type game in (13.4) if

$$\mathbb{E}\left[\langle \nabla_{y_i} L_i^1(y^*, w), y_i - y_i^* \rangle\right]$$
$$+ \mathbb{E}\left[\langle \nabla_{\mathbb{E}[x_i]} L_i^2(\mathbb{E}[x^*]), \mathbb{E}[x_i] - \mathbb{E}[x_i^*] \rangle\right] \geq 0, \qquad (13.9)$$

for all $i \in \mathcal{N}$, and for every $\mathbb{E}[y_i] = 0$, $\mathbb{E}[x_i] \geq 0$, and $\sum_{j \in \mathcal{N}} \mathbb{E}[x_j] \geq 1$.

Lemma 4 *There is at most one mean-field-type variational equilibrium x^{VE}.*

Proof 33 (Lemma 4) *We can restrict the search domain to be bounded by $[0, 10]^n$. Hence, the intersection set will be convex, compact and non-empty. Noting that*

$$\left(\nabla_{y_i} L^1(y, w), \nabla_{\mathbb{E}[x_i]} L^2(\mathbb{E}[x])\right)$$

is strongly monotone operator, the variational inequality in (13.9) has at most one solution in the intersection set.

Lemma 5 *The strategic profile*

$$x^{\mathrm{VE}} = \left(\frac{1}{n}, \ldots, \frac{1}{n}\right) - \vec{N}_{\mathrm{GNE}}$$

is a variational equilibrium of the mean-field-type game.

Proof 34 (Lemma 5) *We directly verify that the strategic profile given by* $x^{\text{VE}} = (1/n, \ldots, 1/n) - \vec{N}_{\text{GNE}}$ *is a variational equilibrium since it satisfies the inequality in* (13.9).

Lemma 6 *The variational equilibrium of the mean-field-type game is the most efficient GNE.*

Proof 35 (Lemma 6) *Notice that*

$$\underset{x^* \in \text{GNE}}{\text{minimize}} \sum_{i \in \mathcal{N}} \mathbb{E}[L_i(x^*)]$$

$$= \underset{y \in \{\vec{N}_{\text{GNE}}\}}{\text{minimize}} \sum_{i \in \mathcal{N}} L_i^1(y, w) + \underset{\mathbb{E}[x] \in \Delta_{n-1}}{\text{minimize}} \sum_{i \in \mathcal{N}} L_i^2(\mathbb{E}[x^*]). \qquad (13.10)$$

Taking the second term from the right hand side in (13.10), *one obtains*

$$\underset{\mathbb{E}[x] \in \Delta_{n-1}}{\text{minimize}} \sum_{i \in \mathcal{N}} \left\{ \sum_{j \in \mathcal{N}} \mathbb{E}[x_j]^2 + 2\mathbb{E}[x_i] + 1 \right\}$$

$$= \underset{\mathbb{E}[x] \in \Delta_{n-1}}{\text{minimize}} \, n \sum_{j \in \mathcal{N}} \mathbb{E}[x_j]^2 + 2 + n,$$

and

$$\mathbb{E}[x^*] = \arg \underset{\mathbb{E}[x] \in \Delta_{n-1}}{\min} \sum_{j \in \mathcal{N}} \mathbb{E}[x_j]^2.$$

Now, for a concave function f, *applying the Jensen's inequality one has*

$$f\left(\frac{1}{n} \sum_{i=1}^{n} x_i\right) \geq \frac{1}{n} \sum_{i=1}^{n} f(x_i).$$

Let $f(x) = -x^2$, *then*

$$-\left(\frac{1}{n}\right)^2 \geq -\frac{1}{n} \sum_{i=1}^{n} x_i^2,$$

$$\frac{1}{n} \leq \sum_{i=1}^{n} x_i^2,$$

showing that

$$\mathbb{E}[x^{\text{VI}}] = \left(\frac{1}{n}, \ldots, \frac{1}{n}\right)$$

minimizes $\sum_{i \in \mathcal{N}} L_i^2(\mathbb{E}[x])$.

Lemma 7 *The most efficient mean-field-type GNE does not coincide with the global optimum in the stochastic case as values $\vec{N}_{\text{global}} \neq \vec{N}_{\text{GNE}}$, where*

$$\left(\vec{N}_{\text{global}}\right)_i = -\frac{w_i - \mathbb{E}[w_i]}{\delta_i}, \tag{13.11a}$$

$$\left(\vec{N}_{\text{GNE}}\right)_i = -\frac{w_i - \mathbb{E}[w_i]}{1 + \varepsilon_{ii}}, \tag{13.11b}$$

and $\delta_i = \sum_{j \in \mathcal{N}}(1 + \varepsilon_{ji})$.

Proof 36 (Lemma 7) *Taking the first term from the right-hand side in (13.10) yields*

$$\underset{y \in \{\vec{N}_{\text{GNE}}\}}{\text{minimize}} \sum_{i \in \mathcal{N}} \left\{ \sum_{j \in \mathcal{N}}(1 + \varepsilon_{ij})\mathbb{E}[y_j^2] + 2\mathbb{E}[y_i(w_i - \mathbb{E}[w_i])] \right\},$$

$$= \underset{y \in \{\vec{N}_{\text{GNE}}\}}{\text{minimize}} \sum_{j \in \mathcal{N}} \left(\sum_{i \in \mathcal{N}}(1 + \varepsilon_{ij}) \right) \mathbb{E}[y_j^2] + 2 \sum_{i \in \mathcal{N}} \mathbb{E}[y_i(w_i - \mathbb{E}[w_i])],$$

$$= \underset{y \in \{\vec{N}_{\text{GNE}}\}}{\text{minimize}} \sum_{j \in \mathcal{N}} \delta_j \left(\mathbb{E}[y_j] + \frac{w_i - \mathbb{E}[w_i]}{\delta_j} \right)^2 - \sum_{j \in \mathcal{N}} \frac{\sigma^2}{\delta_j},$$

completing the proof.

13.1.4   Potential Constrained Mean-Field-Type Game

Extending the potential games definition from the deterministic counterpart, the constrained mean-field-type game is potential if the incentive of all the players can be represented by means of a unique potential function. Thus, let us consider the following constrained potential game problem:

$$\underset{x_i}{\text{minimize}} \ \mathbb{E}\left[P^1(x - \mathbb{E}[x], w)\right] + P^2(\mathbb{E}[x]),$$

subject to

$$\mathbb{E}[x_i] \geq 0,$$

$$\sum_{j \in \mathcal{N}} \mathbb{E}[x_j] \geq 1,$$

where

$$P^1(x - \mathbb{E}[x], w - \mathbb{E}[w]) = \sum_{i \in \mathcal{N}}(1 + \varepsilon_{ii})\text{var}(x_i) + 2 \sum_{i \in \mathcal{N}} \text{cov}(x_i, w_i),$$

$$P^2(\mathbb{E}[x], \mathbb{E}[w]) = \sum_{i \in \mathcal{N}} (\mathbb{E}[x_i] + \mathbb{E}[w_i])^2.$$

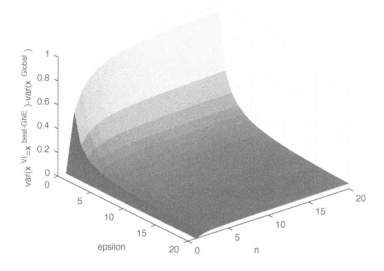

FIGURE 13.1
Gap between var $\left(x^{\mathrm{VI}} = x^{\mathrm{best-GNE}}\right)$ and var $\left(x^{\mathrm{Global}}\right)$ as different parameters $\varepsilon_{ij} = \varepsilon$, for all $i, j \in \mathcal{N}$, and number of players n change for a fixed variance $\sigma = 1$.

One verifies that the functions P^k, $k = \{1, 2\}$, solve

$$\nabla_{\mathbb{E}[x_i]} P^2 (\mathbb{E}[x], \bar{w}) = \nabla_{\mathbb{E}[x_i]} L_i^2 (\mathbb{E}[x], \bar{w}),$$
$$\nabla_{y_i} P^1 (x - \mathbb{E}[x], w - \mathbb{E}[w]) = \nabla_{y_i} L_i^1 (x - \mathbb{E}[x], w - \mathbb{E}[w]).$$

Therefore, the mean-field-type game can be solved by optimizing the single potential functions $P^1(x - \mathbb{E}[x], w)$ and $P^2(\mathbb{E}[x])$ satisfying the established constraints.

13.1.5 Efficiency Analysis

According to Lemma 7, none of the mean-field-type equilibrium provides the same efficiency of the global optimum in the stochastic case. Therefore, we are interested in analyzing the efficiency of the GNE with respect to the global optimum as different parameters change in the game. Then, we analyze how the efficiency changes under four specific scenarios, i.e.,

- Variation of the variance

- Variation of ε-parameters

- Variation of number of players

- Variation of connectivity under graphs

Let Φ denote a risk Key-Performance-Index (KPI) that computes the gap between the variance for the best GNE and the variance under the global solution as follows:

$$\Phi = \mathrm{var}\left(x^{\mathrm{VI}} = x^{\mathrm{best-GNE}}\right) - \mathrm{var}\left(x^{\mathrm{Global}}\right), \tag{13.12}$$

$$= \sigma^2 \sum_{i \in \mathcal{N}} \left(\frac{1}{1+\varepsilon_{ii}} - \frac{1}{1+\varepsilon_{ii} + \sum_{j \in \mathcal{N}\setminus\{i\}}(1+\varepsilon_{ij})}\right).$$

Next, we analyze how the risk KPI is affected by different parameters in the strategic game according to (13.12):

13.1.5.1 Variations of the Variance

It can be seen from (13.12) that the risk key performance index Φ is directly proportional and unbounded with respect to the variance σ of the random variable, i.e., the gap between the variance obtained with the variational equilibrium and the global solution increases as the variance of the noisy signals w_i, for all $i \in \mathcal{N}$, increase.

13.1.5.2 Variations of the ε-parameters

In order to analyze the risk key performance index Φ under the variation of the ε-parameter, let us fix the variance of the random variable, i.e., $\sigma = 1$, and let $\varepsilon_{ij} = \varepsilon$, for all $i,j \in \mathcal{N}$. Figure 13.1 shows the behavior of Φ. It can be seen that the existing gap between the variance for the best GNE and under the global solution vanishes as ϵ increasing, independently of the number n of players.

13.1.5.3 Variations of the Number of Players

In order to analyze the risk key performance index Φ as number of players changes, let us fix the variance $\sigma = 1$ and $\varepsilon_{ij} = \varepsilon$, for all $i,j \in \mathcal{N}$. It can be seen in Figure 13.1 that, different from the trend of the changes over ε, the key performance index Φ increases as the number of players increase and has a limit given by 1.

13.1.5.4 Variations of Connectivity under Graphs

Let $\mathcal{G} = (\mathcal{N}, \mathcal{E})$ be a undirected graph where $\mathcal{E} \subseteq \{(i,j)|i,j \in \mathcal{N}\}$ denotes the set of links representing how players regard others, i.e., (i,j) means that the player $i \in \mathcal{N}$ incorporates the variance of the player $j \in \mathcal{N}$ within its cost function $L_i^1(y,w)$. Thus, the set of links are $\mathcal{E} = \{(i,j)|\varepsilon_{ij} \neq 0\}$. Therefore, notice that the existing gap between the variance for the best GNE and under the global solution behaves similarly as with the number of players n, but with the number of links. Indeed, notice that under graphs, the parameter δ_i in $\bar{N}_{\mathrm{global}} = -(w_i - \mathbb{E}[w_i])/\delta_i$ becomes $\delta_i = (1+\varepsilon_{ii}) + \sum_{j \in \mathcal{N}_i}(1+\varepsilon_{ji})$ being $\mathcal{N}_i = \{j|\varepsilon_{ji} \neq 0\}$. Figure 13.2 presents an example of the decrement of the

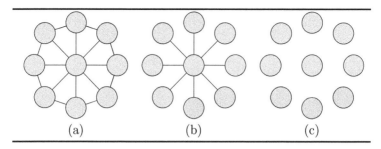

FIGURE 13.2
Different topologies for the comparison among $\delta-$parameters in (13.11a) with $\varepsilon_{ij} \geq 0$, for all $i, j \in \mathcal{N}$.

risk key performance index Φ in (13.12) as the number of links decreases. The KPIs $\Phi_{(a)}$, $\Phi_{(b)}$ and $\Phi_{(c)}$ corresponding to Figures 13.2(a-c), respectively; are as follows:

$$\Phi_{(a)} > \Phi_{(b)} > \Phi_{(c)}.$$

13.1.6 Learning Variational Equilibria

It is desired to design a learning algorithm that converges to a mean-field-type GNE. More precisely, it is desirable to guarantee converge to the efficient GNE given by the mean-field-type variational equilibrium. To this end, consider the following learning dynamics:

$$\dot{y}_i = -\frac{\partial}{\partial y_i} P^1(y, w - \mathbb{E}[w]), \tag{13.13a}$$

$$\mathbb{E}[\dot{x}_i] = -\mathbb{E}[x_i]\left[\frac{\partial}{\partial \mathbb{E}[x_i]} P^2(\mathbb{E}[x], \mathbb{E}[w])\right.$$

$$\left. - \sum_{j \in \mathcal{N}} \mathbb{E}[x_j]\frac{\partial}{\partial \mathbb{E}[x_i]} P^2(\mathbb{E}[x], \mathbb{E}[w])\right], \tag{13.13b}$$

$$\mathbb{E}[x_i(0)] > 0, \forall i \in \mathcal{N}, \tag{13.13c}$$

$$\sum_{j \in \mathcal{N}} \mathbb{E}[x_j(0)] = 1, \tag{13.13d}$$

which converges to the variational equilibrium given by $x^* = y^* + \mathbb{E}[x^*]$. Let us consider the mean-field-type game in (13.4) for two decision-makers. Figure 13.3 presents the random variables w_1 and w_2; and shows the evolution of the strategies x_1 and x_2 under the learning algorithm presented in (13.13). It can be seen the convergence of the trajectories to the mean-field-type variational equilibrium given by $x^{\text{best}-\text{GNE}} = x^{\text{VI}} = (1/2, 1/2)$.

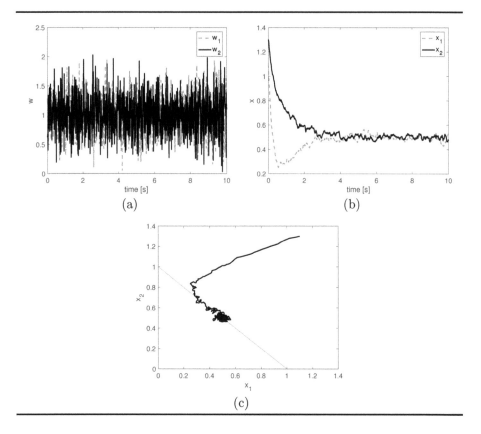

FIGURE 13.3
Evolution of the variables w and x under the learning algorithm in (13.13) for
the constrained MFTG presented in (13.4) with $n = 2$. Figures correspond to:
(a) evolution of w_1 and w_2, and (b)-(c) evolution of x_1 and x_2.

13.2   Model

Based on the motivating example of the previous section we propose a class of
constrained mean-field-type games in the stationary regime. We present some
analysis and several learning algorithms for the converge to the variational
mean-field-type equilibria.

Let us consider a set of $n \geq 2$ ($n \in \mathbb{N}$) decision-makers given by $\mathcal{N} =
\{1, \ldots, n\}$. Each decision-maker is interested in minimizing its own mean-
field-type cost functional denoted by $L_i(x, w, D_w)$, for all $i \in \mathcal{N}$, which can

be divided into two parts, i.e.,

$$L_i(x, w, D_w) = L_i^1(y, w) + L_i^2(\mathbb{E}[x]),$$
$$x = y + \mathbb{E}[x],$$

where $y, x, w \in \mathbb{R}^n$, being x the strategy profile of all decision-makers, and w a random variable with certain distribution denoted by D_w. We consider different type of constraints, i.e.,

$$\begin{cases} \underset{y_i, \mathbb{E}[x_i]}{\text{minimize}} \ \mathbb{E}\left[L_i^1(y, w) + L_i^2(\mathbb{E}[x])\right], \\\\ \text{subject to} \\\\ G \cdot \mathbb{E}[x] \leq g, \end{cases} \tag{13.14}$$

where $G \in \mathbb{R}^{b \times n}$, and $g \in \mathbb{R}^b$ allow describing b inequality constraints. Alternatively,

$$\begin{cases} \underset{y_i, \mathbb{E}[x_i]}{\text{minimize}} \ \mathbb{E}\left[L_i^1(y, w) + L_i^2(\mathbb{E}[x])\right], \\\\ \text{subject to} \\\\ \sum_{j \in \mathcal{N}} \mathbb{E}[x_j] = m, \\\\ 0 \leq \underline{x}_j \leq \mathbb{E}[x_j] \leq \bar{x}_j, \ \forall j \in \mathcal{N} \end{cases} \tag{13.15}$$

where $m \in \mathbb{R}_{>0}$. Finally, the mean-field-type game problem can just consider the same constraints as in (13.13c), i.e.,

$$\begin{cases} \underset{y_i, \mathbb{E}[x_i]}{\text{minimize}} \ \mathbb{E}\left[L_i^1(y, w) + L_i^2(\mathbb{E}[x])\right], \\\\ \text{subject to} \\\\ \sum_{j \in \mathcal{N}} \mathbb{E}[x_j] = m, \\\\ \mathbb{E}[x_j] \geq 0, \ \forall j \in \mathcal{N} \end{cases} \tag{13.16}$$

In the motivation example presented in Section 13.1, we introduced the concepts of mean-field-type generalized Nash equilibrium, and mean-field-type variational equilibrium for the specific stochastic game in (13.4b). Next, we present the definitions of the same concepts, but in general terms corresponding to problem (13.14).

Definition 33 (Mean-field-type best-response strategy) *A strategy profile is said to be feasible if it satisfies the constraints. Any feasible strategy*

$$(x_1, \ldots, x_{i-1}, x_i^* = \mathbb{E}[x_i^*] + y_i^*, x_{i+1}, \ldots, x_n)$$

Mean-Field-Type Games for Engineers

such that

$$y_i^* \in \arg\min_{y_i} \mathbb{E}[L_i^1(y, w)], \ and$$

$$\mathbb{E}[x_i^*] \in \arg\min_{\mathbb{E}[x_i]} L_i^2(\mathbb{E}[x])],$$

satisfying constraints in either (13.14), (13.15), or (13.16), for all $i \in \mathcal{N}$, is a mean-field-type best response of the player $i \in \mathcal{N}$, denoted by $\mathrm{BR}_i(\{x_j\}_{j\in\mathcal{N}\setminus\{i\}})$, against the strategies $x_{-i} = \mathbb{E}[x_{-i}] + y_{-i}$ selected by the other decision-makers $\mathcal{N} \setminus \{i\}$.

The set of mean-field-type generalized Nash equilibrium is defined by using the mean-field-type best-response as shown next.

Definition 34 (Mean-field-type generalized Nash equilibrium) *Any feasible strategy $x^* = \mathbb{E}[x^*] + y^*$ such that $x_i^* \in \mathrm{BR}_i(\{x_j^*\}_{j\in\mathcal{N}\setminus\{i\}})$, for all $i \in \mathcal{N}$, is a mean-field-type generalized Nash equilibrium.*

As observed in Section 13.1 there can be multiple mean-field-type generalized Nash equilibria. One way to refine this set in the differentiable case is with variational mean-field-type game equilibrium, defined next.

Definition 35 (Mean-field-type variational equilibrium) *A strategy profile $x^* = \mathbb{E}[x^*] + y^*$ is a variational equilibrium of the mean-field-type game in (13.4) if it is feasible and*

$$\mathbb{E}\left[\langle \nabla_{y_i} L_i^1(y^*, w), y_i - y_i^* \rangle \right]$$
$$+ \mathbb{E}\left[\langle \nabla_{\mathbb{E}[x_i]} L_i^2(\mathbb{E}[x^*]), \mathbb{E}[x_i] - \mathbb{E}[x_i^*]\rangle \right] \geq 0,$$

for all $i \in \mathcal{N}$ and for every $\mathbb{E}[y_i] = 0$, $G \cdot \mathbb{E}[x] \geq g$.

Definition 36 (Strict monotonicity) *An operator F is strictly monotone if the following inequality holds:*

$$\langle x - y, F(x) - F(y) \rangle > 0, \forall x \neq y$$

Notice that, in one dimension, the strict monotonicity is equivalent to: $x > y$ implies $F(x) > F(y)$ for all $x \neq y$. Strictly monotone operators are of great interest because they provide a very important result in equilibrium seeking theory. The variational inequality $\langle x - x^*, F(x^*) \rangle \geq 0, \forall x$ has at most one solution if F is a strictly monotone operator. To prove statement, first observe that if two elements x_1 and x_2 are solutions, one gets $\langle y - x_1, F(x_1) \rangle \geq 0, \forall y$ and $\langle y - x_2, F(x_2) \rangle \geq 0$. By exchanging the roles of x_1 and x_2 one arrives at

$$\langle (x_1 - x_2), -F(x_1) \rangle \geq 0,$$
$$\langle (x_1 - x_2), F(x_2) \rangle \geq 0 \tag{13.17}$$

By adding the two inequalities one obtains

$$\langle x_1 - x_2, F(x_1) - F(x_2) \rangle \leq 0.$$

By strictly monotonicity of F this means that $x_1 = x_2$.

Proposition 20 *Let L_i^1, L_i^2 be quasi-convex functions in their i^{th} component and jointly continuous. Assume, in addition, that these functions are coercive. Then, there exists at least one GNE of mean-field type under constraints of type (13.14) (13.15) and (13.16). Moreover, there is at most one mean-field-type variational equilibrium if the gradients generates a strictly monotone operator.*

Proof 37 (Proposition 20) *We observe that the set of constraints is non-empty, convex but not compact. We can reduce it to a compact set because the objective functions are coercive. The problem is then reduced to a mean-field type game problem under non-empty, convex and compact constraint set. Under the assumption above, there exists at least one point $\mathbb{E}[x]$ for the deterministic part and one for the stochastic part. Combining together, we obtain the announced result.*

Next we introduce a class of constrained games known as constrained potential games, and under which several learning algorithms to seek constrained Nash equilibria have been studied.

13.3 Learning Algorithms

Definition 37 (Constrained potential mean-field-type games) *Let us consider a mean-field-type game problem whose cost functional is given by*

$$L_i(x, w_i, D_{w_i}) = L_i^1(x - \mathbb{E}[x], w_i - \mathbb{E}[w_i]) + L_i^2(\mathbb{E}[x], \mathbb{E}[w_i]).$$

This is a potential mean-field-type game if there exist functions $P^1(x - \mathbb{E}[x], w - \mathbb{E}[w])$ and $P^2(\mathbb{E}[x], \mathbb{E}[w])$ such that $\nabla_{v^k} P^k = \nabla_{v^k} L_i^k$, for all $i \in \mathcal{N}$, $k \in \{1, 2\}$ with $(v^1, v^2) = (x_i - \mathbb{E}[x_i], \mathbb{E}[x_i])$.

According to Definition 37, some of the equilibrium of the mean-field-type game can be obtained by means of the optimization of the corresponding potential functions. Thus, several learning algorithms can be implemented in the solution of these type of games with convergence guarantees for strictly convex potential functions. In the following section, we discuss the use of learning algorithms to seek the mean-field-type variational equilibrium.

Considering potential games with strictly monotone gradient potential functions, several learning algorithms can be implemented depending on the imposed constraints. We discuss some alternative algorithms in order to seek variational equilibria (see Definition 35). Notice that, if the potential function is convex and the domain unconstrained then any gradient-based learning algorithm is suitable to seek the mean-field-type game equilibrium. The issue here comes from the constraints (13.1). Then, the constrained optimization problem in (13.14) can be solved by means of the dual decomposition algorithm with separable objectives [132]. To this end, let us compute the Lagrangian function, i.e.,

$$\mathcal{L}(y, w, \mathbb{E}[x], \mu) = P^1(y, w - \mathbb{E}[w]) + P^2(\mathbb{E}[x], \mathbb{E}[w]) + \mu^\top (G\mathbb{E}[x] - g).$$

The dual decomposition algorithm consists of

$$\hat{y} \in \arg\min_{y} \ P^1(y, w - \mathbb{E}[w]),$$

$$\hat{x} \in \arg\min_{\mathbb{E}[x]} \ P^2(\mathbb{E}[x], \mathbb{E}[w]) + \mu^\top G\mathbb{E}[x],$$

$$\mu^+ = \max\left\{\mu + \varrho\left(G\hat{x} - g\right), 0\right\},$$

with $\varrho > 0$, which converges to the mean-field-type variational equilibrium given that the potential function $P^1(y, w - \mathbb{E}[w])$ and $P^2(\mathbb{E}[x], \mathbb{E}[w])$ are assumed to be strictly convex functions.

When considering an equality constraint, there are some suitable alternative algorithms that have been mainly designed for solving resource allocation problems. Thus, the problem in (13.15) can be addressed by using the constrained evolutionary dynamics presented in [41], e.g., using the Smith-replicator dynamics given by

$$\dot{y}_i = -\frac{\partial}{\partial y_i} P^1(y, w - \mathbb{E}[w]),$$

$$\mathbb{E}[\dot{x}_i] = -[\hat{x}_i]_+ \sum_{j \in \mathcal{S}} [\check{x}_j]_+ \left[\frac{\partial P^2(\mathbb{E}[x], \mathbb{E}[w])}{\partial \mathbb{E}[x_i]} - \frac{\partial P^2(\mathbb{E}[x], \mathbb{E}[w])}{\partial \mathbb{E}[x_j]}\right]_+$$

$$+ [\check{x}_i]_+ \sum_{j \in \mathcal{S}} [\hat{x}_j]_+ \left[\frac{\partial P^2(\mathbb{E}[x], \mathbb{E}[w])}{\partial \mathbb{E}[x_j]} - \frac{\partial P^2(\mathbb{E}[x], \mathbb{E}[w])}{\partial \mathbb{E}[x_i]}\right]_+,$$

$$\underline{x} \le \mathbb{E}[x(0)] \le \bar{x},$$

$$\sum_{j \in \mathcal{N}} \mathbb{E}[x_j(0)] = m,$$

where $[\cdot]_+ = \max(0, \cdot)$, $\hat{x} = \bar{x} - \mathbb{E}[x]$, $\check{x} = \mathbb{E}[x] - \underline{x}$. This algorithm converges to the variational equilibrium given the potential function $P^1(y, w - \mathbb{E}[w])$ and $P^2(\mathbb{E}[x], \mathbb{E}[w])$ are assumed to be strictly convex functions. Finally, the example in (13.13) consists of a gradient system for an unconstrained optimization problem together with the replicator dynamics [133], which is a gradient algorithm that evolves within an invariant set able to satisfy the constraints imposed in (13.16). In fact, the problem in (13.16) can be addressed by means of any evolutionary dynamics algorithm either centralized or distributed [128, 134], e.g., the distributed Smith dynamics

$$\dot{y}_i = -\frac{\partial}{\partial y_i} P^1(y, w - \mathbb{E}[w]),$$

$$\mathbb{E}[\dot{x}_i] = -\sum_{j\in\mathcal{S}} \mathbb{E}[x_j] \left[\frac{\partial P^2(\mathbb{E}[x], \mathbb{E}[w])}{\partial \mathbb{E}[x_i]} - \frac{\partial P^2(\mathbb{E}[x], \mathbb{E}[w])}{\partial \mathbb{E}[x_j]} \right]_+$$

$$+ \mathbb{E}[x_i] \sum_{j\in\mathcal{S}} \left[\frac{\partial P^2(\mathbb{E}[x], \mathbb{E}[w])}{\partial \mathbb{E}[x_j]} - \frac{\partial P^2(\mathbb{E}[x], \mathbb{E}[w])}{\partial \mathbb{E}[x_i]} \right]_+ ,$$

$$\mathbb{E}[x_i(0)] > 0, \forall i \in \mathcal{N},$$

$$\sum_{j\in\mathcal{N}} \mathbb{E}[x_j(0)] = m,$$

where $[\cdot]_+ = \max(0, \cdot)$, which converges to the variational equilibrium since the potential function $P^1(y, w - \mathbb{E}[w])$ and $P^2(\mathbb{E}[x], \mathbb{E}[w])$ are assumed to be strictly convex functions.

13.4 Equilibrium under Migration Constraints

In this section, we assume that each decision-maker i has a migration incentive or cost $\eta_{i,ll'}$ to move from component (i, l) to (i, l') with choice constraint neighbors of l given by $\mathcal{N}_{il}$. When $\eta_{i,ll'} > 0$, there is a migration incentive for i from l to l'. When $\eta_{i,ll'} < 0$ it becomes a migration cost. When $\eta = 0$ and $\mathcal{N}_{il}$ is the set of all choices of i, then we retrieve the classical mean-field-type equilibrium.

Next, we define a mean-field-type equilibrium with migration incentive/cost. A feasible strategic profile $x^* = \mathbb{E}[x^*] + y^*$ is an equilibrium under migration incentive/cost η of the mean-field-type game in (13.4) if

$$\mathbb{E}[x_{il}^*] > 0 \implies$$
$$\begin{cases} \mathbb{E}\left[\nabla_{y_{il}} L_i^1(y^*, w) \right] = \max_{l'\in\mathcal{N}_{il}} \mathbb{E}\left[\nabla_{y_{il'}} L_i^1(y^*, w) \right], \\ \nabla_{\mathbb{E}[x_{il}]} L_i^2(\mathbb{E}[x^*]) + \eta_{i,ll} = \max_{l'\in\mathcal{N}_{il}} \nabla_{\mathbb{E}[x_{il'}]} L_i^2(\mathbb{E}[x^*]) + \eta_{i,ll'}, \end{cases} \quad (13.18)$$

for every decision-maker $i \in \mathcal{N}$. This type of solution concepts have plenty of applications. It includes

- Cost with $\eta_{ij} > 0$ (i) electricity transmission and distribution cost from the point of production to the end user point, (ii) Pricing of electrical vehicle charging station in smart cities, (iii) transportation and UBER pricing, (iv) cost sharing between cities in water distribution networks,

- Benefit with $\eta_{ij} < 0$: (i) smart city planning with incentive/benefit for the citizens to join specific areas, (ii) mechanism design for cloud servers in blockchain-based networks (validation task delegation).

This equilibrium has three type of constraints:

- Strategy constraint on $\mathbb{E}[x]$

- Choice movement constraint given by $\mathcal{N}_{i,l}$

- Incentive/cost for a change of choice η

This equilibrium can be reformulated as an equilibrium of a modified game with cost functional $\tilde{L}_i^2 = L_i^2 + \sum_{l,l'} \eta_{i,ll'} \tilde{x}_{i,ll'}$ where $x_{il} = \sum_{l'} \tilde{x}_{i,l'l}$. It follows that

$$\nabla_{\mathbb{E}[\tilde{x}_{i,ll}]} \tilde{L}_i^2(\mathbb{E}[x^*]) = \max_{l' \in \mathcal{N}_{il}} \nabla_{\mathbb{E}[\tilde{x}_{ill'}]} \tilde{L}_i^2(\mathbb{E}[x^*]),$$

leading to similar as in Definition 35 with the neighborhood constraint $\mathcal{N}_{il}$.

In this chapter, we have introduced a class of constrained mean-field-type games under stationary regime. These kinds of games consider distribution-dependent cost through the first moment and could take into account both mean and variance terms in the payoff of the decision-makers. We have presented concepts such as mean-field-type equilibrium, mean-field-type variational equilibrium, mean-field-type potential games, and we introduce some learning algorithms to converge to the best Nash equilibrium.

14

Mean-Field-Type Model Predictive Control

We study and design a mean-field-type model predictive control, which is a class of risk-aware control strategy considering within the cost functional not only the mean but also the variance of both the system states and control inputs. We formally present the relationship between the proposed approach and the so-called chance-constrained control techniques.

There are different ways to handle uncertainties in the design of control techniques. On one hand, robust model predictive control (MPC) allow computing the appropriate control inputs considering the worse scenario unknown bounded disturbances by solving a min-max optimization problem. On the other hand, stochastic MPC controllers study the computation of the optimal control inputs considering the probabilistic distributions associated with the disturbances. One of the main risk-aware optimization-based control strategies is the Chance-Constrained Model Predictive Control (CC-MPC). For instance, a CC-MPC approach is studied and applied to a large-scale WDN in [135] whereas the work in [136] presents the design of a robust periodic MPC controller for the same WDN as in [135]. Alternatively, [137] studies stochastic MPC controllers with demands based on Gaussian processes for the same kind of WDN. Moreover, other MPC approaches that combine both the robust and stochastic methods have been considered in [138]. This control approach is widely used in the control of a large variety of engineering applications,

Prior to discussing the risk-aware MPC controllers, we present the general idea of the optimization-based receding-horizon control strategies.

14.1 Problem Statement

Consider a state-space discrete-time system modeled by the following difference equation:

$$x_{k+1} = Ax_k + \bar{A}\mathbb{E}[x_k] + Bu_k + \bar{B}\mathbb{E}[u_k] + B_d d_k + Sw_k, \qquad (14.1)$$

$$x_k \triangleq x_0,$$

where $x \in \mathbb{R}^{n_x}$ denotes the system-state vector, $u \in \mathbb{R}^{n_u}$ is the vector

DOI: 10.1201/9781003098607-14

of control inputs, $[B_d d_k + S w_k] \in \mathbb{R}^{n_x}$ denotes the vector of disturbances, which are divided into two parts. First, $d_k \in \mathbb{R}^{n_d}$ denotes the known nominal disturbance given by the demand of resources, whereas $w_k \sim \mathcal{N}(0,1)$, $w_k \in \mathbb{R}^{n_d}$ denotes the unknown uncertainty associated with the demand.

The state-space matrices are given by $A, \bar{A} \in \mathbb{R}^{n_x \times n_x}$, $B, \bar{B} \in \mathbb{R}^{n_x \times n_u}$, $B_d \in \mathbb{R}^{n_x \times n_d}$, and $S \in \mathbb{R}^{n_x \times n_d}$. In addition, different from other approaches reported in the literature, the system in (14.1) involves mean-field terms associated with the distribution for both the system states and control inputs, i.e., $\mathbb{E}[x] \in \mathbb{R}^{n_x}$ and $\mathbb{E}[u] \in \mathbb{R}^{n_u}$.

The MPC controller is in charged of minimizing a cost functional $g(u)$ throughout a fixed time horizon $[k..k+N]$ with $N \in \mathbb{N}$ by solving the following optimization problem every time instant:

$$\underset{u_{k|k},\dots,u_{k+N-1|k}}{\text{minimize}} \quad \mathbb{E}\left[g(x_{k+N|k}) + \sum_{\ell=k}^{k+N-1} h(x_{\ell|k}, u_{\ell|k}) \right], \qquad (14.2a)$$

subject to

$$x_{\ell+1|k} = A x_{\ell|k} + \bar{A}\mathbb{E}[x_{\ell|k}] + B u_{\ell|k} + \bar{B}\mathbb{E}[u_{\ell|k}] + B_d d_{\ell|k} + S w_{\ell|k}, \qquad (14.2b)$$

$$u_{\ell|k} \in \mathbb{U}, \qquad (14.2c)$$

$$x_{j|k} \in \mathbb{X}, \qquad (14.2d)$$

$$x_{k|k} \triangleq x_k, \qquad (14.2e)$$

for all $\ell \in [k..k+N-1]$, and $j \in [k, k+N]$, where $\mathbb{X}$ and $\mathbb{U}$ correspond to the non-empty feasible sets for the system states and control inputs, respectively. Assuming feasibility of the optimization problem in (14.7), then an optimal sequence of control inputs is obtained at each time instance, which is denoted by

$$\hat{u}_k^* \in \arg \underset{u_{k|k},\dots,u_{k+N-1|k}}{\min} \mathbb{E}\left[g(x_{k+N|k}) + \sum_{\ell=k}^{k+N-1} h(x_{\ell|k}, u_{\ell|k}) \right],$$

under constraints (14.2b)-(14.2d). Therefore, the optimal control inputs $\hat{u}_k^*$ generates a sequence of system states denoted by $\hat{x}_k^*$, given a certain known sequence of disturbances $\hat{d}_k$, i.e.,

$$\hat{u}_k^* \triangleq \left(u_{k|k}^*, \dots, u_{k+N-1|k}^* \right),$$

$$\hat{x}_k^* \triangleq \left(x_{k+1|k}^*, \dots, x_{k+N|k}^* \right),$$

$$\hat{d}_k \triangleq \left(d_{k|k}, \dots, d_{k+N-1|k} \right).$$

Notice that, only the first control input $u_{k|k}^*$ from the sequence $\hat{u}_k^*$ can be applied to the system state in (14.1). Then, following the receding-horizon philosophy, at time step $k+1$, a new system state $x_{k+1|k+1} \neq x_{k+1|k}$ is measured and a new optimal control sequence $\hat{u}_{k+1}^*$ is computed.

14.2 Risk-Aware Model Predictive Control Approaches

We discuss the two main risk-aware control approaches, i.e., the CC-MPC and the proposed mean-field-type model predictive control (MFT-MPC), and the existing relationship between them.

14.2.1 Chance-Constrained Model Predictive Control

The CC-MPC is in charged of minimizing a cost functional $L(x, u)$ subject to several operational and physical constraints, i.e.,

$$L(x, u) = g(x_{k+N|k}) + \sum_{\ell=k}^{k+N-1} h(x_{\ell|k}, u_{\ell|k}), \qquad (14.3)$$

where the function $g : \mathbb{R}^{n_x} \to \mathbb{R}$ corresponds to the terminal cost, and the function $h : \mathbb{R}^{n_x} \times \mathbb{R}^{n_u} \to \mathbb{R}$ corresponds to the running cost. Thus, the optimization problem behind the CC-MPC is given by:

$$\underset{u_{k|k},\ldots,u_{k+N-1|k}}{\text{minimize}} \quad \mathbb{E}\left[L(x, u)\right], \qquad (14.4a)$$

subject to

$$x_{\ell+1|k} = Ax_{\ell|k} + \bar{A}\mathbb{E}[x_{\ell|k}] + Bu_{\ell|k} + \bar{B}\mathbb{E}[u_{\ell|k}] + B_d d_{\ell|k} + Sw_{\ell|k}, \qquad (14.4b)$$

$$\mathbb{P}\left(u_{\ell|k} \in \mathbb{U}\right) \geq 1 - \delta_u, \qquad (14.4c)$$

$$\mathbb{P}\left(x_{j|k} \in \mathbb{X}\right) \geq 1 - \delta_x, \qquad (14.4d)$$

$$x_{k|k} \triangleq x_k, \qquad (14.4e)$$

for all $\ell \in [k..k + N - 1]$, and $j \in [k..k + N]$. In addition, the feasible sets for the control inputs and system states are given by non-empty sets $\mathbb{U}$ and $\mathbb{X}$, respectively. The parameters $\delta_x, \delta_u \in (0, 1)$ determine the risk levels [139]. Notice that, for fixed risk parameters δ_x and δ_u, the feasibility of the problem in (14.4) is compromised as the noise-variance parameter S increases.

14.2.2 Mean-Field-Type Model Predictive Control

We present the novel MFT-MPC controller, and its relationship with the aforementioned CC-MPC controller. The MFT-MPC controller is in charged of minimizing a cost functional $L(x, u)$. On the other hand, we modify $L(x, u)$ to be a risk-aware cost functional for the MFT-MPC controller as follows:

$$L(x, u) = g(\mathbb{E}[x_{k+N|k}]) + g_v(x_{k+N|k}, \mathbb{E}[x_{k+N|k}])$$

$$+ \sum_{\ell=k}^{k+N-1} \left[h(\mathbb{E}[x_{\ell|k}], \mathbb{E}[u_{\ell|k}]) \right]$$

$$+ \sum_{\ell=k}^{k+N-1} \left[h_v(x_{\ell|k}, u_{\ell|k}, \mathbb{E}[x_{\ell|k}], \mathbb{E}[u_{\ell|k}]) \right], \qquad (14.5)$$

where

$$g_v(x_{k+N|k}, \mathbb{E}[x_{k+N|k}]) = \langle \bar{V}(x_{k+N|k} - \mathbb{E}[x_{k+N|k}]), x_{k+N|k} - \mathbb{E}[x_{k+N|k}] \rangle, \tag{14.6a}$$

$$h_v(x_{\ell|k}, u_{\ell|k}, \mathbb{E}[x_{\ell|k}], \mathbb{E}[u_{\ell|k}]) = \langle \bar{R}(u_{\ell|k} - \mathbb{E}[u_{\ell|k}]), u_{\ell|k} - \mathbb{E}[u_{\ell|k}] \rangle$$
$$+ \langle \bar{Q}(x_{\ell|k} - \mathbb{E}[x_{\ell|k}]), x_{\ell|k} - \mathbb{E}[x_{\ell|k}] \rangle, \qquad (14.6b)$$

being $g : \mathbb{R}^{n_x} \to \mathbb{R}$ and $g_v : \mathbb{R}^{n_x} \times \mathbb{R}^{n_x} \to \mathbb{R}$ correspond to the terminal cost, and the functions $h : \mathbb{R}^{n_x} \times \mathbb{R}^{n_u} \to \mathbb{R}$ and $h_v : \mathbb{R}^{n_x} \times \mathbb{R}^{n_u} \times \mathbb{R}^{n_x} \times \mathbb{R}^{n_u} \to \mathbb{R}$ correspond to the running cost. Besides, the prioritization weight matrices are given by $\bar{V}, \bar{Q} \succeq 0$, and $\bar{R} \succ 0$. Thus, the optimization problem behind the MFT-MPC is given by:

$$\underset{u_{k|k}, \ldots, u_{k+N-1|k}}{\text{minimize}} \quad \mathbb{E}\left[L(x, u) \right], \qquad (14.7a)$$

subject to

$$x_{\ell+1|k} = A x_{\ell|k} + \bar{A}\mathbb{E}[x_{\ell|k}] + B u_{\ell|k} + \bar{B}\mathbb{E}[u_{\ell|k}] + B_d d_{\ell|k} + S w_{\ell|k}, \quad (14.7b)$$
$$\mathbb{E}[u_{\ell|k}] \in \mathbb{U}^c, \qquad (14.7c)$$
$$\mathbb{E}[x_{j|k}] \in \mathbb{X}^c, \qquad (14.7d)$$

for all $\ell \in [k..k+N-1]$, and $j \in [k..k+N]$. In addition, the feasible sets for the control inputs and system states are given by non-empty sets $\mathbb{U}^c \subset \mathbb{U}$ and $\mathbb{X}^c \subset \mathbb{X}$, respectively; defined as follows:

$$\mathbb{D}^x = \{ z_x \in \mathbb{R}^{n_x} : \|z_x - x\| \le \varepsilon_x, \ x \in \partial\mathbb{X} \}, \qquad (14.8a)$$
$$\mathbb{X}^c = \overline{(\mathbb{X} \setminus \mathbb{D}^x)}, \qquad (14.8b)$$
$$\mathbb{D}^u = \{ z_u \in \mathbb{R}^{n_u} : \|z_u - u\| \le \varepsilon_u, \ u \in \partial\mathbb{U} \}, \qquad (14.8c)$$
$$\mathbb{U}^c = \overline{(\mathbb{U} \setminus \mathbb{D}^u)}, \qquad (14.8d)$$

where $\varepsilon_x, \varepsilon_u > 0$ (see graphical example in Figure 14.1).

14.2.3 Chance-Constrained vs Mean-Field Type Model Predictive Control

We briefly discuss some features of the MFT-MPC such as its connection with the so-called CC-MPC. The following remark points out the main characteristic of the MFT-MPC.

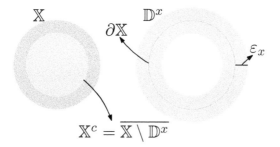

FIGURE 14.1
Graphical example for the sets $\mathbb{X}$, $\mathbb{D}^x$, and $\mathbb{X}^c$.

Remark 30 (Mean-Variance Minimization) *The cost functional in (14.5) involves variance minimization, i.e., the expectation of the cost functional $\mathbb{E}\left[J(x, u)\right]$ considers the following variance terms:*

$$\mathrm{var}_{[\bar{Q}]}(x_{j|k}) = \mathbb{E}\left\langle \bar{Q}(x_{j|k} - \mathbb{E}[x_{j|k}]), x_{j|k} - \mathbb{E}[x_{j|k}]\right\rangle,$$
$$\mathrm{var}_{[\bar{R}]}(u_{\ell|k}) = \mathbb{E}\left\langle \bar{R}(u_{\ell|k} - \mathbb{E}[u_{\ell|k}]), u_{\ell|k} - \mathbb{E}[u_{\ell|k}]\right\rangle,$$

for all time instants $j \in [k..k+N]$ and $\ell \in [k..k+N-1]$. Such terms make the MFT-MPC controller to perform risk minimization.

Proposition 21 presents the relationship between the MFT-MPC and the CC-MPC whose optimization problems are given by (14.7) and (14.2), respectively.

Proposition 21 (Relationship between CC-MPC and MFT-MPC)
The mean-variance minimization $\mathbb{E}[L(x, u)]$ considered in the MFT-MPC (see Problem (14.7) and Remark 30) is related to the CC-MPC (see Problem (14.4)) with the following parameters:

$$\delta_x = \frac{\mathrm{var}[x]}{\varepsilon_x^2},$$
$$\delta_u = \frac{\mathrm{var}[u]}{\varepsilon_u^2},$$

for some $\varepsilon_x, \varepsilon_u > 0$, which determines the non-empty sets $\mathbb{X}^c$ in (14.8b) and $\mathbb{U}^c$ in (14.8d), respectively (see Figure 14.1).

Proof 38 (Proposition 21) *First, we present the relationship between the variance minimization and the chance-constrained optimization. Let us consider a non-empty set $\mathbb{X}^c \subset \mathbb{X}$ as defined in (14.8b), where $\varepsilon_x > 0$. Therefore, let $\mathbb{P}\left(x \in \mathbb{X}^c\right) \leq \mathbb{P}\left(x \in \mathbb{X}\right)$. From the Markov's Inequality [140] we have:*

$$\mathbb{P}\left(y \geq a\right) \leq \frac{\mathbb{E}[y]}{a},$$

being $a > 0$ and y a non-negative random variable. Thus,

$$\mathbb{P}\left(y \leq a\right) \geq 1 - \frac{\mathbb{E}[y]}{a}.$$

Therefore

$$\mathbb{P}\left(\|x - \mathbb{E}[x]\| \leq \varepsilon_x\right) \geq 1 - \frac{\text{var}[x]}{\varepsilon_x^2}. \tag{14.9}$$

Considering the fact $\mathbb{E}[x] \in \mathbb{X}$ according to constraint in (14.7d), combining (14.9) and (14.8b), and taking into consideration that $\mathbb{P}\left(x \in \mathbb{X}^c\right) \leq \mathbb{P}\left(x \in \mathbb{X}\right)$ yields

$$\mathbb{P}\left(x \in \mathbb{X}\right) \geq 1 - \frac{\text{var}[x]}{\varepsilon_x^2}, \tag{14.10}$$

which is associated with the Chebyshev's Inequality [140], and it is the same chance-constraint in (14.4d) taking

$$\delta_x = \frac{\text{var}[x]}{\varepsilon_x^2},$$

showing the announced relationship between CC-MPC and MFT-MPC.

The result in Proposition 21 allows the MFT-MPC controller to adjust the risk parameters in the CC-MPC controller by penalizing the variance-minimization terms as presented in Lemma 8.

Lemma 8 (Risk Parameters) *An increment in the prioritization in $\bar{V}$ and $\bar{Q}$ ($\bar{R}$) in the MFT-MPC corresponds to a reduction of the parameters δ_x (δ_u) in the CC-MPC.*

Proof 39 (Lemma 8) *First, let us consider two different prioritization weights for the penalization over the variance terms in x, i.e., $\bar{Q}_i = \bar{V}_i$, with $i \in \{1, 2\}$, and $\bar{Q}_1 > \bar{Q}_2$. Now, from Problem in (14.7) yields $\text{var}_{[\bar{Q}_1]}(x_k) > \text{var}_{[\bar{Q}_2]}(x_k)$. Finally, from (14.10) and (14.4d), it is concluded that an increment in the prioritization in $\bar{V}$ and $\bar{Q}$ in the MFT-MPC corresponds to a reduction of the parameters δ_x in the CC-MPC. Similar arguments are applied for the control input u.*

Besides, the feasibility of the chance-constraints in the CC-MPC controller are associated with the variance minimization in the MFT-MPC controller as presented next in Remark 31.

Remark 31 *In order to increase the probability $\mathbb{P}(x \in \mathbb{X})$ in the MFT-MPC controller with respect to the CC-MPC controller, it is necessary that, according to Proposition 21*

$$1 - \frac{\text{var}[x]}{\varepsilon_x^2} \geq 1 - \delta_x.$$

Therefore, it is concluded that

$$\text{var}[x] \le \varepsilon_x^2 \delta_x.$$

Same procedure is applied for the constraint over the control input u, i.e., var$[u] \le \varepsilon_u^2 \delta_u$.

14.2.4 Decomposition and Stability

We now discuss the decomposition of the optimization problem behind the MFT-MPC.

Lemma 9 (MFT-MPC Problem Decomposition) *The problem in (14.7) can be decomposed into two parts, i.e., an unconstrained stochastic optimization problem depending $x_k - \mathbb{E}[x_k]$ and $u_k - \mathbb{E}[u_k]$; and a constrained deterministic optimization problem depending on $\mathbb{E}[x_k]$ and $\mathbb{E}[u_k]$.*

Proof 40 (Lemma 9) *First of all, notice that the decomposition is possible due to the fact that*

$$x_k - \mathbb{E}[x_k] \perp \mathbb{E}[x_k],$$
$$u_k - \mathbb{E}[u_k] \perp \mathbb{E}[u_k].$$

Let $y_k = x_k - \mathbb{E}[x_k]$, and $z_k = u_k - \mathbb{E}[u_k]$; then the unconstrained stochastic optimization problem is given by

$$\underset{z_{k|k},\dots,z_{k+N-1|k}}{\text{minimize}} \ \mathbb{E}\left[g_v(y_{k+N|k}) + \sum_{\ell=k}^{k+N-1} h_v(y_{\ell|k}, z_{\ell|k}) \right], \qquad (14.11a)$$

subject to

$$y_{\ell+1|k} = A y_{\ell|k} + B z_{\ell|k} + S w_{\ell|k}, \qquad (14.11b)$$
$$y_{k|k} \triangleq y_k, \qquad (14.11c)$$

for all $\ell \in [k..k+N-1]$. On the other hand, the optimization problem behind the deterministic MPC controller is as follows:

$$\underset{\mathbb{E}[u_{k|k}],\dots,\mathbb{E}[u_{k+N-1|k}]}{\text{minimize}} \ g(\mathbb{E}[x_{k+N|k}]) + \sum_{\ell=k}^{k+N-1} h(\mathbb{E}[x_{\ell|k}], \mathbb{E}[u_{\ell|k}]), \qquad (14.12a)$$

subject to

$$\mathbb{E}[x_{\ell+1|k}] = (A+\bar{A})\mathbb{E}[x_{\ell|k}] + (B+\bar{B})\mathbb{E}[u_{\ell|k}] + B_d d_{\ell|k}, \qquad (14.12b)$$
$$\mathbb{E}[u_{\ell|k}] \in \mathbb{U}^c, \qquad (14.12c)$$
$$\mathbb{E}[x_{j|k}] \in \mathbb{X}^c, \qquad (14.12d)$$

$$x_{k|k} \triangleq x_k, \tag{14.12e}$$

for all $\ell \in [k..k+N-1]$, and $j \in [k..k+N]$. Finally, the optimal control input is obtained from the sum of the solution of the previous optimization problems, i.e.,

$$u^*_{k|k} = z^*_{k|k} + \mathbb{E}[u^*_{k|k}].$$

Completing the orthogonal decomposition.

According to Lemma 9, and different from the problem in (14.4), the feasibility is not compromised as the noise-variance parameter S increases. Thus, the probability that the stochastic state $x \in \mathbb{X}$ and the control inputs $u \in \mathbb{U}$ is increased by augmenting either the variance penalization in the cost functional, or increasing the parameters ε_x and ε_u keeping the sets $\mathbb{X}^c, \mathbb{U}^c \neq \emptyset$.

Now, we are interested in analyzing the stability of the stochastic dynamical system given by:

$$x_{\ell+1|k} = Ax_{\ell|k} + \bar{A}\mathbb{E}[x_{\ell|k}] + Bu_{\ell|k} + \bar{B}\mathbb{E}[u_{\ell|k}] + B_d d_{\ell|k} + Sw_{\ell|k},$$

under the proposed risk-aware MPC controller. To this end by definition and minimization of variance, the system state $x_{\ell|k}$ evolves around its expected value $\mathbb{E}[x_{\ell|k}]$. Therefore, the stability of the system can be analyzed by means of its expectation, i.e., analyzing the stability of the following system:

$$\mathbb{E}[x_{\ell+1|k}] = \mathbb{E}\Big[Ax_{\ell|k} + \bar{A}\mathbb{E}[x_{\ell|k}] + Bu_{\ell|k} + \bar{B}\mathbb{E}[u_{\ell|k}] + B_d d_{\ell|k} + Sw_{\ell|k}\Big],$$
$$= (A+\bar{A})\mathbb{E}[x_{\ell|k}] + (B+\bar{B})\mathbb{E}[u_{\ell|k}] + B_d d_{\ell|k}.$$

Thus, the stability analysis is reduced to the statement made in Remark 32.

Remark 32 *According to Lemma 9, the MFT-MPC is reduced to a constrained deterministic MPC controller (14.12), and a stochastic unconstrained optimization problem (14.11). In this regard, the stability of the MFT-MPC is reduced to the stability of the deterministic dynamics of $\mathbb{E}[x]$, which has been widely studied in the literature. Thus, both functions g_v and h_v are designed to guarantee the closed-loop system stability [141–143].*

We have presented a mean-field-type model predictive control (MFT-MPC) design by incorporating risk minimization by means of variance terms in both the system states and control inputs. We have shown the existing tight relationship between the chance-constrained model predictive control (CC-MPC) and the MFT-MPC controller. We have presented the connection between the variance minimization and the risk parameters in the chance-constraints in a CC-MPC controller.

Moreover, it has been shown that the feasibility of the MFT-MPC can be improved either by augmenting the variance penalization in the cost functionals g_v and h_v to reduce the variance terms $\text{var}[x]$ and $\text{var}[u]$, or by increasing the parameters ε_x and ε_u reducing the feasible sets $\mathbb{X}^c$ and $\mathbb{U}^c$ for the expectation values $\mathbb{E}[x]$ and $\mathbb{E}[u]$, respectively.

15

Data-Driven Mean-Field-Type Games

In the previous chapters we have studied model-based mean-field-type control and game problems, i.e., by assuming knowledge about the dynamical behavior of the system. Nevertheless, there are several situations and applications in which the dynamical model of the states evolution is unknown. The lack of model might be because of the complexity of the stochastic system, or there must be some problems and applications where the system is completely unknown and only information about certain inputs and an output is available.

In this chapter, we study data-driven mean-field-type game problems based on machine learning. To this end, we compute an estimated stochastic model (learning-based modeling) based on massive data representing or providing information about the dynamical behavior of an unknown system. Thus, machine learning techniques enable the possibility to establish appropriate control inputs, which are computed based on learned optimality conditions, which are estimated from data as well by exploiting statistical learning techniques as discussed in [144].

First, we present the problem statement of the data-driven mean-field-type game. Afterward, we discuss about the main machine learning philosophy. Finally, we illustrate the machine-learning-based data-driven mean-field-type game followed by numerical examples. It is important to highlight that the use of machine learning in the stochastic interactive decision-making field is still a research topic and this chapter only intends to present a first approach using linear regression.

DOI: 10.1201/9781003098607-15

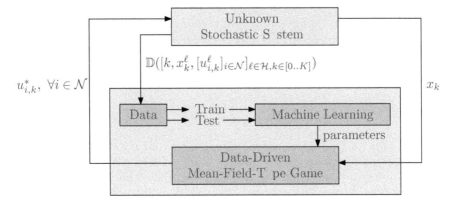

FIGURE 15.1
General scheme for a data-driven mean-field-type game problem by using machine learning.

15.1   Data-Driven Mean-Field-Type Game Problem

The data-driven mean-field-type game technique consists of determining the appropriate (estimated) optimal control inputs of mean-field type for all the $n \geq 2$ $(n \in \mathbb{N})$ decision-makers in the set $\mathcal{N} = \{1, \ldots, n\}$, which are interacting strategically without having knowledge about the dynamical model of the system (unknown system state evolution). Instead, this game-theoretical strategy focuses on the analysis of massive data about the dynamical behavior of the system.

The considered unknown system can be running in either continuous or discrete time. In any of the aforementioned cases, the available information (captured/measured data) about the behavior of the system is sampled. Thus, it is suitable to address the data-driven game problem in the context of discrete-time mean-field-type game problems as it was studied in Chapters 11 and 12.

Figure 15.1 shows the general scheme for a data-driven mean-field-type game problem. The model of the stochastic dynamical system is unknown and a closed-loop is established. The available historical data about the behavior of the system is denoted by $\mathbb{D}$, and its structure is explained in detail in Section 15.3.

Such data $\mathbb{D}$ is used in order to train an estimation of the model for the unknown system by using machine learning techniques. There exists a large variety of machine-learning or statistical learning techniques [144], for example, linear regression, logistic regression, K-nearest neighbors, decision trees, support vector machines, among many others. Depending on the nature of available data and the problem, the most appropriate machine-learning tool

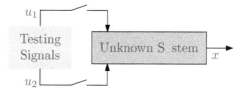

FIGURE 15.2
Input/output configuration for the unknown system in a two-player mean-field-type game problem.

is selected. For instance, in the example presented in this chapter, we use the most basic technique known as linear regression since we assume the dynamical behavior of the system is linear as it has been treated throughout this book.

By means of a system-state feedback x_k and the machine-learning-based parameters, the data-driven mean-field-type game problem can be solved and the optimal control inputs $u_{i,k}^*$, for all $i \in \mathcal{N}$, are computed and applied to the unknown system. Let us denote the unknown system dynamics by

(Discrete-time unknown system):
$$x_{k+1} = b(k, x_k, [u_{i,k}]_{i\in\mathcal{N}}, m_k) + \sigma(k, x_k, [u_{i,k}]_{i\in\mathcal{N}}, m_k)w_k,$$

where m_k denotes the probability measure of x at time k, and w denotes the white noise perturbing the system. In order to capture data from the unknown system, multiple tests are performed. Let $\mathcal{H} = \{1, \ldots, h\}$ denote the set of $h \in \mathbb{N}$ tests. At each test, arbitrary control input signals $u_{i,k}^\ell$, for all $i \in \mathcal{N}$, $\ell \in \mathcal{H}$, $k \in [0..K]$, are used to stimulate the unknown system. In the meantime, the response of the system x_k^ℓ, for all $k \in [0..K]$, is captured and stored. For instance, Figure 15.2 shows the scheme that can be implemented for the recollection of data for a two-decision-maker mean-field-type game problem.

The following Assumption 1 imposes some conditions over the testing signals in order to recollect the information about the dynamical behavior.

Assumption 1 *The testing control inputs $u_{i,k}$ for all $i \in \mathcal{N}$, which are used to capture data for the machine learning purposes (see Figure 15.2), are uncorrelated or have small correlation.*

Notice that, Assumption 1 is quite important in order to distinguish the input signals in the machine learning procedure. Another alternative consists of exciting the unknown system using a unique signal from a decision-makers and to perform multiple tests using the control inputs one by one to identify the control-dependent parameters independently.

Remark 33 *By identifying a suitable model for the unknown system based on*

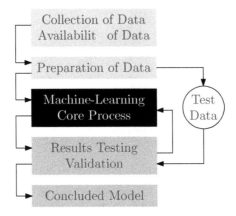

FIGURE 15.3
Machine-learning scheme.

machine learning techniques, it is therefore possible to implement model-based mean-field-type game solutions as introduced in Chapters 11 and 12. Thus, the data-driven mean-field-type game problem is solved.

In the following section, we introduce the general machine-learning philosophy regarding the treatment of data for training purposes.

15.2   Machine Learning Philosophy

Figure 15.3 presents the general scheme for the machine learning process, which consists in three main components as described next:

- **Data Treatment:** this procedure consists of identifying the data availability about the unknown system behavior. It is worthy to mention that the amount of data has considerable implications over the learning process, i.e., the more data and/or more tests h we have, a better learning can be performed.

  On the other hand, in the first step of the machine-learning procedure, it is necessary to prepare and organize data for the analysis. Within the preparation of data, the information is divided into two parts: (i) learning-oriented data, which is used in the machine-learning core; and (ii) testing-oriented data, which is used in order to evaluate the learning output. Then, data $\mathbb{D}$ is split into two sets.

- **Machine-Learning Core:** once data is ready for the analysis, an objective functional is defined. Such objective is commonly established by means of

TABLE 15.1

Summary of available data for machine-learning purposes.

		Initial Data $\mathbb{D}$	
Time	State	Decision-maker 1	Decision-maker 2
k	x_k	$u_{1,k}$	$u_{2,k}$
0	x_0^1	$u_{1,0}^1$	$u_{2,0}^1$
...	...	...	...
K	x_K^1	$u_{1,K}^1$	$u_{2,K}^1$
0	x_0^2	$u_{1,0}^2$	$u_{2,0}^2$
...	...	...	...
K	x_K^2	$u_{1,K}^2$	$u_{2,K}^2$
...	...	...	...
0	x_0^h	$u_{1,0}^h$	$u_{2,0}^h$
...	...	...	...
K	x_K^h	$u_{1,K}^h$	$u_{2,K}^h$

inputs (available from data and measurable in a real implementation) and outputs (main variable of interest). In the mean-field-type game problem, the inputs are given by the decision-makers' strategies or control inputs, and the output is given by the system state.

Learning techniques are implemented in order to obtain an estimated model that could return an appropriate output based on arbitrary inputs previously established onto the unknown system, and also taking into consideration the reaction of the system against the aforementioned inputs. This learning is executed by means of the learning-oriented data.

- **Results/Output:** after performing the learning procedure, it is necessary to quantify the performance of the obtained model. The learned-model quality assessment is determined by means of the testing-oriented data.

   Completing the machine-learning procedure, some estimated parameters are obtained and optimal control inputs can be computed by following the model-based semi-explicit or explicit solutions. Indeed, the evolution of the Riccati equations can be computed by using the estimated model with the learned parameters given by the machine learning procedure.

   In the following section, we present a strategy in order to execute the machine-learning-based estimation to perform the data-driven mean-field-type game.

15.3 Machine-Learning-Based (Linear Regression) Data-Driven Mean-Field-Type games

Let us consider a two-decision-maker scalar-valued state mean-field-type game problem involving an unknown stochastic system, see the scenario presented in Figure 15.2. We follow the machine-learning scheme presented in Figure 15.3 oriented to the data-driven mean-field-type game problem.

15.3.1 Availability of Data

For simplicity, let us consider a set of $h \in \mathbb{N}$ tests performed over the unknown system denoted by $\mathcal{H} = \{1, \ldots, h\}$. These tests are made by applying some control inputs $u_{i,k}$, for all the decision-makers $i \in \mathcal{N}$. Assumption 2 imposes conditions over the initial condition for all the tests.

Assumption 2 *The available data, denoted by $\mathbb{D}$, corresponds to the evolution of the system states with inputs $u_{i,k}$, for all $i \in \mathcal{N}$, from the same initial conditions $x_0 = x_0^\ell$ for all the tests $\ell \in \mathcal{H}$.*

It is defined a discrete-time horizon $[0..K]$ for each test. We assume that all the tests $\ell \in \mathcal{H}$ are made using the same time window. The collection of captured/available data for the test $\ell \in \mathcal{H}$ is given by

$$\mathbb{D}^\ell \triangleq \left(k, x_k^\ell, [u_{i,k}^\ell]_{i \in \mathcal{N}} \right), \quad k \in [0..K].$$

The overall available data set is given by $\mathbb{D} \triangleq \left(\mathbb{D}^\ell \right)_{\ell \in \mathcal{H}}$. Table 15.1 presents a summary of the available data about the behavior of the unknown system.

15.3.2 Preparation of Data

Considering the finite h measured values of x_k for each test at time instant k in the available data $\mathbb{D}$, it is possible to compute a histogram. Moreover, $\mathbb{D}$ allows computing a continuous Kernel-density function for the system state at time instant $t = k$, which is denoted by $m(t, x)$, or the Kernel-density as an histogram $m_{k,x}$. Either by means of the Kernel-density function or the computed histogram, one estimates the expected value trajectories for both the system state and control inputs as shown next.

The estimation of the expected value of the system states $\mathbb{E}x_k$ and the control input $\mathbb{E}u_{i,k}$, which are denoted by $\mathbb{E}y_k$ and $\mathbb{E}\hat{u}_{i,k}$, respectively; are computed by using the available data as follows:

$$\mathbb{E}y_k = \frac{1}{h} \sum_{\ell \in \mathcal{H}} x_k^\ell, \ \forall k \in [0, K], \tag{15.1a}$$

TABLE 15.2
Summary of prepared data for machine-learning purposes.

Preparation of Data (Machine-learning-oriented data)			
Next State	Expected State	Expected Decision-maker 1	Expected Decision-maker 2
x_{k+1}	$\mathbb{E}y_k$	$\mathbb{E}u_{1,k}$	$\mathbb{E}u_{2,k}$
x_1^1	$\mathbb{E}y_0$	$\mathbb{E}u_{1,0}$	$\mathbb{E}u_{2,0}$
...	...	...	...
x_{K+1}^1	$\mathbb{E}y_K$	$\mathbb{E}u_{1,K}$	$\mathbb{E}u_{2,K}$
x_1^2	$\mathbb{E}y_0$	$\mathbb{E}u_{1,0}$	$\mathbb{E}u_{2,0}$
...	...	...	...
x_{K+1}^2	$\mathbb{E}y_K$	$\mathbb{E}u_{1,K}$	$\mathbb{E}u_{2,K}$
...	...	...	...
x_1^h	$\mathbb{E}y_0$	$\mathbb{E}u_{1,0}$	$\mathbb{E}u_{2,0}$
...	...	...	...
x_{K+1}^h	$\mathbb{E}y_K$	$\mathbb{E}u_{1,K}$	$\mathbb{E}u_{2,K}$

$$\mathbb{E}\hat{u}_{i,k} = \frac{1}{h}\sum_{\ell\in\mathcal{H}} u_{i,k}^\ell, \ \forall i \in \mathcal{N}, \ k \in [0,K]. \qquad (15.1b)$$

Table 15.2 shows a summary of the preparation of data prior to performing the model estimation procedure based on machine learning.

15.3.3 Machine-Learning Core

Once data is prepared for the machine learning purposes by computing an estimation of the expected trajectories for the system states and the corresponding control inputs for all the decision-makers, it is necessary to postulate a dynamical structure that the unknown system might follow. In principle, this task is another problem, and machine learning can also be used in order to identify the most appropriate structure within a set of structure families.

In contrast, in this chapter we assume that the dynamical structure of the unknown system is known. Thus, we assume that the following dynamical structure is suitable:

$$x_{k+1} = b_0 + b_1 x_k + \bar{b}_1 \mathbb{E}x_k + \sum_{i\in\mathcal{N}} \left(b_{2i} u_{i,k} + \bar{b}_{2i}\mathbb{E}u_{i,k} \right) + \sigma w_k. \qquad (15.2)$$

Let us define the following deterministic system dynamics:

$$\mathbb{E}x_{k+1}^\ell = b_0^\ell + (b_1^\ell + \bar{b}_1^\ell)\mathbb{E}x_k^\ell + \sum_{i\in\mathcal{N}}(b_{2i}^\ell + \bar{b}_{2i}^\ell)\mathbb{E}u_{i,k}^\ell, \ \ell \in \mathcal{H}, \qquad (15.3)$$

$$= b_0^\ell + \hat{b}_1^\ell \mathbb{E}x_k^\ell + \sum_{i\in\mathcal{N}} \hat{b}_{2i}^\ell \mathbb{E}u_{i,k}^\ell, \ \ell \in \mathcal{H},$$

Remark 34 *The model in* (15.2) *works as a template for the machine-learning-based training purposes. It is important to highlight that the structure of the dynamics is also unknown and it is required to test multiple possibilities in order to learn the appropriate model. In this chapter, we assume that the dynamics of the unknown system is linear.*

In the previous dynamics, $\mathbb{E}x^\ell_{k+1}$, $\mathbb{E}x^\ell_k$ and $\mathbb{E}u^\ell_{i,k}$ are known for all decision-makers $i \in \mathcal{N}$, all time steps $k \in [0..K]$, and all the tests $\ell \in \mathcal{H}$. However, the parameters b_0, $\hat{b}^\ell_1$, and $\hat{b}^\ell_{2i}$, for all $i \in \mathcal{N}$, are unknown and should be trained by means of machine learning. To this end, let us consider the following variables:

$$
\Phi^\ell = \begin{bmatrix}
1 & \mathbb{E}y^\ell_0 & \mathbb{E}u^\ell_{1,0} & \cdots & \mathbb{E}u^\ell_{n,0} \\
1 & \mathbb{E}y^\ell_1 & \mathbb{E}u^\ell_{1,1} & \cdots & \mathbb{E}u^\ell_{n,1} \\
\vdots & \vdots & \vdots & \ddots & \vdots \\
1 & \mathbb{E}y^\ell_K & \mathbb{E}u^\ell_{1,K} & \cdots & \mathbb{E}u^\ell_{n,K}
\end{bmatrix},
$$

$$
b^\ell = \begin{bmatrix} b^\ell_0 & \hat{b}^\ell_1 & \hat{b}^\ell_{2,1} & \cdots & \hat{b}^\ell_{2,n} \end{bmatrix}^\mathsf{T},
$$

$$
\mathbb{E}y^\ell = \begin{bmatrix} \mathbb{E}y^\ell_1 & \mathbb{E}y^\ell_2 & \cdots & \mathbb{E}y^\ell_{K+1} \end{bmatrix}^\mathsf{T},
$$

where $\Phi^\ell \in \mathbb{R}^{K \times n+2}$, $b^\ell \in \mathbb{R}^{n+2}$, and $\mathbb{E}y^\ell \in \mathbb{R}^K$. Therefore, the optimization problem that should be solved is shown below. We pursue to reduce the gap between the known behavior of the system and the proposed structured stochastic model:

$$
\underset{b^\ell}{\text{minimize}} \; \left\langle \Phi^\ell b^\ell - \mathbb{E}y^\ell, \Phi^\ell b^\ell - \mathbb{E}y^\ell \right\rangle, \; \forall \ell \in \mathcal{H}. \tag{15.4}
$$

Once the parameters b^ℓ_0, $\hat{b}^\ell_1$, $\hat{b}^\ell_{2,1}, \ldots \hat{b}^\ell_{2,n}$ are identified in the deterministic equation (15.3), then we proceed by identifying the individual parameters in the stochastic model. Let us consider then the following subtraction:

$$
\begin{aligned}
x^\ell_{k+1} - \mathbb{E}x^\ell_{k+1} &= b^\ell_0 + b^\ell_1 x^\ell_k + \bar{b}^\ell_1 \mathbb{E}x^\ell_k + \sum_{i \in \mathcal{N}} b^\ell_{2i} u^\ell_{i,k} + \bar{b}^\ell_{2i} \mathbb{E}u^\ell_{i,k} \\
&\quad - b^\ell_0 - (b^\ell_1 + \bar{b}^\ell_1)\mathbb{E}x^\ell_k - \sum_{i \in \mathcal{N}}(b^\ell_{2i} + \bar{b}^\ell_{2i})\mathbb{E}u^\ell_{i,k}, \; \ell \in \mathcal{H}, \\
&= b^\ell_1 (x^\ell_k - \mathbb{E}x^\ell_k) + \sum_{i \in \mathcal{N}} b^\ell_{2i}(u^\ell_{i,k} - \mathbb{E}u^\ell_{i,k}), \; \ell \in \mathcal{H}.
\end{aligned}
$$

Table 15.3 presents the necessary data in order to identify the new parameters of the model. Therefore, let us introduce the following variables:

$$
\tilde{\Phi}^\ell = \begin{bmatrix}
(x^\ell_0 - \mathbb{E}y^\ell_0) & (u^\ell_{1,0} - \mathbb{E}u^\ell_{1,0}) & \cdots & (u^\ell_{n,0} - \mathbb{E}u^\ell_{n,0}) \\
(x^\ell_1 - \mathbb{E}y^\ell_1) & (u^\ell_{1,1} - \mathbb{E}u^\ell_{1,1}) & \cdots & (u^\ell_{n,1} - \mathbb{E}u^\ell_{n,1}) \\
\vdots & & \ddots & \vdots \\
(x^\ell_K - \mathbb{E}y^\ell_K) & (u^\ell_{1,K} - \mathbb{E}u^\ell_{1,K}) & \cdots & (u^\ell_{n,K} - \mathbb{E}u^\ell_{n,K})
\end{bmatrix},
$$

TABLE 15.3
Second summary of prepared data for machine-learning purposes.

Preparation of Data (Machine-learning-oriented data)		
Next State Difference	Expected Diff. Decision-maker 1	Expected Diff. Decision-maker 2
$\dfrac{x_{k+1} - \mathbb{E}y_{k+1}}{x_1^1 - \mathbb{E}y_1}$	$\dfrac{u_{1,k} - \mathbb{E}u_{1,k}}{u_{1,0} - \mathbb{E}u_{1,0}}$	$\dfrac{u_{2,k} - \mathbb{E}u_{2,k}}{u_{2,0} - \mathbb{E}u_{2,0}}$
$\ldots$	$\ldots$	$\ldots$
$\dfrac{x_{K+1}^1 - \mathbb{E}y_{K+1}}{x_1^2 - \mathbb{E}y_1}$	$\dfrac{u_{1,K} - \mathbb{E}u_{1,K}}{u_{1,0} - \mathbb{E}u_{1,0}}$	$\dfrac{u_{2,K} - \mathbb{E}u_{2,K}}{u_{2,0} - \mathbb{E}u_{2,0}}$
$\ldots$	$\ldots$	$\ldots$
$x_{K+1}^2 - \mathbb{E}y_{K+1}$	$u_{1,K} - \mathbb{E}u_{1,K}$	$u_{2,K} - \mathbb{E}u_{2,K}$
$\ldots$	$\ldots$	$\ldots$
$x_1^h - \mathbb{E}y_1$	$u_{1,0} - \mathbb{E}u_{1,0}$	$u_{2,0} - \mathbb{E}u_{2,0}$
$\ldots$	$\ldots$	$\ldots$
$x_{K+1}^h - \mathbb{E}y_{K+1}$	$u_{1,K} - \mathbb{E}u_{1,K}$	$u_{2,K} - \mathbb{E}u_{2,K}$

$$\tilde{b}^\ell = \begin{bmatrix} b_1^\ell & b_{2,1}^\ell & \cdots & b_{2,n}^\ell \end{bmatrix}^\top,$$
$$\tilde{y}^\ell = \begin{bmatrix} (x_1^\ell - \mathbb{E}y_1^\ell) & (x_2^\ell - \mathbb{E}y_2^\ell) & \cdots & (x_{K+1}^\ell - \mathbb{E}y_{K+1}^\ell) \end{bmatrix}^\top,$$

and the parameters $b_1^\ell, b_{2,1}^\ell, \ldots, b_{2,n}^\ell$, for all $\ell \in \mathcal{H}$, are found by solving the optimization problem:

$$\underset{\tilde{b}^\ell}{\text{minimize}} \left\langle \tilde{\Phi}^\ell \tilde{b}^\ell - \tilde{y}^\ell, \tilde{\Phi}^\ell \tilde{b}^\ell - \tilde{y}^\ell \right\rangle, \ \forall \ell \in \mathcal{H}, \tag{15.5}$$

where $\tilde{\Phi}^\ell \in \mathbb{R}^{K \times n+1}$, $\tilde{b}^\ell \in \mathbb{R}^{n+1}$, and $\tilde{y}^\ell \in \mathbb{R}^K$. With the optimal solution of the former optimization problem in (15.4), we obtain the estimation for b_0^ℓ. With the solution of the latter problem we have the estimated terms b_1^ℓ, and b_{2j}^ℓ, for all $j \in \mathcal{N}$ and $\ell \in \mathcal{H}$. Moreover, the other parameters for the system are computed as follows:

$$\bar{b}_1^\ell = \hat{b}_1^\ell - b_1^\ell, \ \forall \ell \in \mathcal{H},$$
$$\bar{b}_{2j}^\ell = \hat{b}_{2j}^\ell - b_{2j}^\ell, \ \forall j \in \mathcal{N}, \ \ell \in \mathcal{H}.$$

Finally, it is necessary to estimate the σ term. To this end, we compute the following trajectory denoted by ξ^ℓ:

$$\xi^\ell = x_{k+1}^\ell - b_0^\ell - b_1^\ell x_k^\ell - \bar{b}_1^\ell \mathbb{E}x_k^\ell - \sum_{j \in \mathcal{N}} \left(b_{2j}^\ell u_{j,k}^\ell + \bar{b}_{2j}^\ell \mathbb{E}u_{j,k}^\ell \right),$$

and

$$\sigma^\ell = \sqrt{\text{var}(\xi^\ell)}, \ \forall \ell \in \mathcal{H}.$$

In order to improve the accuracy of the estimation, we obtain as many parameters as tests h from the set $\mathcal{H}$. We compute the average of such terms to obtain a final estimation of the parameters, i.e.,

$$b_0 = \frac{1}{h} \sum_{\ell \in \mathcal{H}} b_0^\ell, \qquad\qquad b_1 = \frac{1}{h} \sum_{\ell \in \mathcal{H}} b_1^\ell, \qquad\qquad \text{(15.6a)}$$

$$\bar{b}_1 = \frac{1}{h} \sum_{\ell \in \mathcal{H}} \bar{b}_1^\ell, \qquad\qquad b_{2j} = \frac{1}{h} \sum_{\ell \in \mathcal{H}} b_{2j}^\ell, \ \forall j \in \mathcal{N}, \qquad \text{(15.6b)}$$

$$\bar{b}_{2j} = \frac{1}{h} \sum_{\ell \in \mathcal{H}} \bar{b}_{2j}^\ell, \ \forall j \in \mathcal{N}, \qquad \sigma = \frac{1}{h} \sum_{\ell \in \mathcal{H}} \sigma^\ell. \qquad \text{(15.6c)}$$

Next section discusses how well the machine-learning-based estimation performs. In case the results are not appropriate, it is necessary to enlarge the training data to have more information about the behavior or the system, or to evaluate if the chosen structure for the dynamical model is appropriate.

15.4   Error and Performance Metrics

After having determined the machine-learning-based parameters, it is indispensable to quantify how well the estimation fits the actual dynamical behavior of the system. To evaluate the performance of the machine-learning output, we use either the test-oriented data or new generated data over the unknown system. This assessment consists of comparing the actual output x_{k+1} and the estimated output that is generated from the learned dynamical model y_{k+1}, i.e., by means of an error $e_k = x_k - y_k$ as follows [144]:

- **Mean Absolute Error:** denoted by MAE and given by

$$\mathrm{MAE}(k, x, y) = \frac{1}{K} \sum_{k \in [0, K] \cap \mathbb{Z}_{\geq 0}} |x_k - y_k|$$

- **Mean Squared Error:** denoted by MSE and given by

$$\mathrm{MSE}(k, x, y) = \frac{1}{K} \sum_{k \in [0, K] \cap \mathbb{Z}_{\geq 0}} (x_k - y_k)^2$$

- **Root Mean Squared Error:** denoted by RMSE and given by

$$\mathrm{RMSE}(k, x, y) = \sqrt{\frac{1}{K} \sum_{k \in [0, K] \cap \mathbb{Z}_{\geq 0}} (x_k - y_k)^2}$$

TABLE 15.4
Initial data $\mathbb{D}$ for machine-learning purposes.

Data	Test		k	State $[x_{k+1}]$	State $[x_k]$	Control input $[u_{1,k}]$	Control input $[u_{2,k}]$
0	0	tests 0	0.0	-0.286958	0.399464	-0.482293	0.184236
1	1	tests 0	1.0	2.829336	-0.286958	-1.562252	1.141813
2	2	tests 0	2.0	7.989195	2.829336	2.261716	0.538175
3	3	tests 0	3.0	13.729748	7.989195	0.376886	1.422338
4	4	tests 0	4.0	16.581772	13.729748	1.386673	1.154714
...	...	...	...	...	...	...	...
99995	99995	tests 99	995.0	-166.167656	-169.362757	-4.046709	0.745054
99996	99996	tests 99	996.0	-170.465694	-166.167656	-6.153590	-0.481266
99997	99997	tests 99	997.0	-169.261156	-170.465694	-5.729804	0.161896
99998	99998	tests 99	998.0	-169.365552	-169.261156	-5.942358	0.498624
99999	99999	tests 99	999.0	-174.892673	-169.365552	-4.110319	-2.363684

TABLE 15.5
Correlation among the decision-makers control-inputs.

	Control input $u_{1,k}$	Control input $u_{2,k}$
Control input $u_{1,k}$	1	-0.00119
Control input $u_{2,k}$	-0.00119	1

15.5 Numerical Example

In order to validate the effectiveness of the proposed machine-learning-based approach, we use a known model (real system). Then, such system is treated as unknown to generate massive data, prepare and organize it. We apply the machine learning technique in order to identify a suitable model. Thus, the learned model can be compared with the real one. Let us consider the "unknown" system with two decision-makers as in (15.2) and with the following parameters:

$$b_0^{\text{real}} = 1, \qquad b_1^{\text{real}} = 0.7,$$
$$\bar{b}_1^{\text{real}} = 0.5, \qquad b_{21}^{\text{real}} = 1.5,$$
$$\bar{b}_{21}^{\text{real}} = 2, \qquad b_{22}^{\text{real}} = 2.5,$$
$$\bar{b}_{22}^{\text{real}} = 3, \qquad \sigma^{\text{real}} = 1.5$$

Now, we perform 100 different tests $\mathcal{H} = \{1, \ldots, 100\}$ in a time windows $[0..1000]$, by applying the following control input signals $u_{1,k}$ and $u_{2,k}$. Table

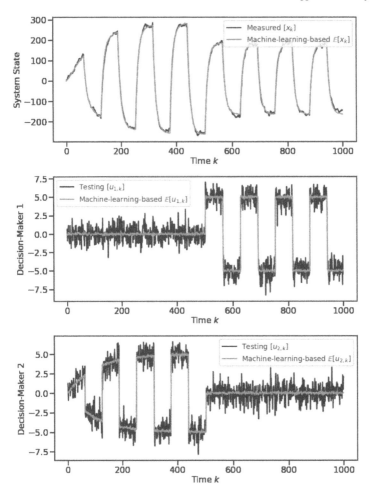

FIGURE 15.4
Machine-learning-based expected values in comparison with a particular test.
This figure corresponds to Test 7.

15.4 shows the data $\mathbb{D}$ about the dynamical behavior of the "unknown" system.
This data is generated by using the control inputs as presented in Figure 15.4
corresponding to the test $7 \in \mathcal{H}$.

According to Assumption 1, it is necessary to guarantee that $u_{1,k}$ and $u_{2,k}$
are highly uncorrelated. To this end, the control inputs are applied to the
system in a sequential manner as shown in Figure 15.4. The corresponding
correlation matrix is presented in Table 15.5.

Continuing with the preparation of data, we compute the empirical expected trajectory for the system state and the control inputs following (15.1)
using all data for the 100 tests from $\mathcal{H}$.

TABLE 15.6
First preparation of data for machine-learning purposes.

State $\mathbb{E}[x_{k+1}]$	State $\mathbb{E}[x_k]$	Control input $\mathbb{E}[u_{1,k}]$	Control input $\mathbb{E}[u_{2,k}]$
0.972331	0.144576	-0.089594	0.062337
2.027029	0.972331	-0.160471	0.088414
3.139136	2.027029	-0.098340	0.038366
4.890869	3.139136	0.060389	0.222316
6.706799	4.890869	0.124079	0.156796
...	...	...	...
-163.861008	-163.545049	-4.990320	-0.040719
-164.096483	-163.861008	-5.114689	-0.024985
-164.217920	-164.096483	-5.069535	-0.014337
-164.451868	-164.217920	-4.905714	-0.010529
-164.500045	-164.451868	-4.892615	0.022669

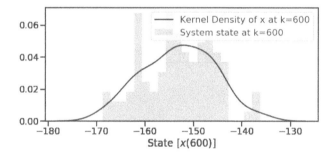

FIGURE 15.5
Data-based probability measure of x at $k = 600$.

From data, and by grouping according to the discrete time instant k, one obtains a kernel density estimation for either the system state or the control inputs. As an example, Figure 15.5 presents the kernel density function m_k (histogram) and $m(t, x)$ (continuous density estimation) for time instant $k = 600$. Taking into consideration the data $\mathbb{D}$ in Tables 15.4 and 15.6, and defining as inputs

$$\mathbb{E}x_k^\ell, \text{ and } \mathbb{E}u_{i,k}^\ell, \ \forall i \in \mathcal{N},$$

and as output

$$\mathbb{E}x_{k+1}^\ell,$$

the parameters b_0^ℓ, $\hat{b}_1^\ell$, and $\hat{b}_{2i}^\ell$, for all $i \in \mathcal{N}$ are learned. Afterward, the difference variables

$$(x_k^\ell - \mathbb{E}x_k^\ell), \text{ and } (u_{i,k}^\ell - \mathbb{E}u_{i,k}^\ell), \ \forall i \in \mathcal{N},$$

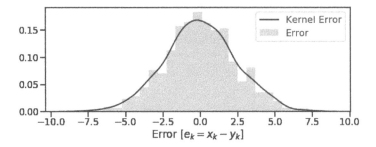

FIGURE 15.6
Error distribution with $e_k = x_k - y_k$.

TABLE 15.7
Error metrics between a real trajectory and machine-learning-based estimated
trajectories

Error Metrics	Error Value
Mean Absolute Error	2.0182
Mean Squared Error	6.2670
Root Mean Squared Error	2.5034

are taken as inputs; and
$$(x_{k+1}^\ell - \mathbb{E}x_{k+1}^\ell)$$
is taken as an output in order to learn the parameters b_1^ℓ and b_{2i}^ℓ, for all
$i \in \mathcal{N}$. This previously described learning procedure is performed for all the
tests $\ell \in \mathcal{H}$. Thus, one obtains a distribution for all the learned parameters
as presented in Figure 15.7.

Finally, following the estimation based on the mean value of the parameters
distributions as in (15.6), once obtains the parameters:

$$b_0^{\text{learned}} = 0.9929, \qquad b_1^{\text{learned}} = 0.8996,$$
$$\bar{b}_1^{\text{learned}} = 0.00072, \qquad b_{21}^{\text{learned}} = 1.4945,$$
$$\bar{b}_{21}^{\text{learned}} = 1.9968, \qquad b_{22}^{\text{learned}} = 2.5059,$$
$$\bar{b}_{22}^{\text{learned}} = 1.9770, \qquad \sigma^{\text{learned}} = 1.6932,$$

In order to evaluate the machine-learning-based parameter accuracy, the
test-oriented data may be used to establish a comparison, or a new trajectory
can be compared. Figure 15.6 presents the distribution of the error between
the trajectories from a real test and the machine-learning-based estimated
trajectory. Hence, error quantification is presented in Table 15.7.

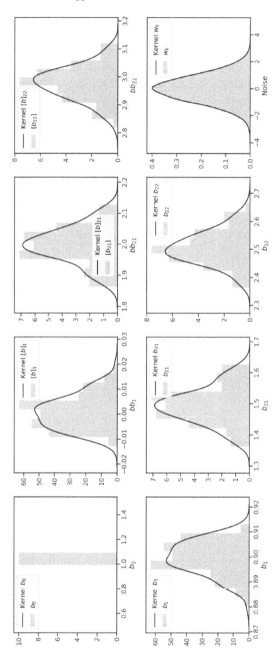

FIGURE 15.7
Machine-learning-based parameters.

This chapter has introduced simple machine-learning techniques (linear regression) in order to solve a data-driven mean-field-type game problem. Notice that, once a machine-learning-based model is obtained from massive data describing the dynamical behavior of the system, then any game theoretical solution can be implemented, e.g., non-cooperative, cooperative, co-opetitive, sequential, mutual support, among others.

Hence, we have presented a simple scalar-valued numerical example. However, the machine-learning technique is easily extensible to vector and even matrix-valued cases.

16

Applications

After having studied all the risk-aware approaches by means of the mean-field-type game theory in both continuous and discrete-time context, this chapter is devoted to present several engineering applications. In particular we address the following applications:

- Water distribution problem

- Micro-grid energy storage problem in a smart grid

- Continuous stirred tank reactor problem

- Mechanism design problem

- Evacuation building problem

- Coronavirus spread problem

16.1 Water Distribution Systems

This section is devoted to discuss the water distribution networks. These kinds of systems involve uncertainties associated mainly to the existing demands of the water resource. Moreover, the uncertainties might also be related to imperfect measurements about the reservoir levels. We address two different examples. First, we study a five-tank benchmark involving some constraints and two decision-makers modifying two different sets of control inputs, which are given by flow. On the other hand, we present the Barcelona drinking water network and we design a risk-aware model predictive controller.

16.1.1 Five-Tank Water System

We first apply the presented results in Chapter 10 for a constrained engineering problem. Let us consider the multi-variable system presented in Figure 16.1, which consists of five tanks (reservoirs) involving two decision-makers $\mathcal{N} = \{1, 2\}$. This system is a modification of the reported four-tank benchmark

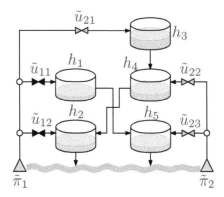

FIGURE 16.1
Illustrative example. Five-tank benchmark involving two players and two coupled input constraints. Players 1 and 2 correspond to the black and gray colors,
respectively.

in [145] by including a new tank and a stochastic process into the problem.
The control inputs corresponding to the decision-makers are given by:

$$\tilde{u}_1 = (\tilde{u}_{11} \ \tilde{u}_{12})^\top,$$
$$\tilde{u}_2 = (\tilde{u}_{21} \ \tilde{u}_{22} \ \tilde{u}_{23})^\top.$$

There is coupling between the decision-makers for the imposed constraints.
Next, we introduce linear dynamics corresponding to the system.

Let h_i denotes the level of the ith tank, whose dynamics are simply expressed by a mass balance equations as inflows and outflows, i.e.,

$$\dot{h}_i = q_i^{\text{in}} - q_i^{\text{out}}, \ \forall i \in \{1, \ldots, 5\},$$

where the outflow is a function of the current tank level h_i. In this case, we
consider a linear relationship with respect to h_i, i.e., $q_i^{\text{out}} = \varrho_i h_i$, and the
inflow q_i^{in} is given by either a control input, or by the outflow from another
previous tank. Thus, the whole deterministic system can be modeled by the
following differential equation:

$$\begin{pmatrix} \dot{h}_1 \\ \dot{h}_2 \\ \dot{h}_3 \\ \dot{h}_4 \\ \dot{h}_5 \end{pmatrix} = \underbrace{\begin{pmatrix} -\varrho_1 & 0 & 0 & 0 & 0 \\ 0 & -\varrho_2 & 0 & \varrho_4 & 0 \\ 0 & 0 & -\varrho_3 & 0 & 0 \\ 0 & 0 & \varrho_3 & -\varrho_4 & 0 \\ 0 & 0 & 0 & 0 & -\varrho_5 \end{pmatrix}}_{A} \begin{pmatrix} h_1 \\ h_2 \\ h_3 \\ h_4 \\ h_5 \end{pmatrix}$$

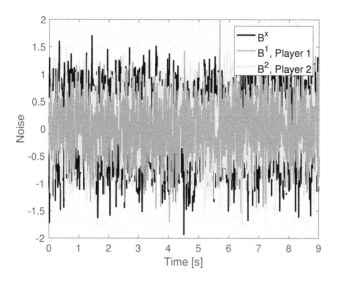

FIGURE 16.2
Brownian motions.

$$+ \underbrace{\begin{pmatrix} 1 & 0 \\ 0 & 1 \\ 0 & 0 \\ 0 & 0 \\ 0 & 0 \end{pmatrix}}_{B_1} \tilde{u}_1 + \underbrace{\begin{pmatrix} 0 & 0 & 0 \\ 0 & 0 & 0 \\ 1 & 0 & 0 \\ 0 & 1 & 0 \\ 0 & 0 & 1 \end{pmatrix}}_{B_2} \tilde{u}_2,$$

where $\varrho_i > 0$, for all $i \in \{1, \ldots, 5\}$, and dynamics can be compacted as

$$\dot{h} = Ah + \sum_{j \in \mathcal{N}} B_j \tilde{u}_j.$$

Once the system dynamics have been presented, we proceed next to transform the problem into the suitable form including the auxiliary variables to consider multiple coupled constraints.

It is desired to stabilize all the tank levels to the references $h^d = (2 \quad 5 \quad 1 \quad 2 \quad 3)^\top$ while satisfying the coupled input constraints

$$\begin{cases} \tilde{u}_{11} + \tilde{u}_{12} + \tilde{u}_{21} = \tilde{\pi}_1, \\ \tilde{u}_{22} + \tilde{u}_{23} = \tilde{\pi}_2, \end{cases} \tag{16.1}$$

with the fixed resources given by $\tilde{\pi}_1 = 7$, and $\tilde{\pi}_2 = 10$. Thus, the considered cost functional for each decision-maker is given by $L_i(x, u)$ in (10.4) (Chapter 10) where $x = h - h^d$ denotes the error between the system state and the

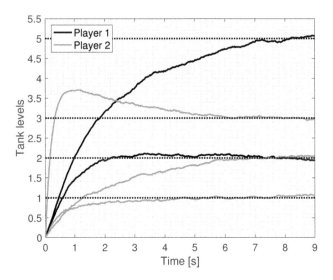

FIGURE 16.3
Evolution of the system states.

desired levels. Furthermore, we consider the following prioritization using the weight parameters:

$$Q_i = 1000\mathbf{I}_{[5]}, \qquad \bar{Q}_i = 0.5Q_i,$$
$$R_1 = \mathbf{I}_{[2]}, \qquad R_2 = \mathbf{I}_{[3]}, \qquad \bar{R}_i = 0.5R_i,$$

for all $i \in \{1, 2\}$. It yields that the error dynamics are

$$(\dot{h} - \dot{h}^d) = A(h - h^d) + \sum_{j \in \mathcal{N}} B_j(\tilde{u}_j - \tilde{u}_j^d).$$

Therefore, taking the error dynamics with

$$u_j = \tilde{u}_j - \tilde{u}_j^d,$$

for all $j \in \mathcal{N}$, and adding a Brownian noise B_r^x to it in order to model a stochastic behavior, which can be associated with an imperfect model or to a noisy inflow/outflow, we obtain the dynamics

$$dx = \left(Ax + \sum_{j \in \mathcal{N}} B_j u_j \right) dt + S dB_r^x,$$

with $S = 0.5\mathbf{I}_{[5]}$. Figure 16.2 shows the considered noise corresponding to the system state and to the two decision-makers. Hence, it can be seen that according to the proposed approach, $\mathcal{C} = \{1, 2\}$ and with the vectors

$$u^1 = (u_{11} \ u_{12} \ u_{21})^{\top},$$

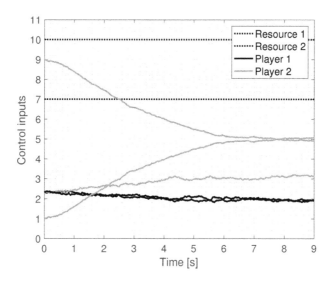

FIGURE 16.4
Evolution of the optimal control inputs for the players.

$$u^2 = (u_{22} \ u_{23})^\top.$$

Getting (10.2) and (10.3). Thus, the same model as in (10.6) is obtained with

$$z = (x_1 \ \dots \ x_5 \ u_1 \ \dots \ u_5)^\top,$$

$\tilde{\bar{A}} = \mathbf{0}$, and the new control inputs

$$v_1 = (v_{11} \ v_{12})^\top,$$
$$v_2 = (v_{21} \ v_{22} \ v_{23})^\top.$$

Also, considering the following matrices:

$$\tilde{B}_1 = \begin{pmatrix} 0 & 0 \\ 2 & -1 \\ -1 & 2 \\ -1 & -1 \\ 0 & 0 \\ 0 & 0 \end{pmatrix}, \quad \tilde{B}_2 = \begin{pmatrix} 0 & 0 & 0 \\ -1 & 0 & 0 \\ -1 & 0 & 0 \\ 2 & 0 & 0 \\ 0 & 1 & -1 \\ 0 & -1 & 1 \end{pmatrix},$$

the parameters $\beta^1 = \beta^2 = 0.5$, and the following constraints:

$$\left\langle (u_{11} \ u_{12} \ u_{21})^\top, \mathbb{1}_{[3]} \right\rangle = \left\langle u^1, \mathbb{1}_{[3]} \right\rangle = 0,$$
$$\left\langle (u_{22} \ u_{23})^\top, \mathbb{1}_{[2]} \right\rangle = \left\langle u^2, \mathbb{1}_{[2]} \right\rangle = 0.$$

Using the transformed dynamics, the mean-field-type game problem can be stated and solved by means of the proposed approach. Figure 16.3 presents the evolution of the system states where we can observe the variance reduction. On the other hand, Figure 16.4 shows the evolution of the optimal flows. Notice that, the imposed constraints are satisfied for all the time. Figure 16.4 also presents the sum of the flows along the time, which are equal to 10 and 7 as desired.

The role of the mean-field-type game approach consists in taking into consideration risk terms that mitigates the effects of the perturbations presented in Figure 16.2. Next, we continue presenting engineering application with the Barcelona drinking water system.

16.1.2   Barcelona Drinking Water Distribution Network

We present an engineering large-scale application in order to illustrate the performance of the mean-field-type MPC controller and the influence that the prioritization weights have over the variance minimization. We apply the mean-field-type MPC controller to the Large-Scale WDN of Barcelona (see Figure 16.5, taken from [146]), which has been widely used as a large-scale case study for the analysis of optimization-based controllers, e.g., [146, 147]. On the other hand, a detailed model of this network has been reported in [136] where matrices for the system are presented. The author in [135] shows the periodicity of the demands. In this case study $n_x = 17$, $n_u = 61$, $n_d = 25$, and $c = 11$. We present a comparison of the performance of two stochastic MFT-MPC controllers considering different noise-variance parameter S, and with respect to the fully deterministic case, i.e., with $S = 0$.

We present the novel MFT-MPC controller applied to the WDN by considering the following cost functionals:

$$g(\xi_{k+N|k}) = \langle V\xi_{k+N|k}, \xi_{k+N|k} \rangle,$$
$$h(\xi_{\ell|k}, u_{\ell|k}) = \langle Q\xi_{\ell|k}, \xi_{\ell|k} \rangle + \langle Y(\alpha_1 + \alpha_{2\ell}), u_{\ell|k} \rangle + \langle R\Delta u_{\ell|k}, \Delta u_{\ell|k} \rangle,$$

and functionals g_v and h_v as

$$g_v(x_{k+N|k}, \mathbb{E}[x_{k+N|k}]) = \langle \bar{V}(x_{k+N|k} - \mathbb{E}[x_{k+N|k}]), x_{k+N|k} - \mathbb{E}[x_{k+N|k}] \rangle,$$
$$h_v(x_{\ell|k}, u_{\ell|k}, \mathbb{E}[x_{\ell|k}], \mathbb{E}[u_{\ell|k}]) = \langle \bar{R}(u_{\ell|k} - \mathbb{E}[u_{\ell|k}]), u_{\ell|k} - \mathbb{E}[u_{\ell|k}] \rangle$$
$$+ \langle \bar{Q}(x_{\ell|k} - \mathbb{E}[x_{\ell|k}]), x_{\ell|k} - \mathbb{E}[x_{\ell|k}] \rangle,$$

where $\xi \in \mathbb{R}^{n_x}$ denotes an auxiliary deterministic variable used to establish a constraint over the systems states $x \in \mathbb{R}^{n_x}$ above a security level that is denoted by $x_s \in \mathbb{R}^{n_x}$, and $\alpha_1 \in \mathbb{R}^{n_u}$ penalizes the cost of water and $\alpha_{2k} \in \mathbb{R}^{n_u}$ is a time-varying parameter associated with the costs of energy to operate the actuators in the DWN.

The costs functional also penalizes the slew rate in terms of

$$\Delta u_{k+\ell} = u_{k+\ell} - u_{k+\ell-1}, \ \forall \ell \in [0, N-1] \cap \mathbb{Z}_{>0}$$

in order to guarantee a smooth operation of the actuators in the system. Thus, the optimization problem behind the MFT-MPC is given by:

$$\underset{u_{k|k}, \ldots, u_{k+N-1|k}}{\text{minimize}} \quad \mathbb{E}\left[L(u)\right], \tag{16.2a}$$

subject to

$$x_{\ell+1|k} = Ax_{\ell|k} + Bu_{\ell|k} + B_d d_{\ell|k} + Sw_{\ell|k}, \tag{16.2b}$$
$$\mathbb{E}[u_{\ell|k}] \in \mathbb{U}_k^c, \tag{16.2c}$$
$$\mathbb{E}[x_{j|k}] \in \mathbb{X}_k^c, \tag{16.2d}$$
$$x_{k|k} \triangleq x_k, \tag{16.2e}$$

for all $\ell \in [k..k+N-1]$, and $j \in [k..k+N]$. In addition, the feasible sets for the control inputs and system states are given by

$$\mathbb{U}_k = \mathbb{U}^1 \cap \mathbb{U}_k^2, \tag{16.3a}$$
$$\mathbb{U}^1 \triangleq \left\{\mathbb{E}[u] \in \mathbb{R}^{n_u} \,\middle|\, \underline{u} \leq \mathbb{E}[u] \leq \bar{u}\right\}, \tag{16.3b}$$
$$\mathbb{U}_k^2 \triangleq \left\{\mathbb{E}[u] \in \mathbb{R}^{n_u} \,\middle|\, E_u \mathbb{E}[u] + E_d \mathbb{E}[B_d d_k + Sw_k] \leq \phi\right\}, \tag{16.3c}$$
$$\mathbb{X}_k = \mathbb{X}^1 \cap \mathbb{X}_k^2, \tag{16.3d}$$
$$\mathbb{X}^1 \triangleq \left\{\mathbb{E}[x] \in \mathbb{R}^{n_x} \,\middle|\, \underline{x} \leq \mathbb{E}[x] \leq \bar{x}\right\}, \tag{16.3e}$$
$$\mathbb{X}_k^2 \triangleq \left\{\mathbb{E}[x] \in \mathbb{R}^{n_x} \,\middle|\, \mathbb{E}[x] \geq x_s - \xi_k\right\}, \tag{16.3f}$$

where $E_u \in \mathbb{R}^{c \times n_u}$ and $E_d \in \mathbb{R}^{c \times n_d}$ allow describing $c \in \mathbb{N}$ constraints associated with the required mass balance in the WDN for all the joint nodes, and $\phi \in \mathbb{R}_{>0}^c$ establishes a tolerance for the mass balance constraint assigned by the company in charged of the operation of the network. Besides, notice that, ϕ also prevents $\mathbb{U}^c$ to be empty. In all the cases, the sets $\mathbb{X}^c$ and $\mathbb{U}^c$ have been conveniently reduced by a ten percent with respect to $\mathbb{X}$ and $\mathbb{U}$. Such reduction is associated with the parameters S we are considering as explained next.

We consider three scenarios involving different noise-variance terms as follows:

(1) **MFT-MPC:** Considering the following weight parameters:

$$Y = I_{n_u}, \quad R = 0.001 I_{n_u}, \quad Q = 0.01 I_{n_x}, \quad \bar{Q} = 0.01 I_{n_x}, \quad \bar{R} = 0.05 I_{n_u},$$

and a variance term $S = S_1$ for the noise w given by $S_1 = 0.1 \cdot \text{diag}(\hat{d})$, where $\hat{d} = \max\{d_k : k \in [1..24]\}$.

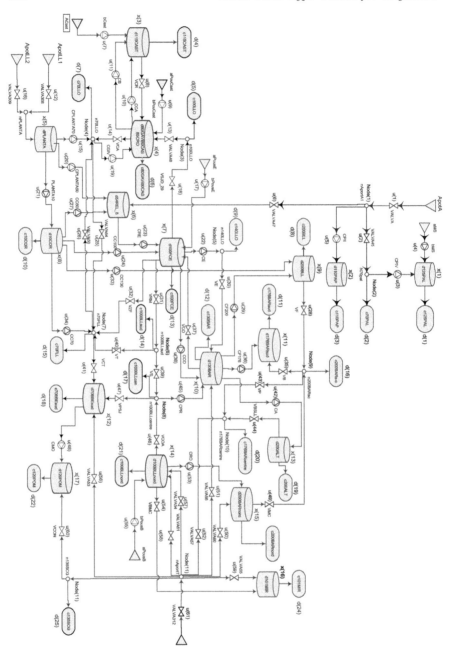

FIGURE 16.5
Water distribution network.

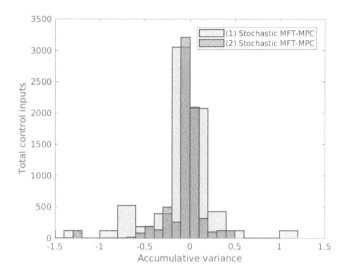

FIGURE 16.6
Control inputs variance comparison for the two different MFT-MPC controllers.

(2) MFT-MPC: Considering the same parameters $Y, R, Q, \bar{Q}$, and $\bar{R}$ as in Scenario 1, and with noise-variance term $S = S_2$ where $S_1 = 1.4S_2$.

(3) Deterministic MPC: Considering the same parameters $Y, R, Q, \bar{Q}$, and $\bar{R}$ as in Scenario 1, and with noise-variance term $S = 0$.

Figure 16.7 presents the evolution of the system states x_4, x_{11}, and x_{12}, and the evolution of the optimal control inputs u_{36}, u_{42}, and u_{54}. These system states have been selected since the corresponding tanks involve several inputs and outputs, nodes and demands. For instance, state x_4 involves the control inputs u_9, u_{13}, u_{19}, u_{14}, u_{11}, u_{10}, and u_8; the nodes "n100LLO" and "n70LLO", and the demands d_5, d_6, and d_7. The results presented in Figure 16.7 correspond to the deterministic MPC, and the MFT-MPC controllers corresponding to Scenarios 1 and 2 with different noise-variance parameters. It can be seen that the variance of the system states are similar for the two different MFT-MPC controllers.

However, the control inputs variance for the first MFT-MPC with S_1 and for the second MFT-MPC with S_2 are different. Regarding the variance minimization, Figure 16.6 shows the comparison between the accumulative control inputs to evidence the variance for the two MFT-MPC controllers. It is observed that the variance with S_1 is greater than the variance with S_2 as expected due to the fact $S_1 > S_2$.

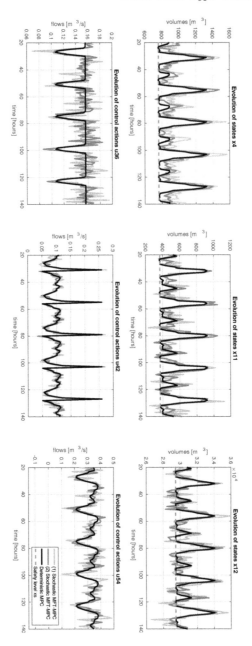

FIGURE 16.7
Results corresponding to the proposed stochastic MFT-MPC controllers for
Scenarios 1 and 2; and behavior of the deterministic MPC controller.

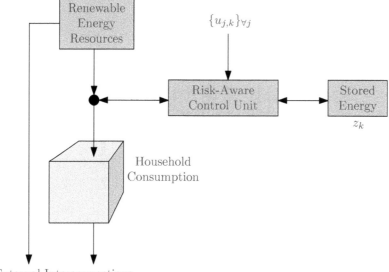

FIGURE 16.8
General scheme of the micro-grid involving energy storage. (Adapted from [1].)

16.2 Micro-grid Energy Storage

Let us consider a simple smart micro-grid model inspired from the work presented in [1]. The main objective it to manage, in an optimal manner, the energy storage in the micro-grid. The micro-grid is composed of several components among which we can find:

- The household consumption

- The power generation to be stored

- The energy storage device

- Both internal and external power flows, and

- The control unit

In this case, we focus on the design of a risk-aware control unit able to minimize the variance associated with the risk caused by the uncertainties in the model.

Figure 16.8 shows the scheme of the storage energy problem and all the associated components. The power is generated by using some renewable resources and, depending on the household consumption, so energy is stored. In

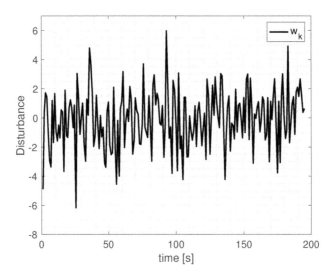

FIGURE 16.9
Evolution of the noise applied to the system.

order to control such storage, several controlled power flows should be determined as explained in the following section.

16.2.1   Model

Let us consider the following system dynamics corresponding to the micro-grid energy storage application as presented in [1]. The energy storage device is modeled as a reservoir with a certain level of performance.

Besides, for the sake of simplicity, we assume that both the charge and discharge are without loss. We consider that the stored energy in the device has a decay factor a, which will depend on the implemented technology. Let z denotes the amount of energy stored in the micro-grid and let μ denote the desired reference. The control inputs u_j, for all $j \in \mathcal{N}$, denote the power flows modifying the stored energy, and w_k represents the uncertainty affecting the storage device. In summary, the model can be simplified as follows:

$$z_{k+1} = az_k + \sum_{j \in \mathcal{N}} b_{2j} u_{j,k} + \sigma w_k,$$

Moreover, let $\mu \in \mathbb{R}$ denote a desired reference for the system state $z \in \mathbb{R}$. Now, let us express the system dynamics in terms of the error $x = z - \mu$, i.e.,

$$x_{k+1} = az_k + \sum_{j \in \mathcal{N}} b_{2j} u_{j,k} - \mu + \sigma w_k,$$

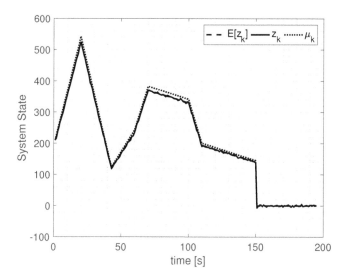

FIGURE 16.10
Evolution of the system state.

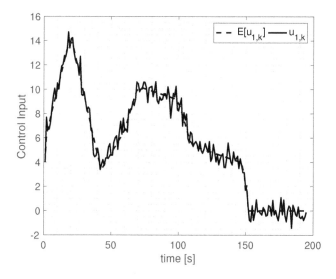

FIGURE 16.11
Evolution of the control input for the first player and its expectation, i.e., $u_{1,k}$ and $\mathbb{E}[u_{1,k}]$.

$$= a(x_k + \mu) + \sum_{j \in \mathcal{N}} b_{2j} u_{j,k} - \mu + \sigma w_k,$$

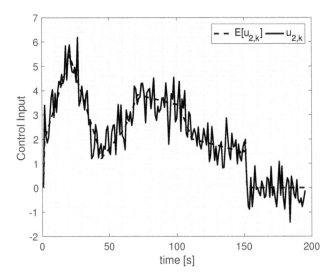

FIGURE 16.12
Evolution of the control input for the second player and its expectation, i.e., $u_{2,k}$ and $\mathbb{E}[u_{2,k}]$.

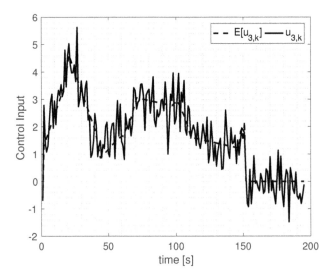

FIGURE 16.13
Evolution of the control input for the third player and its expectation, i.e., $u_{3,k}$ and $\mathbb{E}[u_{3,k}]$.

$$= (a-1)\mu + ax_k + \sum_{j \in \mathcal{N}} b_{2j}u_{j,k} + \sigma w_k. \tag{16.4}$$

These dynamics in (16.4) are of the same form as in (11.1) (see Chapter 11) with $b_0 = (a-1)\mu$, $b_1 = a$, and $\bar{b}_1 = \bar{b}_{2j} = 0$, for all $j \in \mathcal{N}$.

Let us consider a three-decision-maker game problem with the following parameters for the system states:

$$a = 0.95,$$
$$b_{21} = 1.1,$$
$$b_{22} = 1.07,$$
$$b_{23} = 1.03,$$
$$\sigma = 2,$$

and the following parameters for the cost function

$$N = 200,$$
$$q_1 = \bar{q}_1 = 90,$$
$$q_2 = \bar{q}_2 = 92,$$
$$q_3 = \bar{q}_3 = 100,$$
$$r_j = \bar{r}_j = 5, \ \forall j \in \mathcal{N}.$$

Figure 16.9 presents the evolution of the considered noise perturbing the system.

16.2.2 Numerical Results

We solve and implement a non-cooperative mean-field-type game under the aforementioned parameters. On the other hand, Figure 16.10 shows the evolution of the system state z tracking an arbitrary time-varying desired reference μ. Besides, the same figure presents the evolution of the expected state $\mathbb{E}[z]$ showing the high variance reduction over the system state according to the selected prioritization weights $q_j + \bar{q}_j$, for all $j \in \mathcal{N}$, i.e., $\mathbb{E}[(z - \mathbb{E}[z])^2]$. Figures 16.11, 16.12 and 16.13 present the evolution of the optimal control inputs and their expected values for u_1, u_2, and u_3, respectively. It can be seen that the variance reduction for the optimal control inputs is lower than the variance reduction assigned to the system states due to the prioritization-weights selection, i.e., $r_j = \bar{r}_j = 5$, for all $j \in \mathcal{N}$.

Finally, notice that the same problem can be addressed by using multiple storage energy devices. To this end, results from either the vector-valued or matrix-valued setup may be used and implemented. Furthermore, multiple solution concepts can be studied under this scenario, i.e., cooperative, co-opetitive, adversarial, among others.

For the next engineering application consisting of a reactor we will use and applied a continuous-time mean-field-type approach.

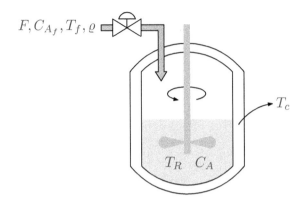

FIGURE 16.14
Continuous stirred tank reactor.

16.3 Continuous Stirred Tank Reactor

In this section we use a gain-scheduling technique to applied the linear-quadratic mean-field-type control into a non-linear system, i.e., we proposed the design of a gain-schedule mean-field-type control (GS-MFTC). Let us consider a stochastic version of the CSTR model presented in [148], as shown in Figure 16.14, i.e.,

$$dC_A = \left(\frac{F}{V}(C_{A_f} - C_A) - C_A k_0 \exp\left(-\frac{E}{R_a T_R}\right) \right) d\hat{t} + \tilde{\sigma}_1 dB_1, \qquad (16.5a)$$

$$dT_R = \left(\frac{F}{V}(T_f - T_R) + \frac{(-\Delta H)}{\rho C_P} C_A k_0 \exp\left(-\frac{E}{R_a T_R}\right) \right. $$
$$\left. + \frac{h_0 A_0}{\varrho C_P V}(T_C - T_R) \right) d\hat{t} + \tilde{\sigma}_2 dB_2, \qquad (16.5b)$$

The variables of model in (16.5) are described in Table 16.1. In Lemma 10, we present a detailed deduction of the stochastic version of the non-linear model as in [148] by applying Itô's formula.

Lemma 10 (Stochastic Non-linear Model) *The stochastic CSTR model in (16.5) can be expressed as follows:*

$$dx_1 = \left(-x_1 + (1 - x_1) D_a \exp\left(\frac{x_2 \gamma}{\gamma + x_2}\right) \right) dt - \sigma_1 dB_1, \qquad (16.6a)$$

$$dx_2 = \left(-x_2 + \hat{B} D_a (1 - x_1) \exp\left(\frac{x_2 \gamma}{\gamma + x_2}\right) - (x_2 - x_c) u \right) dt$$

TABLE 16.1
Description of the variables in the model (16.5).

Variable	Description	Units
A_0	heat transfer area	m^2
C_{A_f}	feed concentration	gmol/m^3
C_P	specific heat capacity	J/kg
E	activation energy	J/gmol
F	volumetric flow rate	m^3/min
h_0	heat transfer coefficient	J/(s m^2 K)
k_0	Arrhenious pre-exponential constant	1/min
R_a	gas constant	J/gmol K
T_C	coolant temperature	K
T_f	feed temperature	K
V	reactor volume	m^3
ΔH	heat of reaction	J/gmol
ϱ	reactant density	kg/m^3

$$+ \sigma_2 dB_2, \tag{16.6b}$$

where

$$x_1 = \frac{C_{A_f} - C_A}{C_{A_f}},$$

$$x_2 = \frac{T_R - T_f}{T_f}\gamma,$$

$$u = \frac{h_0 A_0}{\varrho C_P F},$$

$$t = \hat{t}\frac{F}{V},$$

$$\gamma = \frac{E}{R_a T_f},$$

$$D_a = \frac{V}{F}k_0 \exp(-\gamma),$$

$$\sigma_1 = \frac{\tilde{\sigma}_1}{C_{A_f}},$$

$$\sigma_2 = \frac{\tilde{\sigma}_2 \gamma}{T_f},$$

$$\hat{B} = \frac{(-\Delta H)C_{A_f}\gamma}{\varrho C_P T_f},$$

$$x_c = \frac{T_C - T_f}{T_f}\gamma,$$

are the respective variables.

Proof 41 (Lemma 10) *The deterministic version of the model is presented in [148]. Different from [148], here we present a detail deduction of the corresponding stochastic model using the Itô's formula. Notice that,*

$$\frac{E}{(R_a T_R)} = \frac{\gamma T_f}{T_R}.$$

Thus,

$$-\frac{T_f\gamma}{T_R} = -\frac{T_f\gamma}{T_R + T_f - T_f} = \frac{-\gamma^2}{\gamma + \frac{T_R - T_f}{T_f}\gamma} = -\gamma + \frac{x_2\gamma}{\gamma + x_2}.$$

The model in (16.5) is re-written as follows:

$$dC_A = \left(C_{A_f} - C_A - D_a C_A \exp\left(\frac{x_2\gamma}{\gamma + x_2}\right) \right) dt + \tilde{\sigma}_1 dB_1,$$

$$dT_R = \left(T_f - T_R + \frac{(-\Delta H)}{\rho C_P} D_a C_A \exp\left(\frac{x_2\gamma}{\gamma + x_2}\right) + \frac{h_0 A_0}{\varrho C_P F}(T_C - T_R) \right) dt$$
$$+ \tilde{\sigma}_2 dB_2.$$

We apply Itô's formula to the function x_1, i.e.,

$$dx_1 = \left(-\frac{C_{A_f} - C_A}{C_{A_f}} + \frac{D_a C_A}{C_{A_f}} \exp\left(\frac{x_2\gamma}{\gamma + x_2}\right) \right) dt - \frac{\tilde{\sigma}_1}{C_{A_f}} dB_1,$$

adding and subtracting $D_a \exp\left((x_2\gamma)/(\gamma + x_2)\right)$ yields

$$dx_1 = \left(-x_1 + (1 - x_1)D_a \exp\left(\frac{x_2\gamma}{\gamma + x_2}\right) \right) dt - \frac{\tilde{\sigma}_1}{C_{A_f}} dB_1.$$

Similarly, we apply Itô's formula to the function x_2, i.e.,

$$dx_2 = \left(\frac{T_f - T_R}{T_f}\gamma + \frac{(-\Delta H)\gamma}{\rho C_P T_f} D_a C_A \exp\left(\frac{x_2\gamma}{\gamma + x_2}\right) \right.$$
$$\left. + \frac{h_0 A_0}{\varrho C_P F T_f}(T_C - T_R)\gamma \right) dt + \frac{\tilde{\sigma}_2\gamma}{T_f} dB_2,$$

adding and subtracting

$$\frac{h_0 A_0}{\varrho C_P F},$$

$$\frac{(-\Delta H)C_{A_f}}{\varrho C_P T_f} D_a \exp\left(\frac{x_2 \gamma}{\gamma + x_2}\right),$$

it finally yields

$$\mathrm{d}x_2 = \left(-x_2 + \hat{B}D_a(1-x_1)\exp\left(\frac{x_2\gamma}{\gamma+x_2}\right) - (x_2-x_c)u\right)\mathrm{d}t + \frac{\tilde{\sigma}_2\gamma}{T_f}\mathrm{d}B_2,$$

completing the deduction of the non-linear dimensionless model.

16.3.1 Linearization-Based Scheduling and Risk-Aware Control Problem

We present the scheduling method based on Jacobian linearization of the non-linear system dynamics. Let $\Theta = \{1, \ldots, n\}$ denote the set of possible operating points, where $\theta \in \Theta$[1] corresponds to the current operating point given by $x^\theta \in \mathbb{R}^{\ell_x}$.

Consider a general stochastic non-linear system given by

$$\mathrm{d}x = D_r(x, \mathbb{E}[x], u, \mathbb{E}[u])\mathrm{d}t + D_f \mathrm{d}B, \tag{16.7a}$$
$$x(0) := x_0, \tag{16.7b}$$

where $x \in \mathbb{R}^{\ell_x}$ and $u \in \mathbb{R}^{\ell_u}$ denote the system states and control inputs, respectively; and $\mathbb{E}[x] \in \mathbb{R}^{\ell_x}$ and $\mathbb{E}[u] \in \mathbb{R}^{\ell_u}$ denote their expectation values, respectively. Hence, $D_r : \mathbb{R}^{\ell_x} \times \mathbb{R}^{\ell_x} \times \mathbb{R}^{\ell_u} \times \mathbb{R}^{\ell_u} \to \mathbb{R}^{\ell_x}$ is twice differentiable and corresponds to the drift, and $D_f \in \mathbb{R}$ to a constant diffusion, where $B \in \mathbb{R}^{\ell_x}$ denotes a standard Brownian motion. The objective is to stabilize the non-linear system in (16.7) around an operating point x^θ. Let

$$\tilde{x} = x - x^\theta,$$
$$\mathbb{E}[\tilde{x}] = \mathbb{E}[x] - \mathbb{E}[x^\theta],$$
$$\tilde{u} = u - u^\theta,$$
$$\mathbb{E}[\tilde{u}] = \mathbb{E}[u] - \mathbb{E}[u^\theta],$$

be the error terms for system states and control inputs. Then, the cost functional is given by

$$L(\tilde{x}, \mathbb{E}[\tilde{x}], \tilde{u}, \mathbb{E}[\tilde{u}])) = \langle Q(T)\tilde{x}(T), \tilde{x}(T)\rangle + \langle \bar{Q}(T)\mathbb{E}[\tilde{x}(T)], \mathbb{E}[\tilde{x}(T)]\rangle$$
$$+ \int_0^T \langle Q\tilde{x}, \tilde{x}\rangle + \langle \bar{Q}\mathbb{E}[\tilde{x}], \mathbb{E}[\tilde{x}]\rangle \,\mathrm{d}t$$
$$+ \int_0^T \langle R\tilde{u}, \tilde{u}\rangle + \langle \bar{R}\mathbb{E}[\tilde{u}], \mathbb{E}[\tilde{u}]\rangle \,\mathrm{d}t, \tag{16.8}$$

---

[1]We advise not to confuse the operation points with the set of Poisson jumps used in other chapters given that we use the same notation.

where Q, $Q + \bar{Q} \succ 0$, and R, $R + \bar{R} \succeq 0$ are symmetric matrices, and the corresponding optimization problem for the non-linear mean-field-type controller is as follows:

$$\underset{u,\mathbb{E}[u]}{\text{minimize}} \ \mathbb{E}\left[L\left(\tilde{x}, \mathbb{E}[\tilde{x}], \tilde{u}, \mathbb{E}[\tilde{u}]\right)\right], \tag{16.9a}$$

$$\text{subject to (16.7).} \tag{16.9b}$$

Lemma 11 (Risk Minimization) *The expected value of the cost functional, i.e.,*

$$\mathbb{E}L\left(\tilde{x}, \mathbb{E}[\tilde{x}], \tilde{u}, \mathbb{E}[\tilde{u}]\right),$$

involves variance terms for both the system states and control inputs, i.e.,

$$\mathbb{E}\left[\langle Q\left(\tilde{x} - \mathbb{E}[\tilde{x}]\right), \left(\tilde{x} - \mathbb{E}[\tilde{x}]\right)\rangle\right],$$
$$\mathbb{E}\left[\langle R\left(\tilde{u} - \mathbb{E}[\tilde{u}]\right), \left(\tilde{u} - \mathbb{E}[\tilde{u}]\right)\rangle\right].$$

Therefore, (16.9) is a risk-minimization problem.

Proof 42 (Lemma 11) *The proof directly follows after applying the orthogonal decomposition given by $(\tilde{x} - \mathbb{E}[\tilde{x}]) \perp \mathbb{E}[\tilde{x}]$ and $(\tilde{u} - \mathbb{E}[\tilde{u}]) \perp \mathbb{E}[\tilde{u}]$.*

Let us consider the following linear system dynamics computed from (16.7):

$$d\tilde{x} = \left(A^\theta \tilde{x} + \bar{A}^\theta \mathbb{E}[\tilde{x}] + B^\theta \tilde{u} + \bar{B}^\theta \mathbb{E}[\tilde{u}]\right) dt + \sigma^\theta dB, \tag{16.10a}$$

$$\tilde{x}(0) := \tilde{x}_0, \tag{16.10b}$$

where the state-space matrices are obtained from a Jacobian linearization given by

$$A^\theta = \nabla_x D_r(x, \mathbb{E}[x], u, \mathbb{E}[u])\Big|_{(x^\theta, \mathbb{E}[x^\theta], u^\theta, \mathbb{E}[u^\theta])}, \tag{16.11a}$$

$$\bar{A}^\theta = \nabla_{\mathbb{E}[x]} D_r(x, \mathbb{E}[x], u, \mathbb{E}[u])\Big|_{(x^\theta, \mathbb{E}[x^\theta], u^\theta, \mathbb{E}[u^\theta])}, \tag{16.11b}$$

$$B^\theta = \nabla_u D_r(x, \mathbb{E}[x], u, \mathbb{E}[u])\Big|_{(x^\theta, \mathbb{E}[x^\theta], u^\theta, \mathbb{E}[u^\theta])}, \tag{16.11c}$$

$$\bar{B}^\theta = \nabla_{\mathbb{E}[u]} D_r(x, \mathbb{E}[x], u, \mathbb{E}[u])\Big|_{(x^\theta, \mathbb{E}[x^\theta], u^\theta, \mathbb{E}[u^\theta])}, \tag{16.11d}$$

$$\sigma^\theta = D_f. \tag{16.11e}$$

Once the stochastic model in (16.10) is obtained, then a LQ mean-field-type control can be designed to operate around the operating point x^θ as discussed in the next section. Thus, the appropriate gains are assigned to the feedback state and its expectation conveniently as the operating points vary along the time.

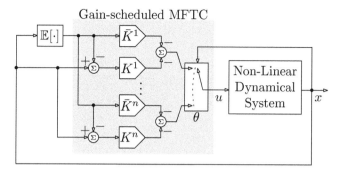

FIGURE 16.15
Gain-scheduled Mean-Field-Type Control Diagram with n operation points and $\theta \in \{1, \ldots, n\}$.

16.3.2 Gain-Scheduled Mean-Field-Type Control

We consider two different approaches for the risk-aware control design based on mean-field-type theory. First, we show the risk-neutral approach, which involves the first and second moment terms, i.e., the mean and variance of the system state and the control inputs. On the second hand, we consider the risk-sensitive approach, which involves also higher order terms.

In order to design the Risk-Neutral and Risk-Sensitive Gain-Scheduled Mean-Field-Type Control, it is necessary to identify the appropriate gains $K^{\theta}, \bar{K}^{\theta}, \theta \in \Theta$. The diagram of the GS-MFTC is presented in Figure 16.15 with n operating points. whereas Figure 16.16 shows the diagram corresponding to the mean-field-free case.

16.3.2.1 Design

We solve the infinite horizon Risk-Neutral Mean-Field-Type Control whose respective cost functional $L^{\infty}(\tilde{x}, \mathbb{E}[\tilde{x}], \tilde{u}, \mathbb{E}[\tilde{u}])$ is given by

$$L^{\infty}(\tilde{x}, \mathbb{E}[\tilde{x}], \tilde{u}, \mathbb{E}[\tilde{u}]) = \int_0^{\infty} \langle Q(\tilde{x} - \mathbb{E}[\tilde{x}]), \tilde{x} - \mathbb{E}[\tilde{x}] \rangle + \langle (Q + \bar{Q})\mathbb{E}[\tilde{x}], \mathbb{E}[\tilde{x}] \rangle \, dt$$
$$+ \int_0^{\infty} \langle R(\tilde{u} - \mathbb{E}[\tilde{u}]), \tilde{u} - \mathbb{E}[\tilde{u}] \rangle + \langle (R + \bar{R})\mathbb{E}[\tilde{u}], \mathbb{E}[\tilde{u}] \rangle \, dt,$$

and the corresponding optimization problem is given by

$$\underset{u, \mathbb{E}[u]}{\text{minimize}} \ \mathbb{E}\left[L^{\infty}\left(\tilde{x}, \mathbb{E}[\tilde{x}], \tilde{u}, \mathbb{E}[\tilde{u}]\right)\right], \tag{16.12a}$$

subject to

$$d\tilde{x} = (A^{\theta}(\tilde{x} - \mathbb{E}[\tilde{x}]) + (A^{\theta} + \bar{A}^{\theta})\mathbb{E}[\tilde{x}] + B^{\theta}(\tilde{u} - \mathbb{E}[\tilde{u}])$$
$$+ (B^{\theta} + \bar{B}^{\theta})\mathbb{E}[\tilde{u}])dt + \sigma^{\theta}dB, \tag{16.12b}$$
$$\tilde{x}(0) = x_0. \tag{16.12c}$$

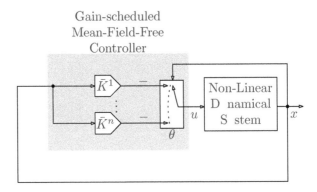

FIGURE 16.16
Gain-scheduled mean-field-free control diagram with n operation points and $\theta \in \{1, \ldots, n\}$.

The solution of the infinite horizon problem is presented in Proposition 22.

Proposition 22 (Infinite-Horizon Risk-Neutral) *The feedback optimal control inputs of the mean-field-type control and the optimal cost, which solve Problem in* (16.12), *are given by:*

$$\tilde{u}^* = -\bar{K}^\theta \mathbb{E}[\tilde{x}] - K^\theta(\tilde{x} - \mathbb{E}[\tilde{x}]),$$

with

$$\bar{K}^\theta = (R + \bar{R})^{-1}(B^\theta + \bar{B}^\theta)^\top \bar{P}^\infty,$$
$$K^\theta = R^{-1}(B^\theta)^\top P^\infty,$$

where the following equations:

$$0 = Q + P^\infty(A^\theta)^\top + A^\theta P^\infty - P^\infty B^\theta R^{-1}(B^\theta)^\top P^\infty,$$
$$0 = Q + \bar{Q} + \bar{P}^\infty(A^\theta + \bar{A}^\theta)^\top + (A^\theta + \bar{A}^\theta)\bar{P}^\infty$$
$$- \bar{P}^\infty(B^\theta + \bar{B}^\theta)(R + \bar{R})^{-1}(B^\theta + \bar{B}^\theta)^\top \bar{P}^\infty,$$

are solved by P^∞ and $\bar{P}^\infty$.

This result is obtained directly by solving the steady-state equilibrium of the Riccati equations presented in Chapter 9. The result presented in Proposition 22 shows a control structure involving two main gains, i.e., $\bar{K}^\theta$ and K^θ. The risk-aware control scheme is the one presented Figure 16.15. On the other hand, we also solve the Infinite-Horizon Risk-Sensitive Problem, i.e.,

$$\underset{u, \mathbb{E}[u]}{\text{minimize}} \ \frac{1}{\lambda} \log \mathbb{E}\left[\lambda e^{J^\infty(\tilde{x}, \mathbb{E}[\tilde{x}], \tilde{u}, \mathbb{E}[\tilde{u}])}\right], \tag{16.13}$$

subject to (16.12b) and (16.12c). The solution of the infinite horizon problem is presented in Proposition 23.

Proposition 23 (Infinite-Horizon Risk-Sensitive) *The feedback optimal control inputs of the mean-field-type control and the optimal cost, which solve Problem in* (16.3.2.1), *are given by:*

$$\tilde{u}^* = -\bar{K}^\theta \mathbb{E}[\tilde{x}] - K^\theta(\tilde{x} - \mathbb{E}[\tilde{x}]),$$

with

$$\bar{K}^\theta = (R + \bar{R})^{-1}(B^\theta + \bar{B}^\theta)^\top \bar{P}^\infty,$$
$$K^\theta = R^{-1}(B^\theta)^\top P^\infty,$$

where the following equations:

$$0 = Q + P^\infty(A^\theta)^\top + A^\theta P^\infty - P^\infty \left[B^\theta R^{-1}(B^\theta)^\top - 2\lambda\sigma^\theta(\sigma^\theta)^\top \right] P^\infty,$$
$$0 = Q + \bar{Q} + \bar{P}^\infty(A^\theta + \bar{A}^\theta)^\top + (A^\theta + \bar{A}^\theta)\bar{P}^\infty$$
$$\quad - \bar{P}^\infty(B^\theta + \bar{B}^\theta)(R + \bar{R})^{-1}(B^\theta + \bar{B}^\theta)^\top \bar{P}^\infty,$$

are solved by P^∞ *and* $\bar{P}^\infty$.

This result is obtained directly by solving the steady-state equilibrium of the Riccati equations presented in Chapter 9.

16.3.2.2 Local Stability of the Operating Points

Here, we analyze the closed-loop stability of the GS-MFTC around the corresponding operating points $\tilde{x}^\theta, \tilde{u}^\theta$, for all $\theta \in \Theta$. Thus, let us consider the following transformation of the state-space model in (16.12b):

$$d(\tilde{x} - \mathbb{E}[\tilde{x}]) = \left(A^\theta(\tilde{x} - \mathbb{E}[\tilde{x}]) + B^\theta(\tilde{u} - \mathbb{E}[\tilde{u}]) \right) dt + \sigma^\theta dB.$$

Therefore, replacing the optimal control input from either Proposition 22 or Proposition 23 yields

$$\tilde{u}^* - \mathbb{E}[\tilde{u}^*] = -K^\theta(\tilde{x} - \mathbb{E}[\tilde{x}]),$$
$$\mathbb{E}[\tilde{u}^*] = -\bar{K}^\theta \mathbb{E}[\tilde{x}],$$

and

$$d(\dot{x} - \mathbb{E}[x]) = \left(A^\theta - B^\theta K^\theta \right) (\tilde{x} - \mathbb{E}[\tilde{x}]) dt + \sigma^\theta dB.$$

Considering the variable $z = [(\tilde{x} - \mathbb{E}[\tilde{x}])^\top \quad \mathbb{E}[\tilde{x}]^\top]^\top$ with $(\tilde{x} - \mathbb{E}[\tilde{x}]) \perp \mathbb{E}[\tilde{x}]$, it follows:

$$dz = A_z^\theta z dt + S_z^\theta dB,$$

where

$$A_z^\theta = \begin{bmatrix} A^\theta - B^\theta K^\theta & 0 \\ 0 & A^\theta + \bar{A}^\theta - (B^\theta + \bar{B}^\theta)\bar{K}^\theta \end{bmatrix},$$

TABLE 16.2
Value of the variables for the case study.

Variable	Value	Units
A_0	8.1755	m^2
C_{A_f}	2114.5	$gmol/m^3$
C_P	3571.3	J/kg
E	75361.14	$J/gmol$
F	0.1605	m^3/min
h_0	2.5552e4	$J/(s\ m^2\ K)$
k_0	2.8267e11	$1/min$
R_a	8.3174	$J/gmol\ K$
T_C	279	K
T_f	295.22	K
V	2.4069	m^3
ΔH	−0.9e5	$J/gmol$
ϱ	1000	kg/m^3

$$S_z^\theta = \begin{bmatrix} \sigma^\theta \\ 0 \end{bmatrix}.$$

Thus, the stability of the operating point is determined by the matrix A_z^θ. The respective operating point associated with $\theta \in \Theta$ is locally stable if the real component of the spectrum of A_z^θ is negative, i.e., $\Re\{\lambda(A_z^\theta)\}_i < 0$, for all components $i \in \{1,\ldots,2\ell_x\}$.

16.3.3   Risk-Aware Numerical Illustrative Example

Let us consider the system parameters presented in Table 16.2, which have been chosen inspired by the case study presented in [149]. Besides, we consider the following three operating points, i.e., $[C_A^\theta \quad T_R^\theta]^\top$ where $\theta \in \Theta = \{1,2,3\}$:

$$[C_A^1 \quad T_R^1]^\top = [1.388e3 \quad 304.8390]^\top, \tag{16.14a}$$
$$[C_A^2 \quad T_R^2]^\top = [919.6546 \quad 314.4581]^\top, \tag{16.14b}$$
$$[C_A^3 \quad T_R^3]^\top = [521.3688 \quad 324.0771]^\top, \tag{16.14c}$$

corresponding to the following operating system states x^θ and control inputs u^θ, $\theta \in \Theta = \{1,2,3\}$:

$$[x_1^1 \quad x_2^1]^\top = [0.3436 \quad 1]^\top,$$
$$[x_1^2 \quad x_2^2]^\top = [0.5651 \quad 2]^\top,$$
$$[x_1^3 \quad x_2^3]^\top = [0.7534 \quad 3]^\top,$$

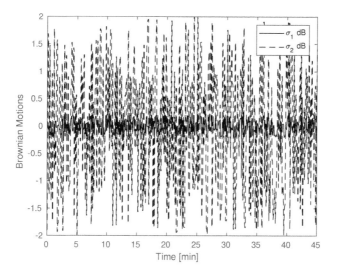

FIGURE 16.17
Noise Brownian motions for both the reactant concentration and the reactor temperature.

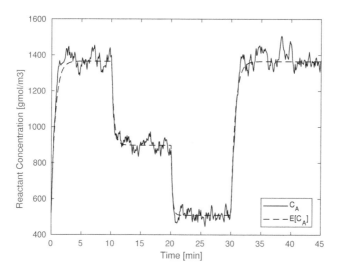

FIGURE 16.18
Performance of the GS-MFTC. Evolution of the reactant concentration C_A and its expectation $\mathbb{E}[C_A]$ tracking the reference C_A^{ref}.

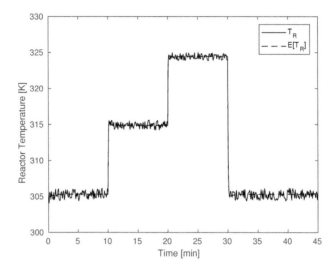

FIGURE 16.19
Performance of the GS-MFTC. Evolution of the reactor temperature T_R and its expectation $\mathbb{E}[T_R]$ tracking the reference T_R^{ref}.

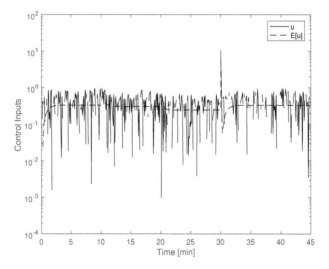

FIGURE 16.20
Evolution of the optimal control input and its expectation.

$$u^1 = 0.3363,$$

$$u^2 = 0.3066,$$
$$u^3 = 0.2505.$$

The parameters for the cost functional are

$$Q = 50I_2,$$
$$\bar{Q} = 80I_2,$$
$$R = 1,$$
$$\bar{R} = 2.$$

Therefore, the state-space matrices $A^\theta, \bar{A}^\theta, B^\theta, \bar{B}^\theta, \sigma^\theta, \theta \in \Theta$, corresponding to the operating points are computed from (16.6) and (16.11), i.e.,

$$\bar{A}^\theta = 0,$$
$$\bar{B}^\theta = 0,$$

$$A^\theta_{11} = \left[-D_a\psi(x_2) - 1 \right]\Big|_{x_2^\theta},$$

$$A^\theta_{12} = \left[-D_a\psi(x_2)\left(\frac{\gamma}{\gamma + x_2} - \frac{\gamma x_2}{(\gamma + x_2)^2} \right)(x_1 - 1) \right]\Big|_{x_1^\theta, x_2^\theta},$$

$$A^\theta_{21} = \left[-\hat{B}D_a\psi(x_2) \right]\Big|_{x_2^\theta},$$

$$A^\theta_{22} = \left[-u - \hat{B}D_a\psi(x_2)\left(\frac{\gamma}{\gamma + x_2} - \frac{\gamma x_2}{(\gamma + x_2)^2} \right)(x_1 - 1) - 1 \right]\Big|_{u^\theta, x_2^\theta},$$

$$B^\theta = \begin{bmatrix} 0 \\ x_c - x_2 \end{bmatrix}\Big|_{x_2^\theta}, \quad \sigma^\theta = \begin{bmatrix} \sigma_1 \\ \sigma_2 \end{bmatrix}, \quad \forall \theta \in \Theta,$$

where $\sigma_1 = 0.2$, $\sigma_2 = 2$, and

$$\psi(x_2) = \exp\left(\frac{x_2\gamma}{\gamma + x_2} \right).$$

Now, by using the operating points and by applying the result in Proposition 22, the mean-field-type control gains $K^\theta, \bar{K}^\theta$, for all $\theta \in \Theta$, are computed.

Thus, it is obtained that

$$
P^\infty = \begin{cases}
\begin{bmatrix} 16.6049 & -0.1199 \\ -0.1199 & 2.6932 \end{bmatrix}, & \text{if } \theta = 1, \\[2em]
\begin{bmatrix} 11.5604 & -0.3106 \\ -0.3106 & 2.0222 \end{bmatrix}, & \text{if } \theta = 2, \\[2em]
\begin{bmatrix} 7.6961 & -0.6024 \\ -0.6024 & 1.6017 \end{bmatrix}, & \text{if } \theta = 3,
\end{cases}
$$

$$
\bar{P}^\infty = \begin{cases}
\begin{bmatrix} 43.3166 & -0.4108 \\ -0.4108 & 7.5331 \end{bmatrix}, & \text{if } \theta = 1, \\[2em]
\begin{bmatrix} 30.3773 & -0.9678 \\ -0.9678 & 5.6677 \end{bmatrix}, & \text{if } \theta = 2, \\[2em]
\begin{bmatrix} 20.6323 & -1.8181 \\ -1.8181 & 4.4904 \end{bmatrix}, & \text{if } \theta = 3,
\end{cases}
$$

and

$$
\begin{aligned}
K^1 &= [0.3220 \quad -7.2346], \\
\bar{K}^1 &= [0.3678 \quad -6.7453], \\
K^2 &= [1.1449 \quad -7.4541], \\
\bar{K}^2 &= [1.1892 \quad -6.9642], \\
K^3 &= [2.8232 \quad -7.5058], \\
\bar{K}^3 &= [2.8400 \quad -7.0144].
\end{aligned}
$$

In order to verify the local stability of the operating points, we compute the matrices A_z^θ, and the corresponding spectrum $\lambda(A_z^\theta)$, for all $\theta \in \Theta$ as follows:

$$
A_z^1 = \begin{bmatrix}
-1.5234 & 0.3222 & 0 & 0 \\
-2.0344 & -18.9852 & 0 & 0 \\
0 & 0 & -1.5234 & 0.3222 \\
0 & 0 & -1.9114 & -17.6707
\end{bmatrix},
$$

$$
A_z^2 = \begin{bmatrix}
-2.2992 & 0.4980 & 0 & 0 \\
-2.9772 & -26.0254 & 0 & 0 \\
0 & 0 & -2.2992 & 0.4980 \\
0 & 0 & -2.8138 & -24.2192
\end{bmatrix},
$$

$$A_z^3 = \begin{bmatrix} -4.0557 & 0.6253 & 0 & 0 \\ -3.6975 & -32.9605 & 0 & 0 \\ 0 & 0 & -4.0557 & 0.6253 \\ 0 & 0 & -3.6189 & -30.6577 \end{bmatrix},$$

and with the following respective spectrum:

$$\lambda(A_z^1) = \{-1.5610, -18.9476, -1.5616, -17.6324\},$$
$$\lambda(A_z^2) = \{-2.3619, -25.9628, -2.3633, -24.1551\},$$
$$\lambda(A_z^3) = \{-4.1359, -32.8803, -4.1410, -30.5723\},$$

showing the local stability of the considered operating points with the designed GS-MFTC.

Next, we present the performance of the GS-MFTC over the non-linear model in (16.6), and tracking a reference signals $C_A^{\text{ref}}, T_R^{\text{ref}}$ changing their value among the different considered operating points C_A^θ, T_R^θ, for all $\theta \in \Theta$ according to (16.14) (see Figure 16.15) as follows:

$$C_A^{\text{ref}} = \begin{cases} C_A^1 & \text{if } t \in [0,10] \cup (30,45], \\ C_A^2 & \text{if } t \in (10,20], \\ C_A^3 & \text{if } t \in (20,30], \end{cases} \tag{16.15}$$

and

$$T_R^{\text{ref}} = \begin{cases} T_R^1 & \text{if } t \in [0,10] \cup (30,45], \\ T_R^2 & \text{if } t \in (10,20], \\ T_R^3 & \text{if } t \in (20,30], \end{cases} \tag{16.16}$$

Figure 16.17 shows the noise perturbing the stochastic CSTR. The evolution of the reactant concentration is presented in Figure 16.18. It can be observed that, the reactant concentration tracks the desired reference established in (16.15). Similarly, Figure 16.19 shows the evolution of the reactor temperature according to the desired references presented in (16.16), and Figure 16.20 shows the evolution of the optimal control input u^* and its expectation $\mathbb{E}[u^*]$.

16.4 Mechanism Design in Evolutionary Games

Consider a large population of agents selecting an available strategy from the set $\mathcal{S} = \{1, \ldots, n\}$, being $n \geq 2$ ($n \in \mathbb{N}$). The scalar $x_i \in \mathbb{R}_{\geq 0}$ corresponds to the proportion of agents selecting the strategy $i \in \mathcal{S}$. Hence, $x \in \mathbb{R}^n$ describes a population state or a strategic distribution among the strategies. The set representing the possible population states, such that the population mass remains constant, is given by unitary simplex $\Delta_+ = \Delta \cap \mathbb{R}_{\geq 0}$, where

$$\Delta := \left\{ x \in \mathbb{R}^n : \sum_{i \in \mathcal{S}} x_i = 1 \right\}.$$

In order to maximize their individual utilities, agents make decisions based on the values of the fitness functions $f_i : \Delta_+ \to \mathbb{R}$, which map the current population states to rewards associated with each of the strategies $i \in \mathcal{S}$. Therefore, the fitness functional of the entire population is given by $f : \Delta_+ \to \mathbb{R}^n$, where each component corresponds to f_i. Here, it is important to advise not to confuse these fitness functions with the ansatz of the optimal costs in the mean-field-type games.[2]

Definition 38 (Population Game) *A population game in the simplex Δ_+ is characterized by its corresponding fitness function whose mapping is given by $f : \Delta_+ \to \mathbb{R}^n$.*

Since $x \in \Delta_+$, the population state x can be equivalently interpreted as a *mixed strategy* used by the player, i.e., players select the action $i \in \mathcal{S}$ with probability x_i [36].

The standard main objective in a population game is to reach a Nash equilibrium introduced next.

Definition 39 (Population-Game Equilibria) *[128, page 24] A population state $x^* \in \Delta$ is said to be a Nash equilibrium if each used strategy earns the maximum payoff, i.e.,*

$$\text{NE}(f) = \left\{ x^* \in \Delta : \forall i \in \mathcal{S}, x_i^* > 0 \implies f_i(x^*) \geq f_j(x^*), \forall j \in \mathcal{S} \right\}$$

corresponds to the Nash equilibria.

The set of Nash equilibria in a population game f can be reached by implemented a class of evolutionary dynamics [128]. In this paper we focus mainly on a class of evolutionary dynamics with imperfect fitness observation, which further motivates the development of equilibrium selection techniques.

In particular, by identifying the flow of information between strategies by an undirected and connected graph $\mathcal{G} = (\mathcal{S}, \mathcal{E})$, where $\mathcal{E} \subseteq \{(i, j) | i, j \in \mathcal{S}\}$ is the set of links allowing migration among strategies $\mathcal{S}$, and $\mathcal{N}_i = \{j | (i, j) \in \mathcal{E}\}$ is the set of neighbors of the strategy $i \in \mathcal{S}$. We consider a type of projection dynamics, given by

$$x \in \Delta_+, \quad \dot{x} = -x + P_K(x + L_p f(x)), \tag{16.17}$$

where L_p is the Laplacian matrix of $\mathcal{G}$, and $P_K(x) = \operatorname{argmin}_{u \in K} |x - u|_2$ is

---

[2]We make this remark since we use the same notation for the ansatz and the fitness functions.

the Lipschitz continuous Euclidean projection on the set K, which in this case corresponds to the Cartesian product $K = [0, 1] \times [0, 1] \times \ldots [0, 1]$. Note that by construction the dynamics (16.17) can be implemented in a distributed way, and whenever $x \in \text{int}(\Delta_+)$ the dynamics (16.17) simplify to

$$\dot{x} = L_p f(x), \tag{16.18}$$

whose i^{th} component is given by

$$\dot{x}_i = |\mathcal{N}_i| f_i(x) - \sum_{j \in \mathcal{N}_i} f_j(x), \ \forall i \in \mathcal{S}. \tag{16.19}$$

Indeed, the dynamics (16.17) can be seen as a distributed version of the Lipschitz projection evolutionary dynamics considered in [150].

Remark 35 *Since the projection $P_K(\cdot)$ in box constraints can be implemented in a distributed way, and since L_p is a Laplacian matrix, the dynamics (16.17) can be fully implemented in a decentralized way.*

In order to incorporate randomness and incentives into the dynamics (16.17), let us consider a vector of "perturbed" and "controlled" fitness functions, given by

$$\tilde{f}(x, u, \xi) = f(x) + g_1(u) + g_2(\xi), \tag{16.20}$$

where ξ is a Gaussian random variable, $g_1 : \mathbb{R}^p \to \mathbb{R}^n$ and $g_2 : \mathbb{R}^n \to \mathbb{R}^n$ are continuously differentiable functions, and $u \in \mathbb{R}^p$ is a control input.

The interpretation of ξ is presented in Figures 16.21(b) and 16.21(d), where the population selecting a strategy has imperfect information (noisy measurements) about the fitness functions corresponding to the other strategies. Moreover, Figures 16.21(c) and 16.21(d) show the interpretation of the control input u that allows manipulating the evolution of the population states as desired (mechanism inducing a desired behavior).

Under this modification we obtain a class of *controlled evolutionary dynamics with imperfect fitness observation* given by

$$x \in \Delta_+, \ \dot{x} = -x + P_K \left(x + L_p \tilde{f}(x, u, \xi) \right), \ K = \mathbb{R}_+^n. \tag{16.21}$$

Note that in system (16.21) the noise affects the perception of the fitness functions, even when $u = 0$. This behavior is illustrated in Figure 16.22 by means of two population-game examples, and its scheme is presented in Figures 16.21(a)-16.21(b), where a game with two stable equilibrium points, e.g., the Zeeman's game, is simulated under additive Gaussian noise acting on the fitness, and with input u set to zero. In Figure 16.21(b), the population selecting the first strategy has an imperfect observation of the fitness obtained by those selecting the second strategy because of the noise ξ_2, and vice versa. In this case, the variance of the solutions is reflected in the emerging asymptotic behavior. On the other hand, when $u \neq 0$, the input signal can be seen as

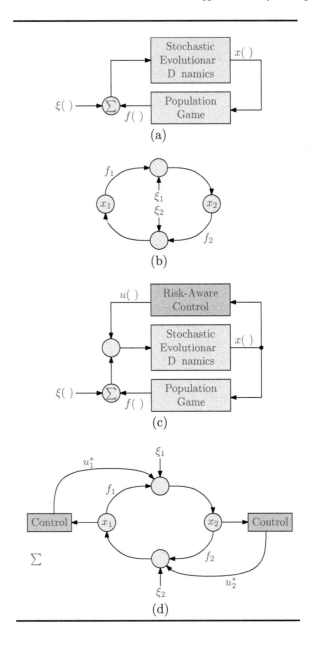

FIGURE 16.21
(a)–(b) Evolutionary dynamics with imperfect fitness observation for a population game, and two-strategy population game. (c)–(d) Closed-loop of the multi-layer game for the equilibrium selection and two-strategy population game.

an incentive mechanism designed to induce particular Nash equilibria even though there is imperfect information about the measured fitness functions. Of special interest for us are the cases where $u = \alpha(x)$, i.e., an actual feedback law mapping the population states to incentive inputs.

16.4.1 A Risk-Aware Approach to the Equilibrium Selection

We show that the Nash equilibrium selection problem can be conveniently modeled as a risk-aware control problem by using mean-field-type theory that simultaneously minimizes the variance, for which explicit solutions can be found whenever the trajectories of (16.21) remain in the interior of the positive orthant. To make this problem mathematically tractable we focus on the case where $f(x)$, $g_1(u)$ and $g_2(\xi)$ are linear functions, such that the controller fitness function (16.20) is given by

$$\tilde{f}(x, u, \xi) = Dx + Wu + \tilde{S}\xi, \tag{16.22}$$

for some matrices $D \in \mathbb{R}^{n \times n}$, $W \in \mathbb{R}^{n \times m}$ and $\tilde{S} > 0$. Then, it follows that in the interior of the simplex the dynamics (16.21) are given by

$$\dot{x} = L_p Dx + L_p Wu + L_p \tilde{S}\xi, \tag{16.23}$$

where, bye taking $L_p \tilde{S}\xi dt = SdB$, yields the following stochastic differential equation:

$$dx = (L_p Dx + L_p Wu)\, dt + SdB. \tag{16.24}$$

Lemma 12 *Under the fitness selection in (16.20), the stochastic dynamics in (16.24) renders the Δ almost surely invariant.*

We consider two different equilibrium selection problems:

- Designing feedback laws $u = \alpha(x)$ that induce a desired Nash equilibrium x^d, and

- Designing feedback laws that induce an unknown Nash equilibrium x^* that satisfies a given optimality criterion.

These two problems are addressed by designing appropriate incentives that modify the fitness functions while minimizing the variance given by $\mathrm{var}(x - x^d)$. The general scheme of the proposed novel approach is shown in Figures 16.21(c)-16.21(d). Besides, notice that, under both scenarios it is necessary to know the matrices L_p, D, and W.

16.4.1.1 Known Desired Nash Equilibrium

Let $x^d \in \mathrm{int}(\Delta)$ be a desired Nash equilibrium. The dynamics in (16.23) in the error coordinates $x^e = x - x^d$ and $u^e = u - u^d$, yields

$$\dot{x} - \dot{x}^d = L_p Dx - L_p Dx^d + L_p W(u - u^d) + L_p \tilde{S}\xi, \tag{16.25a}$$

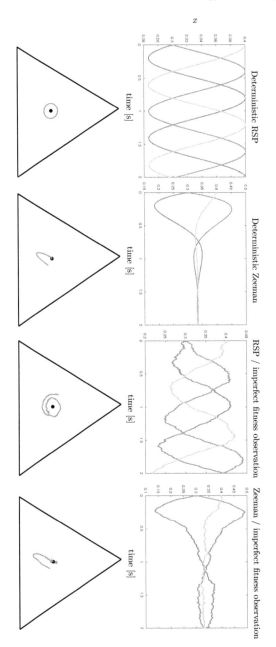

FIGURE 16.22
Projection dynamics behavior for both the RSP and Zeeman game with and without noisy fitness functions.

$$dx^e = (Ax^e + Bu^e)dt + SdB, \tag{16.25b}$$

where $A = L_p D$ and $B = L_p W$. Taking the expectation we obtain the dynamics of the mean error, given by

$$\mathbb{E}[\dot{x}^e] = Ax^e + Bu^e. \tag{16.26}$$

In order to converge to x^d, we can formulate the Nash equilibrium selection problem as a stochastic control problem of the form

$$\begin{cases} \underset{u^e}{\text{minimize}} \ \mathbb{E}[L(u^e)], \\[2mm] \text{subject to} \\[2mm] (16.25b), \text{ and } x(0) \in \Delta, \end{cases} \tag{16.27}$$

where the cost functional can be selected either as the classic cost

$$L(u^e) = \langle Q(T)x^e(T), x^e(T) \rangle + \int_0^T \langle Qx^e, x^e \rangle \, dt + \int_0^T \langle Ru^e, u^e \rangle \, dt,$$

with symmetric positive semidefinite matrices Q and R; or, alternatively, as the following mean-variance cost function:

$$L(u^e) = \langle Q(T)\tilde{x}(T), \tilde{x}(T) \rangle + \left\langle \bar{Q}(T)\mathbb{E}[x^e(T)], \mathbb{E}[x^e(T)] \right\rangle$$
$$+ \int_0^T \left[\langle Q\tilde{x}, \tilde{x} \rangle + \left\langle \bar{Q}\mathbb{E}[x^e], \mathbb{E}[x^e] \right\rangle \right] dt$$
$$+ \int_0^T \left[\langle R\tilde{u}, \tilde{u} \rangle + \left\langle \bar{R}\mathbb{E}[u^e], \mathbb{E}[u^e] \right\rangle \right] dt. \tag{16.28}$$

where

$$\tilde{u} = u^e - \mathbb{E}[u^e], \tag{16.29a}$$
$$\tilde{x} = x^e - \mathbb{E}[x^e], \tag{16.29b}$$
$$\tilde{Q} = Q + \bar{Q}, \tag{16.29c}$$
$$\tilde{R} = R + \bar{R}. \tag{16.29d}$$

Thus, the expected cost $\mathbb{E}[L(u^e)]$ involves the variance terms

$$\mathbb{E} \langle Q\tilde{x}, \tilde{x} \rangle = \text{var}_Q[x^e], \tag{16.30a}$$
$$\mathbb{E} \langle R\tilde{u}, \tilde{u} \rangle = \text{var}_R[u^e]. \tag{16.30b}$$

To solve the optimization problem presented in (16.27), it is necessary to have knowledge about the desired Nash equilibrium x^d, which, in general, is not available. Next, we discuss some alternatives for the equilibrium selection problem with unknown desired equilibrium.

16.4.1.2   Unknown Desired Nash Equilibrium

In most applications, the set of desired Nash equilibria is unknown. Therefore, the previous approach cannot be directly implemented. In this regard, since it is not possible to regulate the population states towards a specified equilibrium, other criterion should be implemented in the equilibrium selection. Potential criteria include points that maximize the fitness functions, that minimize of the variance, or the combination of both by incorporating weights that assign different prioritization over the n strategies. Based on this, the equilibrium selection problem is now modeled as the following optimization problem:

$$\left\{ \begin{array}{l} \underset{u}{\text{minimize}} \ \mathbb{E}[L(u)], \\[1.5em] \text{subject to} \\[1.5em] (16.24), \ \text{and} \ x(0) \in \Delta. \end{array} \right. \tag{16.31}$$

where the new cost function corresponding to the stochastic control problem for the equilibrium selection can be taken as

$$L(u) = -\langle Q(T)Dx(T), Dx(T)\rangle - \int_0^T \langle QDx, Dx\rangle \, \mathrm{d}t + \int_0^T \langle Ru, u\rangle \, \mathrm{d}t,$$

or as

$$L(u) = \langle Q(T)(\tilde{x}(T)), \tilde{x}(T)\rangle - \left\langle \tilde{Q}(T)D\mathbb{E}[x(T)], D\mathbb{E}[x(T)]\right\rangle$$
$$+ \int_0^T \langle Q\tilde{x}, \tilde{x}\rangle - \left\langle \tilde{Q}D\mathbb{E}[x], D\mathbb{E}[x]\right\rangle \mathrm{d}t$$
$$+ \int_0^T \langle R\tilde{u}, \tilde{u}\rangle + \left\langle \tilde{R}\mathbb{E}[u], \mathbb{E}[u]\right\rangle \mathrm{d}t, \tag{16.32}$$

for the risk-aware control (mean-field-type control), where

$$\tilde{u} = u - \mathbb{E}[u], \quad \tilde{x} = x - \mathbb{E}[x],$$

and $\tilde{Q}, \tilde{R}$ as in (16.29). The latter cost in (16.32) involves also the terms in (16.30), i.e., $\mathrm{var}_Q[x]$, and $\mathrm{var}_R[u]$.

Remark 36 *Even though the original population game may have a unique Nash equilibrium, the equilibrium selection can still be performed by means of Problem in (16.31) due to the fact the matrices $Q, \tilde{Q}, R,$ and $\tilde{R}$ can induce a new Nash equilibrium.*

The previous discussion revealed that if the trajectories of the evolutionary dynamics evolve in the interior of the simplex for all the time, the Nash equilibrium selection problem can be formalized as a stochastic optimization problem, even though the desired Nash point is unknown. We now show in the next section how to construct explicit solutions for both problems in the case where such solutions exist.

16.4.2   Risk-Aware Control Design

Let us consider a stochastic system with the general form

$$\mathrm{d}x = D_r(x, u)\mathrm{d}t + S\mathrm{d}B \tag{16.33}$$

where $D_r(x, u)$ denotes the state-and-control-input dependent drift component and S the state-and-control-input independent diffusion, i.e.,

$$\mathrm{d}x = \left(Ax + \bar{A}\mathbb{E}[x] + Bu + \bar{B}\bar{u}\right)\mathrm{d}t + S\mathrm{d}B. \tag{16.34}$$

The dynamics in (16.34) can be written in terms of $\tilde{x}$ and $\tilde{u}$ given in
(16.29). The following Proposition 24 presents a explicit characterization of
the solutions of general-field-type control problems.

Proposition 24 *Suppose that the following general risk-aware control problem admits a solution:*

$$\underset{u}{\text{minimize}}\ \mathbb{E}\left[\langle Q(T)\tilde{x}(T), \tilde{x}(T)\rangle + \left\langle \tilde{Q}(T)\mathbb{E}[x(T)], \mathbb{E}[x(T)]\right\rangle\right]$$

$$+ \mathbb{E}\int_0^T \langle Q\tilde{x}, \tilde{x}\rangle + \left\langle \tilde{Q}\mathbb{E}[x], \mathbb{E}[x]\right\rangle \mathrm{d}t$$

$$+ \mathbb{E}\int_0^T \langle R\tilde{u}, \tilde{u}\rangle + \left\langle \tilde{R}\mathbb{E}[u], \mathbb{E}[u]\right\rangle \mathrm{d}t,$$

subject to

$$\mathrm{d}x = \left(A\tilde{x} + (A + \bar{A})\mathbb{E}[x] + B\tilde{u} + (B + \bar{B})\mathbb{E}[u]\right)\mathrm{d}t + S\mathrm{d}B,$$
$$x(0) \in \Delta,$$

*where $\tilde{x} = x - \mathbb{E}[x]$, $\tilde{u} = u - \mathbb{E}[u]$, and $Q \succ 0$, $Q + \tilde{Q} \succ 0$, and $R \succeq 0$,
$R + \bar{R} \succeq 0$ are symmetric matrices. The feedback optimal control inputs of the
risk-aware controller are given by:*

$$u^* - \mathbb{E}[u]^* = -\frac{1}{2}R^{-1}B^\top(P^\top + P)(x - \mathbb{E}[x]), \tag{16.35a}$$

$$\mathbb{E}[u]^* = -\frac{1}{2}(R + \bar{R})^{-1}(B + \bar{B})^\top(\bar{P}^\top + \bar{P})\mathbb{E}[x]. \tag{16.35b}$$

The optimal cost is given by

$$L(u^*) = \langle P(0)(x(0) - \mathbb{E}[x(0)]), x(0) - \mathbb{E}[x(0)]\rangle$$
$$+ \left\langle \bar{P}(0)\mathbb{E}[x(0)], \mathbb{E}[x(0)]\right\rangle + \delta(0),$$

where P, $\bar{P}$, and δ solve the following ordinary differential equations:

$$\dot{P} = -Q - PA^\top - AP + PBR^{-1}B^\top P,$$
$$\dot{\bar{P}} = -Q - \bar{Q} - \bar{P}(A + \bar{A})^\top - (A + \bar{A})\bar{P} + \bar{P}(B + \bar{B})(R + \bar{R})^{-1}(B + \bar{B})^\top\bar{P},$$
$$\dot{\delta} = -\left\langle (P^\top + P)S, S\right\rangle,$$

*with boundary terminal conditions $P(T) = Q(T)$, $\bar{P}(T) = Q(T) + \bar{Q}(T)$, and
$\delta(T) = 0$.*

The result in Proposition 24 was generated for a risk-aware control, which corresponds to the risk-aware control since it minimizes variance terms. Other results omitting the variance can be directly retrieved. For instance, Corollary 6 below presents the results for the deterministic and mean-field-free control scenario.

Corollary 6 *Let us consider a deterministic and mean-field-free control problem, whose cost is given by the following function:*

$$\underset{u \in \mathcal{U}}{\text{minimize}} \ \langle \bar{Q}(T)x(T), x(T) \rangle + \int_0^T \langle \bar{Q}x, x \rangle + \langle \bar{R}u, u \rangle \, \mathrm{d}t,$$

subject to

$$\mathrm{d}x = \left(\bar{A}x + \bar{B}u \right) \mathrm{d}t,$$
$$x(0) \in \Delta,$$

where $\bar{Q} \succ 0$, and $\bar{R} \succeq 0$ are symmetric matrices. The feedback optimal control inputs of the mean-field-free control and the optimal cost are given by:

$$u^* = -\frac{1}{2}\bar{R}^{-1}\bar{B}^\top(\bar{P}^\top + \bar{P})x,$$
$$L(u^*) = \langle \bar{P}(0)x(0), x(0) \rangle,$$

where $\bar{P}$ solves the following ordinary differential equations:

$$\dot{\bar{P}} = -\bar{Q} - \bar{P}\bar{A}^\top - \bar{A}\bar{P} + \bar{P}\bar{B}(\bar{R})^{-1}\bar{B}^\top\bar{P},$$

with boundary terminal condition $\bar{P}(T) = \bar{Q}(T)$.

Proof 43 (Corollary 6) *This result is obtained straight-forwardly from Proposition 24 by considering that $x(t) = \mathbb{E}[x(t)]$, $u(t) = \mathbb{E}[u(t)]$, and matrices Q, R, A, and B being null.*

Remark 37 *The results of Proposition 24 and Corollary 6 are developed for the stochastic dynamics (16.24), which correspond to the evolutionary dynamics (16.17) for the case when $x \in int(\Delta_+)$. This allows us to exploit the structure of the dynamics in the interior of the simplex in order to obtain a mathematically tractable problem for the design the control input u. Note, however, that without any further penalty function there is no guarantee that the closed-loop solutions induced by u will remain in $int(\Delta_+)$. However, in order to enforce this restriction, an additional penalty term that grows unbounded as $x \to \partial\Delta_+$ can be added to the cost functions $L(u)$. This type of penalty functions are standard in optimal control problems where additional constraints need to be satisfied by the trajectories of the closed-loop system, e.g., see [151], [152].*

Remark 38 *The equilibrium selection consists of risk-aware problems since the costs functions in (16.8) and (16.32) incorporate variance terms for both the population states and the control input. This can be directly observed by computing the expectation of the cost functions, which generates variance terms.*

We can now proceed to apply the feedback law (16.35) in order to induce particular Nash equilibria in the evolutionary dynamics with imperfect fitness observation.

Remark 39 *The equilibrium selection under other nonlinear population dynamics such as replicator or Smith dynamics would result in a non-LQ-like optimization. Nevertheless, the direct method can potentially be used since it has already been shown in the literature that it is not limited to the linear-quadratic cases [71].*

We have studied the risk-aware control design for the equilibrium selection, i.e., considering that incentives are applied to drive the population to a desired state. Nevertheless, the risk-aware game approach should be also investigated. For instances, assume that the agents belonging to different strategies aim to converge to individual desired proportions x_i^d, $i \in \mathcal{S}$. In addition, let $x_1^d, \ldots, x_n^d$ imply a conflicting population state, i.e., $x^d \notin \Delta$. Thus, each strategic proportion modifies individually its fitness function competing against others while minimizing the efforts (energy).

16.4.3 Illustrative Example

To illustrate the performance of the equilibrium selection in the evolutionary dynamics with imperfect fitness observation by means of a risk-aware control, we consider two different classic population games with fitness functions of the form $f(x) = Dx$, i.e.,

1. The classic Rock-Paper-Scissor Game (RPS) is characterized by the matrix:

$$D = \begin{pmatrix} 0 & 1 & -1 \\ -1 & 0 & 1 \\ 1 & -1 & 0 \end{pmatrix}, \qquad (16.36)$$

2. The classic Zeeman's Game is characterized by the matrix:

$$D = \begin{pmatrix} 0 & 6 & -4 \\ -3 & 0 & 5 \\ -1 & 3 & 0 \end{pmatrix}. \qquad (16.37)$$

For an initial condition belonging to the simplex $x(0) = (0.3, 0.4, 0.3)$, the population states under the deterministic projection dynamics converge to a limit cycle for the RPS game, and to $x^* = (1/3, 1/3, 1/3)$ for the Zeeman's

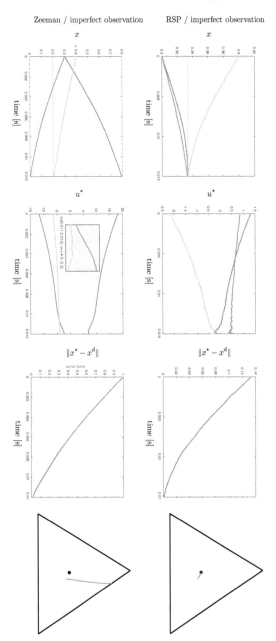

FIGURE 16.23
Behavior of the risk-aware controller over the projection dynamics for both the RSP and Zeeman game with imperfect fitness observation.

game, as presented in Figure 16.22. Moreover, when considering the stochastic projection dynamics, the trajectories are perturbed as shown in Figure 16.22. The objective is to control the projection dynamics with imperfect fitness observation by means of a risk-aware controller to reach the following desired equilibria x^d:

1. RPS Game: $x^d = (1/3, 1/3, 1/3)$; see Figure 16.22.
2. Zeeman's Game: $x^d = (4/5, 0, 1/5)$. This equilibrium point is not stable. Thus, even though the initial population state is $x = x^d$, the trajectories of the states would be repealed by this equilibrium; see Figure 16.22.

Remark 40 *The trajectories under the projection dynamics do not converge to the desired equilibrium point $x^d = (1/3, 1/3, 1/3)$ in the RPS Game, neither to the desired equilibrium point $x^d = (4/5, 0, 1/5)$ in the Zeeman's Game, as shown in Figure 16.22.*

Figure 16.23 presents the evolution of the population states when adopting the risk-aware controller in charged of the equilibrium selection problem (see Figure 16.21(c)). It can be seen that the trajectories are driven to the desired points with a monotone decreasing error $\|x - x^d\|$.

We have presented an equilibrium selection mechanism based on a risk-aware controller that is able to consider both mean and variance minimization within the cost functional and system dynamics. Besides, we have discussed the equilibrium selection problem for cases where the desired equilibrium is known, and when it is unknown. We showed that this equilibrium selection can be seen as a mechanism design where the control input becomes the appropriate incentive to induce a wanted equilibrium. From a different perspective, the equilibrium selection can be interpreted as a creation of incentives a population via external mechanisms, where there exists an upper layer entity modifying rewards in a convenient way to guarantee a desired emerging population state.

16.5 Multi-level Building Evacuation with Smoke

The exposition of this engineering application is mainly divided into two parts. First, we propose a simple Markov chain-based evacuation and smoke model over graphs. Second, we design a risk-aware evacuation controller that provides movement advice for each discretized area in order to minimize the risk(variance).

The model describes the move of a population mass and the smoke propagation throughout a two-dimensional space to be evacuated, and includes several evacuation zones, multiple smoke exits, and also various smoke sources. The model allows reducing the complexity of already reported models in the literature, i.e., we avoid computing complex PDEs offered by other models (e.g., Navier-Stokes approach [153]). We deduce the Master/Kolmogorov-Forward equation from the Markov chain and show that it suitably captures the behavior seen in evacuation procedures, and we formally validate specific features, e.g., the fact the model illustrates the movement of the whole mass to the evacuation areas and that the population mass can exclusively exit through the evacuation zones. The population mass moves to the evacuation areas as quick as possible while avoiding crowd and smoke, while the smoke spreads covering the whole space. To this end, the decision-makers estimate the security level associated with each one of the adjacency zones to which they can move. In a real situation, decision-makers cannot quickly compute the insecurity level for all the direction at every step. Consequently, we consider the insecurity levels to be stochastic incorporating risk terms into the evacuation problem.

The solution of the mean-field-type control problem is computed in a semi-explicit way by setting the problem using mean-field-type theory to reduce the evacuation time. We examine the effect of both the smoke and the risk-awareness. The solution of the mean-field-type control is obtained from solving the backward-forward partial integro-differential system involving both the Hamilton-Jacobi-Bellman equation for mean-field-type terms and the Fokker-Planck-Kolmogorov equation for the evolution of the probability density of the states. To this end, we propose an appropriate guess functional and follow a simple terms completion procedure to optimize the Hamiltonian as we have been studying throughout this book.

16.5.1 Markov-Chain-Based Motion Model for Evacuation over Graphs

Consider a continuous-time Markov chain $\{X_t\}_{t \geq 0}$ with discrete state space $\mathcal{A} = \{1, \ldots, n\}$. Let $p_{k,t|j}$ denote the transition probability from state j to k. From the Chapman-Kolmogorov equation, we obtain that the probability $p_{k,t|j} = \mathbb{P}(X_t = k | X_0 = j)$ can be expressed in terms of the product of probabilities in an intermediate step as follows: $p_{k,t|j} = \sum_{\ell \in \mathcal{A}} p_{k,t|\ell} p_{\ell,t'|j}$ with $\sum_{k \in \mathcal{A}} p_{k,t|\ell} = 1$, for all t. Let $\varrho_{\ell k}$ denote the jump intensity from state ℓ to k with $p_{k,t+\delta|\ell} \approx \delta \varrho_{\ell k}$, and $\sum_{\ell \in \mathcal{A}} \varrho_{k\ell} = 0$. The *Kolmogorov-Forward Equa-*

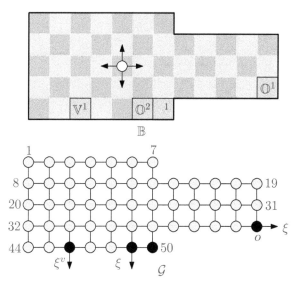

FIGURE 16.24
Representation of the space $\mathbb{B}$, its respective discretization into n regions, and an example graph $\mathcal{G}$, and $\mathcal{V} = \{46\}$, $\mathcal{O} = \{43, 49\}$, and $\mathcal{F} = \{50\}$.

tion/Master Equation is given by

$$\dot{p}_k = \sum_{\ell \in \mathcal{A}} \varrho_{\ell k} p_\ell, \quad \forall k \in \mathcal{A}, \tag{16.38}$$

$$\dot{p}_k = \sum_{\ell \in \mathcal{A} \backslash \{k\}} \varrho_{\ell k} p_\ell - p_k \sum_{\ell \in \mathcal{A} \backslash \{k\}} \varrho_{k \ell}, \quad \forall k \in \mathcal{A}. \tag{16.39}$$

They describe the time evolution of the probabilities of occupying a certain state. The *Master Equation* is used throughout this section to model the evacuation procedure.

Consider a population of decision-makers staying in a two-dimensional space $\mathbb{B} \subset \mathbb{R}^{d \times d}$ (space to be evacuated). When there is an emergency, the whole population makes decisions to move throughout $\mathbb{B}$ in order to reach areas allowing them to evacuate the building. The decision-makers take into consideration several factors to evacuate such as crowd/obstacles avoidance, shortest path, among others. There are some considerations under the evacuation problem as described next. Consider a discretization of $\mathbb{B}$ into a grid composed of $n \in \mathbb{N}$ two-dimensional areas. The set of discretized regions is denoted by $\mathcal{A} = \{1, \ldots, n\}$ (corresponding to the chain discrete state space). Moreover, let $\mathcal{O} \subset \mathcal{A}$ denote the set of selected areas through which it is possible to evacuate the space $\mathbb{B}$, e.g., main doors; and let $\mathcal{V} \subset \mathcal{A}$ be the set of areas where there are smoke exits, e.g., windows. Finally, we consider the set of smoke sources as $\mathcal{F} \subset \mathcal{A}$, which is mainly given by the origin of the fire.

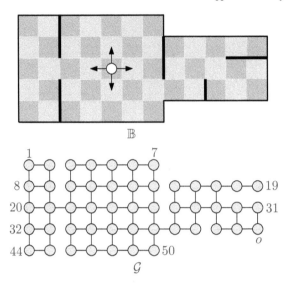

FIGURE 16.25
Representation of spacial constraints such as walls and obstacles in the graph $\mathcal{G}^f$.

Therefore, each discretized region is denoted by $\mathbb{B}^\ell \subset \mathbb{B}$, for all $\ell \in \mathcal{A}$. The evacuation areas are denoted by $\mathbb{O}^i \subset \mathbb{B}$, for all $i \in \mathcal{O}$; the areas where smoke can flow out are denoted by $\mathbb{V}^j \subset \mathbb{B}$, for all $j \in \mathcal{V}$; and the smoke-source regions are given by $\mathbb{F}^k \subset \mathbb{B}$.

There are several constraints throughout $\mathbb{B}$ to move. Such motion constraints are captured by using a graph representation. Let $\mathcal{G} \triangleq (\mathcal{A}, \mathcal{E})$ be a connected graph describing how the decision-maker can move throughout $\mathbb{B}$ where $\mathcal{E} = \{(i,j) : i,j \in \mathcal{A}\}$ is the set of links representing the allowed transitions among the regions in $\mathbb{B}$, i.e., $(i,j) \in \mathcal{E}$ denotes that a proportion of decision-makers in the two-dimensional region i can move to the region j and vice versa. The adjacency matrix of the graph $\mathcal{G}$ is denoted by $A^d \in \{1,0\}^{n \times n}$ and it provides information about the connectivity. Let $A^d = [a^d_{ij}]$ where $a^d_{ij} = 1$ if $(i,j) \in \mathcal{E}$ and $a^d_{ij} = 0$ if $(i,j) \notin \mathcal{E}$. Finally, consider the set of neighbors of a node $i \in \mathcal{A}$ given by $\mathcal{N}_i = \{j : (i,j) \in \mathcal{E}\}$ where $\mathcal{N}_i \neq \emptyset$ provided that $\mathcal{G}$ is connected. It is assumed that the decision-makers composes a mass denoted by m and they are rational in the sense that they can select a region where to stay along the time t. Decision-makers move from a node to another in the discretization of $\mathbb{B}$ following a jump intensity.

Notice that in our model, since the decision-makers can only move to connected regions, then the probability to "jump" in the physical space is null. For instance, the probability to instantaneously move from, an arbitrary region located in the center of $\mathbb{B}$ given by $\ell \in \mathcal{A}$ with $\ell \notin \mathcal{N}_j$, for all $j \in \mathcal{O}$, to the region $\mathbb{O}^j$ is zero. We can write (16.38) as $\dot{p} = M^\top p$ where M is the

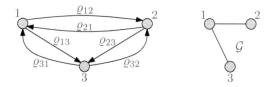

FIGURE 16.26
Relationship between possible jump intensities in the Markov chain and the
links $\mathcal{E}$ in a connected graph $\mathcal{G}$.

jump-intensity matrix. We have that

$$p_{j,t+\delta|\ell} \approx \delta \varrho_{\ell j} = \delta a^d_{\ell j} \varrho_{\ell j} = 0, \ \forall j \in \mathcal{O}, \ \ell \notin \mathcal{N}_j,$$

since $(\ell, j) \notin \mathcal{E}$, and

$$\dot{p} = M^\top p = [M^\top \circ A^d]p.$$

As an example, consider three-state Markov chain with transition rates
$\varrho_{23} = \varrho_{32} = 0$. Then, the relationship between the Markov chain and the
corresponding graph $\mathcal{G}$ is the one presented in Figure 16.26, which illustrates
a Markov chain over graphs. Next, we divide the evacuation modeling into
two parts. First, we present the motion model to move the population mass
to the evacuation areas $\mathbb{O}^\ell$ (Section 16.5.1.1), and then the actual evacuation
to empty the areas $\mathbb{O}^\ell$, for all $\ell \in \mathcal{O}$ (Section 16.5.1.2). Another advantage of
the graph-based model is highlighted in Remark 43 below.

Remark 41 *As another advantage of the proposed model, besides reducing
the complexity of other models to consider crowd dynamics and/or smoke dynamics, is that the consideration of spacial constraints such as walls and/or
obstacles is straight-forward. These constraints are captured by means of the
jump intensities, or equivalently, through the appropriate design of the connected graph representing the discretized space. Figure 16.25 presents an illustrative example in the presence of multiple spacial constraints given by walls.*

16.5.1.1 Reaching Evacuation Areas

Before evacuating the whole area $\mathbb{B}$, it is necessary to ensure that the population mass is driven to the evacuation areas $\mathbb{O}^\ell$, for all $\ell \in \mathcal{O}$.

Remark 42 *The evacuation problem consists of moving the whole population
distribution to the nodes $\mathcal{O}^\ell$, for all $\ell \in \mathcal{O}$. This is a planning problem since
given an initial mass distribution $x_0 \in \mathbb{R}^n$ at time $t = 0$, it is desired to obtain
a population distribution such that $\sum_{\ell \in \mathcal{O}} x_\ell(T) = m$ in a terminal time T
called as evacuation time. Figures 16.27 and 16.28 illustrates the planning
problem.*

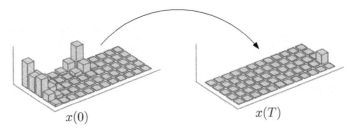

FIGURE 16.27
Mass of people motion with an initial distribution $x(0)$ evacuating the whole space $\mathbb{B}$ within time T and a unique exit area.

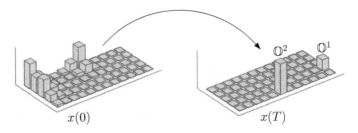

FIGURE 16.28
Mass of people motion with an initial distribution $x(0)$ evacuating the whole space $\mathbb{B}$ within time T and with two exit areas $\mathbb{O}^1$ and $\mathbb{O}^2$.

The probability of occupying the state $i \in \mathcal{A}$ is denoted by $p_i \in \mathbb{R}^+$. Then, $x_i \in \mathbb{R}^+$, and $x_i = p_i m$ denotes the mass of decision-makers occupying the area $i \in \mathcal{A}$. Thus, $x \in \mathbb{X}$ provides information about the mass distribution along the space $\mathbb{B}$ and

$$\mathbb{X} = \left\{ x \in \mathbb{R}^n : \sum_{i \in \mathcal{A}} x_i = m, \ x_i \geq 0, \ \forall i \in \mathcal{A} \right\},$$

with the *Master Equation*

$$\dot{x}_k = \sum_{\ell \in \mathcal{A}\setminus\{k\}} \varrho_{\ell k} x_\ell - x_k \sum_{\ell \in \mathcal{A}\setminus\{k\}} \varrho_{k\ell}, \ \forall k \in \mathcal{A}, \qquad (16.40)$$

and initial distribution $x_k(0) \in \mathbb{X}$. The same reasoning is applied in the evolutionary games community as in Remark 43.

Remark 43 *The Master Equation in (16.40) is associated with the Mean Dynamics used in evolutionary games and population dynamics field as in [41]. To this end, a new term $\varrho_{\ell\ell} = 0$ is added to and subtracted from (16.40), which does not modify the evolution of the population state x_k, for all $k \in \mathcal{A}$.*

Alternatively from the *Master Equation* in (16.40), we have the *Kolmogorov-Forward Equation* as follows:

$$\dot{x}_k = \sum_{\ell \in \mathcal{A}} \varrho_{\ell k} x_\ell, \quad \forall k \in \mathcal{A}, \tag{16.41}$$

with initial condition $x_k(0) \in \mathbb{X}$. Now, the mapping $g_i : \mathbb{R}^+ \to \mathbb{R}_{\leq 0}$, for all $i \in \mathcal{A}$, gets a population mass x_i and returns a non-positive number associated with the level of insecurity of being in the ith state at time t, which is going to serve to define the appropriate jump intensities in the Markov chain describing the evolution of the population mass during the evacuation. The function g_i, for all $i \in \mathcal{A}$, takes into consideration the following aspects:

1. Distance from the discretized region of $\mathbb{B}$, corresponding to $i \in \mathcal{A}$, to evacuation regions $\mathbb{O}^j$, for all $j \in \mathcal{O}$.

2. The crowd at the discretized region of $\mathbb{B}$ corresponding to $i \in \mathcal{A}$.

3. Level of smoke in the discretized region of $\mathbb{B}$ corresponding to $i \in \mathcal{A}$.

For simplicity, we select $g_i = -\psi_i x_i$, where $\psi_i \in \mathbb{R}_{>0}$ takes into consideration the distance and smoke, whereas x_i allows to take into consideration the crowd. Therefore, $\psi_\ell > \varepsilon > 0$, for all $\ell \in \mathcal{O}$, is a small number reflecting a small insecurity level since it corresponds to the evacuation regions to exit the space $\mathbb{B}$.

Remark 44 *The function $g : \mathbb{R}^n_{\geq 0} \to \mathbb{R}^n_{\leq 0}$ returns a non-positive number since there is no safe place in $\mathbb{B}$, i.e., the function g captures the fact that $\mathbb{B}$ needs to be evacuated.*

The specific behavior of the population is determined by the jump intensities ϱ_{ij}, for all $i, j \in \mathcal{A}$. Let

$$\varrho_{\ell k} = \begin{cases} a^d_{k\ell} \dfrac{\max(g_k(x_k) - g_\ell(x_\ell), 0)}{x_\ell}, & \text{if } x_\ell > 0, \\ 0, & \text{otherwise} \end{cases}$$

be the jump intensity making decision-makers move from the area i to j. Decision-makers move if and only if the security level in $g_j(x_j)$ is greater, i.e., if such move helps to evacuate the area $\mathbb{B}$. The dynamics to move the population mass from $\mathbb{B}$ to $\mathbb{O}^\ell$, for all $\ell \in \mathcal{O}$, is

$$\dot{x}_k = \sum_{\ell \in \mathcal{A} \setminus \{k\}} a^d_{k\ell} \left(g_k(x_k) - g_\ell(x_\ell) \right), \quad \forall k \in \mathcal{A},$$

$$= g_k(x_k) \sum_{\ell \in \mathcal{A}} a^d_{k\ell} - \sum_{\ell \in \mathcal{A}} a^d_{k\ell} g_\ell(x_\ell), \quad \forall k \in \mathcal{A},$$

$$\dot{x} = \left(\text{diag}(\mathbf{1}^\top A^d) - A^d \right) g(x). \tag{16.42}$$

Lemma 13 *Dynamics in (16.42) describes the evolution of the population mass within $\mathbb{B}$ without evacuating the regions $\mathbb{O}^\ell$, for all $\ell \in \mathcal{O}$.*

Proof 44 (Lemma 13) *The population mass inside $\mathbb{B}$ remains constant, i.e., there is no evacuation of the regions $\mathbb{O}^\ell$, for all $\ell \in \mathcal{O}$. This is straightforwardly observed from the fact that in $\sum_{k \in \mathcal{A}} \dot{x}_k$ we have*

$$\sum_{k \in \mathcal{A}} \sum_{\ell \in \mathcal{A}} a_{k\ell}^d g_k(x_k) = \sum_{k \in \mathcal{A}} \sum_{\ell \in \mathcal{A}} a_{k\ell}^d g_\ell(x_\ell),$$

since $a_{k\ell}^d = a_{\ell k}^d$.

Lemma 14 *Let $\bar{\psi} \in \min\{\psi_\ell\}_{\ell \in \mathcal{A}\setminus\mathcal{O}}$ denote the minimum insecurity level for all the areas different from the evacuation regions. The equilibrium point x^* for the dynamics (16.42) is given by a population distribution that concentrate people in the evacuation regions $\mathbb{O}^\ell$, for all $\ell \in \mathcal{O}$ since $\psi_o = \varepsilon$, for all $o \in \mathcal{O}$, where $\varepsilon \ll \bar{\psi}$.*

Proof 45 (Lemma 14) *Consider the evacuation dynamics (16.42) with $x^\top \mathbf{1} = m$, for all time t, where $\dot{x} = 0$ only if $g(X) \in \operatorname{span}(\mathbf{1})$, i.e., when $g_i(x_i) = g_j(x_j)$, for all $i, j \in \mathcal{A}$. Thus, taking the insecurity functions $g(x) = -\psi x$ where the matrix $\psi \succ 0$ with*

$$\psi = \operatorname{diag}([\psi_1 \quad \cdots \quad \psi_n]^\top).$$

Then, one obtains that in the worse case

$$\frac{1}{|n - \mathcal{O}|} \sum_{i \in \mathcal{A}\setminus\mathcal{O}} g_i(x_i) = \frac{1}{|\mathcal{O}|} \sum_{o \in \mathcal{O}} g_o(x_o),$$

$$-\frac{1}{|n - \mathcal{O}|} \sum_{i \in \mathcal{A}\setminus\mathcal{O}} \bar{\psi} x_i = -\frac{1}{|\mathcal{O}|} \sum_{o \in \mathcal{O}} \varepsilon x_o,$$

and replacing the functions describing insecurity levels yields

$$\sum_{o \in \mathcal{O}} x_o = \frac{|\mathcal{O}|}{|n - \mathcal{O}|} \frac{\bar{\psi}}{\varepsilon} \sum_{i \in \mathcal{A}\setminus\mathcal{O}} x_i = \frac{\theta}{\varepsilon}\left(m - \sum_{o \in \mathcal{O}} x_o \right),$$

with $\theta = \psi|\mathcal{O}| \cdot |n - \mathcal{O}|^{-1}$, and from which the relationship between m and the $\sum_{o \in \mathcal{O}} x_o$ is obtained that

$$\sum_{o \in \mathcal{O}} x_o \left(\frac{\varepsilon + \theta\bar{\psi}}{\varepsilon} \right) = \theta \frac{\bar{\psi}}{\varepsilon} m,$$

where

$$\sum_{o \in \mathcal{O}} x_o = \frac{\theta}{\varepsilon + \theta} m \approx m,$$

since $\varepsilon \approx 0$, completing the proof.

Lemma 15 *The dynamics in (16.42) drive the whole mass m distributed on* $\mathbb{B}$ *to the evacuation regions* $\mathbb{O}^\ell$, *for all* $\ell \in \mathcal{O}$, *i.e.,* $x \to x^*$ *as time* $t \to \infty$.

Proof 46 (Lemma 15) *Let us consider the following Lyapunov candidate*

$$V_1(x) = \frac{1}{2} \sum_{\ell \in \mathcal{A}} (x_\ell - x_\ell^*)^2 \,.$$

We have

$$\begin{aligned}
\dot{V}_1(x) &= \sum_{\ell \in \mathcal{A}} (x_\ell - x_\ell^*)\, \dot{x}_\ell, \\
&= (x - x^*)^\top \dot{x}, \\
&= -(x - x^*)^\top \left(\mathrm{diag}(\mathbf{1}^\top A^d) - A^d \right) \psi x, \\
&= -x^\top \left(\mathrm{diag}(\mathbf{1}^\top A^d) - A^d \right) \psi x, \\
&\leq 0, \quad \left(since \ \left[\left(\mathrm{diag}(\mathbf{1}^\top A^d) - A^d \right) \psi \right] \succeq 0 \right)
\end{aligned}$$

where equality holds only at x^*, *completing the proof.*

16.5.1.2 Evacuation of the Whole Area

Once the population mass is moved to the evacuation areas $\mathbb{O}^\ell$, for all $\ell \in \mathcal{O}$, by means of the Markov chain modeling, then we consider an outflow rate denoted by $\xi^x \in \mathbb{R}_{>0}$ from $\mathbb{B}$ through $\mathbb{O}^\ell$, making the mass $m \to 0$ as $t \to \infty$. To this end, we consider a modified version of the *Master Equation* as follows:

$$\dot{x}_k = \sum_{\ell \in \mathcal{A}\setminus\{k\}} a_{k\ell}^d \left(g_k(x_k, s_k) - g_\ell(x_\ell, s_\ell) \right) - \xi^x x_k I_{\mathbb{O}}(k), \qquad (16.43)$$

for all $k \in \mathcal{A}$, where $I_{\mathbb{O}}$ denotes the indicator function $I_{\mathbb{O}} : \mathcal{A} \to \{0,1\}$ with $I_{\mathbb{O}}(k) = 0$ if $k \in \mathcal{A} \setminus \mathcal{O}$ and $I_{\mathbb{O}}(k) = 1$ if $k \in \mathcal{O}$. By adding the null term $(g_k(x_k, s_k) - g_k(x_k, s_k))$ to the sum in the latter expression, it yields in a compacted form that

$$\dot{x} = \left(\mathrm{diag}(\mathbf{1}^\top A^d) - A^d \right) g(x, s) - \xi^x H x, \qquad (16.44)$$

where $H \in \{0,1\}^{n \times n}$ being $H = \mathrm{diag}(I_{\mathbb{O}}(1), \ldots, I_{\mathbb{O}}(n))$. Notice that, the latter equation is directly obtained from the *Density-Dependent Master Equation*, i.e.,

$$\dot{x}_k = \sum_{\ell \in \mathcal{A}\setminus\{k\}} a_{\ell k}^d \varrho_{\ell k} x_\ell - x_k \sum_{\ell \in \mathcal{A}\setminus\{k\}} a_{k\ell}^d \varrho_{k\ell} - \xi^x x_k I_{\mathbb{O}}(k), \qquad (16.45)$$

for all $k \in \mathcal{A}$, which is a subtle modification of the *Master Equation* in (16.39). The role that (16.39) plays in the evolutionary games field has been highlighted in Remark 43. Moreover, when the population mass is allowed to change along the time, then the expression in (16.45) is also related to evolutionary games as pointed out in Remark 45.

Remark 45 *When either death or birth is considered within the population, the mass m is not longer constant obtaining the density-dependent evolutionary dynamics. For instance, in [154], the density-dependent replicator dynamics are studied by considering dynamics over m, and this behavior can be extended to any evolutionary dynamics.*

At first impression, the density-independent and density-dependent evolutionary dynamics are not related to each other since one of them satisfy invariance of the simplex $\mathbb{X}$ whereas the other does not. Nevertheless, the following result shows that the density-dependent dynamics could be considered as density-independent dynamics by incorporating a sink for the death factor.

Proposition 25 *The particular density-dependent dynamics (16.44) is equivalent to the density-independent dynamics (16.40) by introducing a new sink strategy denoted by y, i.e., $\mathcal{A}^y = \mathcal{A} \cup \{y\}$ and $\mathcal{E}^y = \mathcal{E} \cup \{(o,y)\}$, for all $o \in \mathcal{O}$; and with corresponding joining jump intensity given by the outflow in the building, i.e., $\varrho_{oy} = \xi^x$ and $\varrho_{yo} = 0$, for all $o \in \mathcal{O}$.*

Proof 47 (Proposition 25) *For the sake of clarity, let us write the dynamics (16.44) in terms of inflow and outflow. First, let us consider the dynamics corresponding to the nodes $\mathcal{A} \setminus \mathcal{O}$, i.e.,*

$$\dot{x}_k = \sum_{\ell \in \mathcal{A} \setminus \{k\}} a_{\ell k}^d \varrho_{\ell k} x_\ell - x_k \sum_{\ell \in \mathcal{A} \setminus \{k\}} a_{k\ell}^d \varrho_{k\ell}, \forall k \in \mathcal{A} \setminus \mathcal{O},$$

since $(\ell, y) \notin \mathcal{E}$, for all $\ell \in \mathcal{A} \setminus \mathcal{O}$, the sum set can be changed into $\mathcal{A}^y$ as follows:

$$\dot{x}_k = \sum_{\ell \in \mathcal{A}^y \setminus \{k\}} a_{\ell k}^d \varrho_{\ell k} x_\ell - x_k \sum_{\ell \in \mathcal{A}^y \setminus \{k\}} a_{k\ell}^d \varrho_{k\ell}, \forall k \in \mathcal{A} \setminus \mathcal{O}.$$

Now, we analyze the dynamics for all $o \in \mathcal{O}$ corresponding to the evacuation areas $\mathbb{O}^o$, i.e.,

$$\dot{x}_o = \sum_{\ell \in \mathcal{A} \setminus \{o\}} a_{\ell o}^d \varrho_{\ell o} x_\ell - x_o \sum_{\ell \in \mathcal{A} \setminus \{o\}} a_{o\ell}^d \varrho_{o\ell} - \xi^x x_o, \forall o \in \mathcal{O},$$

since $\varrho_{yo} = 0$, for all $o \in \mathcal{O}$, these dynamics can be re-written changing the sum set into $\mathcal{A}^y$ as follows:

$$\dot{x}_o = \sum_{\ell \in \mathcal{A}^y \setminus \{o\}} a_{\ell o}^d \varrho_{\ell o} x_\ell - x_o \overbrace{\left(\sum_{\ell \in \mathcal{A} \setminus \{o\}} a_{o\ell}^d \varrho_{o\ell} - \xi^x \right)}^{\sum_{\ell \in \mathcal{A}^y \setminus \{o\}} a_{o\ell}^d \varrho_{o\ell}}.$$

Finally, considering the dynamics for the fictitious strategy $y \in \mathcal{A}^y$ yields

$$\dot{x}_y = \xi^x x_o = \varrho_{oy} x_o - x_y \varrho_{yo},$$

and $a_{y\ell}^d = 0$, for all $\ell \in \mathcal{A} \setminus \mathcal{O}$, since y is only connected to all $o \in \mathcal{O}$, the sum over all the other terms is zero and

$$\dot{x}_y = \sum_{\ell \in \mathcal{A}^y \setminus \{y\}} a_{\ell y}^d \varrho_{\ell y} x_\ell - x_y \sum_{\ell \in \mathcal{A}^y \setminus \{y\}} a_{y\ell}^d \varrho_{y\ell},$$

showing the announced equivalence.

Proposition 26 *The model* (16.44) *captures the population mass evacuation from* $\mathbb{B}$, *i.e.,* $m \to 0$ *as* $t \to \infty$.

Proof 48 (Proposition 26) *Consider the Lyapunov function* $V(m) = m^2$. *Therefore,* $\dot{V}(m) = 2m\dot{m}$, *and*

$$\dot{V}(m) = 2m\mathbf{1}^\top \left[(\mathrm{diag}(\mathbf{1}^\top A^d) - A^d) g(x,s) - \xi^x Hx \right],$$

where $\mathbf{1}^\top \mathrm{diag}(\mathbf{1}^\top A^d) = \mathbf{1}^\top A^d$, *and*

$$\dot{V}(m) = -2m\xi^x \mathbf{1}^\top Hx = -2m\xi^x \sum_{o \in \mathcal{O}} x_o \leq 0,$$

with $m \geq 0$. *Since all the population mass is driven to the areas* $\mathbb{O}^o$, *for all* $o \in \mathcal{O}$, *then* $x_o \geq 0$ *along the time and* $\dot{V}(m) = 0$ *holds only if* $m = 0$ *completing the proof.*

Proposition 27 *Evacuation is only allowed through the safety area* $\mathbb{O}$ *under dynamics* (16.44).

Proof 49 (Proposition 27) *This proof is straight-forward by considering the dynamics* $\dot{m}$, *i.e.,*

$$\dot{m} = -\xi^x \mathbf{1}^\top Hx = -\xi^x \sum_{o \in \mathcal{O}} x_o,$$

where $\dot{m} = 0$ *if* $\xi^x = 0$. *Therefore, evacuation is only allowed through the evacuation areas* $\mathbb{O}^o$, *for all* $o \in \mathcal{O}$. *This is also related to the result presented in Lemma 13.*

16.5.1.3 Jump Intensities for Evacuation

The level of insecurity is determined by three factors

$$g_i(x_i, s_i) = - \left(\varepsilon_i^d + \varepsilon_i^c + s_i \right) x_i$$

as follows:

- *Distance:* The parameter $\varepsilon_i^d \in \mathbb{R}_{<0}$ penalizes the distance from the ith state to the any $\mathbb{O}^\ell$.

- *Crowd:* There is a parameter $\varepsilon^c \in \mathbb{R}_{<0}$ that penalizes the crowd located at the state $i \in \mathcal{A} \setminus \{o\}$.

- *Smoke:* The parameter $\varepsilon^s \in \mathbb{R}_{<0}$ penalizes the amount of smoke around the state $i \in \mathcal{A} \setminus \{o\}$.

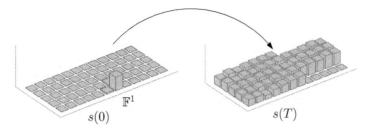

FIGURE 16.29
Smoke spread given a fire source $s(0)$ and covering the whole space $\mathbb{B}$ within time T, i.e., smoke distribution $s(T)$.

16.5.2   Markov-Chain-Based Modeling for Smoke Motion

We take into account the role of smoke in the evacuation problem since it can considerably modify the trajectories selected by the decision-makers. Indeed, the evacuation trajectories vary not only depending on the initial condition for the population distribution, but also on the origin of the smoke (fire). The smoke dynamics is normally described by the *Navier-Stokes* equations [153]. Thus, the evolution of the smoke dynamics is described by means of partial differential equations. In contrast, we model the evolution of the smoke distribution by using the Markov chain-based motion model as we did for the decision-makers. Differently, smoke spreads throughout the space covering the whole space $\mathbb{B}$ as time passes. Moreover, let us consider areas denoted by $\mathbb{V}^i$ with $i \in \mathcal{V}$, where there are exits for the smoke, e.g., windows, and with an exit rate given by $\xi^v \in \mathbb{R}^+$. Thus, we also model how the smoke evacuates the area $\mathbb{B}$.

Let $s_i \in \mathbb{R}^+$ denote the proportion of smoke around the state $i \in \mathcal{A}$. Similar to the motion for the decision-makers, the smoke propagates following a connected graph representing the possible spread in the physical space, i.e.,

$$\dot{s}_k = \sum_{\ell \in \mathcal{A}\setminus\{k\}} \varrho_{\ell k} s_\ell - s_k \sum_{\ell \in \mathcal{A}\setminus\{k\}} \varrho_{k\ell} - \xi^v s_k I_{\mathbb{V}}(k), \qquad (16.46)$$

for all $k \in \mathcal{A}$, where $\sum_{i \in \mathcal{A}} s_i(0) = m^s$. It is considered that the smoke propagates until achieving an uniform distribution throughout the space $\mathbb{B}$. Then, let $h_i(s_i) : \mathbb{R} \to \mathbb{R}$ denote the "incentives" smoke has to propagate, e.g., we consider $h_i(s_i) = -s_i$ reflecting a propagation covering the whole space as shown in the following Lemma:

Lemma 16 *The smoke is distributed in an evenhanded or homogeneous manner throughout the space when there are no exits, i.e., $\mathcal{V} = \emptyset$ (See Figure 16.29).*

Proof 50 (Lemma 16) *We follow the same procedure presented in Lemma 14 showing that the dynamics (16.46) correspond to a consensus algorithm.*

Lemma 17 *The smoke mass is reduced as there are exits throughout the space* $\mathbb{B}$.

Proof 51 (Lemma 17) *This result is straight-forward by applying the same reasoning applied in Proposition 26.*

16.5.3 Mean-Field-Type Control for the Evacuation

Let us consider the evacuation dynamics in (16.44) with uncertainties associated with the perception of the safety levels at each discretized region of $\mathbb{B}$ as follows:

$$\dot{x} = \left(\text{diag}(1^\top A^d) - A^d\right)(g(x,s) + \omega) - \xi^x H x, \tag{16.47}$$

where, according to the jump intensities

$$g_i(x_i, s_i) = -\left(\varepsilon_i^d + \varepsilon_i^c + s_i\right) x_i,$$

and

$$g(x,s) = -\left(E^d + E^c + S\right) x,$$

with $E^d = \text{diag}(\varepsilon^d)$, $E^c = \text{diag}(\varepsilon^c)$, and $S = \text{diag}(s)$. Besides, ω denotes a white noise where $\omega dt = dW$ denoting W a standard Brownian motion. We propose to design a mean-field-type controller in charged of minimizing the risk in the evacuation problem by means of a control input signal Bu where $u \in \mathbb{R}^n$ denotes a the mean-field-type control input and $B \in \mathbb{R}^{n \times n}$, i.e.,

$$\dot{x} = \left(\text{diag}(1^\top A^d) - A^d\right)(g(x,s) + \omega) - \xi^x H x + Bu. \tag{16.48}$$

The control input can provide suggestions about the security levels for all the discretization areas. It is worthy to point out that B has the same graph-constrained structure, and we do not allow the control input to modify the security levels for the evacuation areas $\mathbb{O}^\ell$, for all $\ell \in \mathcal{O}$. Thus, let us consider the indicator function $I_{\mathbb{B}\setminus\mathbb{O}} : \mathcal{A} \rightarrow \{0,1\}$, where $I_{\mathbb{B}\setminus\mathbb{O}}(k) = 0$ if $k \in \mathcal{O}$, and $I_{\mathbb{B}\setminus\mathbb{O}}(k) = 1$ if $k \in \mathcal{A} \setminus \mathcal{O}$. Hence, $B = \left(\text{diag}(1^\top A^d) - A^d\right) \text{diag}(I_{\mathbb{B}\setminus\mathbb{O}}(1), \dots, I_{\mathbb{B}\setminus\mathbb{O}}(n))$.

Remark 46 *The control input only affects the transition rates by means of the security functions but not the evacuation rate. If there is no evacuation rates from $\mathbb{B}$, i.e., either $\xi^x = 0$, $H = \mathbf{0}$ or $\mathcal{O} = \{\emptyset\}$, then the population mass inside $\mathbb{B}$ remains constant under the dynamics (16.48).*

Remark 46 shows that the control input satisfies the evacuation constraints, i.e., it is only possible to evacuate through $\mathbb{O}^\ell$, for all $\ell \in \mathcal{O}$, and it is only possible to provide advise to move to adjacency areas satisfying the graph $\mathcal{G}$. The control input $u_i \in \mathbb{R}$ can be seen or interpreted as an ideal advice that, the population mass in the space corresponding to $i \in \mathcal{A}$, receives in order to optimize the evacuation. Thus, (16.47) is rewritten incorporating a control action u as a stochastic differential equation as follows:

$$\mathrm{d}x = (Ax(t) + Bu(t)) \, \mathrm{d}t + \sigma \mathrm{d}W(t),$$

where

$$A = -\overbrace{\left(\mathrm{diag}(1^{\top}A^d) - A^d\right)}^{\text{movement constraints}} \overbrace{\left(E^d + E^c + S - \xi^x H\right)}^{\psi \text{ safety perception}},$$
$$\sigma = \left(\mathrm{diag}(1^{\top}A^d) - A^d\right).$$

The mean-field-type control takes into consideration not only the mean of the variables of interest but also risk quantities expressed in terms of the variance.

Let $\phi(t, dx)$ be the probability measure of x. Let us consider the following compacted notation for the variance terms:

$$\mathrm{var}(x(t)) = \mathbb{E}\langle Q(t)(x(t) - \mathbb{E}[x(t)]), x(t) - \mathbb{E}[x(t)]\rangle,$$
$$\mathrm{var}(u(t)) = \mathbb{E}\langle R(t)(u(t) - \mathbb{E}[u(t)]), u(t) - \mathbb{E}[u(t)]\rangle,$$

where the expectation values are computed by means of the probability measure $\phi(t, dt)$, i.e.,

$$\mathbb{E}[x_i] = \int_{\mathbb{R}^n} x_i \, \phi(t, \mathrm{d}x_i, \mathrm{d}x_{-i}),$$

$$\mathbb{E}[u_i] = \int_{\mathbb{R}^n} u_i(\phi, x)\phi(t, \mathrm{d}x_i, \mathrm{d}x_{-i}).$$

For the sake of simplicity and abusing of notation, we denote the expected value of the vectors x and u as

$$\mathbb{E}[x] = \int x \, \phi(t, \mathrm{d}x),$$

$$\mathbb{E}[u] = \int u(t, x, \phi)\phi(t, \mathrm{d}x).$$

Then, we define both the terminal cost $h(x, \phi)$ and running cost $l(x, \phi, u)$ in (16.49) and (16.50), respectively.

$$h(x, \phi) = \left\langle \tilde{Q}(T) \int y\phi(T, \mathrm{d}y), \int y\phi(T, \mathrm{d}y) \right\rangle$$
$$+ \left\langle Q(T)\tilde{x}(T) \left(x(T) - \int y\phi(T, \mathrm{d}y)\right), x(T) - \int y\phi(T, \mathrm{d}y) \right\rangle,$$

$$(16.49)$$

$$l(x,\phi,u) = \left\langle \tilde{Q}(t) \int y\phi(t,dy), \int y\phi(t,dy) \right\rangle$$

$$+ \left\langle Q(t) \left(x(t) - \int y\phi(t,dy) \right), x(t) - \int y\phi(t,dy) \right\rangle$$

$$+ \left\langle \tilde{R}(t) \int u(t,y,\phi)\phi(t,dy), \int u(t,y,\phi)\phi(t,dy) \right\rangle$$

$$+ \left\langle R(t) \left(u(t) - \int u(t,y,\phi)\phi(t,dy) \right), u(t) - \int u(t,y,\phi)\phi(t,dy) \right\rangle,$$

$$(16.50)$$

The mean-field-type control problem is as follows:

$$\underset{u\in\mathbb{R}^n}{\text{minimize}} \ \mathbb{E}[L(x,\phi,u)] = \mathbb{E}\left[h(x,\phi) + \int_0^T l(x,\phi,u)dt \right], \qquad (16.51a)$$

subject to

$$dx(t) = (Ax(t) + Bu(t))\,dt + \sigma dW(t), \qquad (16.51b)$$

$$x(0) \triangleq x_0 \text{ (given initial population distribution).} \qquad (16.51c)$$

The optimal cost is denoted by

$$V(t,\phi) = \underset{u}{\text{minimize}} \ \mathbb{E}\left[L(x,\phi,u) \right].$$

Proposition 28 *Let $Q(t) \succ 0$, $Q(t) + \bar{Q}(t) \succ 0$, and $R(t) \succeq 0$, $R(t) + \bar{R}(t) \succeq 0$, for all t, be symmetric matrices. The feedback optimal control input $u^*(t)$ of the risk-aware controller is given by:*

$$u^*(t) = -\frac{(R(t) + \bar{R}(t))^{-1}B^\top(\bar{P}(t)^\top + \bar{P}(t))}{2}\int y\phi(t,dy)$$

$$- \frac{R(t)^{-1}B^\top(P(t)^\top + P(t))}{2}\left(x(t) - \int y\phi(t,dy) \right).$$

where P, $\bar{P}$, and δ solve the following ordinary differential equations:

$$\dot{P}(t) = -Q(t) - P(t)A^\top - AP(t) + P(t)BR(t)^{-1}B^\top P(t),$$

$$\dot{\bar{P}}(t) = -Q(t) - \bar{Q}(t) - \bar{P}(t)A^\top - A\bar{P}(t)$$

$$+ \bar{P}(t)B(R(t) + \bar{R}(t))^{-1}B^\top \bar{P}(t),$$

$$\dot{\delta}(t) = -\left\langle (P(t)^\top + P(t))\sigma, \sigma \right\rangle,$$

with terminal boundary conditions $P(T) = Q(T)$, $\bar{P}(T) = Q(T) + \bar{Q}(T)$, and $\delta(T) = 0$.

Proof 52 (Proposition 28) *This proof follows the same reasoning as the direct method procedure. Here, we highlight the HJB solution introduced in Chapter 1, Section 1.4. The solution of problem in (16.51) is given by solving the Backward-Forward Partial Integro-Differential System given by [26]*

$$0 = V_t(t, \phi) + H(x, \phi, V_{x\phi}, V_{xx\phi}), \tag{16.52a}$$

$$V(T, \phi) = \int_x h(x, \phi)\phi(T, dx), \tag{16.52b}$$

$$\frac{\partial}{\partial t}\phi = -\sum_{i=1}^{n} \frac{\partial}{\partial x_i} [Ax(t) + Bu(t)]_i \, \phi$$

$$+ \frac{1}{2}\sum_{i=1}^{n}\sum_{j=1}^{n} \frac{\partial^2}{\partial x_i \partial x_j} [\sigma_{ij}\phi], \tag{16.52c}$$

$$\phi(0) = \phi_0. \tag{16.52d}$$

where (16.52a) constitutes the Hamilton-Jacobi-Bellman equation for the mean-field-type control with terminal condition (16.52b), and with Hamiltonian $H(x, \phi, V_{x\phi}, V_{xx\phi})$. Hence, (16.52c) is the Fokker-Planck-Kolmogorov equation describing the evolution of the joint probability density with boundary condition (16.52d). Thus,

$$H(x, \phi, V_{x\phi}, V_{xx\phi}) = \underset{u}{\text{minimize}} \int l(x, \phi, u) + \left\langle \frac{\partial^2}{\partial x \partial \phi} V(t, \phi), Ax(t) + Bu(t) \right\rangle$$

$$+ \frac{1}{2} \left\langle \frac{\partial^3}{\partial x^2 \partial \phi} \sigma V(t, \phi), \sigma \right\rangle \phi(t, dx).$$

We propose the following guess functional $V(t, \phi)$ for the solution:

$$\left\langle P(t) \left(x(t) - \int y\phi(t, dy) \right), x(t) - \int y\phi(t, dy) \right\rangle$$

$$+ \left\langle \bar{P}(t) \int y\phi(t, dy), \int y\phi(t, dy) \right\rangle + \delta(t) \int \phi(t, dy). \tag{16.53}$$

Therefore,

$$\frac{\partial}{\partial t} V(t, \phi) = \left\langle \dot{P}(t) \left(x(t) - \int y\phi(t, dy) \right), x(t) - \int y\phi(t, dy) \right\rangle$$

$$+ \left\langle \dot{\bar{P}}(t) \int y\phi(t, dy), \int y\phi(t, dy) \right\rangle + \dot{\delta}(t)$$

$$- \left\langle (P(t)^\top + P(t)) \left(x(t) - \int y\phi(t, dy) \right), \int (Ax(t) + Bu(t)) \phi(t, dy) \right\rangle$$

$$+ \left\langle (\bar{P}(t)^\top + \bar{P}) \int y\phi(t, dy), \int (Ax(t) + Bu(t)) \phi(t, dy) \right\rangle,$$

and

$$\frac{\partial^2}{\partial x \partial \phi} V(t, \phi) = (P(t)^\top + P(t)) \left(x - \int y \phi(t, dy) \right),$$

$$\frac{\partial^3}{\partial x^2 \partial \phi} V(t, \phi) = (P(t)^\top + P(t)).$$

The Hamiltonian is strictly convex in both u and

$$u - \int u(t, y, \phi) \phi(t, dy).$$

Thus, we apply the terms completion and follow a similar procedure for the terms in $\int u(t, y, \phi) \phi(t, dy)$, obtaining the optimal control input. Moreover, replacing back the terms completion in (16.52a), the announced Riccati equations are obtained and the boundary conditions are set by (16.52b), completing the proof.

Remark 47 *This proof can be also developed by applying the direct method using the guess functional $V(t, \phi)$ in (16.53).*

16.5.4 Single-Level Numerical Results

Consider the single-level (single-population) example presented in Figure 16.24 involving 50 states in the discretized space $\mathbb{B}$, i.e., $\mathcal{A} = \{1, \ldots, 50\}$, where the evacuation areas $\mathbb{O}^1$, and $\mathbb{O}^2$ correspond to the states $\mathcal{O} = \{43, 49\} \subset \mathcal{A}$. We analyze three different scenarios. First, we consider the control-free scenario without smoke. Under this scenario, people prefer to evacuate by using the nearest exit, which according to the initial population distribution, it is $\{49\} \in \mathcal{O}$ for most of them (see Table 16.3).

As a second scenario, the free-control case is performed, where it can be observed that the population moves throughout $\mathbb{B}$ reaching the evacuation areas. Under this scenario we consider smoke, whose propagation is presented in Figure 16.30. Figure 16.31 shows the percentage reduction of the population inside the area $\mathbb{B}$ along the time, which is a reduction of 39% after 500 seconds.

Table 16.3 shows the effect of the smoke in the evacuation. The source of the smoke is close to evacuation region $\{49\} \in \mathcal{O}$ modifying the behavior observed in Scenario 1. The risk(variance) behavior can be seen in Figure 16.33 when there is no controller. The noisy behavior is caused by the uncertainties related to the measurements to compute the security-level functions and the smoke propagation, and such volatility belongs to an approximated range $[-0.5, 0.5]$. Figure 16.32 shows details about the evolution of the population during the evacuation.

On the other hand, the risk-aware controller is implemented in order to

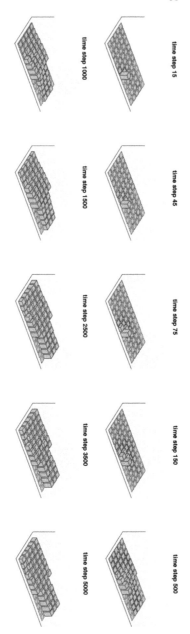

FIGURE 16.30
Evolution of the smoke throughout the area $\mathbb{B}$ for 8.3 minutes and with time steps 0.1 seconds.

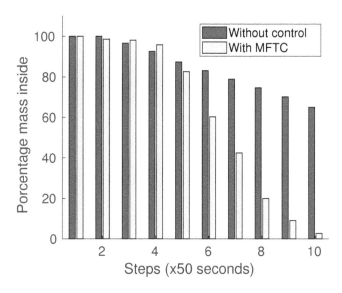

FIGURE 16.31
Evacuation comparison with and without mean-field-type controller.

TABLE 16.3
Summary of results for three different scenarios with and without MFTC/smoke.

Scn.	MFTC	Smoke	Smoke Source	Total Evac.	Evacuation $\{43\} \in \mathcal{A}$	Evacuation $\{49\} \in \mathcal{A}$
1	NO	NO	-	99%	21.5455%	78.4545%
2	NO	YES	$\{50\}$	39%	47.4304%	52.5696%
3	YES	YES	$\{50\}$	90%	44.9810%	55.0190%

provide people an advise to move along the space $\mathbb{B}$ to maximize the evacuation rates. Figure 16.31 shows that, when adopting the risk-aware controller, the evacuation is considerably enhanced, i.e., after 500 seconds, more than 90% percent of the population mass is evacuated. It is important to recall Remark 46 and highlight that the control inputs do not modify the population mass when there is no evacuation areas, i.e., the control input only modifies the perception security-level functions. Figure 16.33 shows that the risk(variance) is also reduced, and now the volatility belongs to an approximated range $[-2.5, 2.5]$. Table 16.3 presents the improvement in the evacuated mass when minimizing the risk.

We have proposed an evacuation model by means of Markov chains and the discretization of the space. To this end, we have presented the Master/Kolmogorov-Forward equation –which is obtained from the Chapman-

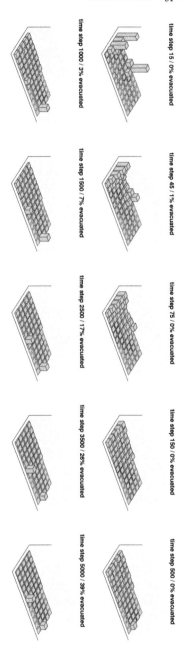

FIGURE 16.32
Evolution of the population mass throughout the area $\mathbb{B}$ without risk minimization for 8.3 minutes and with time steps 0.1 seconds.

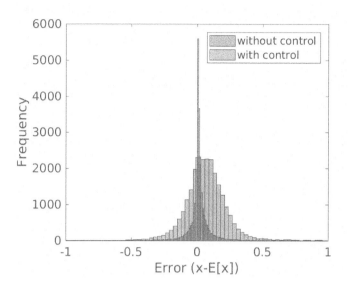

FIGURE 16.33
Variance comparison with and without mean-field-type controller.

Kolmogorov equation– and that uses jump intensities to represent the probability of being occupying a certain state. The appropriate selection of the jump intensities leads to a suitable evacuation model. We have shown that the behavior of both, people evacuation and smoke propagation, can be captured by the simple Markov-chain-based modeling instead of using complex PDEs such as in the Navier-Stokes model. We have discussed both the density-independent and density-dependent approaches where death and/or birth is allowed, and have formally established the relationship between them. More further work related to this approach, we refer the reader to [31].

We have presented the effects of the smoke and the effect of uncertainties (risk-awareness) in the evacuation process. We have analyzed the model by using invariant set analysis, Lyapunov stability analysis, and have solved the risk-aware (variance-aware) problem semi-explicitly by means of mean-field-type theory. To this end, we presented the backward-forward partial integro-differential system to be solved including the respective Hamilton-Jacobi-Bellman and Fokker-Planck-Kolmogorov equations.

The system has been solved by postulating the appropriate guess functional and following a simple terms completion procedure to optimize the Hamiltonian, which is convex with respect to the control inputs. The direct method can be implemented to solve the problem by using the same guess functional presented in this application.

16.6   Coronavirus Propagation Control

At the end of 2019, coronavirus disease (COVID-19) appeared in Wuhan, China. After very few months, by April 2020, already millions of infected cases have been reported in the entire globe affecting hundreds of countries throughout all the continents. This COVID-19 pandemic has left more than a million deaths. In order to reduce the propagation of the virus and achieve an effective monitoring of its progress, governments have imposed some rules/policies such as canceling both international and domestic flights, imposing lock-downs, decreasing the congestion in the public transportation systems, and/or recommending social distance/isolation. All these decisions have a tremendous economic implications and, in some cases, are not effective enough [155]. There is a quite urgent necessity in studying how to model and control the spread of the virus. Several studies have been already reported worldwide as presented next.

Regarding the behavior of the virus, some researchers have studied its response against different temperature conditions in order to identify in which regions the virus could spread in an easier manner [156]; and in [157], the relationship between the exposition with the pollution and the COVID-19 infection is studied. Other studies are oriented to identify how the immune system mechanisms and response against the COVID-19 can be improved [158]. In contrast, there are other reported analysis focusing on the economic impact of the virus. For instance, in [159], it is studied if the COVID-19 affects the stock market and how the amount of infected and death cases have a negative implication over the stock returns. In [160], the economic and social impact of the COVID-19 in the financial markets and institutions is analyzed by considering either a direct or indirect way. Finally, as a third main approach, researchers are working on the appropriate modeling for the propagation of the COVID-19 by using, adapting and/or enhancing pandemic mathematics. For instance, different types of SIS and SIR stochastic epidemic models are presented in [161], the Auto-Regressive Integrated Moving Average models, which are used to predict the epidemiological tendency of the virus as in [162], where the study is mainly focused on the Italian, Spanish and French cases. Hence, some deterministic and stochastic SI models considering a transmission rate [163], or the so-called Kolmogorov-forward equation as it is made in [164] by means of several methods oriented to stochastic disease dynamics modeling. Other models based on ordinary differential equations have been presented for specific regions as in [165], where the Italian case is considered. Finally, and quite related to the topic addressed in this book, the work in [166] analyzes the COVID-19 from a mean-field-type game-theoretical perspective. As an example, and to motivate the importance of modeling and studying how to control the propagation of a pandemic, Figure 16.34 shows the quick propagation of the virus in Colombia when some domestic flights are allowed throughout the country during just 400 iterations representing possible interactions in the population.

16.6.1 Single-Player Problem

To model the propagation of the COVID-19 we consideration that the individuals can belong to any of the following possible states:

Exposed: in this state the individual is healthy and exposed to get the disease

Quarantined: in this state, we assume that the individual is healthy and is in quarantine in order to reduce the probability to get infected

Infected: in this state the individual is sick by the COVID-19 and is susceptible to either passing away or recovering from the disease

Recovered: in this state, the individual is healthy after having suffered from the disease

Death: this state indicates that the individual has passed away because of the COVID-19

Let us consider a Markov process $\{X(t)\}_{t \geq 0}$ with the discrete state space $\mathcal{S}$ given by the possible states previously introduced, i.e., $\mathcal{S} = \{E, Q, I, R, D\}$ corresponding to Exposed, Quarantined, Infected, Recovered, and Death, respectively. Let $\mathbb{P}(X(t) = s) \in [0, 1]$ denote the probability that the state $X(t)$ is $s \in \mathcal{S}$ at time t. Hence, let $m(t) = [\mathbb{P}(X(t) = s)]_{s \in \mathcal{S}} \in [0, 1]^{|\mathcal{S}|}$ denote the probability distribution of $X(t)$, from which an histogram may be constructed.

The government and/or authorities impose some laws and/or policies to control the propagation of the virus, e.g., close the borders, cancel domestic and international flights, close restaurants, suspend concerts, impose compulsory quarantine, etc. These decisions are given by means of control inputs denoted by $u \in \mathbb{R}^{|\mathcal{S}|}$ and are normally economically costly.

Let $q_{ss'}(m(t), u(t)) \geq 0$ denote the transition rate (also known as jump intensities) from state $s \in \mathcal{S}$ to state $s' \in \mathcal{S}$ such that

$$q_{ss} = - \sum_{s' \in \mathcal{S} \setminus \{s\}} q_{ss'}.$$

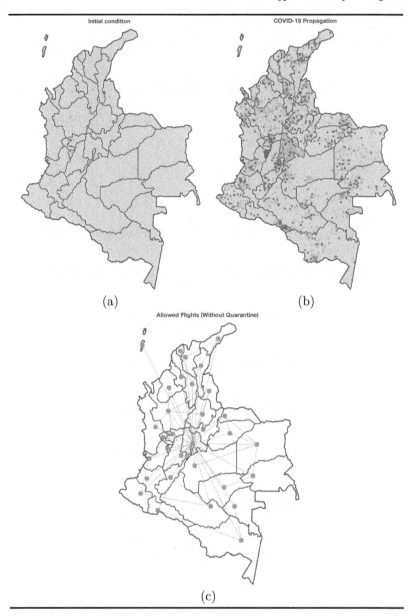

FIGURE 16.34
Propagation of the virus in Colombia with migration constraints among the
32 departments without quarantine prevention. Dots in the map represent
infected/dead people: (a) initial condition, (b) spread of the virus after 400
iterations (interactions), (c) allowed air traffic.

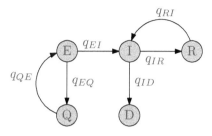

FIGURE 16.35
Transition rates among the finite states $\mathcal{S}$.

Figure 16.35 shows the transition rates among the different states in the COVID-19 modeling. The transition rates $q_{ss'} : [0,1]^{|\mathcal{S}|} \times \mathbb{U} \to \mathbb{R}_{\geq 0}$, for all $s, s' \in \mathcal{S}$, are functions depending on the probabilities of being occupying a state from $\mathcal{S}$ at certain time t, and the imposed policies and/or laws form the government to mitigate the propagation of the virus. Next, we introduce two different approaches. First, we present the atomic case that models an individual changing from one state to another according to the transition rates. Second, we present a mean-field-type approach that models the probabilities of being occupying a certain state. If a large (or infinite) number of individuals is considered, then the probabilities can be interpreted as a proportion of individuals and the mean-field term is given by the distribution throughout the finite states $\mathcal{S}$.

16.6.1.1   Control Problem of Mean-Field Type

The authorities are interested in having control over the propagation of the virus by imposing laws/policies by means of the signals $u \in \mathbb{R}^{|\mathcal{S}|}$, which directly affect the transition rates among the states $\mathcal{S}$ in the Markov chain as shown in Figure 16.35. At the same time, it is important to guarantee that the designed policies do not considerably affect the economy of the country/region under study.

There is therefore a dilemma, or in other words, there are two contradictory objectives that have to be appropriately balanced. On one hand, even though it is well known that, so far, the isolation is the best strategy to mitigate the propagation of the virus, it implies the freeze of the market affecting the whole population from the economical point of view.

Consider the Kolmogorov-forward equation, which describes the evolution of the probabilities of being occupying a certain state along the time. In fact, when considering a large population with finite number of individuals, these probabilities can be interpreted as the proportion of individuals occupying a certain state in $\mathcal{S}$. The problem is given by

$$\min_{u \in \mathcal{U}} L(m(t), u(t)), \tag{16.54a}$$

subject to

$$\frac{d}{dt}m_s(t) = \sum_{s'\in\mathcal{S}\setminus\{s\}} q_{s's}(m(t),u(t))\, m_{s'}(t) - \sum_{s'\in\mathcal{S}\setminus\{s\}} q_{ss'}(m(t),u(t))\, m_s(t),$$

$$\forall s \in \mathcal{S}, \tag{16.54b}$$

$$q_{ss}(m(t),u(t)) = -\sum_{s'\in\mathcal{S}\setminus\{s\}} q_{ss'}(m(t),u(t)), \tag{16.54c}$$

$$m(0) = m_0. \tag{16.54d}$$

Once the control problem has been introduced. The mitigation of the COVID-19 can be studied by considering multiple decision-makers. When there are multiple entities deciding, then we call the problem to be a game or a strategic interaction. Next, we address the game theoretical perspective.

16.6.2   Multiple-Decision-Maker Problem

The problem with several involved players (decision-makers)) in the interactive decision is analyzed, e.g., multiple regions, states, countries or continents. The set of players is denoted by $\mathcal{P} = \{1,\ldots,n\}$, and the system dynamics describing the evolution of the COVID-19 for the different players are coupled to each other. The evolution of the COVID-19 for each player is modeled by means of a Markov chain with a finite number of states $\mathcal{S}^p = \{E^p, Q^p, I^p, R^p, D^p\}$, for all $p \in \mathcal{P}$.

The transit of an individual among the regions, which are given by players, is made by means of a connected undirected graph, i.e., exposed, infected and recovered individuals can travel from one region to another following a constrained graph denoted by $\mathcal{G}^E = (\mathcal{V}^E, \mathcal{E}^E)$, $\mathcal{G}^I = (\mathcal{V}^I, \mathcal{E}^I)$, and $\mathcal{G}^R = (\mathcal{V}^R, \mathcal{E}^R)$, respectively. The nodes in the graphs and sets of edges are given by

$$\mathcal{V}^E = \{E^p\}_{p\in\mathcal{P}}, \qquad \mathcal{E}^E \subseteq \{(i,j): i,j \in \mathcal{V}^E\}, \tag{16.55a}$$

$$\mathcal{V}^I = \{I^p\}_{p\in\mathcal{P}}, \qquad \mathcal{E}^I \subseteq \{(i,j): i,j \in \mathcal{V}^I\}, \tag{16.55b}$$

$$\mathcal{V}^R = \{R^p\}_{p\in\mathcal{P}}, \qquad \mathcal{E}^R \subseteq \{(i,j): i,j \in \mathcal{V}^R\}. \tag{16.55c}$$

Thus, there is a new bigger aggregated Markov process emerging denoted by $\{\tilde{X}(t)\}_{t\geq 0}$, where the set of finite states is given by $\tilde{\mathcal{S}} = \bigcup_{p\in\mathcal{P}} \mathcal{S}^p$. Besides, there are some migration constraints representing the air-traffic restrictions among the players. Let $A \in [0,1]^{\sum_{p\in\mathcal{P}}|\mathcal{S}^p|}$ be the adjacency matrix for the aggregated graph $\mathcal{G} = \bigcup_{s\in\mathcal{S}} \mathcal{G}^s$ with $\mathcal{V} = \bigcup_{p\in\mathcal{P}} \mathcal{S}^p$ and $\mathcal{E} \subset \bigcup_{s\in\mathcal{S}} \mathcal{E}^s$.

The transition rates (jump intensities) in the overall Markov process $\{\tilde{X}(t)\}_{t\geq 0}$ satisfy the following condition. For two states $s \in \mathcal{S}^p$, and $s \in \mathcal{S}^r$, with $p \neq r$,

$$q_{ss'} \begin{cases} = 0 & \text{if } (s,s') \notin \mathcal{E}, \\ > 0 & \text{otherwise.} \end{cases}$$

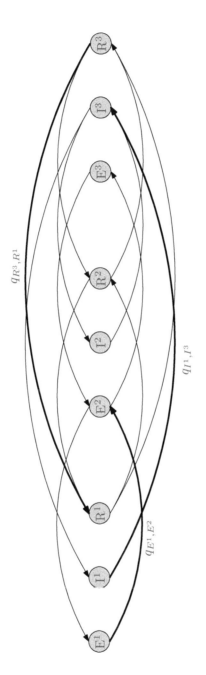

FIGURE 16.36
Transition rates among three different players, i.e., $\mathcal{P} = \{1, 2, 3\}$ corresponding to the example in (16.55).

Figure 16.36 presents an example involving three players $\mathcal{P} = \{1,2,3\}$ where the graphs $\mathcal{G}^E$, $\mathcal{G}^I$, and $\mathcal{G}^R$ are characterized by

$$\mathcal{E}^E \subseteq \{(E^1, E^2), (E^2, E^3)\},$$
$$\mathcal{E}^I \subseteq \{(I^1, I^3), (I^2, I^3)\},$$
$$\mathcal{E}^R \subseteq \{(R^1, R^2), (R^2, R^3), (R^3, R^1)\}.$$

In addition, each player imposes local policies to have control over the COVID-19 and also assigns heterogeneous prioritization to the factors composing the cost functional, i.e., the balance between the minimization of the virus and the economical crisis.

There are several possible cases in the strategic interaction. First scenario consists in a non-cooperative game in which the players pursue to minimize their own cost functional without caring about the economical crisis or virus spread of other players. On the other hand, the players can cooperate to each other in order to overcome the COVID-19, i.e., all the players optimize jointly the required policies to mitigate the virus. Finally, players can establish coalitions in a co-opetitive game, where some players cooperate to each other but are indifferent about the situation of some others. These scenarios are analyzed next.

16.6.2.1 Non-cooperative Games

Each player establishes policies to minimize its own cost functional denoted by $L^p(m^p(t), u^p(t))$, where m^p is the probability measure corresponding to the player p, for all $p \in \mathcal{P}$, i.e.,

$$L^p(\{m^p(t)\}_{p\in\mathcal{P}}, u^p(t)) = h^p\left(\{m^p(T)\}_{p\in\mathcal{P}}\right) + \int_0^T \ell^p\left(\{m^p(t)\}_{p\in\mathcal{P}}, u^p(t)\right) dt,$$

where $h^p : [0,1]^{\sum_{p\in\mathcal{P}} |\mathcal{S}^p|} \times \bigcup_{p\in\mathcal{P}} \mathcal{S}^p \to \mathbb{R}$ denotes the terminal cost, and $\ell^p : [0,1]^{\sum_{p\in\mathcal{P}} |\mathcal{S}^p|} \times \bigcup_{p\in\mathcal{P}} \mathcal{S}^p \times \mathbb{U}^p \to \mathbb{R}$ denotes the running cost. The non-atomic game problem is

$$\underset{u^p \in \mathcal{U}^p}{\text{minimize}} \; L^p(\{m^r(t)\}_{r\in\mathcal{P}}, u^p(t)), \tag{16.56a}$$

subject to

$$\frac{\mathrm{d}}{\mathrm{d}t} m_s(t) = \sum_{s'\in\mathcal{S}\backslash\{s\}} q_{s's}(\{m^p(t), u^p(t)\}_{p\in\mathcal{P}}) \, m_{s'}(t)$$
$$- \sum_{s'\in\mathcal{S}\backslash\{s\}} q_{ss'}(\{m^p(t), u^p(t)\}_{p\in\mathcal{P}}) \, m_s(t), \; \forall s \in \bigcup_{p\in\mathcal{P}} \mathcal{S}^p, \tag{16.56b}$$

$$q_{ss}(\{m^p(t), u^p(t)\}_{p\in\mathcal{P}}) = - \sum_{s' \bigcup_{p\in\mathcal{P}} \mathcal{S}^p\backslash\{s\}} q_{ss'}(\{m^p(t), u^p(t)\}_{p\in\mathcal{P}}), \tag{16.56c}$$

$$m^p(0) = m_0^p, \quad \forall p \in \mathcal{P}. \tag{16.56d}$$

The solution of the problem in (16.56) is given by a Nash equilibrium as it has been studied throughout the book.

Bibliography

[1] A. Ouammi, H. Dagdougui, and R. Sacile. Optimal control of power flows and energy local storages in a network of microgrids modeled as a system of systems. *IEEE Transactions on Control Systems Technology*, 23(1):128–138, 2015.

[2] H. M. Markowitz. Portfolio selection. *The Journal of Finance*, 7(1):77–91, 1952.

[3] H. M. Markowitz. The utility of wealth. *The Journal of Political Economy (Cowles Foundation Paper 57)*, 2:151–158, 1952.

[4] H. M. Markowitz. *Portfolio Selection: Efficient Diversification of Investments*. John Wiley & Sons, Inc. Chapman & Hall, Ltd. 1959.

[5] Robert Aumann. *Collected papers*, volume 1. The MIT Press, Cambridge, MA, 2000.

[6] T. E. Duncan and B. Pasik-Duncan. Solvable stochastic differential games in rank one compact symmetric spaces. *International Journal of Control*, 91(11):2445–2450, 2018.

[7] T. E. Duncan. Linear exponential quadratic stochastic differential games. *IEEE Transactions on Automatic Control*, 61(9):2550–2552, 2016.

[8] T. E. Duncan and B. Pasik-Duncan. Linear-quadratic fractional Gaussian control. *SIAM Journal on Control and Optimization*, 51(6):4504–4519, 2013.

[9] T. E. Duncan, B. Maslowski, and B. Pasik-Duncan. Linear-exponential-quadratic control for stochastic equations in a Hilbert space. *SIAM Journal on Control and Optimization*, 50(1):507–531, 2012.

[10] T. E. Duncan. Linear-exponential-quadratic Gaussian control. *IEEE Transactions on Automatic Control*, 58(11):2910–2911, 2013.

[11] T. E. Duncan. Linear-quadratic stochastic differential games with general noise processes. In F. El Ouardighi and K. Kogan, editors, *Economics and Management Science: Essays in Honor of Charles S. Tapiero. Operations Research and Management Series.*, volume 198. Springer Intern. Springer, 2014.

[12] P. E. Caines. Mean field games. In John Baillieul and Tariq Samad, editors, *Encyclopedia of Systems and Control*, pages 706–712. Springer London, London, 2015.

[13] M. Bardi. Explicit solutions of some linear-quadratic mean field games. *American Institute of Mathematical Sciences*, 7(2):243–261, 2012.

[14] M. Bardi and F. S. Priuli. Linear-quadratic n-person and mean-field games with ergodic cost. *SIAM Journal on Control and Optimization*, 52(5):3022–3052, 2014.

[15] H. Tembine, D. Bauso, and T. Basar. Robust linear quadratic mean-field games in crowd-seeking social networks. In *Proceedings of the 52nd IEEE Conference on Decision and Control (CDC)*, pages 3134–3139, Firenze, Italy, 2013.

[16] H. Tembine, Q. Zhu, and T. Basar. Risk-sensitive mean-field games. *IEEE Transactions on Automatic Control*, 59(4):835–850, 2014.

[17] A. Bensoussan, J. Frehse, and S.C.P. Yam. *Mean Field Games and Mean Field Type Control Theory, SpringerBriefs in Mathematics*. Springer, 2013.

[18] A. Bensoussan. Explicit solutions of linear quadratic differential games. In H. Yan, G. Yin, and Q. Zhang, editors, *International Series in Operations Research and Management Science*, volume 94. Springer, 2006.

[19] T. E. Duncan and B. Pasik-Duncan. A direct method for solving stochastic control problems. *Communications in Information and Systems*, 12(1):1–14, 2012.

[20] D. L. Lukes and D. L. Russell. A global theory of linear-quadratic differential games. *Journal of Mathematical Analysis and Applications*, 33(1):96–123, 1971.

[21] J. C. Engwerda. On the open-loop Nash equilibrium in lq-games. *Journal of Economic Dynamics and Control*, 22(5):729–762, 1998.

[22] E. J. Dockner, S. Jorgensen, N. Van Long, and G. Sorger. *Differential Games in Economics and Management Science*. Cambridge University Press, Business and Economics, 2000.

[23] A. Bensoussan, J. Frehse, and S. C. P. Yam. On the interpretation of the master equation. *Stochastic Processes and their Applications*, 127(7):2093–2137, 2017.

[24] B. Djehiche B, H. Tembine, and R. Tempone. A stochastic maximum principle for risk-sensitive mean-field type control. *IEEE Transactions on Automatic Control*, 60(10):2640–2649, 2015.

[25] H. Tembine. *Distributed Strategic Learning for Wireless Engineers*. CRC Press, Taylor & Francis, 2012.

[26] H. Tembine. Risk-sensitive mean-field-type games with l^p-norm drifts. *Automatica*, 59(2015):224–237, 2015.

[27] B. Djehiche, A. Tcheukam, and H. Tembine. Mean-field-type games in engineering. *AIMS Electronics and Electrical Engineering*, 1(1):18–73, 2017.

[28] H. Tembine. Nonasymptotic mean-field games. *IEEE Transactions on Cybernetics*, 44(12):2744–2756, 2014.

[29] B. Djehiche, T. Basar, and H. Tembine. *Mean-Field-Type Game Theory*. Springer, 2020. Under Preparation.

[30] H. Tembine. Energy-constrained mean-field games in wireless networks. *Strategic Behavior and the Environment*, 4(2):187–211, 2014.

[31] J. Barreiro-Gomez, S. E. Choutri, and H. Tembine. Risk-awareness in multi-level building evacuation with smoke: Burj Khalifa case study. *Automatica*, 129, 109625 2021.

[32] B. Djehiche, J. Barreiro-Gomez, and H. Tembine. Electricity price dynamics in the smart grid: A mean-field-type game perspective. In *23rd International Symposium on Mathematical Theory of Networks and Systems (MTNS)*, pages 631–636, Hong Kong, China, 2018.

[33] H. M. Markowitz. *Harry Markowitz: Selected Works*. World Scientific-nobel Laureate Series. World Scientific Publishing Company, 2009.

[34] N. Nisan, T. Roughgarden, É. Tardos, and V. V. Vazirani. *Algorithmic Game Theory*. Cambridge University Press, New York, NY, USA, 2007.

[35] J. R. Marden and J. S. Shamma. *Game-Theoretic Learning in Distributed Control*, pages 1–36. Springer International Publishing, Cham, 2018.

[36] S. Lasaulce and H. Tembine. *Game Theory and Learning for Wireless Networks: Fundamentals and Applications*. Academic Press, 2011.

[37] J. Barreiro-Gomez and H. Tembine. Mean-field-type model predictive control: An application to water distribution networks. *IEEE Access*, 7(2019):135332–135339, 2019.

[38] J. Barreiro-Gomez, C. Ocampo-Martinez, and N. Quijano. Dynamical tuning for MPC using population games: A water supply network application. *ISA Transactions*, 69(2017):175–186, 2017.

[39] N. Quijano, C. Ocampo-Martinez, J. Barreiro-Gomez, G. Obando, A. Pantoja, and E. Mojica-Nava. The role of population games and evolutionary dynamics in distributed control systems. *IEEE Control Systems*, 37(1):70–97, 2017.

[40] J. Barreiro-Gomez, F. Dörfler, and H. Tembine. Distributed and robust population games with applications to optimal frequency control in power systems. In *American Control Conference (ACC)*, pages 5762–5767, Milwaukee, 2018.

[41] J. Barreiro-Gomez and H. Tembine. Constrained evolutionary games by using a mixture of imitation dynamics. *Automatica*, 97(2018):254–262, 2018.

[42] J. I. Poveda, P. N. Brown, J. R. Marden, and A. R. Teel. A class of distributed adaptive pricing mechanisms for societal systems with limited information. In *Proceedings of the 56th IEEE Conference on Decision and Control (CDC)*, pages 1490–1495, Melbourne, Australia, 2017.

[43] J. Barreiro-Gomez and H. Tembine. Blockchain token economics: A mean-field-type game perspective. *IEEE Access*, 2019. DOI: 10.1109/ACCESS.2019.2917517.

[44] G. Obando, A. Pantoja, and N. Quijano. Building temperature control based on population dynamics. *IEEE Transactions on Control Systems Technology*, 22(1):404–412, 2014.

[45] J. M. Lasry and P. L. Lions. Mean field games. *Japanese Journal of Mathematics*, 2(1):229–260, 2007.

[46] A. Bensoussan, B. Djehiche, H. Tembine, and P. Yam. Risk-sensitive mean-field-type control. In *Proceedings of the 56th IEEE Conference on Decision and Control (CDC)*, pages 33–38, Melbourne, Australia, 2017. DOI: 10.1109/CDC.2017.8263639.

[47] D. Andersson and B. Djehiche. A maximum principle for sdes of mean-field type. *Applied Mathematics & Optimization*, 63(2011):341–356, 2011.

[48] R. Buckdahn, B. Djehiche, and J. Li. A general stochastic maximum principle for SDEs of mean-field type. *Applied Mathematics & Optimization*, 64(2011):197–216, 2011.

[49] J. J. Absalom Hosking. A stochastic maximum principle for a stochastic differential game of a mean-field type. *Applied Mathematics & Optimization*, 66(2012):415–454, 2012.

[50] R. Elliott, X. Li, and Y. Ni. Discrete time mean-field stochastic linear-quadratic optimal control problems. *Automatica*, 49(2013):3222–3233, 2013.

[51] Y. Shen and T. K. Siu. The maximum principle for a jump-diffusion mean-field model and its application to the mean–variance problem. *Nonlinear Analysis*, 86(2013):58–73, 2013.

[52] M. Laurière and O. Pironneau. Dynamic programming for mean-field type control. *Comptes Rendus de l'Académie des Sciences Paris, Series I*, 352(2014):707–713, 2014.

[53] W. Guangchen, W. Zhen, and Z. Chenghui. Maximum principles for partially observed mean-field stochastic systems with application to financial engineering. In *Proceedings of the 33rd Chinese Control Conference*, pages 5357–5362, Nanjing, China, 2014.

[54] G. Wang, C. Zhang, and W. Zhang. Stochastic maximum principle for mean-field type optimal control under partial information. *IEEE Transactions on Automatic Control*, 59(2):522–528, 2014.

[55] A. Bensoussan, S.C.P. Yam, and Z. Zhang. Well-posedness of mean-field type forward–backward stochastic differential equations. *Stochastic Processes and their Applications*, 125(2015):3327–3354, 2015.

[56] R. Carmona and F. Delarue. Mean field forward-backward stochastic differential equations. *Electron. Commun. Probab.*, 18(2013):1–15, 2013.

[57] L. Ma, T. Zhang, W. Zhang, and B. Chen. Finite horizon mean-field stochastic H_2/H_∞ control for continuous-time systems with (x, v)-dependent noise. *Journal of The Franklin Institute*, 352(2015):5393–5414, 2015.

[58] Y. Ni, J. Zhang, and X. Li. Indefinite mean-field stochastic linear-quadratic optimal control. *IEEE Transactions on Automatic Control*, 60(7):1786–1800, 2015.

[59] B. Djehiche and H. Tembine. Risk-sensitive mean-field type control under partial observation. In F. E. Benth and G. D. Nunno, editors, *Stochastics of Environmental and Financial Economics*, pages 243–263. Springer, Oslo, Norway, 2016.

[60] H. Ma and B. Liu. Maximum principle for partially observed risk-sensitive optimal control problems of mean-field type. *European Journal of Control*, 32(2016):16–23, 2016.

[61] W. Guangchen, W. Zhen, and Z. Chenghui. A partially observed optimal control problem for mean-field type forward-backward stochastic system. In *Proceedings of the 35th Chinese Control Conference*, pages 1781–1786, Chengdu, China, 2016.

[62] H. Ma and B. Liu. Linear-quadratic optimal control problem for partially observed forward-bacxkward stochastic differential equations of mean-field type. *Asian Journal of Control*, 18(6):2146–2157, 2016.

[63] M. Laurière and O. Pironneau. Dynamic programming for mean-field type control. *J Optim Theory Appl*, 169(2016):902–924, 2016.

[64] P. J. Graber. Linear quadratic mean field type control and mean field games with common noise, with application to production of an exhaustible resource. *Applied Mathematics & Optimization*, 74(3):459–486, 2016.

[65] A. K. Cissé and H. Tembine. Cooperative mean-field type games. In *Proceedings of the 19th World Congress The International Federation of Automatic Control*, pages 8995–9000, Cape Town, South Africa, 2014.

[66] H. Tembine. Uncertainty quantification in mean-field-type teams and games. In *Proceedings of the IEEE Control Conference on Decision and Control (CDC)*, pages 4418–4423, Osaka, Japan, 2015.

[67] B. Djehiche and M. Huang. A characterization of sub-game perfect equilibria for sdes of mean-field type. *Dynamic Games and Applications*, 6(2016):55–81, 2016.

[68] A. Aurell. Mean-field type games between two players driven by backward stochastic differential equations. *Games*, 9(88):1–26, 2018.

[69] B. Djehiche and S. Hamadène. Optimal control and zero-sum stochastic differential game problems of mean-field type. *Applied Mathematics & Optimization*, (2018):1–28, 2018.

[70] H. Tembine T. E. Duncan. Linear-quadratic mean-field-type games: A direct method. *Games Journal*, 7(2018):1–18, 2018.

[71] J. Barreiro-Gomez, T. E. Duncan, B. Pasik-Duncan, and H. Tembine. Semi-explicit solutions to some non-linear non-quadratic mean-field-type games: A direct method. *IEEE Transactions on Automatic Control*, 65(6):2582–2697, 2020.

[72] J. Barreiro-Gomez, T. E. Duncan, and H. Tembine. Discrete-time linear-quadratic mean-field-type repeated games: Perfect, incomplete, and imperfect information. *Automatica*, 112(2020):108647, 2020.

[73] A. Bensoussan, B. Djehiche, H. Tembine, and S. C. P. Yam. Mean-field-type games with jump and regime switching. *Dynamic Games and Applications*, 10(1): 1–39, 2020.

[74] J. Barreiro-Gomez, T. E. Duncan, and H. Tembine. Linear-quadratic mean-field-type games: Jump-diffusion process with regime switching. *IEEE Transactions on Automatic Control*, 64(10):4329–4336, 2019.

[75] J. Barreiro-Gomez, T. E. Duncan, and H. Tembine. Co-opetitive linear-quadratic mean-field-type games. *IEEE Transactions on Cybernetics*, 2019. DOI: 10.1109/TCYB.2019.2901006.

[76] J. Gao and H. Tembine. Distributed mean-field-type filters for big data assimilation. In *2016 IEEE 18th International Conference on High Performance Computing and Communications; IEEE 14th International Conference on Smart City; IEEE 2nd International Conference on Data Science and Systems*, pages 1446–1453, 2016.

[77] A. T. Siwe and H. Tembine. Network security as public good: A mean-field-type game theory approach. In *2016 13th International Multi-Conference on Systems, Signals & Devices (SSD)*, pages 601–606, Leipzig, Germany, 2016.

[78] H. Tembine. Mean-field-type optimization for demand-supply management under operational constraints in smart grid. *Energy Systems*, 7(2016):333–356, 2016.

[79] A. Aurell and B. Djehiche. Mean-field type modeling of nonlocal crowd aversion in pedestrian crowd dynamics. *SIAM Journal on Control and Optimization*, 56(1):434–455, 2018.

[80] J. Barreiro-Gomez, T. E. Duncan, and H. Tembine. Linear-quadratic mean-field-type games with multiple input constraints. *IEEE Control Systems Letters*, 3(3):511–516, 2019.

[81] A. Aurell and B. Djehiche. Modeling tagged pedestrian motion: A mean-field type game approach. *Transportation Research Part B*, 121(2019):168–183, 2019.

[82] J. Gao and H. Tembine. Distributed mean-field-type filters for traffic networks. *IEEE Transactions on Intelligent Transportation Systems*, 20(2):507–521, 2019.

[83] J. Barreiro-Gomez, T. E. Duncan, and H. Tembine. Linear-quadratic mean-field-type games-based stochastic model predictive control: A microgrid energy storage application. In *Proceedings of the American Control Conference*, pages 3224–3229, Philadelphia, PA, 2019.

[84] A. Bensoussan, J. Frehse, and S.C.P. Yam. *Mean Field Games and Mean Field Type Control Theory*, volume 1. Springer Briefs in Mathematics, New York, 2013.

[85] D. A. Gomes and J. Saúde. Mean field games models—A brief survey. *Dynamic Games and Applications*, 4(2):110–154, Jun 2014.

[86] P. E. Caines, M. Huang, and R. Malhame. Mean field games. In T. Başar and G. Zaccour, editors, *Handbook of Dynamic Game Theory*, pages 1–28. Springer International Publishing AG, 2017.

[87] A. Wald. On some systems of equations of mathematical economics. *Econometrica*, 19:368–403, 1951.

[88] J. M. von Neumann. *Theory of Games and Economic Behavior*. Princeton University Press, 1953.

[89] J. Barreiro-Gomez and H. Tembine. A matlab-based mean-field-type games toolbox: Continuous-time version. *IEEE Access*, 7:126500–126514, 2019.

[90] M. Deutsch. A theory of cooperation and competition. *Human Relations*, 2:129–151, 1949.

[91] S. J. Chión, V. Charles, and M. Tavana. Impact of incentive schemes and personality-tradeoffs on two-agent coopetition with numerical estimations. *Measurement*, 125(2018):182–195, 2018.

[92] J. Barreiro-Gomez, C. Ocampo-Martinez, N. Quijano, and J. M. Maestre. Non-centralized control for flow-based distribution networks: A game-theoretical insight. *Journal of the Franklin Institute*, 354(14):5771–5796, 2017.

[93] P. Skowron, K. Rzadca, and A. Datta. Cooperation and competition when bidding for complex projects: Centralized and decentralized perspectives. *IEEE Intelligent Systems*, 32(1):17–23, 2017.

[94] H. V. Stackelberg. The theory of the market economy, trans. by aj peacock, london, william hodge. *Originally published as Grundlagen der Theoretischen Volkswirtschaftlehre*, 1948.

[95] M. Simaan and J. B. Cruz. On the Stackelberg strategy in nonzero-sum games. *Journal of Optimization Theory and Applications*, 11(5):533–555, 1973.

[96] A. Bagchi and T. Basar. Stackelberg strategies in linear-quadratic stochastic differential games. *Journal of Optimization Theory and Applications*, 35(3):443–464, 1981.

[97] A. Bensoussan, S. Chen, and S. P. Sethi. The maximum principle for global solutions of stochastic Stackelberg differential games. *SIAM Journal on Control and Optimization*, 53(4):1956–1981, 2015.

[98] L. Pan and J. Yong. A differential game with multi-level of hierarchy. *Journal of Mathematical Analysis and Applications*, 161(2):522–544, 1991.

[99] M. Simaan and J. Cruz. A stackelberg solution for games with many players. *IEEE Transactions on Automatic Control*, 18(3):322–324, 1973.

[100] J. Cruz. Leader-follower strategies for multilevel systems. *IEEE Transactions on Automatic Control*, 23(2):244–255, 1978.

[101] B. Gardner and J. Cruz. Feedback Stackelberg strategy for m-level hierarchical games. *IEEE Transactions on Automatic Control*, 23(3):489–491, 1978.

[102] T. Basar and H. Selbuz. Closed-loop Stackelberg strategies with applications in the optimal control of multilevel systems. *IEEE Transactions on Automatic Control*, 24(2):166–179, 1979.

[103] Y. Lin, X. Jiang, and W. Zhang. An open-loop Stackelberg strategy for the linear quadratic mean-field stochastic differential game. *IEEE Transactions on Automatic Control*, 64(1):97–110, 2019.

[104] K. Du and Z. Wu. Linear-quadratic Stackelberg game for mean-field backward stochastic differential system and application. *Mathematical Problems in Engineering*, Article ID 1798585, 17 pages, 2019.

[105] J. Moon and T. Basar. Linear-quadratic stochastic differential Stackelberg games with a high population of followers. In *Proceedings of the 54th IEEE Conference on Decision and Control*, pages 2270–2275, Osaka, Japan, 2015.

[106] A. Bensoussan, M. H. M. Chau, and S. C. P. Yam. Mean-field Stackelberg games: Aggregation of delayed instructions. *SIAM Journal on Control Optimization*, 53(4):2237–2266, 2015.

[107] A. Bensoussan, M.H.M. Chau, Y. Lai, and S.C.P. Yam. Linear-quadratic mean field Stackelberg games with state and control delays. *SIAM Journal on Control and Optimization*, 55(4):2748–2781, 2017.

[108] A. Y. Averboukh. Stackelberg solution for first-order mean-field game with a major player. *Izvestiya Instituta Matematiki i Informatiki. Udmurtskij*, 52, 2018.

[109] J. Moon and T. Basar. Linear quadratic mean field Stackelberg differential games. *Automatica*, 97(2018):200–213, 2018.

[110] J. Shi, G. Wang, and J. Xiong. Leader-follower stochastic differential game with asymmetric information and applications. *Automatica*, 63(2016):60–73, 2016.

[111] M. Nourian, P. Caines, R. P. Malhamé, and M. Huang. Mean field LQG control in leader-follower stochastic multi-agent systems: Likelihood ratio based adaptation. *IEEE Transactions on Automatic Control*, 57(11):2801–2816, 2012.

[112] H. Cai and G. Hu. Distributed tracking control of an interconnected leader-follower multiagent system. *IEEE Transactions on Automatic Control*, 62(7):3494–3501, 2017.

[113] Y. Li, D. Shi, and T. Chen. False data injection attacks on networked control systems: A Stackelberg game analysis. *IEEE Transactions on Automatic Control*, 63(10):3503–3509, 2018.

[114] J. Barreiro-Gomez, C. Ocampo-Martinez, and N. Quijano. Partitioning for large-scale systems: A sequential distributed MPC design. In *20th IFAC World Congress*, pages 8838–8843, Toulouse, France, 2017.

[115] Z. El Oula Frihi, J. Barreiro-Gomez, S. E. Choutri, B. Djehiche, and H. Tembine. Stackelberg mean-field-type games with polynomial cost. In *21th IFAC World Congress*, Berlin, Germany, 2020.

[116] Z. El Oula Frihi, J. Barreiro-Gomez, S. E. Choutri, and H. Tembine. Hierarchical structures and leadership design inmean-field-type games with polynomial cost. *Games*, 11(30):1–26, 2020.

[117] C. Berge. Théorie générale des jeux à n personnes [general theory of n-person games]. *Paris: Gauthier-Villars*, 1957.

[118] O. Musy, A. Potter, and T. Tazdait. A new theorem to find Berge equilibria. *International Game Theory Review*, 14(1250005), 2012.

[119] B. Crettez. On sugden's mutually beneficial practice and Berge equilibrium. *International Review of Economics*, 128, 2017.

[120] B. Crettez. A new sufficient condition for a Berge equilibrium to be a Berge-Vaisman equilibrium. *Journal of Quantitative Economics*, 15(3), 451-459, 2017.

[121] K. Keskin and H. C. Saglam. On the existence of Berge equilibrium: An order theoretic approach. *International Game Theory Review*, 17(03), 2015.

[122] H. W. Corley. A mixed cooperative dual to the Nash equilibrium. *Game Theory*, 1–7, 2015.

[123] H. W. Corley and P. Kwain. An algorithm for computing all Berge equilibria. *Game Theory*, 1–2, 2015.

[124] A. Pottier and R. Nessah. Berge-Vaisman and Nash equilibria: Transformation of games. *International Game Theory Review*, 16(04), 2014.

[125] M. Larbani and V.I. Zhukovskii. Berge equilibrium in normal form static games: a literature review. *Izv. IMI udGU*, 49:80–110, 2017.

[126] V.I. Zhukovskii, K.N. Kudryavtsev, and A.S. Gorbatov. Berge equilibrium in cournot model of oligopoly. *Vestnik Udmurtskogo Universiteta, Matematika. Mekhanika*, 2015.

[127] J.O. Hairault and T. Sopraseuth. *Exchange Rate Dynamics*, volume 1. Routledge Taylor & Francis Group, NewYork, 2004.

[128] W. H. Sandholm. *Population Games and Evolutionary Dynamics*. Cambridge, MA, MIT Press, 2010.

[129] J. Martinez-Piazuelo, G. Diaz-Garcia, N. Quijano, and L. F. Giraldo. Discrete-time distributed population dynamics for optimization and control. *IEEE Transactions on Systems, Man and Cybernetics: Systems*, 2021.

[130] A. Nagurney and D. Zhang. Projected dynamical systems in the formulation, stability analysis, and computation of fixed demand traffic network equilibria. *Transportation Science*, 31:147–158, 1997.

[131] J. B. Rosen. Existence and uniqueness of equilibrium points for concave n-person games. *Econometrica*, 33(3):520–534, 1965.

[132] S. Boyd and L. Vandenberghe. *Convex Optimization*. Cambridge University Press, 2004.

[133] P. D. Taylor and L. B. Jonker. Evolutionary stable strategies and game dynamics. *Mathematical Biosciences*, 40(1):145–156, 1978.

[134] J. Barreiro-Gomez, G.Obando, and N. Quijano. Distributed population dynamics: Optimization and control applications. *IEEE Transactions on Systems, Man, and Cybernetics: Systems*, 47(2):304–314, 2017.

[135] J. M. Grosso, C. Ocampo-Martinez, V. Puig, and B. Joseph. Chance-constrained model predictive control for drinking water networks. *Journal of Process Control*, 24(2014):504–516, 2014.

[136] M. Pereira, D. Muñoz de la Peña, D. Limon, I. Alvarado, and T. Alamo. Application to a drinking water network of robust periodic MPC. *Control Engineering Practice*, 57(2016):50–60, 2016.

[137] Y. Wang, C. Ocampo-Martinez, and V. Puig. Stochastic model predictive control based on Gaussian processes applied to drinking water networks. *IET Control Theory & Applications*, 10(8):947–955, 2016.

[138] P. Sopasakis, D. Herceg, A. Bemporad, and P. Patrinos. Risk-averse model predictive control. *Automatica*, 100(2019):281–288, 2019.

[139] Y. Yang and C. Sutanto. Chance-constrained optimization for non-convex programs using scenario-based methods. *ISA Transactions*, 90(2019):157–168, 2019.

[140] B. K. Ghosh. Probability inequalities related to Markov's theorem. *The American Statistician*, 56(3):186–190, 2002.

[141] D. Mayne. Model predictive control: Recent developments and future promise. *Automatica*, 50(2014):2967–2986, 2014.

[142] J. Rawlings and D. Mayne. *Model Predictive Control Theory and Design*. Nob Hill Pub, Llc, 2009.

[143] P. D. Christofides, R. Scattolini, D. Muñoz de la Peña, and J. Liu. Distributed model predictive control: A tutorial review and future research directions. *Computers & Chemical Engineering*, 51(5):21–41, 2013.

[144] G. James, D. Witten, T. Hastie, and R. Tibshirani. *An Introduction to Statistical Learning with Applications in R*. Springer, 2014.

[145] K. H. Johansson. The quadruple-tank process: A multivariable laboratory process with an adjustable zero. *IEEE Transactions on Control Systems Technology*, 8(3):456–465, 2000.

[146] J. Barreiro-Gomez. *The Role of Population Games in the Design of Optimization-Based Controllers*. Springer editorial, 2019.

[147] J. Grosso. *Economic and Robust Operation of Generalised Flow-based Networks*. Doctoral dissertation. Universidad Politècnica de Catalunya. Automatic Control Department., 2015.

[148] C. Liou and T. Hsiue. Exact linearization and control of a continuous stirred tank reactor. *Journal of the Chinese Institute of Engineers*, 18(6):825–833, 1995.

[149] G. Wang, S. Peng, and H. Huang. A sliding observer for nonlinear process control. *Chemical Engineering Science*, 52(5):787–805, 1997.

[150] T. L. Friesz, D. Bernstein, N. J. Mehta, R. L. Tobin, and S. Ganjalizadeh. Day-today dynamic network disequilibria and idealized traveler information systems. *Operations Research*, 42:1120–1136, 1994.

[151] P. Malisani, F. Chaplais, and N. Petit. Design of penalty functions for optimal control of linear dynamical systems under state and input constraints. In *Proceedings of 50st IEEE Conference on Decision and Control and European Control Conference (CDC-ECC)*, pages 6697–6704, Orlando, FL, 2011.

[152] P. Malisani, F. Chaplais, and N. Petit. An interior penalty method for optimal control problems with state and input constraints of nonlinear systems. *Optimal Control Applications and Methods*, 37(2016):3–33, 2016.

[153] B. Merci and T. Beji. *Fluid Mechanics Aspects of Fire and Smoke Dynamics in Enclosures*, volume 1. Taylor & Francis, 2013.

[154] R. Cressman and V. Křivan. Migration dynamics for the ideal free distribution. *The American Naturalist*, 168(3):384–397, 2006.

[155] J. Hellewell, S. Abbott, A. Gimma, N. I. Bosse, C. I. Jarvis, T. W. Russell, J. D. Munday, A. J. Kucharski, W. J. Edmunds, S. Funk, and R. M. Eggo. Feasibility of controlling covid-19 outbreaks by isolation of cases and contacts. *Lancet Glob Health*, 8:488–496, 2020.

[156] M. F. Bashir, B. Ma, Bilal, B. Komal, M. A. Bashir, D. Tan, and M. Bashir. Correlation between climate indicators and covid-19 pandemic in New York, USA. *Science of the Total Environment*, 278(2020):138835, 2020.

[157] Y. Zhu, J. Xie, F. Huang, and L. Cao. Association between short-term exposure to air pollution and covid-19 infection: Evidence from China. *Science of the Total Environment*, 727(2020):138704, 2020.

[158] F. Taghizadeh-Hesary and H. Akbari. The powerful immune system against powerful covid-19: A hypothesis. *Medical Hypotheses*, 20(2020):1–7, 2020. DOI: https://doi.org/10.1016/j.mehy.2020.109762.

[159] A. M. Al-Awadhi, K. Alsaifi, A. Al-Awadhi, and S. Alhammadi. Death and contagious infectious diseases: Impact of the covid-19 virus on stock market returns. *Journal of Behavioral and Experimental Finance*, 27(2020):100326, 2020.

[160] J. W. Goodell. Covid-19 and finance: Agendas for future research. *Finance Research Letters*, 20(2020):1–12, 2020. DOI: https://doi.org/10.1016/j.frl.2020.101512.

[161] L. J. S. Allen. An introduction to stochastic epidemic models. In *Mathematical epidemiology*, pages 81–130. Springer editorial, 2008.

[162] Z. Ceylan. Estimation of covid-19 prevalence in italy, spain, and france. *Science of the Total Environment*, 20(2020):1–24, 2020. DOI: https://doi.org/10.1016/j.scitotenv.2020.138817.

[163] J. A Jacquez and P. O'Neill. Reproduction numbers and thresholds in stochastic epidemic models i. homogeneous populations. *Mathematical Biosciences*, 107(2):161–186, 1991.

[164] M. J. Keeling and J. V. Ross. On methods for studying stochastic disease dynamics. *Journal of the Royal Society Interface*, 5(19):171–181, 2008.

[165] G. Giordano, F. Blanchini, R. Bruno, P. Colaneri, A. Di Filippo, A. Di Matteo, and M. Colaneri. Modelling the covid-19 epidemic and implementation of population-wide interventions in Italy. *Nature Medicine*. DOI: https://doi.org/10.1038/s41591-020-0883-7.

[166] H. Tembine. COVID-19: Data-driven mean-field-type game perspective. *Games*, 11(4):51, 2020.

Index

For Product Safety Concerns and Information please contact our EU
representative GPSR@taylorandfrancis.com
Taylor & Francis Verlag GmbH, Kaufingerstraße 24, 80331 München, Germany

www.ingramcontent.com/pod-product-compliance
Ingram Content Group UK Ltd.
Pitfield, Milton Keynes, MK11 3LW, UK
UKHW021445080625
459435UK00011B/377